PRAISE FOR

RIGHTFUL HERITAGE

"Engrossing. . . . [An] enjoyably exhaustive new biography."
—Clay Risen, *New York Times Book Review*

"Douglas Brinkley's high-spirited and admirably thorough new book on FDR . . . show[s] how strands of Roosevelt's life united to shape approaches to preservation." —Dennis Drabelle, *Washington Post*

"Brinkley makes wonderfully clear, Roosevelt was among the first national leaders to see from the start that forestation meant water and soil stabilization, in good times and bad. What a president! What a book!" —Nigel Hamilton, *Boston Globe*

"Brinkley is one of the nation's most acclaimed and popular historians with a work ethic that rivals Paul Bunyan's. *Rightful Heritage* is a massive book with rich characters. . . . Brinkley draws vivid portraits of the multiple players involved in the era's environmental movement."
—*USA Today*

"[A] brightly written, highly useful argument, especially in a time when the public domain is under siege." —*Kirkus Reviews*

"Douglas Brinkley is the most distinguished student of America's environmental history. His books on Teddy Roosevelt, Alaska, Hurricane Katrina, and now Franklin D. Roosevelt are masterful studies of the country's natural history. His new work on FDR's extraordinary role in preserving the nation's natural wonders is a landmark achievement. It is the fullest, most compelling study of FDR's part in perpetuating our natural treasures. It is a must-read for anyone interested in the environment." —Robert Dallek

"*Rightful Heritage* is a marvelous book in every sense; one of Douglas Brinkley's very best. Franklin Delano Roosevelt's achievements in conservation helped preserve the nation's natural bounty, but also made it accessible to the citizenry as never before. An easily forgotten legacy of the New Deal, those achievements presaged bold efforts at global conservation during World War II. By telling this grand story so well, Brinkley provokes readers to appreciate how, with great leadership and sufficient political will, the national government can perform wonders of its own." —Sean Wilentz

"Following his definitive look at Teddy Roosevelt's passion for wilderness conservation, Doug Brinkley now brings us a colorful, exciting narrative of how his cousin FDR carried on the cause of being a warrior for the environment. By highlighting FDR's drive for protecting our land, Brinkley gives us a wonderful, timely new perspective on FDR; his wife, Eleanor; and the dedicated environmentalists around them." —Walter Isaacson

"With this engaging book, Douglas Brinkley recovers one of Franklin Roosevelt's long-overlooked legacies: his stewardship of America's natural resources. This is a vivid history of an important subject, and we are lucky that Brinkley has turned his attention to the Roosevelt we do not generally associate with the preservation of our environment." —Jon Meacham

"Douglas Brinkley's stunningly researched and compellingly written *Rightful Heritage* tells the story of a little-known White House love affair—FDR's with the American wilderness. In our search for compassionate and clearheaded leaders to guide us through and beyond our current environmental crisis, Brinkley's vividly detailed account of Roosevelt's pioneering preservationism, which ensured that enormous swaths of landscape would remain 'forever wild,' may serve as a much-needed beacon and bible." —Megan Marshall

"*Rightful Heritage* is an irresistibly powerful and beautiful tale of America's fraught love affair with its land, told by one of our most gifted historians. From the white pines of the Adirondacks to the black volcanoes of Hawaii, Brinkley follows FDR on an astonishing journey that, despite war, depression, and political infighting, somehow preserved what is most precious to us." —Candice Millard

"Few contemporary observers, perhaps, would label Roosevelt an environmentalist par excellence. Yet, in a majestic new book, *Rightful Heritage*, historian Douglas Brinkley argues that the president's fervor for, and concern about, the natural world constituted a major thread of his life and work." —Peter Tonguette, *Columbus Dispatch*

"*Rightful Heritage* is an excellent read for those who are interested in a nontraditional narrative about Franklin D. Roosevelt or who are passionate about wildlife preservation. . . . Brinkley guides readers through the U.S. president's life from boyhood through the presidency, with particular focus paid to the creation and work of the Civilian Conservation Corps." —Jennifer Ball, *Deseret News*

"A valuable addition to the many books of Roosevelt's influence and importance. . . . Take it along to trace his presence in whatever national park you are visiting." —Robert Hoover, *Dallas Morning News*

"Brinkley's book adds significantly to knowledge of FDR as both man and president, and ranks among the best books on this major historical figure." —*Publishers Weekly* (starred review)

"Brinkley crafts a detailed study that will attract legions of faithful readers. Scholars will savor the author's meticulous annotations in addition to endnotes highlighting a lesser-studied aspect of Franklin's legacy of governmental action." —*Library Journal*

RIGHTFUL HERITAGE

ALSO BY DOUGLAS BRINKLEY

The Quiet World: Saving Alaska's Wilderness Kingdom, 1879–1960

The Wilderness Warrior: Theodore Roosevelt and
the Crusade for America

The Great Deluge: Hurricane Katrina, New Orleans,
and the Mississippi Gulf Coast

Wheels for the World: Henry Ford, His Company,
and a Century of Progress, 1903–2003

The Western Paradox: Bernard DeVoto Conservation Reader
(editor, with Patricia Nelson Limerick)

The Atlantic Charter

Driven Patriot: The Life and Times of James Forrestal

FDR and the Creation of the U.N.
(with Townsend Hoopes)

RIGHTFUL HERITAGE

Franklin D. Roosevelt and the Land of America

DOUGLAS BRINKLEY

HARPER PERENNIAL

NEW YORK • LONDON • TORONTO • SYDNEY • NEW DELHI • AUCKLAND

A hardcover edition of this book was published in 2016 by Harper, an imprint of HarperCollins Publishers.

HarperCollins books may be purchased for educational, business, or sales promotional use. For information, please e-mail the Special Markets Department at SPsales@harpercollins.com.

FIRST HARPER PERENNIAL EDITION PUBLISHED 2017.

Designed by Leah Carlson-Stanisic

Library of Congress Cataloging-in-Publication Data has been applied for.

ISBN 978-0-06-208925-0 (pbk.)

17 18 19 20 21 LSC 10 9 8 7 6 5 4 3 2 1

TO MY BLESSED CHILDREN:
BENTON, JOHNNY, AND CASSADY

I see an America whose rivers and valleys and lakes—
hills and streams and plains—the mountains over our land
and nature's wealth deep under the earth—are protected
as the rightful heritage of all the people.

—FRANKLIN D. ROOSEVELT,
NOVEMBER 2, 1940, CLEVELAND, OHIO

Each town should have a park, or rather a primitive forest,
of five hundred or a thousand acres, where a stick should
never be cut for fuel, a common possession forever,
for instruction and recreation.

—HENRY DAVID THOREAU, *JOURNAL*, OCTOBER 15, 1859

CONTENTS

PART THREE: CONSERVATION EXPANSION, 1937–1939

PART FOUR: WORLD WAR II AND GLOBAL CONSERVATION, 1940–1945

PART ONE

THE EDUCATION OF A HUDSON RIVER CONSERVATIONIST, 1882–1932

CHAPTER ONE

"ALL THAT IS IN ME GOES BACK TO THE HUDSON"

I

There was never a eureka moment that transformed Franklin D. Roosevelt into a dyed-in-the-wool forest conservationist. His passion for the natural world was instead an emotion, inclination, and inherited disposition. Throughout his life he firmly believed that natural surroundings influenced a person's health, social behavior, and mood for the better. "Franklin Delano Roosevelt's family owned land in and around Poughkeepsie and along the banks of the Hudson River for four generations," his wife, Eleanor Roosevelt, wrote, "but even before that his Roosevelt ancestors lived just a bit farther down the Hudson River . . . so the river in all of its aspects and the countryside as a whole were familiar and deeply rooted in my husband's consciousness."[1]

If tourists with an active imagination started at the Hudson River's headwaters in a lake high in the Adirondacks, and traveled 315 miles south to New York Bay, they'd see a grand cavalcade of American history unfold. To Roosevelt, the Hudson, a tidal river where Atlantic seawater flooded far upstream, was the great wellspring of the nation. The river was the first in America to be explored by Europeans, its waters beheld the first steamships to operate successfully, and its banks were the birthplace of the nation's first lucrative railroad.[2] A connoisseur of Hudson River Valley history, Roosevelt concurred with Dutch explorer Henry Hudson, who in 1609 found the river's leafy shores and bluffs "as pleasant a land as one can tread upon."[3]

America's only four-term president was born on January 30, 1882, in the provincial village of Hyde Park, New York, amid rich Dutchess

County farm country. His parents, James and Sara Roosevelt, were both from prominent Hudson Valley families, on opposite sides of the river. The tranquil beauty of the Hudson River stitched the families together: the Roosevelts on the east side (Hyde Park) and the Delanos on the west (Newburgh). In 1880, James Roosevelt was a well-preserved widower of fifty-four—with one child, a son named James, who was known as "Rosey"—when he married Sara Delano, a woman twenty-six years his junior. Franklin, their only child together, was named after Sara's favorite uncle.

James Roosevelt graduated from Union College in 1847 and Harvard Law School four years later. He increased his inherited fortune through investments in real estate, bituminous coal, banking, and railroads. Fancying himself a country squire, he had the leisure of spending quality time with little Franklin, exploring nature, teaching him to honor trees as the noblest and weightiest of all living organisms, to respect the ethics of land stewardship, and to be a country gentleman, too. Franklin's nickname for his father was "Popsy," a term of endearment. "Mr. James passed on intact his own fondness for the outdoors," historian Geoffrey C. Ward explained. "He loved trees, for example, knew their varieties, would allow none to be cut unless they were dead or diseased, and he made sure his son felt the same way."[4]

In 1867 James Roosevelt had purchased an estate in Hyde Park. He later acquired additional acreage adjacent to it on both sides of the Albany Post Road.[5] Like most upscale properties in the mid-Hudson region, the Springwood estate—simply called "Hyde Park" by most Roosevelts—had Hudson River access as well as fruit orchards, vegetable gardens, a stocked pond, wide fields, and thick woodlands. It would be FDR's home throughout his life. Covering one square mile, the over six-hundred-acre estate formed a long, slender parallelogram, with its west side bordering the mile-wide river.[6] Though the rambling Main House had unpretentious amenities, it boasted a marvelous view of the scenic Hudson. From his third-floor bedroom little Franklin could look west across the river to see the hardwood forests of the Catskill Mountains. The Hudson River Valley became increasingly sanctified to many Americans after the Civil War, as magazines like *Scribner's* and the *Atlan-*

tic Monthly depicted the region's pastoral charms; so did Currier and Ives in such lithographs as "Hudson River–Crow's Nest," and "On the Hudson."[7]

Springwood, Franklin Roosevelt's lifelong home in Hyde Park, New York, was photographed from the air in 1932, looking to the southwest. The house is nestled among the trees (*center*). Franklin sailed and ice-boated on the Hudson River (*at top*), while his mother took pride in the formal rose gardens (*right*).

Whenever Franklin played at Springwood, collecting leaves in the hemlock forest, tobogganing down snowy hills, catching minnows in Elbow Creek, or chasing rabbits along the riverfront, he was living what Leo Marx called—in his study of agrarian romanticism, *The Machine in the Garden*—"the pastoral ideal."[8] Photos exist of Franklin riding his Welsh pony (Debby), playing with the family border collie (Monk), and leading his pet goat into a "secret garden." When the Hudson froze there were marvelous sleigh rides down the river to Newburgh. Locals made a livelihood selling blocks of ice to New York City and beyond.[9] In the spring, when the snow melted, Franklin would wear rubber hip boots and sail little boats on the Hudson.[10] In a flight of imagination, he would pretend hemlocks were ships, hanging sheets on their thick branches to make them resemble the mast of the USS *Constitution* (he had a small replica of the famous War of 1812 frigate in his bedroom trunk). He could only imagine the aquatic universe under the rippling surface of the river—the eels, water plants, and detritus from ships past. Memories of these Hyde Park outdoor adventures turned FDR, as an adult, into a tireless booster of the Boy Scouts of America; he believed that all young men deserved to experience the joys of nature just as he had done.[11]

As an only child tutored at home, Franklin had only a few playmates in Hyde Park: namely Edmond Rogers, Mary Newbold, and his niece Helen Roosevelt. Instead, during his leisure hours, Franklin roamed the estate's grounds, nurtured by chittering squirrels and drumming doves. Wandering around the terraced hills that rose from the Hudson River to Springwood's Main House, tracking robins and vireos, turning horses out to pasture, he developed an almost tangible intimacy with the land, never taking the staggering beauty of the Hudson River Valley for granted.[12] "The land existed for FDR in time as well as space," historian John Sears explained. "It had a history, and that history was organically connected to the present."[13] Dutchess County had once been inhabited by Algonquian-speaking Wappinger Indians, one of the northernmost of the Delaware peoples. Young Franklin often scoured his property for artifacts and relics.

Whenever Franklin arrived home after one of his Indian relic hunts, invariably scuffed up, his mother, Sara Delano Roosevelt, would ask him scoldingly and in alarm, "Where have you been?"

"Oh, out in the woods," was the usual answer.

"What doing?" Sara would ask.

"Climbing trees."[14]

In the 1880s there was a backlash in some circles against rapacious industrialists who mauled landscapes, like those of the Hudson River Valley and the Adirondacks, for timber and minerals. Painter Thomas Cole, in his seminal "Essay on American Scenery," bemoaned what he called the "ravages of the axe," which were destroying pastoral landscapes.[15] Deeply concerned about deforestation, the Roosevelts wanted state regulations and industry safeguards in place to fight wanton extraction. While much of nineteenth-century New York City smelled of smoke, coal, and oil, FDR's upstate environment during his childhood was pine-scented. As landscape perservationsists his parents sought to keep the Hudson River Valley green for future generations.

Since most of the land in New York where reforestation was especially needed during the Gilded Age was in the hands of private owners, the most practical solution to saving forests was acquiring more public lands. In 1885, the state launched two major forest preserves, which covered nearly a million acres in the Adirondacks and, on a much smaller scale, the Catskills. Unfortunately, the law was weak. In 1894, when FDR was twelve, the state superseded it with an amendment mandating that all state-owned forests be "forever kept as wild." After that, forest preserves were regulated to protect nature, first and foremost. Further, they were part of a movement in New York that allowed them to exist amid privately owned parcels in which commercial development would be largely restricted by the state. As these innovative "parks" coalesced, they were located across massive tracts in the Catskills and Adirondacks. Chopping timber in the "forever wild" forest preserves in the parks was banned outright, because of the amendment. On the private lands in the parks, it was overseen by the state to ensure conservation was at least considered. These developments in forestry, followed closely by the Roosevelt family while Franklin was growing up, gave birth to the modern wilderness preservation movement.[16]

The U.S. Department of Agriculture (USDA) joined the state of New York in visionary protection of American forestlands. At the prodding of grassroots activists and conservation-minded presidents such as Benjamin Harrison and Grover Cleveland, Congress preserved vast forests in the West. On the intellectual front, George Perkins Marsh's 1864 environmental manifesto *Man and Nature; Or, Physical Geography as Modified by Human Action*, which warned of the perils of deforestation, was influential among American conservationists. Picking up Marsh's torch was Charles Sprague Sargent, the director of Harvard University's Arnold Arboretum, who worried about shortsighted forestry practices. From 1887 to 1897 Sargent published his *Report on the Forests of North America*, which called for immediate reform of destructive timber-management practices throughout the nation. The Roosevelts of Hyde Park happily subscribed to Sargent's *Garden and Forest* magazine, a weekly aimed at fostering awareness of and interest in woodlots, horticulture, landscape design, and scenic preservation.[17]

Sara's father, Warren Delano, a Republican, had hired the building and landscape architect Andrew Jackson Downing, whose portfolio included the White House and U.S. Capitol grounds, to redesign his estate, Algonac (a name derived from an Algonquin Indian term meaning "hill and river"). Situated on the bluffs two miles above the town of Newburgh, the mansion left a lifelong impression on FDR. Algonac had originally been a modest brick-and-stucco house, but Downing enlarged it to forty rooms and situated it toward the south to capture views of the Hudson River, as well as the surrounding mountains. Two towers and a large veranda were also added. It appeared to visitors that Algonac was almost an organic part of the Orange County, New York, landscape, not an intrusion on it. But Warren Delano, the patriarchal millionaire, wasn't what later generations would consider an environmentalist. He owned coal mines near Johnstown, Pennsylvania, and copper mines in Maryland and Tennessee. (As a consequence, there were towns named Delano in both states.)[18] The inside of the manor had heavy carved redwood furniture, teakwood screens, and Buddhist temple bells. On the grounds there were a dozen species of trees. One of FDR's boyhood loves was embarking on the twenty-mile trip to Algonac, crossing the Hudson on the Beacon-Newburgh ferry. The compound there was so bucolic that the Roosevelts and Delanos used "Algonac" as a code word for "good news."[19]

Downing, who made his home along the Hudson all his life, was an early influence on urban parks. He promoted his philosophy of landscape architecture, designed to be more authentically American than manicured English gardens, in *The Horticulturist*, a magazine he founded, and in *A Treatise on the Theory and Practice of Landscape Gardening, Adapted to North America* (1841). His credo of living in harmony with nature took root throughout the Hudson River Valley.

Downing went from delivering plants to wealthy clients—the "River Families," as the local landed gentry were known—to being invited into their homes for grand dinner parties. Even after his death at the age of thirty-six, his popularity among River Families like the Roosevelts and Delanos remained strong.

Both James and Sara Roosevelt concurred with the popular Jacksonian novelist Catherine Sedgwick, who believed that "no American in good report, whether he be rich or poor," should "build a house or lay out a garden without consulting Downing's work."[20] Downing had a deep influence on Franklin's philosophy of living in harmony with nature. The Roosevelts adhered almost religiously to his principle of maintaining woodlands and making the natural setting important.[21] "We believe," Downing wrote, "in the bettering influence of beautiful cottages and country houses—in the improvement in human nature necessarily resulting to all classes."[22] Using local wood and stone was Downing's way of having his homes sink into the natural landscape. What James and Sara Roosevelt took from Downing—and passed on to Franklin—was the concept that Springwood should be a "middle landscape," a "rural Arcadia . . . aesthetically balanced between the extremes of wilderness and city."[23]

James Roosevelt carefully designed Springwood's grounds with refined simplicity, letting nature set the terms, in what Downing termed the "accessible perfect seclusion" from New York City.[24] He considered his prime riverfront acreage his greatest asset; the land would bring solace to his son Franklin for the rest of his life. With more than six hundred contiguous acres to work with—some of them pastureland and wetlands—James adopted an informal arrangement of landscape elements. On the grounds surrounding the house, he included a variety of tree species, laid out an irregular placement of plants, and created some gravel roads. Maintenance of Springwood's rose gardens, half-domesticated forests, wildflower meadows, pleasure grounds, and farmland was time-consuming, and executing Downing-style improvements required commitment—all the more so in the early twentieth century, when Franklin would expand the family estate to 1,424 acres.

James Roosevelt was convinced that fresh air was a curative. He taught Franklin that growing crops and trees was the key to lifelong happiness. He always maintained a vegetable patch, and Franklin continued the farm-to-table tradition when he came of age. Likewise, James dabbled in scientific forestry, growing trees on select Springwood parcels for eventual sale. Every Christmas, he gave evergreen trees to his friends, the domestic staff, and the groundskeepers.[25] FDR kept this fine family tradition going even during the Second World War.

James Roosevelt was also an accomplished equestrian; he and Franklin often rode horses side by side around Dutchess County. Everything from saddle care to dismounting like a gentleman was learned. Before Franklin was born, James had bred some very fast trotting horses for the racetrack. The stables James constructed at Springwood for the trotters boasted a three-story tower topped with a copper weather vane (with his initials in bold relief).[26] Sometimes, for variety, Franklin and James would ride twenty miles together from Hyde Park to visit the Delanos in Newburgh. During FDR's childhood, his father promoted the notion that "fresh air" built strong men. "He instilled in his son," biographer Kenneth S. Davis wrote, "the manly virtues."[27]

Sara Delano Roosevelt, FDR's mother, raised in Newburgh, was strong willed and attractive, with brown hair and bright eyes. She had a lifelong passion for learning. In girlhood she cruised to Hong Kong and traveled Europe extensively. Sir Walter Raleigh's six-volume *History of the World* was her favorite literary work.[28] Fluent in German and French, she learned to paint portraits and landscapes from Hudson River Valley artist Frederic Edwin Church, who lived in nearby Greenpoint at his estate, Olana, overlooking the Hudson. And she loved the fresh air, often saying, "All weather is good weather."[29]

For generations, the seafaring Delanos had made fortunes in the trade of opium, sugar, and tea from China; Sara traveled there for the first time at age eight. The Delanos could trace their roots back to the *Mayflower* and to French Huguenot immigrants of the seventeenth century. In 1817 Captain Amasa Delano of Massachusetts wrote the first quasi-scientific tract about the Galápagos Islands, decades before Charles Darwin's expedition on the *Beagle*. One of Sara's grandfathers had been a New Bedford whaling captain; in the early nineteenth century, he built a home in Fairhaven, Massachusetts, within a day's sail of Buzzards Bay and Naushon Island.[30] "Despite Sara's far-flung travels," biographer Jan Pottker wrote, "her schooling abroad, and her frequent trips to Manhattan, she preferred the country to city life."[31]

Sara was a careful mother who instinctively played the role of Hyde Park matriarch.[32] Blessed with a gift for homemaking and enabled by a loyal staff, she indulged Franklin's every whim.[33] Any playground scratch or bruise her boy suffered was dramatized into a life-threatening medical crisis. He was raised to become a Dutchess County gentleman—the most fitting trajectory for a patrician lad. All of her husband's distant cousin Theodore Roosevelt's sweat-and-blood rhetoric, as typified in his Darwinian philosophy of the "strenuous life," left her stone cold. "My son Franklin is a Delano," she insisted, "not a Roosevelt at all."[34]

There were seminal differences between the Oyster Bay (Long Island) and Hudson River Valley branches of the Roosevelt family. The Oyster Bay clan was Republican; those in the mid-Hudson, like James, were Democrats. Though Theodore Roosevelt was only FDR's fifth cousin, he cast a long and triumphant shadow. To FDR, he was always "Uncle Theodore," his idol and counselor. Both the Oyster Bay and the Hudson River clans believed passionately in the conservation of natural resources. They shared the idea that depleted soil, polluted water, clear-cut forests, and dwindling wildlife could all be *restored* to their former glory with proper scientific management. Although the reform-minded TR articulated the sentiment best, the entire family was appalled at the crime, poverty, and urban blight that had turned so many New York City neighborhoods into hellholes of squalor. All the Roosevelts believed cities needed to imitate the virtues of small towns like Oyster Bay and Hyde Park, by maintaining tree-lined streets and building handsome parks. While not antiurban per se, the Roosevelts championed "regional cities" where cosmopolitanism could intersect with community values and nature to form an *ideal*.

Sara Roosevelt idealized the Hudson River Valley with a blend of unabashed sentimentality and plain reasoning. There was something magical, she knew, about the 825 square miles of Dutchess County with its many pristine pockets of wild greenery and thick woods. With the exception of the city of Poughkeepsie, life in the county was quietly pastoral and agricultural. In 1938, while Franklin Roosevelt was in the White House, Sara Roosevelt wrote a nostalgic foreword to a locally published book, *Crum Elbow Folks*, that pined for the "pat of horse's feet in the gray sand dust." Like that of the nature essayist John Burroughs, who lived across the Hudson from Springwood in West Park, her favorite bird was the hermit thrush (*Catharus guttatus*)—"the shyest of songsters"—because it "poured forth" glorious woodland melodies. Sara bemoaned the plight of the fox grape, a musky-scented vine that, she recalled, "filled the air with its perfume," while in autumn "the little bunches of bright blue grapes were hung thick among the branches."[35] She maintained that the grape was "ruthlessly destroyed" by misguided farmers who had no aesthetic appreciation for the vine.

Both James and Sara, as good Democrats, thought Grover Cleveland of Buffalo, New York, was a model leader. They supported his campaigns for governor and president with ardor, admiring his bravery in confronting the Democratic machine known as Tammany Hall. Cleveland served two separate terms as president, winning in 1884 and 1892, but losing in 1888. James and Sara felt President Cleveland had used his veto power impressively to prevent special interests from gaining a foothold in his government and successfully kept the country out of international entanglements. At one point, James took Franklin to Washington to meet President Cleveland at the Executive Mansion (it was renamed the White House by Theodore Roosevelt in 1901). "I'm making a strange wish for you, little man, a wish that I suppose no one else would make," Cleveland said to Franklin, placing his hand on the young man's head. "I wish for you that you may never be president of the United States."[36] At the time, in the mid-1890s, Franklin's distant cousin, Theodore, was beginning his path to the White House, as president of the board of the New York City Police Commissioners.

Young Franklin was proud that another distant cousin, Robert Barnwell Roosevelt was considered the preeminent pisciculturist of New York during Reconstruction and the Gilded Age. While RBR, a Democrat, had an amazing résumé—including a term representing New York in Congress, a stint as ambassador to the Netherlands under President Cleveland and political prominence as treasurer of the Democratic National Committee—wildlife protection was his true calling. He wrote stylish essays about the art of angling and about the biology of trout, perch, and shad. A devoted follower of Charles Darwin, RBR documented the evolution of eels and frogs. He railed against industrial outfits that dumped contaminants into the Hudson River, thereby preventing Atlantic sturgeon (*Acipenser oxyrinchus oxyrinchus*) from successfully reproducing.[37] The prolific RBR wrote esteemed books including *Game Fish of the Northern States of America and British Provinces* (1862), *The Game-Birds of the Coasts and Lakes of Northern States of America* (1866), and *Superior Fishing* (1884).[38] Every June, he celebrated the northern migration of striped bass (*Morone saxatilis*) from his Long Island home to the coastal waters of New England. Working with fish culturist Seth Green of Caledonia, New York, RBR helped establish state-run fish hatcheries in Cold Spring Harbor and Rochester.

RBR was one of the early "riverkeepers" of the Hudson. By the late 1890s the waterway was being destroyed by runoff from quarrying; its water was often too toxic to drink. Unregulated factories had turned quiet backwaters and tributaries of the great river into dumping grounds. Oil spills from ships on the Hudson, particularly in the summertime, caused "petroleum vapor conflagration"; this, in turn, triggered respiratory illnesses.[39] Pollution came not only from industrial pollutants but also from human waste. In an era before modern sanitation, pits under privies were supposed to be emptied regularly by "night soil" workers, but sometimes the waste made its way into the Hudson and other rivers in New York—especially as the population grew. The water became contaminated and the incidence of cholera and other communicable diseases soared. After a cholera outbreak killed thousands in 1832, the Old Croton Aqueduct was built alongside the Hudson, running through the towns of Westchester County before entering the Bronx at Van Cortlandt Park. By the 1890s, New York City received 90 percent of its drinking water via this aqueduct.[40]

Prominent River Families like the Harrimans, Osborns, Borgs, Luddingtons, and Morgenthaus, who believed the river helped define them, fought to make water from the Hudson healthy.[41] Like the Roosevelts, these wealthy families considered the river America's Rhine—placid, restrained, and seldom reaching flood stage. Together they labored to save their beloved Hudson from ruin.

II

In the summer of 1883, Sara, James, and Franklin went north on a holiday. Their destination was Campobello Island in the Canadian province of New Brunswick, situated where the Saint Croix River meets the Bay of Fundy (only a moat of water separated Campobello from the fishing hamlet of Lubec, Maine, the easternmost point of the United States).[42] Once the train deposited the Roosevelts in Eastport, Maine, the family took a ferry across the border to get to Campobello, a fifteen-square-mile island.[43] The Roosevelts stayed at Tyn-y-coed (House in the Woods), a secluded inn with wonderful waterfront views and a "reputation for fresh air."[44]

The Roosevelts had such a grand time at "Campo" that they returned there year after year. Eventually they bought a four-acre parcel on the conifer-dotted island, where they built a sprawling, gable-roofed, maroon house of thirty-four rooms. They were finally able to stay in the mansion, dubbed "Granny's Cottage," beginning in 1895.[45] Every summer James Roosevelt would dock his sailing yacht at the island. With Franklin as his first mate, they'd navigate the windy Passamaquoddy Bay, an inlet of the Bay of Fundy characterized by rocks, vicious currents, and drastic tides. When Franklin was six, a photograph was taken of him clutching the steering wheel, seeming at least "momentarily in command" of the vessel.[46]

Franklin Roosevelt's favorite ships were schooners with two or more masts. These rakish vessels had sailed in great numbers in the Atlantic Ocean from 1830 to 1920. By the time Franklin turned ten, he had become a solid sailor with an understanding of how to "change tack," remain flexible in changing winds, and navigate capably even in fierce chop. Everyone was impressed with how easily he read nautical maps and understood weather conditions at sea. For Franklin's sixteenth birthday, his father gave him a twenty-one-foot knockabout; he named the boat *New Moon*.[47]

In late summer the Roosevelts traditionally left Campobello for a month's vacation in the Saint Regis area of the Adirondack Park. From the favored lodge grounds Franklin studied mountainsides replete with hemlock, pine, and spruce trees.[48] Even the decaying tree trunks were of interest to him. Families with vast wealth and good taste—the Rockefellers, Astors, Vanderbilts, and Roosevelts among them—built or leased "camps" high in the Adirondacks to escape the summertime heat and the congestion of their city lives. These camps weren't mere tents in the forest any more than the marble palaces of Newport were "cottages." The stateliest variety, which the Roosevelts favored, were rustic mansions built with timber and stone. The wealthy Adirondacks set shared a belief that small-scale family logging, mindful of reckless forestry practices such as clear-cutting, were vastly preferable to giant outfits like the Hudson River Pulp Company, then using new technology for processing paper pulp from ground wood.[49]

Shuttling around the Adirondacks, New York City, Hyde Park, Campobello Island, and western Europe became commonplace for young Franklin. By the time he turned fourteen, he had already made seven voyages across the Atlantic Ocean, on trips ranging in duration from two to four months.[50] On one such holiday, in May 1891, James, suffering from heart problems, brought his family to Bad Nauheim, a German spa. As people had done since the time of imperial Rome, he sought relief at the world-famous hydrotherapy facilities. While his father "took the cure," Franklin went sightseeing in such nearby cities as Cologne and Heidelberg, and attended a German-language school. He particularly enjoyed exploring the well-tended German forests—managed successfully for centuries—that surrounded Bad Nauheim. "The interesting thing to me, as a boy even," Roosevelt later recalled, "was that the people in that town didn't have to pay taxes. They were supported by their own forest."[51]

III

Barely a day went by when Franklin didn't talk about the world of birds. At ten years old he started dabbling in oology, the collecting of eggs and nests. There is a well-circulated story about young Franklin racing into a family Easter party holding a blue-speckled robin's egg in his hand as if it were a Tiffany jewel. James Roosevelt eventually discovered drawers full of nests and eggs hidden in his son's bedroom. Displeased, he ordered Franklin never to rob a nest of more than one egg. That wildlife conservation lesson stuck. So did Franklin's love of birds. As a boy, he began a very grown-up course of study, reading copiously, making field notes, and demonstrating to others his ability to organize in his own mind all that he was learning. Soon the boy gained his own reputation, independent of his family, as a local authority on birds.[52]

On occasion, Franklin gave impromptu lectures at Springwood and Campobello for family members, neighbors, and household servants on subjects such as the Atlantic Flyway (although this term for the bird migration route stretching from Canada to the Caribbean wasn't officially used until 1947). "Many people do not know what a great variety of birds we have," he wrote in his first ornithological essay. "They can always point out a robin but probably could not tell the difference between a Fox Sparrow and a Song Sparrow and think that a nuthatch is a woodpecker."[53]

When Warren Delano of Newburgh heard his grandson hold forth on "The Shore Birds of Maine," he gifted Franklin a lifetime membership in the American Museum of Natural History (AMNH) on Manhattan's Upper West Side. And he introduced Franklin to the organization's esteemed president, vertebrate paleontologist Dr. Henry Fairfield Osborn.[54] Not only was Osborn a great advocate for Hudson River and Jamaica Bay (between the boroughs of Brooklyn and Queens) ecological preservation, he also headed the Save-the-Redwoods League in California.

The Roosevelts had other connections to important conservationists. Their close friend George Bird Grinnell, editor of *Forest and Stream*, had established the first Audubon Society in 1886, just four years after Franklin's birth. The Audubon Society's declared mission was to outlaw the mass slaughter of wild birds that weren't fit for human consumption; the vandalizing of nests and stealing of eggs; and the use of feathers in fashion or as ornaments.[55] Women's fashion of the period dictated that sophisticates wear hats adorned with exotic plumage from herons and egrets. Whole flocks of migratory waterfowl were being shot in Florida, Georgia, and Louisiana by hunters eager to supply New York milliners with feathers.[56]

On Franklin's eleventh birthday, James Roosevelt gave his boy a handsome pellet gun for the purpose of collecting bird trophies. It wasn't long before his mother was able to record that his first shot struck a crow (*Corvus brachyrhynchos*).[57] The hobby stuck. Wandering around his family's woodlands, he learned that different species of birds had their own favorite kinds of trees. The cedar waxwing (*Bombycilla cedrorum*), for example, gravitated toward hawthorns, while the red-cockaded woodpecker (*Picoides borealis*) claimed longleaf and slash pines. One day near the hamlet of Staatsburg, Franklin studied a Cooper's hawk (*Accipiter cooperii*) that flew right up to him "and appeared to be tame"; the unafraid raptor had probably migrated from Canada and never before encountered humans.[58] Obsessed with bird checklists, Franklin shot and classified three hundred species native to Dutchess County.[59] Most of the bodies were carefully preserved. Family members—Roosevelts and Delanos alike—joked that Franklin was himself a magpie, a collector of *everything* related to ornithology. Learning the complete taxonomy of species, he painstakingly wrote Latin labels for each specimen to place near its claws. "It was not long before the big mahogany cabinet in the library acquired a collection of brand new inhabitants," Sara Roosevelt recalled in her memoir, *My Boy Franklin*. "There was an oriole, a heron, a robin, a woodpecker, and even a hawk, but the winter wren was missing."[60]

Whether Franklin Roosevelt would ever snag a winter wren (*Trog-*

lodytes hiemalis) became a popular dinner-table topic at Springwood. It was a challenge that the self-styled ornithologist took up with gleeful determination. One afternoon he nonchalantly walked into the main house looking for his fowling piece. "There's a winter wren way up in one of the big trees down there," he said confidently. "I want to get him." His mother chuckled at his boyish naïveté. "And do you think that wren is going to oblige you by staying there while you come in and get your gun to go back out and shoot him?" Franklin was undaunted. "Oh yes," he replied, "he'll wait."

An amused Sara Roosevelt watched her son race across the lawn, prepared to tease him for coming home empty-handed. But to her imperishable surprise Franklin returned to the house a single shot later with the dead winter wren in hand.[61]

The majority of Roosevelt's specimens were from Springwood and Crumworld Forest, the neighboring estate belonging to Colonel Archibald Rogers, which may have been the best natural environment in the Hyde Park area for bird-watching. Rogers had worked with the department of forestry at Cornell University to turn his Hyde Park property into an outdoor aviary consisting of a combination of shady tree groves, thick underbrush, and specialized plantings. It was Rogers who encouraged the Cornell Agriculture Experiment Station to help him gather data for a series of Dutchess County residential reforestation projects. Rogers had completed his house in 1889 after buying five smaller estates to form his impressive grounds, and every spring he had trees planted by the thousands. He encouraged the Roosevelts to develop a scientific forestry plan for Springwood. Even after automobiles became ubiquitous, the colonel preferred to travel on horseback to avoid scaring birds. His primary ambition in life was to be the kind of land steward George Washington would have warmly embraced as a neighbor in Mount Vernon, Virginia.

Owing to Franklin's enthusiasm, ornithological pursuits were built into the Roosevelts' European itineraries. While in London one year, Franklin wanted to make an excursion to Osberton, the Nottinghamshire seat of Cecil Foljambe (earl of Liverpool, a friend of the Roosevelts), to study his famous collection. When James canceled the Nottinghamshire trip for business reasons, Franklin was inconsolable.

"Mummy, can't I go without you?" he pleaded.

"You mean you'd visit people you'd never met?" she asked, astounded.

"I'd go anywhere to see those birds!" he answered.

Tired of Franklin's pestering, James and Sara agreed to let their son take the two-hour train ride alone. Foljambe embraced Franklin as if he were kin. Bursting with sophisticated enthusiasm, he showed the young American his world-class collection of birds from the faraway Amazon and Arctic. Franklin considered the experience a high-water mark in his European education.

Over time, Franklin grew into a decent taxidermist. But, as Sara noted, cutting out the insides of an owl or bluebird often turned him "green." More and more, warnings about the lethal effects of arsenic—a chemical commonly used in taxidermy—gave her pause. A public health campaign was under way to outlaw arsenic. Under parental sway, Franklin eventually farmed his specimens out to professionals in Poughkeepsie and New York City. A number of Franklin's preserved birds were accepted by the American Museum of Natural History—the first serious accomplishment in the future president's storied career.[62]

On March 3, 1895, Sara brought Franklin to Manhattan for a meeting of the Linnaean Society of New York. Named after naturalist Carl Linnaeus, the eighteenth-century Swedish naturalist who laid down a lasting foundation for the categorization and naming of species, the organization provided a lively outlet for the study of natural history and ornithology. Among its founding members were nature essayist John Burroughs, editor George Bird Grinnell, and botanist Eugene Bicknell, after whom Bicknell's thrush (*Catharus bicknelli*) was named. Theodore Roosevelt had been a dues-paying member of the Linnaean Society since 1878. As part of Franklin's education, Sara took her son to hear Dr. William Libbey III, a professor of physical geography at Princeton University and director of its esteemed Elizabeth Marsh Museum, deliver a lecture about Hawaii at AMNH. Franklin, fascinated, took careful notes. Once back at Springwood,

Sara helped her son polish his jottings into a full-fledged essay worthy of submission to the Linnaean Society; his piece was ambitious for a boy his age.[63]

Franklin, age ten, practiced archery while on vacation with his family in Germany in 1892. He and his parents visited the resort of Bad Nauheim in south-central Germany every summer during his boyhood. Throughout the surrounding Watterau Valley, he was exposed to the idea of careful forest management, something barely known in America at the time.

Franklin wrote an able description of Maui's Iao Valley, indigenous plant life, and the kukui nut tree (*Aleurites moluccana*), which yielded "great quantities of oil for lamps." But it was Hawaii's active volcanoes—in an area that President Woodrow Wilson would preserve as Hawaii Volcanoes National Park in 1915—that set his imagination aflame. "The volcano of Kilauea is the highest in the world, being over 14,000 feet high, as high as Mt. Blanc," Roosevelt wrote in careful cursive. "Near the volcano are many cracks in the soil from which sulphurous steam comes out. At one end of the crater is the Burning Lake or Lake of Fire, in which Prof. Libbey threw a log of wood and proceeded to run for his life, as the log of wood with a quantity of molten lava was thrown high into the air. The whole surface of the lake was bubbling up and quantities of steam rose from it. Around the crater are several underground passages, in which are huge lava stalactites which sometimes fall around and break with a fearful crash."[64]

Inclement weather never curtailed Roosevelt's ornithological pursuits around Dutchess County. Combating the whistling winter winds, he started keeping "Bird Diaries," written in his elegant penmanship, early in 1896. He marked the date a bird was seen, the weather and temperature at the time of the sighting, the number of specimens he counted, and any other notable traits and characteristics he deemed relevant. Birds were by nature difficult to count, but Franklin tried his hardest. Here's a sample entry from the first Bird Diary:

Wea. Fine *Mon. Feb. 10, 1896* *Ther. 30°*
SHOT
1 fine red male Pine Grosbeak & saw 1 other.
Also, 1 Blue Jay.
SAW
1 flock of about 50 Pine Grosbeaks.
Also, another flock of about 25 individuals.
Also, 14 single Grosbeaks at other times.
Chickadees, Nuthatches, Juncos, Jays, Crows, and Downy Woodpeckers.
Sent Grosbeak to W. W. Harts & Co. New York.[65]

Just a few days after Franklin Roosevelt entered those observations in his field diary, he returned to New York City for a tour of the American Museum of Natural History with the head ornithologist, Dr. Frank M. Chapman, a dear friend of Theodore Roosevelt. Always eager to talk about birding, Chapman was an expert guide. Charged with educational outreach for the museum, he viewed Franklin as a promising protégé. In 1894 Chapman had become the associate editor of the *Auk*, an organ of the American Ornithologists' Union (AOU).[66]

Modeled on the British Ornithological Union, the AOU was created with a primary mission similar to that of the National Audubon Society: preventing bird extinction in North America.[67] Two friends of the Roosevelt family—Dr. Elliott Coues of Washington, D.C.; and ornithologist Dr. C. Hart Merriam of Locust Grove, New York—were among its founders in 1883. The chairman of the AOU was William

Brewster, who became the curator of the Museum of Comparative Zoology at Harvard University in 1885. The AOU led in the creation of waterfowl sanctuaries throughout Florida, Mississippi, Louisiana, and Texas. In 1886, the AOU's Committee on the Protection of Birds drafted a "model law," which was adopted later that year by the government in New York. Making non-game birds safe from hunting, while defining what species would be considered game birds, the new law was the opening salvo in the modern wildlife-protection movement.

But even more significantly, the AOU, in conjunction with the Smithsonian Institution, persuaded the U.S. Department of Agriculture to establish the Division of Economic Ornithology and Mammalogy under the leadership of Dr. Merriam in 1886. While ostensibly this division was funded to help farmers deal with pests, like the English sparrow (*Passer domesticus*), Merriam used his connection to the AOU to begin conducting field surveys and distributing studies of birds, mammals, and other biotic communities.[68]

By the first decade of the twentieth century the Division of Economic Ornithology and Mammalogy had evolved into the Bureau of Biological Survey (the precursor of today's U.S. Fish and Wildlife Service).[69]

Young Franklin was enthralled by the naturalists he was able to meet—and they were impressed by him. In addition to leading the American Museum of Natural History and helping to edit the *Auk*, Dr. Chapman would travel as the British West Indies and Mexico in search of rare species. However, it was his homespun expertise on the "common" birds of the Hudson River Valley—like the robin—that brought him the most acclaim.[70] In the process, as the *New York Times* observed, Dr. Chapman became the most influential man since John James Audubon in getting Americans "interested in birds."[71]

Roosevelt was a willing follower and Dr. Chapman offered the boy an associate membership in the AOU. Poring over Chapman's *Birds of Eastern North America* (first published in 1895) became a ritual for FDR. His bird list grew rapidly throughout late 1896 and 1897, filled with tallies of specimens "shot & stuffed or skinned by F. D. Roosevelt." One banner day Franklin acquired both a scarlet tanager (*Pi-*

ranga olivacea) and an indigo bunting (*Passerina cyanea*)—an admirable ornithological feat.

Joining Franklin in his devotion to birding was a fellow River Family boy, Maunsell Crosby. Five years younger than FDR and raised at Grasmere, an estate just outside the village of Rhinebeck, the boy was the scion of the Livingston family, which had played an important role in the founding of the United States. Philip Livingston had signed the Declaration of Independence, William Livingston had helped draft the Constitution, and Robert Livingston had administered the oath of office to George Washington. Inspired by Dr. Chapman, Maunsell decided early on to become an ornithologist, and he, too, had been granted associate membership in the AOU. Like FDR, he attended Linnaean Society meetings in New York City. In coming years Crosby would conduct the first Audubon "Christmas count" in Dutchess County.[72]

For use in ornithological study by others, Franklin's birds needed to be tagged for identification. In May 1896, soon after he started his Bird Diaries, Franklin noted, "I am to send about 1 dozen Grosbeaks to Museum of Natural History for local collections."[73] It was in fact ten Dutchess County pine grosbeaks (*Pinicola enucleator*) that Franklin eventually donated to the museum. These robin-sized birds, cute like finches and with a slightly forked tail, usually foraged for food in the trees at Springwood. Permanent residents of the Hudson Valley, they made a mellow *teu-teu-teu* sound that Franklin found soothing. Because of their wide distribution, the pine grosbeaks weren't considered rare. Nevertheless, Franklin amassed a good mixture of regional specimens—male and female—for Chapman's shop to study, considering ornithology "one of my chief avocations."[74] That year he also wrote a short article on birds for a children's magazine, the *Foursome* (unfortunately much of his piece had been plagiarized).[75]

Roosevelt often romped around Hyde Park with his half brother, Rosey. Twenty-eight years older than Franklin, Rosey lived only a mile down the Post Road from Springwood. Close despite their age difference, he and Franklin both eagerly anticipated the migration of bird flocks, which occurred every winter and spring. "Shot a Pine

Finch," FDR wrote on an outing with his half brother. "The bird was alone in a small pine tree and he appeared very shy. I had trouble shooting him."[76]

IV

While tutors supervised Franklin's early education, James Roosevelt remained a strong influence on him, especially when it came to land stewardship and forest conservation. Cardiac problems, however, kept James from engaging in strenuous planting or pruning at Springwood. Even though the actual upkeep of Springwood took a backseat

Franklin, thirteen, and his father, James, sixty-seven, posed together in 1895. James, formerly the president of two railroads, had retired early in favor of the country life in Hyde Park. Despite James's flagging health, father and son were playmates, as well as friends. James taught Franklin to ride at the age of four and to sail at six.

to his New York City investment interests in coal and railroads, James continued to instill in Franklin his conviction that proper land management was the best way to protect nature and have a fulfilled life. FDR embraced this belief as his own.

A high-water mark in James Roosevelt's public life came in 1892, when he was chosen as an alternate commissioner to the World's Columbian Exposition in Chicago, organized to celebrate the four-hundredth anniversary of Christopher Columbus's voyage to the

New World. The exposition grounds were designed by Frederick Law Olmsted and planner Daniel Burnham on a six-hundred-acre site; it included a wooded island park and the Midway Plaisance, a mile-long amusement park. The fair, sometimes referred to as the "White City" because of its classical-style architecture, opened in May 1893 and drew an astounding twenty-six million visitors before it closed late that October.

The highlight of the Exposition for FDR—who attended with his Hyde Park friend Edmund Rogers, the colonel's son, arriving via James's private railcar—was watching Native Americans pick up pennies using long whips and studying the taxidermy displayed in dioramas.[77] With wide-eyed wonder, Franklin could see more than fifty thousand specimens of flora and fauna displayed at the Exposition (these became the nucleus of Chicago's Field Museum of Natural History).

An exhibit of trees native to New York State was Franklin's favorite attraction at the fair. The caption under an elegant photograph of a stand of sweet gums explained how, in 1802, Alexander Hamilton had brought *Liquidambar styraciflua* saplings from Mount Vernon in Virginia to the upper end of Manhattan Island to plant himself. The plot in Manhattan where America's first secretary of the treasury planted the sweet gums became known as Hamilton Grange.[78] This display made Roosevelt keenly aware of the symbolism behind planting "historical seeds" at Springwood in the coming years.

Studying natural history—not just ornithology—was Franklin's pastime when his parents took him on another European trip in 1896. That summer, the Roosevelts visited half a dozen cities—and Franklin dashed off to as many natural history museums as he could. In London, when he learned that the Prince of Wales (later, King Edward VII) was presiding over the opening of a new ornithology exhibit at the South Kensington Museum, he was especially excited. Admission to the event, however, was by invitation only. Undeterred, Franklin and his private tutor, Arthur Dumper, artfully crashed the soiree. Roosevelt slipped his American Museum of Natural History membership card to Dumper, who, in turn, presented it to security in lieu of a proper invitation. The guard carefully studied the document, deeming it a valid credential. As Roosevelt later wrote, he and Dumper were thereafter accorded the courtesies due true scientists.[79]

Once the family returned from abroad, Franklin prepared to enter Groton School in Massachusetts. Reverend Dr. Endicott Peabody, the boarding school's headmaster, was an Anglican minister who had been educated at the prestigious British schools Cheltenham and Cambridge. Founded in 1884, Groton was situated along the Nashua River not far from Boston. Peabody's aim was to make Groton a preparatory school on par with British public schools and prepare the sons of the wealthy and prominent for an Ivy League education. Every night following evening prayers, Peabody shook the hand of each boy in a ritual that became known as the "go-by" as they wandered off to bed.

Because Franklin was only fourteen when he started at Groton, there is a mistaken tendency to see him as a blank slate, overparented but well tutored in foreign languages and history. But Franklin—carrying 140 pounds on his nearly six-foot frame—wasn't an enigma or clay to be easily molded. Quite simply, Roosevelt was already what novelist Wallace Stegner called "a placed person," fully belonging to the Dutchess County countryside.[80] As historian David Schuyler noted, Roosevelt, like others from the region, saw the Hudson River Valley as a "sanctified landscape" that represented "a place of transcendent importance to a regional and national cultural identity."[81] No matter what longitude or latitude Roosevelt happened to be in, his inner compass was always pointed toward the Hudson. Over the years he developed the philosophy that Hyde Park, with its strong cultural identity, was a model for other American villages to emulate. Cardinals never seemed so red to Roosevelt, or trees so elegant and full-bodied as on the Springwood grounds. "All that is in me," FDR said, "goes back to the Hudson."[82]

CHAPTER TWO

"I JUST WISH I COULD BE AT HOME TO HELP MARK THE TREES"

I

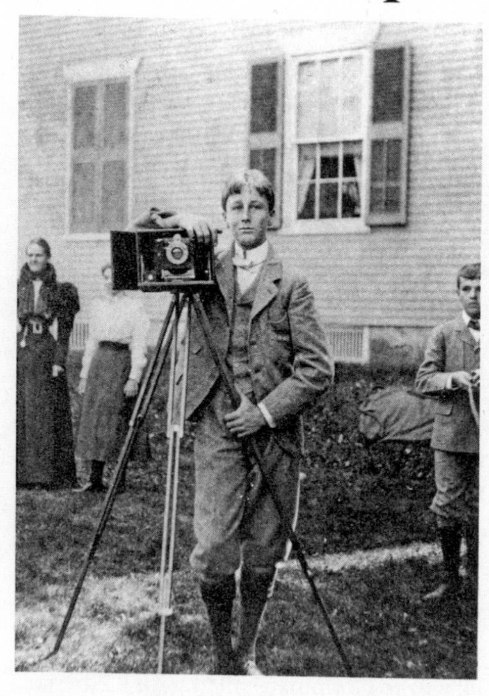

Posing with a camera, Franklin shows off during a stop at the home of his mother's relatives in the sea town of Fairhaven, Massachusetts, in 1897. The fifteen-year-old was on his way to the northern part of the state to enter Groton—the first school he ever attended, after years of private tutoring.

Fourteen-year-old Franklin Roosevelt was homesick during his first semester at Groton School in the fall of 1896. Wedded to country living, he missed the way the blue light shifted hues by the hour on the Hudson, the sight of the Catskill Mountains deep on the western horizon, and the string of Hudson River barges traveling past Springwood. But he adapted to boarding school, falling under the spell of the thirty-nine-year-old headmaster Reverend Dr. Endicott

Peabody. With his slight build and lack of experience with sports, Roosevelt wasn't much of an athlete. Sailing, collecting stamps, and ornithology remained his hobbies.[1] Roosevelt was susceptible to infection and, exposed to boys his age en masse for the first time, fell ill with debilitating conditions such as scarlet fever, measles, and mumps. Although he wasn't an academic star, he enjoyed Greek, Latin, history, literature, and science. Overall, his grades hovered in the "just average," B– range. Nevertheless, Dr. Peabody deemed him a "model Grotonian" and a "champion of its precepts."[2]

Not everyone agreed with Peabody's sterling assessment. In the eyes of his Groton peers, Franklin Roosevelt was handsome and likable, if a little delicate and over-mothered. He had a quick, infectious laugh that was accentuated by his wide and mischievous smile. One classmate recalled that Franklin "developed an independent, cocky manner and at times became very argumentative and sarcastic." The same classmate remembered that FDR "always liked to take the side opposite to that maintained by those with whom he was talking."[3] What nobody disputed was his exuberance of spirit and his easy acclimation to Spartan dormitory living.

Franklin's longing for the Hudson River Valley was a constant theme in his letters home. He made the best of having the Nashua River nearby, regularly prowling around the thickly forested hillsides along its banks in search of winged visitors. "The lovely birds are beginning to arrive," Roosevelt wrote to his parents one spring day. "So far Song Sparrows have come in full force and I have heard several Bluebirds and seen one Robin. It is too bad the holidays come so early this year as only a few birds will have arrived and very few things will be out."[4]

Most parents mailed packages to Grotonians filled with medicine, baked goods, jams, and cheeses. But FDR's parents sent him *Scientific American* magazine, W. M. Gibson's *A Rambler's Calendar of 52 Weeks Among Insects, Birds, and Flowers*, and books by John Burroughs. During his sophomore year, Sara mailed him the recently published two-volume *Audubon and His Journals*. He was spellbound. "I have wanted *Audubon's Journal* ever since it came out and it is the nicest present you could possibly give me," he wrote in appreciation. "I shall spend every moment on it, but I don't really think I could give such a lovely book to the library as I should very much like to have it at home on my ornithological book-shelf!"[5]

Franklin lost himself completely in the world of Audubon's indefatigable treks through Florida, Louisiana, and Kentucky in the 1830s. Next, he asked Sara to send his Dutchess County bird specimens—including his favorite barn owl. "I am very glad my sparrow-hawk came from Rowland and many thanks for sending it," he reported in his next letter home, "and also paying my A. O. U. subscription [for the *Auk*]."[6] *Audubon and His Journals* spurred Roosevelt to re-form the defunct Groton natural history club, adopting guidelines similar to those of the American Ornithologists' Union. "Today a very important event in the history of the school took place, at least I hope so," Roosevelt wrote home in May. "Four or five of us boys have with the approval of the rector formed a natural history society. You know a large room has been set apart in the new school-building for a museum, and there are at present no collections to put in it, so we are to do systematic work."[7]

Roosevelt took a shine to Edward S. Morse, a visiting professor at Groton who taught him about ants, butterflies, and other insects. Morse was one of the founders and the curator of the esteemed Peabody Academy of Science in Groton, as well as editor of *American Naturalist*. With the rector's permission, Morse led students on a trip to a Middlesex County forest to learn how to collect field specimens using brooms, vials, cheesecloth, nets, and glass jars. This romp led Franklin to engage even more with the Natural History Society. "It will be most interesting as well as instructive, and we are all enthusiastic," Roosevelt wrote to his parents. "Our idea is not to let everyone into the Society but only those who care for natural history and who will take an interest and really work."[8]

Morse took Roosevelt and his birder friends at Groton on another outing, this time to collect insects, leaves, and soil samples. Franklin learned about the Native American tribes—the Nashua and Wachuset—that had lived along the Nashua River, hunting, fishing, and growing crops on small plots along the river. They were the original custodians of wild New England. "Professor Morse started by turning over old logs and stones, etc.," Roosevelt wrote to his parents. "He also gave us a good idea about what to collect, how to preserve and label it and other things about the new museum."[9]

While at Groton, FDR heard about a lecture that Dr. Chapman had given at the Linnaean Society of New York in which he argued that the newest generation of bird enthusiasts—which included Franklin Roosevelt and Maunsell Crosby—should retire their guns and taxidermy kits in favor of the camera.[10] The "Chapman doctrine" made a deep impression on FDR, and he mostly stopped shooting birds in order to study them; however, he did occasionally continue to hunt ducks and geese from blinds.

Because Franklin so enjoyed Groton's Natural History Society, he agreed to manage the school's summer camp on Squam Lake in New Hampshire.[11] Serving as part of the faculty, Roosevelt was to teach poor boys from urban areas, aged ten to fourteen, about forestry, bird-watching, camping, swimming, and canoeing. According to a pamphlet published by the Groton School, the camp's purpose was to teach the poor city lads "the differences of environment" in New England.[12] Franklin quickly formed a kinship with many of the first-time campers, taking them out to explore the lake's thirty named islands (and numerous unnamed islets) and watch the common loons (*Gavia immer*) dive underwater for fish. This first visit to Squam Lake also inaugurated his lifelong interest in protecting the rugged White Mountains of New Hampshire and Maine. The most famous peak, 6,288-foot Mount Washington—part of a line of summits known as the Presidential Range—seemed to Roosevelt to be the Mount Rainier of New England. During his time at Squam Lake, Roosevelt pondered why the forest-lush East Coast didn't have large national parks like Yellowstone and Yosemite where American young people could camp in the pristine wilderness.

While Franklin was at Groton, public health became inextricably linked to environmental degradation. Giant oil storage tanks, chemical plants, factory smokestacks, and power lines were at war with the natural world. Steel magnate Andrew Carnegie shocked a meeting of the Pittsburgh Chamber of Commerce by pleading for someone to get rid of the "smoke nuisance" that was causing citizens to flee from western Pennsylvania in search of fresh air. Carnegie, an industrialist-conservationist, was asking the federal government to establish antipollution laws. During Franklin's junior year, Congress passed the Rivers and Harbors Act of 1899, which aimed to preserve navigable waters by outlawing dumping or dredging without permission from the Army Corps of Engineers. Around the same time, the Missouri supreme court ruled in *Missouri v. Illinois and the Sanitary District of Chicago* that Chicago had to maintain a proper drainage canal for sewage: "It is a question of the first magnitude whether the destiny of the great rivers is to be the sewers of the cities along their banks or to be protected against everything which threatens their purity."[13]

The highlight of FDR's time at boarding school came when thirty-eight-year-old Theodore Roosevelt visited Groton in June 1897. TR, who had served as superintendent of the New York City police commission until earlier that year, regaled the Groton boys with law-and-order stories from the Bowery and Hell's Kitchen.

Having just finished his four-volume *The Winning of the West* (1889–1896), TR had adopted wilderness conservation as another of his favored public policy causes. With vivid and commanding personalities, only the two "Johns of the Mountains"—John Muir and John Burroughs—embodied the spirit of the great American outdoors more vividly than TR. Shortly after FDR was born, TR, determined to save North American hoofed mammals from extinction, had co-founded both the Boone and Crockett Club and the American Bison Society. That afternoon at Groton, TR invited Franklin to spend Independence Day picnicking with his family at Sagamore Hill, his estate in Oyster Bay, Long Island; an appreciative FDR accepted the offer without hesitation. That long July weekend with his hero at Sagamore Hill, chatting nonstop about North American plants and animals, proved eye opening for FDR.

In 1887, when Franklin was five years old, TR had published a chapbook, or booklet, called *Summer Birds of the Adirondacks*. Following the picnic, inspired by "Uncle Ted," FDR started collecting texts about North American birds as if they were postage stamps. "I have not got [Edgar Alexander] Mearns' *Birds of the Hudson Highlands* but for several years have tried to get it, as it is very good," Franklin wrote to his parents after talking with Theodore. "In fact, I wrote to Dr. Mearns myself but he could not let me have a copy. He was I believe a great friend of Mr. Arthur Pell as he mentions Mr. P in another pamphlet I have of his, *The Vertebrate Fauna of the Highlands*."[14]

As part of a TR ritual, all of the children were told to scurry down a sand dune to Oyster Bay. "It was awful steep," FDR recalled. "The sand went down with you and you were darned lucky if you didn't end [up] halfway down going head over heels." Climbing back up was even harder. For every two steps upward there would be a step backward. But the kids kept at it. The lesson: never give up.[15]

The weekend at Oyster Bay encouraged FDR to emulate his distant cousin's ardent conservation stance. Once TR became a folkloric Spanish-American War hero in July 1898, FDR took to wearing gold-rimmed pince-nez just like his Uncle Ted's.[16] Historian Blanche Wiesen Cook noted that this affectation made Franklin "look a wee bit silly."[17] Nevertheless, being related to the famous Rough Rider was a boon for Franklin at Groton, a "mark of distinction," a one-up on peers.[18]

For all of their shared passion for the natural world, TR and FDR had noteworthy differences in their respective conservation philosophies. Young FDR preferred soaking up pastoral settings, like those Frederic Church painted, while TR wanted to get lost in the wilderness like Natty Bumppo. TR enjoyed bivouacking in the untamed Rocky Mountains and Dakota Badlands; FDR, by contrast, wanted to take daylong hikes along the Hudson and absorb the biological essence of all he encountered. TR had a bloodlust for big-game trophies, while Franklin largely wanted to protect animals, even hedgehogs and chipmunks, from human-inflicted harm. TR liked having dozens of exotic pets, including a badger and macaw; FDR preferred a dog. TR got terribly seasick; FDR was a first-class salt. Overall, however, FDR's conservation convictions—big forests and wildlife protection—followed the pattern laid out by his illustrious relative.

In the spring of 1900, during Franklin's final semester at Groton, Congressman John Lacey of Iowa introduced America's first serious wildlife protection bill. The Lacey Act would prohibit the interstate transportation of unlawfully killed game animals. Worried that the once ubiquitous passenger pigeon (*Ectopistes migratorius*) was on the verge of extinction, Lacey took to the House floor to protest the "slaughter and destruction" of wild creatures. That year, with progressivism cresting, Congress passed the Lacey Act to sponsor wildlife protection efforts.[19] That same year Theodore Roosevelt, who was now governor of New York (he served in the years 1899 and 1900), signed the Davis Palisades Act, which established Palisades Interstate Park in New York and New Jersey.

The Palisades are the majestic cliffs that stretch along the west side of the lower Hudson River. Under pressure from quarrying operations, they had already been clear-cut of trees and seemed doomed to end up in rubble supporting Manhattan's new skyscrapers. The Roosevelts were among the many New Yorkers who supported the fight to save the Palisades, an effort that finally succeeded during TR's governorship. During his term, short as it was, Roosevelt was also a cheerleader for the continuing effort to maintain New York's mountain regions through public-private parks. Speaking before the state legislature, Governor Roosevelt argued strongly that the Adirondacks and Catskills "should be great parks kept in perpetuity for the benefit and enjoyment of our people."[20]

During Franklin's Groton years, some girls from nearby Boston-area finishing schools joked that the F.D. in his name stood for "feather duster," an insult for a boy whom they considered soft and pampered. In his final term of Groton, when other boys were writing home about friends or the girls they had met, Franklin waxed poetic when he wrote to his mother about red-winged blackbirds and robins. "I had a most delightful experience yesterday," he wrote about his visit to the Cambridge home museum of William Brewster, president of the AOU and owner of "the finest private collection of American birds in the world."[21]

After FDR graduated from Groton in 1900, college was the obvious next step in his life. James Roosevelt was opposed to Franklin's wish to attend the U.S. Naval Academy in Annapolis. An enthusiasm for sailing at Campobello, he reasoned, didn't translate into spending one's whole life in a naval officer's uniform. Following his family's wish, Franklin attended Harvard to pursue a career in business or law, even though his primary interests were in naval history and natural history. For all of his love of forestry, FDR never considered a career in the field. The U.S. Bureau of Forestry hadn't even been created until 1898. As FDR applied for college in 1900, though, Yale was launching a strong new school of forestry and the Society of American Foresters was formed. Change was coming, but meanwhile, the nation faced rampant deforestation, soil erosion, and waterway pollution.

II

Franklin Roosevelt adjusted well to life at Harvard. His awkward years at Groton behind him, he was ever more social. Curious by nature, he made friends easily and applied himself fairly well to his classes. Because of his aristocratic upbringing and friendly countenance, no one yet realized he was a master at reading people. Just as his first semester was winding down in early December 1900, Franklin had written to his seventy-two-year-old "Papa" about perhaps getting a "change of air" in South Carolina to help restore his deteriorating health.[22] But James died a few days later, of endocarditis, in New York City.[23] A bereft FDR honored his father as the most honest, decent person to ever grace the Hudson River Valley.

The will left Springwood to Sara, a widow at forty-six, but it was understood that the responsibility of managing the lands fell to Franklin. "I am grateful to have had Franklin here these first dreadful days," she wrote. "I try to keep busy, but it is all hard. . . . I had all of F's birds out to dust and air."[24] Not long after James Roosevelt died, Sara read her old diaries to conjure up fond memories of her esteemed husband. The effect was to remind her of the joy James had gotten from the Hudson River, the Adirondacks, and Campobello. "I remember what a delight all the beauty of nature was to him," Sara wrote to Franklin, "and how he could enjoy it even when he fell ill."[25]

While on break from Harvard that Christmas season of 1900, Franklin carefully inspected the Springwood grounds with his mother. With James gone, the obligation of maintaining the estate seemed daunting. The Main House, solid with its thick walls of stone, was in relatively good shape; there were just a few rotted beams and exasperating leaks. Franklin was thus able to focus elsewhere, learning what he could about soil management, in hopes of arresting erosion and planting trees at Springwood.

Turning on the green shaded lamp on his desk, he pored over agronomist journals and almanacs in search of cogent information about property maintenance. It sickened him to learn about how virgin stands of hemlock, birch, and spruce trees in Dutchess County had been heavily logged and then burned, leaving the landscape strewn with slash (even though there were laws against it).

If FDR had a quasi–father figure after James's death, it was Uncle Frederic Delano, his mother's brother. After graduating from Harvard in 1885, Frederic began a career as an apprentice machinist with the Chicago, Burlington, and Quincy (CBQ) railroad. But city planning was his passion. Intellectually, he was a child of the Downing and Olmsted school and the burgeoning "city beautiful" movement. While living in Chicago, he served on the Chicago Planning Commission and became famous for his highly publicized efforts to plant trees around every building and keep the shoreline of Lake Michigan as pristine as possible. Young Franklin, whose own favorite president was Thomas Jefferson, relished the fact that Uncle Frederic was a patron of all things having to do with George Washington. As a historical preservationist, Delano helped conserve three sites associated with Washington: Morristown, New Jersey; Valley Forge, Pennsylvania; and Newburgh, New York. And he worked with landscape experts Daniel Burnham, Frederick Law Olmsted, and Augustus Saint-Gaudens on the landmark McMillan Plan of 1902 for *The Improvement of the Park Systems of the District of Columbia*.

Trees were to Delano the great givers of life. A large eastern cottonwood (*Populus deltoides*) growing in the hamlet of Balmville (in the town of Newburgh) was to Delano a "living witness" to the Hudson River Valley's glorious past. Whenever young Franklin visited his uncle at the Delano family estate (Algonac), they would pause at the historic tree. The Balmville tree, in fact, was the oldest example of this species on record in the United States, probably dating from 1699.

According to a nineteenth-century fable, it sprang from the riding crop of George Washington during his encampment in Newburgh in the early 1780s. But dendrologists, immune to folklore, knew that the gnarled tree began life decades before Washington was born. Situated in a glen at what had once been a Native American thoroughfare and a colonial-era crossroads, the Balmville tree—eighty-five feet tall, with a circumference of approximately twenty-five feet—was beloved by George Washington, Andrew Jackson Downing, Frederic Delano, and FDR. What they all admired was the tree's indefatigable will to survive.[26]

As an adult, FDR would often drive around Newburgh, shut off the engine, and ponder life at the Balmville tree. It became a shrine for him.[27] (The record-setting tree at last came down in August 2015.)

It was also Uncle Frederic who nurtured FDR's passion for planting violets at Springwood. Throughout the mid-Hudson, a consortium of violet growers maintained greenhouses, work sheds, and tank houses outfitted with windmills (to guarantee the water supply). Just off the Albany Post Road in Hyde Park, seasonal laborers could be seen picking violets, bunching them into bouquets of one hundred, and wrapping them in wax paper before dropping them into corrugated boxes for shipment. Violets were a huge cash crop in the Hudson River Valley, much as tulips had been in the Netherlands in the seventeenth century.[28]

With help from Uncle Fred, FDR learned how to transplant trees in May if the earth was loose and the best ways to ensure that soil retained its moisture. He learned the agricultural arts of pruning, mulching, spacing, watering, and harvesting from his loyal groundskeeper, William Plog. While about 90 percent of Dutchess County had been logged, the regular rainfall granted the mid-Hudson wonderful recuperative powers. Taking charge of the acreage at Springwood, Roosevelt adopted five rules to prevent the gullying of plowland: (1) create terraced slopes with rock walls and embankments; (2) refrain from removing natural brush or sod from the Hudson's watercourse; (3) plow on contours (not up and down slopes); (4) nurture fallow land vigilantly; and (5) while irrigating, guarantee that water will not drain to form unwelcome gullies. In the process, FDR developed a keen understanding of such soil conservation practices as crop rotation and cover cropping.[29]

Even back at Harvard, FDR was becoming something of an apprentice farmer in order to fill his father's shoes. From one of his science books, he learned that all natural resources, except subterranean minerals, were soil-based; therefore, the destruction of soil could lead only to human despair and environmental degradation. All land needed to be managed carefully so that its soil could stay healthy. With patience and financial investment, Franklin came to believe, abused land could be brought back from ruin. Devastated forests could thrive again as second or third growth after extensive replanting efforts—and a few years' wait—were undertaken. "I just wish I could be at home," he lamented to his mother in early 1901, "to help mark the trees."[30]

Just as James Roosevelt had wanted, Franklin prepared to become a lawyer or an investment banker. A compulsive joiner, FDR took to the Dickey, the Fly Club, the Hasty Pudding, the Memorial Society, the Political Club, the Signet Society, the Social Service Society, the St. Paul's Society, and the Yacht Club. He had helped found Harvard's Political Society, and had been elected secretary of the university Glee Club. But FDR's great distinction at Harvard was working his way up the ranks of the newspaper, the *Harvard Crimson*, from assistant managing editor to managing editor to president.[31] "I must say frankly that I remember my own adventures as an editor," he later recalled, "rather more clearly than I do my routine work as a student."[32]

During the summer of 1901 Franklin, accompanied by his mother, crossed the Atlantic for England and Norway, where he marveled at the beautiful fjords. While in Paris, they learned that President William McKinley had been shot in Buffalo by anarchist Leon Czolgosz, but had initially survived the assassination attempt. His vice president, Theodore Roosevelt, rushed to Buffalo from the Adirondacks, where he had been hiking in the deep woods. By the time Sara and Franklin returned, docking in Hoboken, New Jersey, Theodore had been sworn in as America's twenty-sixth president.

From the "bully pulpit," Theodore Roosevelt proclaimed that natural resource management was his primary public policy concern; under his aegis, *conservation*—of forests, soil, waterways, and wildlife—became the new watchword. The chief of the new U.S. Forest Service, Gifford Pinchot, defined TR's expansive progressive philosophy: "The conservation of natural resources is the key to the future. It is the key to the safety and prosperity of the American people, and of the people of the world, for all time to come. The very existence of our Nation, and of all the rest, depends on conserving the resources which are the foundation of its life. This is why conservation is the greatest material question of all."[33]

Gifford Pinchot was born on August 11, 1865, in Simsbury, Connecticut. His father, James W. Pinchot, was a wealthy dry goods merchant who owned a large, forested estate near Milford, Pennsylvania. Like the Roosevelts, Gifford toured Europe frequently, admiring the Continent's well-managed woodlands. Enamored of forest ecology, Pinchot, on graduating from Yale University in 1889, completed postgraduate work at the French National School of Forestry in Nancy. Pinchot returned to Pennsylvania in 1890 with a missionary zeal to start a forestry management revolution throughout the United States. Thanks to family connections, he was hired by railroad tycoon George Vanderbilt to manage Biltmore, his twenty-thousand-acre estate in Asheville, North Carolina. Pinchot's restoration work at Biltmore proved that a forest could be maintained as cropland and be properly preserved while yielding an annual profit.[34]

The well-traveled Pinchot transformed Biltmore into the "cradle of American forestry" and gave a public face to the honorable calling of silviculture, the study of trees ("what they are and how they grow and how they are protected, handled, harvested, and reproduced").[35]

Just before leaving office early in 1897, President Grover Cleveland acted on Pinchot's recommendation that new forest preserves be established in the West, designating twenty-one million acres for the cause. Westerners with an eye on timbering those public lands, blamed Pinchot for Cleveland's "locking up" of timber resources, and the term Pinchotism became synonymous with the land-grabbing policies of the "feds." Yet dozens of private timber companies and tree growers wanted to consult with Pinchot about their milling operations. Alongside fellow silviculturist Henry S. Graves, Pinchot was a driving force behind the establishment of the Yale School of Forestry in 1900.[36] His books *The White Pine* (1896) and *The Adirondack Spruce* (1898) became essential texts on how to manage North America's eastern forests.

Theodore Roosevelt treated Gifford Pinchot like a son. Both men believed that trees growing above a certain elevation—around 2,500 feet—shouldn't be felled, because they were critical to watersheds. Together they wrestled, bird-watched, hiked the Adirondacks, and scorned lumber interests that plundered forests. Four years into TR's presidency, Pinchot was hired as the chief of the revamped U.S. Forest Service, a post he held from 1905 to 1910.[37] They aggressively established more than one hundred national forests, saving huge tracts of woodlands in the Pacific Northwest, Alaska, and the Rocky Mountain states. On a single day in 1908 TR, on Pinchot's recommendation, established forty-five new national forests in eleven western states. But in 1910, Pinchot would be forced out of the Forest Service—for insubordination—by President William Howard Taft.

FDR's ever-increasing enthusiasm for proper forestry was stoked by his admiration for Pinchot. While at Harvard, FDR marveled at how Pinchot turned Grey Towers, his family home overlooking the Delaware River in Milford, Pennsylvania, into a world-class "tree nursery." (In the twenty-first century, the town of Milford billed itself as the "Birthplace of the American Conservation Movement" because of the connection with Pinchot.) Although they were members of different political parties, FDR always credited Pinchot with setting him "on the conservation road."[38]

Three busy and fun years at Harvard had passed by quickly for FDR. His classmates thought he was a "good fellow" and a regular sport, neither petty nor pompous, yet he was known for his aristocratic air. While he was excluded from the elite Porcellian Club—a huge slight because his father and TR had both been members—his magical surname marked him as someone to watch. The way he signed his letters—F. D. R.—made peers think he was already an accomplished businessman and power player.[39]

Of all the term papers FDR wrote at Harvard, "The Roosevelt Family," for History 101, is the most historically revealing, particularly because of Franklin's exaggerated insistence that his Dutch ancestors were, in fact, egalitarian. "One reason—perhaps the chief—of the virility of the Roosevelts is this very democratic spirit," Roosevelt wrote. "They have never felt that because they were born in good position they could put their hands in their pockets and succeed. They have felt, rather, that being born in a good position, there was no excuse for them if they did not do their duty by the community, and it is because this idea was instilled in them from their birth that they have in nearly every case proved good citizens."[40]

Roosevelt graduated in June 1903, earning an AB in history. He had not blossomed into an intellectual at Harvard; the activity that affected him most from his college days onward was his obsessive stamp collecting.[41] This was no mere hobby: Roosevelt gained vast knowledge about American history and the world at large from stamps. "One thing I have always specialized in ever since I started collecting postage stamps at the age of ten years is geography," Roosevelt said, "and especially the geography of the United States."[42] Rarely a day passed at Harvard when he didn't fiddle around with his stamp books. In the coming years Roosevelt would accept honorary memberships in a host of philately clubs including the Masonic Stamp Club, Washington Philatelic Society, Fort Orange Stamp Club of Albany, and Empire State Philatelic Association. And his passion for forest conservation also grew. In 1929 the Harvard alumni association sent FDR a questionnaire for the twenty-fifth reunion of his graduating class. Bragging about his silviculture prowess, the squire of Hyde Park explained that he'd "rather plant trees than cut them down."[43]

After graduation, Franklin went to Europe for the summer. On returning to New York City, he lived in a well-furnished apartment his mother had rented for him and prepared to study at Columbia Law School. Legal studies in general bored him, but the grind of law was preferable to being a ne'er-do-well living off his monthly trust fund stipend of $1,000. Although most of his new friends in New York were from the privileged classes of the Main Line in Philadelphia, Fifth Avenue in Manhattan, and Back Bay in Boston, he was more socially comfortable with people who had grown up in the scenic Hudson River Valley.

III

LEFT TO RIGHT: Franklin, Sara, and Eleanor Roosevelt at Algonac, Sara's girlhood home, which overlooked the Hudson River from the west. The picture was taken on May 7, 1905, five years after the death of James Roosevelt. Franklin and Eleanor had been married six weeks previously and were soon to leave on a three-month honeymoon in Europe.

In the summer of 1902, on a train from Manhattan to Rhinebeck, Franklin Roosevelt bumped into Anna Eleanor Roosevelt, his fifth cousin once removed. Within eighteen months, Franklin and Eleanor were engaged. The widowed Sara thought that it was wrongheaded for two Roosevelts to get married—no matter how distant their relation. She didn't want to lose Franklin and thought them both too young to wed. Her attempts to derail the relationship—she even took FDR on a Caribbean vacation to distract him—proved futile. On St. Patrick's Day in 1905, Franklin and Eleanor were married in New York by Reverend Endicott Peabody. The twenty-year-old bride's "Uncle Ted," President Theodore Roosevelt—who had trounced his Democratic rival Alton B. Parker in the 1904 presidential election—was supportive of the union. "I am fond of Eleanor as if she were my daughter; and I like you, and trust you, and believe in you," TR wrote to Franklin just before the wedding. "No other success in life—not the Presidency, or anything else—begins to compare with the joy and happiness that come in and from the love of the true man and the true woman. . . . Golden years open before you."[44]

Franklin and Eleanor made an exceedingly good match. Born on October 11, 1884, in New York City, Eleanor Roosevelt spent much of her girlhood upriver from Hyde Park in Tivoli, at her grandmother's mansion. The Catskills were her favored range. Sometimes when Eleanor was a child she visited Springwood for picnics or afternoon tea. Her father, Elliott Roosevelt—Theodore's brother—was a great outdoorsman, world traveler, and big-game hunter. Elliott was afflicted with depression and alcoholism; he was always on the run, getting into trouble with booze, drugs, and women. When Elliott got one of the family's servants pregnant, TR cursed his wayward brother as a "flagrant man swine."[45]

Eleanor never held her father's considerable faults against him. She thought when he rhapsodized about the changing seasons on Slide Mountain or the wild turkeys in the Berkshires, it was pure poetry. In 1932 she would edit a selection of his letters about wilderness adventures in a volume called *Hunting Big Game in the Eighties*. Eleanor's father had passed on to his daughter an appreciation for rural living. "The quiet of the night in the country was such a contrast [to] the continuing sounds of any city, that just the opening of the windows, and listening to the occasional creaking of a branch or the distant cracking of the ice in the brook, was restful in itself," Eleanor wrote in 1936. "Those who never sink into this peace of nature lose a tremendous well of strength, for there is something healing and life-giving in the mere atmosphere surrounding a country house."[46]

Although Eleanor Roosevelt grew up in privileged surroundings, embraced by New York high society, she endured a tragic childhood. Her mother, Anna Rebecca Hall, died of diphtheria on December 7, 1892. Her father committed suicide two years later. Eleanor had every right to hate her disappointing father, but she did not. "With my father I was perfectly happy," she wrote in her autobiography *This Is My Story*. "I loved his voice. . . . Above all I loved the way he treated me."[47] Eleanor, alert and curious, was cared for by her maternal grandmother, Mary Livingston Ludlow Hall, in the river town of Tivoli. Feeling abandoned and depressed, judging herself an "ugly duckling," Eleanor found solace in learning.[48] At age fifteen, Eleanor was sent to Allenswood Academy in London, England. The headmistress, Marie Souvestre, taught her to be intellectually brave and to challenge the status quo when necessary.

Unsympathetic observers of the Roosevelts' marriage have tended to note that they had little in common, which simply was not the case. Both were proud of their bucolic Hudson River Valley roots. Eleanor was Franklin's steadfast companion, sharing a love for the flora and fauna of New York. Added to that foundation was a shared affection for the blue-green rippling ridges of the Catskills and the Shawangunks, a group of mountains just to their south. Decades later, when Eleanor traveled to southern California in 1942 to bask in the Los Angeles sunshine, she noted that while she enjoyed Malibu and Beverly Hills, the great clay banks of the Hudson River were still first in her heart. "Nature is not so kind [in the Hudson Valley], winters are hard, summers are sometimes too hot, sometimes too cold, the lot of the farmer and gardener is always a gamble, and yet I like the change of the seasons," she wrote. "I would miss never having a landscape covered by snow. The coming of spring seems to be more wonderful because of the extremes that lie before it and beyond it. No coloring in the world seems to me more brilliant than an autumn hillside, with scarlet and gold maple and russet oak leaves mixed in with the evergreen of pine and hemlocks."[49]

Franklin and Eleanor's first honeymoon was spent at Hyde Park. In June, the newlyweds sailed across the Atlantic for a three-month honeymoon around Europe, starting and ending in England. In the

Swiss Alps Franklin snapped photographs of peaks sheathed in ice and enormous skies, as if on assignment for *National Geographic.* Franklin and Eleanor dutifully wrote to Sara with updates on the Dolomites, Tyrol, and the Black Forest. Franklin rhapsodized about the mountaintop panoramas and the treeless meadows dotted with edelweiss. Eleanor didn't share her husband's constant pursuit of outdoor activities in all kinds of strange weather. A slight tension developed between them in the Alps. Franklin abandoned Eleanor for a day to go hiking with Kitty Gandy, who owned a hat shop in New York City. "I got up at the UnChristian hour of 7 and started at 8 with Miss Gandy to climb the Faloria, about 4,000 feet above Cortina," Franklin wrote to his mother. "It took us nearly four hours up but the view was well worth the pull, and gave an idea of the wonderful colors of the Dolomites—pink, and yellow rocks, and white slopes of pure limestone—and the clouds were magnificent."[50] Eleanor understandably felt left out, but the hurt was fleeting on the dream holiday.

A highlight of Franklin and Eleanor's European trip—at least for Franklin—was a return visit to Nottinghamshire to see the mounted exotic birds owned by Cecil Foljambe. For her part, Eleanor was charmed by both her husband's strange obsession with avian taxidermy and Foljambe's ancient oak trees. FDR also hiked in Sherwood Forest during his trip to England, imagining he could hear the twang of Robin Hood's bow. In Germany, he marveled anew at the well-managed national woodlands.[51] He wrote enthusiastically to Sara about the German forests, how the "mist rising after the soaking of the last few days . . . was lovely and showed most of the Black Forest. . . . The moisture of all the trees and undergrowth and the bright sun made it very picturesque."[52]

Roosevelt intuitively surmised that forests were a sentimental matter for Germans. The Germans hadn't reclaimed land from the sea on a huge scale like the Dutch, or erected cities in swamps as the Russians and Italians had done in Saint Petersburg and Venice, respectively, but they took marvelous care of their beloved forestlands. Germany was the landscape that most influenced FDR's views on land improvement in the United States. The German people grappled effectively with the inherent tension between industrial progress and land conservation. The Germans were determined to improve soil, cultivate burned-out land, drain swamps, and take care of their impressive woodlands. Much later in the century, ecologists would deem that draining wetlands wasn't usually a good idea, but the Germans' efforts were impressive for the time.[53]

It can be argued that Roosevelt's ascent into the world of practical politics was presaged by his studies of German forestry and reclamation practices.[54] The whole German "community forest" tradition seemed like an inspired public policy for America to adopt. "Today, [a German] must cut only in the manner scientifically worked out which is calculated to serve the ends of the community and not his ends," Roosevelt enthusiastically explained to a friend. "They passed beyond the liberty of the individual to do as he pleased with his own property and found it was necessary to check this liberty for the benefit of the freedom of the whole people."[55]

Franklin marveled that German citizens could be issued tree licenses and receive tax incentives for maintaining community forests.[56] By contrast, it was nearly impossible to establish large community reserves in the American East because so many of his home region's forests were on private land.[57] Convinced that New York state had something to learn from Germany about community forestry, Franklin, on returning home, lobbied family friends to place cut-over lands under state guardianship so that they could be replanted to encourage the renewal of trees. Early in 1906, with the help of Charles Mitchell of Poughkeepsie, FDR began transforming Springwood into a model tree farm.[58] Franklin, as Eleanor put it, assumed responsibility "over the wooded part of the place."[59]

Conservation had become a crusade for young progressives like FDR. On June 8, 1906, President Theodore Roosevelt famously signed the Antiquities Act into law. This audacious piece of legislation allowed a president to declare federal protection for landscapes with archaeological, scientific, and environmental value. In the last three years of his White House tenure, TR would wield this privilege and bypass Congress to save such wonders as the Grand Canyon in Arizona, Muir Woods in California, Devils Tower in Wyoming, and

Mount Olympus in Washington (from lands within the boundaries of Olympic National Forest).

Seeing, meeting, and networking around New York, FDR, conversant in forestry science, talked enthusiastically with farmers about the Adirondack Preserve, the White Mountains of New Hampshire, and the Black Forest of Germany during TR's White House years. These three protected landscapes informed his views of what constituted smart conservation. Roosevelt warned friends that the destruction of forests directly contributed to the poisoning of water supplies—and no large city could afford to sully its safe, reliable supply of water. Not only did trees provide fruit, sap, nuts, firewood, shade, and a weapon against erosion, but their roots tapped into underground water and brought it to the surface—revitalizing the surrounding environment. Trees were more than just aesthetically pleasing to Franklin Roosevelt, they were God's greatest utilitarian invention.

CHAPTER THREE

"HE KNEW EVERY TREE, EVERY ROCK, AND EVERY STREAM"

I

In 1904, in the public debate over whether the Catskill Forest Preserve should be extended into a Park (like that in the Adirondacks) or include only land owned by the state, Franklin Roosevelt argued for the park. Families with deeds dating back to the seventeenth century still lived in Catskills villages like Woodstock and Tannersville. The private land in the affected counties provided residents and idlers with quaint villages surrounded by scenic mountains, verdant forests, and idyllic brooks. FDR wanted to allow private land ownership in Catskill Park and then to enact tough timber laws that would regulate limited cuts.

Usually willing to modify nature for the sake of civilization, Roosevelt envisioned the Catskills as a protected landscape where woodlands and residential villages could coexist while all obeyed conservation laws. In the end, park status was adopted. As of 1912, Catskill Park included 576,120 acres, including 92,000 acres of Forest Preserve. The blend of public and private acreage—untouched wilderness as well as small villages, scientifically managed forests, and small farms—would serve as FDR's rural planning model in which a conservation ethic was prioritized.[1]

Roosevelt, concerned that timber syndicates were felling trees at an unsustainable rate, believed, like Pinchot, that true forestry was the "art of using a forest without destroying it."[2]

At Springwood, FDR tried hard to put his conservation beliefs into practice. Photographer Margaret Bourke-White, in *Life* magazine, captured the essence of Springwood as an "old shoe of a place—

worn, scuffed and scratched, polished into shape, fitting the owner well; but the woodlands were *modernly* managed."[3] William Plog continued to serve as groundskeeper of the estate, and had become part of the extended family. Trimming the hemlock and taking care of the rose garden were among his daily chores. He was paid $45 per month to oversee the property and keep the trees free of disease, fungi, and unwelcome insects. Plog improved the vegetable garden, tended the greenhouse, and helped maintain an irrigation pond. The unspoken promise was that with James deceased, Plog, as surrogate, would teach Franklin that soil was more than a jumble of clays, sands, and silts. Together they laid fieldstones and marked trees that needed to be thinned. In 1947, when asked in an interview for an oral history what Franklin most liked to do at Springwood, Plog immediately responded, "Trek around through the woods."[4]

Roosevelt's neighbors feared that diseases and insects would destroy local woodlands. A devastating blight that struck American chestnut trees (*Castanea dentata*) in 1906 worried Dutchess County farmers. As they battled such attacks, rural farmers turned to professional foresters and so did FDR. Taking tips from Plog and Pinchot, he hoped to raise disease-resistant trees and then share his knowledge with the people he called "forest neighbors."

Determined to turn a modest profit from his Hyde Park forestry operations, he milled cut trees into wooden cross ties for sale to the New York Central Railroad. Excess timber planks were sold to furniture companies. Roosevelt also marketed Christmas trees to Dutchess County residents.[5] A longtime professor at the New York State College of Forestry, Nelson C. Brown, wrote in the journal *American Forests* that FDR hoped to turn his primeval grove of Springwood hemlocks, whose "pristine beauty is unmarred by the ax," into a "museum of what our original forests looked like when the sturdy Dutch forefathers first settled these shores."[6]

The expansion at Springwood—of the grounds, buildings, and tree plots—was partly a result of Franklin and Eleanor starting their family. In 1906, Eleanor gave birth to their first child, Anna, followed by son James in 1907. After a second son, Franklin, died in infancy in 1909, the next Roosevelt child, Elliott, was born in 1910 just as FDR was embarking on his first political campaign, for the state senate. Nevertheless, Anna, James, and Elliott all had childhood memories of hiking along nature trails with their father and embracing certain trees as places of respite.[7] "He knew every tree, every rock and stream," Eleanor Roosevelt recalled, "and never forgot the people who had worked there when he was small."[8]

Between 1907 and 1910, however, Franklin and Eleanor lived mostly in Manhattan, in a town house on East Sixty-Fifth Street that Sara Roosevelt, who lived next door, had purchased for them. To Eleanor's annoyance, Sara knocked out a wall and built a connecting hallway to have access to her beloved Franklin.[9] Even though Franklin was in Manhattan for most of that period, taking classes at Columbia and then practicing law, his interest in conifers (particularly those that grew at the fortieth parallel, the Springwood latitude) intensified.

Having passed the New York bar examination, Roosevelt joined Carter Ledyard & Milburn, a firm at 54 Wall Street, in 1907. However, for Roosevelt, with his many and varied interests, being a Manhattan lawyer was tedious—he admitted he simply "had no aptitude for law."[10] Resisting work he considered boring, Roosevelt, a pastoralist, was always yearning to go fishing, bird-watching, or golfing. His best legal work was assisting old friends from Harvard days beat back charges of public drunkenness. With his passion for naval history, he also occasionally dabbled in admiralty law, taking on a few ferryboat cases and disputes over port-of-entry fees.

In the summer of 1907 Franklin Roosevelt took Eleanor and Anna, who had just turned one, to Campobello. Once again his letters to Sara were filled with reports of picnicking at Raccoon Beach, sailing the Bay of Fundy on "foggy" mornings, and exploring three nearby islands called "the Wolves," collecting limpet shells and gull feathers.[11] Feeling the high winds and strong currents made Roosevelt ponder whether the possibility of the "tidal power" from the Bay of Fundy, both the American and the Canadian sides, could produce electricity for New England. Seeing the double-crested cormorant (*Phalacrocorax auritus*) and the blue-winged teal (*Anas discors*) elated him. "He always said he was shortsighted when he passed people in the streets and didn't recognize people," Eleanor recalled, "but he could always point to a bird and tell me what it was."[12] Owen Winston, a friend of Roosevelt's, came to visit at Campobello that summer, as did Eleanor's brother, Hall, to bird-watch. "Franklin and I, and Hall Roosevelt, went off trying to find cormorants' nests," Winston recalled. "We were fairly successful and got ourselves fairly messed up."[13] Hall recalled that there wasn't a shorebird from New Brunswick to New Jersey that FDR couldn't identify at a glance.

A year after their marriage, Franklin and Eleanor (*seated on the ground, center*) spent the summer at the Roosevelt home on Campobello Island, where they were photographed at a picnic. Sara is seated in a chair (*right*). Franklin's parents had long enjoyed trips to the island, which lay in Canadian waters, taking full advantage of its unexcelled sailing. They built their own lodge there in 1885.

Back in New York City, Roosevelt admitted to his friends at the firm that his goal was to be elected to the state legislature, then secure a high-profile job in the Navy Department, and eventually be elected governor of New York State. This desire for political power seemed to come out of nowhere. He felt strongly that the governorship was a springboard to the White House. It had, after all, been the key to the presidency for his father's favorite Democratic politician, Grover Cleveland, and had also been integral to TR's political ascendance. And just as TR had made a name for himself in conservation circles protecting endangered mammals like buffalo and elk in the American West, FDR hoped to make his conservation mark as a soil and forest protectionist in the Northeast.

Hoping to raise his public profile in New York, FDR accepted an invitation to serve on the Hudson-Fulton Celebration Commission, which was planning a tercentennial celebration in the fall of 1909 to commemorate Henry Hudson's "discovery" of the Hudson River. Believing that the Hudson always sparkled, he helped organize the celebration's maritime events, working closely with his uncle Frederic Delano; they focused on the building of accurate replicas of Hudson's *Half Moon* and Robert Fulton's *Clermont*. FDR saw the extravaganza as a way to highlight the conservation and preservation achievements of the Hudson Valley.[14]

The previous summer, in June 1908, Roosevelt had embarked on a weeklong inspection of the coal holdings owned by his uncle Warren Delano III in the Cumberland Mountains of Kentucky. The poverty and needless environmental destruction he saw there sickened him; heavy logging had turned the forests into a degraded wasteland, and mining had scarred the land. It's probable that FDR's propensity for hydroelectric dams stemmed from this unsettling trip to rural Kentucky. There *had* to be other ways to generate energy that did not involve blowing up mountains or making workers gravely ill. With mounting admiration, Roosevelt viewed public waterpower and wind power as the future of generating electricity and raising the overall standard of living.[15]

Just months before FDR was born in 1882, the world's first hydroelectric power station had started operations along the Fox River in Appleton, Wisconsin. By the time Franklin married Eleanor, he believed that hydroelectric power would soon be a vital energy source. Meanwhile, President Theodore Roosevelt had already promoted the construction of hydroelectric dams in the American West: Arizona's Roosevelt Dam, named in his honor, would be completed in 1911, allowing Phoenix to grow. If dams could be built in the arid West, FDR believed, verdant farms could spring up all over, from eastern Washington to the Colorado Plateau. Roosevelt knew that the potential economic gains for the Yakima Valley in Washington, the Gallatin Valley in Montana, and the upper parts of Wyoming and Idaho—where water from the Shoshone project was already working agricultural miracles—were enormous. Although dam-crazed FDR, like most politicians of the William Howard Taft years, never questioned why the U.S. Army Corps of Engineers didn't also advance building sewage treatment facilities.[16]

And then there was the question of clean air. To FDR's dismay, Thomas Jefferson's dream of an agrarian nation had been jettisoned to make way for smokestacks, cast-iron-framed buildings, and urban ghettos—what John Burroughs collectively called, after visiting Pittsburgh, "the devil's laboratory."[17] The combined effect of factory and automobile exhaust was poisoning the air.

Roosevelt loved cars, pollution aside. Cruising along unmarked rural roads in Dutchess County in his rented red Maxwell Touring Car became high sport. On these backcountry drives, Roosevelt contemplated the interconnectedness of soil, forests, and water. "As a whole," Roosevelt said in 1912, "we are beginning to realize that it is necessary to the health and happiness of the whole people of the State that individuals and lumber companies should not go into wooded areas like the Adirondacks and the Catskills and cut them off root and branch for the benefit of their own pocket."[18]

Whenever Roosevelt sought solitude, away from life at Springwood, he would drive up Charlie Hill Road on Silver Mountain, near the town of Millerton. The summit view to the northwest overlooked the fertile farmlands of northern Dutchess County, Stissing Mountain, and the distant blue-mist Catskills. This was one of Roosevelt's favorite meditative spots. His frequent companion on these jaunts was his friend Maunsell Crosby. In subsequent years Crosby, an accomplished birder, would go on collecting expeditions to Central and South America for the American Museum of Natural History. Roosevelt worked with Crosby on the annual Audubon-sponsored Christmas bird count in Dutchess County. A favorite outing for them each spring involved watching hawks and eagles congregate on Cruger's Island (now part of a 1,600-acre preserve owned by New York State);

only a few other places on the Atlantic Flyway hosted so many raptors at once.[19]

II

After much debate in early 1910, Franklin Roosevelt decided to run for the state senate in New York's Twenty-Sixth District, an area that included Hyde Park. On the stump Roosevelt proved a mellifluous speaker, enunciating clearly and never groping for the correct word. Always in perpetual motion, he campaigned on anti-Tammany measures like eliminating graft and political patronage from government. In a relaxed, confident way, he enjoyed talking about forestry fundamentals with growers of trees, fruits, and vegetables in the Hudson Valley. At the drop of a hat, Roosevelt shared with fellow tree farmers logbooks about planting white pine on his western and northern slopes, and tulip poplar from stock held at a nursery. Part of his agronomist pitch to farmers in the district was that planting trees was, in essence, an insurance policy for the future of their families. A thriving hemlock, he asserted, was akin to "interest" accruing in the bank. Not only did trees stabilize stream banks and curb erosion, but within ten years' time fortunes would be made from timber harvests. Paraphrasing Theodore Roosevelt, FDR would say, "A people without children would face a hopeless future; a country without trees is almost as helpless."[20]

With Congressman Richard Connell, an upstate Democrat, regularly at his side, Roosevelt campaigned from Cold Spring and Carmel (in Putnam County), through Fishkill Landing and Poughquag (in Dutchess County), and on to Germantown, Kinderhook, and Hudson (in Columbia County) to advocate economic justice for farmers. In every town where he stopped, he called everyone he met "friend."[21] Championing the "city beautiful" movement was part of his spiel. "Humboldt, the great traveler, once said: 'You can tell the character of people in a house by looking at the outside,' " Roosevelt told a group of supporters at the Pleasant Valley Public Library. "This is even more true of a community. And I think I can truthfully say that of all the villages of Dutchess County—and I have been in pretty nearly every one—there are very few that appear as favorably as Pleasant Valley."[22]

While campaigning for the state senate, Roosevelt spoke regularly of the two different New Yorks: the city and upstate. Even though he enjoyed living on East Sixty-Fifth Street in Manhattan, he never failed to remind his potential constituents that city dwellers would not have potable water without the upstate forests.[23] Whenever possible, he emphasized to rural audiences his kinship with ex-president Theodore Roosevelt, sometimes even using TR's characteristic words—"delighted" and "bully"—during his speeches.[24] At the same time, FDR was relieved when his distant cousin didn't cite him by name on a campaign stop for the GOP in Dutchess County. TR's main advice to Franklin about campaigning was simply to address head-on the economic problems farmers were experiencing.

To better understand the struggles of rural voters, FDR began keeping a "farm journal," essentially an agronomist's forestry science log, of his plantings and land management accomplishments. Roosevelt's farm journal was decidedly lacking in philosophizing, but one can still detect an undercurrent of Andrew Jackson Downing's influence in its fact-filled pages. Pencil sketches of farm lots were entered. Every area of the Springwood grounds was given a name derived from some naturally occurring feature: River Wood Lot, Locust Pasture, Swamp Pasture, and the like. Roosevelt's hand-drawn maps and detailed notes about everything from raking leaves to fertilizer make the journal an irreplaceable source for understanding how devoted he was to imbuing Springwood with a real land ethic.[25]

FDR used a Dutchess County farmer, Moses Smith, as his political sounding board in the 1910s. Smith lived at Woodlawns, the first parcel of land Franklin owned independently of his mother. Of medium height, with a great blade of a nose and a tangle of jet-black hair hiding under an ever-present hat, Smith leased Springwood acreage from Roosevelt to grow crops—everything from corn to oats to pumpkins. Both men, according to Smith's son, "loved to feel" that they were "using the land to good advantage."[26] Eleanor Roosevelt, in her memoir *I Remember Hyde Park*, recalled how Sara had wanted their meals at Springwood to be full of fresh produce, including "the earliest possible peas, picked when very young, and . . . chickens, eggs, butter, cream, milk."[27]

Roosevelt and Smith regularly talked shop about seeds, climate conditions, pitch pine, and the alluring camphor-like scent of Norway spruce when crushed between two fingers. Roosevelt came to believe that if stubborn Moses Smith, a yeoman farmer, could be persuaded on adopting a soil conservation protocol, then so could any farmer in his district. Smith would often scratch his head while philosophizing to Roosevelt—a quirk Roosevelt found endearing enough to imitate. No matter how famous FDR became, he treated Smith as a fellow farmer trundling crops to the city markets. Roosevelt was impressed with the way Smith bore down on the handles of a plow and drove the blade deep into the ground, as the horses pulled it.

On November 8, 1910, all of Franklin Roosevelt's campaign oratory, shoe-leather hustle, constant smiling, and farm-to-farm handshaking paid off. He defeated his Republican rival by a slim margin. Just days after the victory, the twenty-nine-year-old Roosevelt was asked to chair the senate's Forest, Fish, and Game Committee. He accepted, later joking that the boys in Albany—because of his family name—"couldn't think of anything else" for him to do.[28]

To the general public, the name *Roosevelt* was already synonymous with making conservation a prominent issue in American politics. "It was a post that was supposed to be a sinecure, one of no importance, because in those days there was no such thing as the Conservation Department. . . . There was practically no interest in what you and I know today as conservation in its broadest sense," Roosevelt would tell a West Virginia audience in 1944. "But . . . I was very keen . . . on getting the people of the kind State interested in preventing soil erosion in the Adirondacks."[29]

Franklin and Eleanor quickly rented two floors of the H. King Sturdee house at 248 State Street in Albany. But FDR returned to Hyde Park on weekends to perform constituent work. Although Roosevelt was director of the First National Bank of Poughkeepsie, the Eagle Engine Company, and the Rescue Hook and Ladder Company of Hyde Park, he was careful to avoid possible conflicts of interest: as an anti–Tammany Hall Democrat, he could not afford even the slightest intimation that he was a "bought man." When G. O. Shields of the League of American Sportsmen for the Preservation of Wildlife prodded FDR to introduce an anti–automatic and pump gun bill in Albany, his friend bristled. "I am, I need not tell you, much interested in the preservation of wild life," FDR telegraphed back, "but feel that as chairman of the Forest, Fish, and Game Committee in the Senate I should not be the one to introduce legislation of this kind."[30]

Treated as a wet-behind-the-ears rookie, Roosevelt nevertheless impressed colleagues in Albany with his wide knowledge of conservation, dendrology, and silviculture. Within days, he had befriended Forest, Fish, and Game Commissioner Thomas Mott Osborne, who was likewise an opponent of Tammany and of its boss, Charles F. Murphy. Determined to make conservation his strong suit, Roosevelt made speeches about reforestation, liberally borrowing ideas and syntax from Gifford Pinchot.[31] "When at Hyde Park tomorrow," Franklin wrote to Eleanor, whom he addressed affectionately as "Lamb" or "Babs," "I will go over the locations for planting the 8,000 trees and also see how they are getting on with the clearing of the new pasture."[32]

As chair of the Forest, Fish, and Game Committee, Senator Roosevelt trod carefully when choosing sides on controversial issues. He did heartily defend the Shea-White Plumage Act. The bill, passed the year before he took office, was an extension of the Audubon plumage laws that outlawed the sale of feathers from many bird species. Shea-White further prohibited the importation of feathers, skins, or carcasses of protected bird species.[33] And when a consortium of New York City grocers—eager to secure reliable sources of duck—lobbied mightily for a longer hunting season, Senator Roosevelt countered by sponsoring a bill aimed at reducing bag limits to reverse the diminishment of waterfowl. Backed by the New York Zoological Society and the Audubon Society, Roosevelt argued forcefully that the struggling waterfowl population needed increased protection—not a longer open season. Nevertheless, Roosevelt didn't persuade enough fellow senators to take his pro–duck protection position. "I regret to say," he wrote to an ally, "that Senate Bill No. 9, permitting duck shooting on Long Island, has passed the Senate in spite of much opposition on my part and that of others, by just the necessary number of votes."[34] Roosevelt's efforts weren't entirely in vain, however: the bill failed in the state assembly, and so did not become law.

Roosevelt's commitment to bird protection laws is abundantly clear in letters he wrote to fellow New Yorkers. Showing off, he often brought up his longtime association with the AOU, and his ornithological bona fides in general, when arguing for the protection of small shorebirds and for a moratorium on shooting wood ducks. "I am sorry that I cannot agree with you about the wood duck," Roosevelt wrote to Egbert Bagg of Utica, New York. "It may be true that a man who jumps ducks from a boat cannot tell a wood duck but you must realize that the duck is a species in danger of extermination today. The prohibition against shooting them nevertheless saves many birds every year."[35]

That year, the New York State College of Forestry was reestablished at Syracuse University. Situated on twelve acres near a hill known as Mount Olympus, the College of Forestry afforded visitors exquisite views of the Onondaga Valley. FDR maintained a lifetime attachment to the school, often consulting with its top-flight silviculture experts. Conservation had finally earned a language of its own, a lingua franca that centered on plumage laws, wildfire control, bag limits, and scientific forestry, in which FDR had unusual fluency for a Democratic politician. Roosevelt accepted as his own the definition of conservation offered by the Camp-Fire Club of America—founded in 1904 by William Temple Hornaday to promote big-game management, commonsense hunting laws, habitat restoration efforts, and conservation education—"To protect and use wisely our natural resources, while assuring their preservation."[36]

FDR's interest in American trees also had a historical component. Somewhere, Roosevelt had read that George Washington planted poplars and oaks all over northern Virginia. Washington's diaries and letters from Mount Vernon indeed reveal that the first president was a gifted arborist, and in 1932 an article in *American Forests* noted that Washington tended to his trees "as if they were almost human." This impressed FDR mightily. On more than one occasion, Washington's storied career was linked to trees: like when he famously took command of the Continental Army on July 3, 1775, under a towering elm tree in Cambridge, Massachusetts. Writing to a cousin in 1782, Washington speculated about how "the change of environment"—that is, the replanting of a tree from forest to field—affected the health of a locust or pine. Under the code that George Washington lived by, a planter should be judged, for better or worse, on the health of his woodlands. Washington personally planted at least thirty-seven species of trees on his Virginia acreage. In many ways, Washington was America's first high-profile forestry conservationist.[37]

Roosevelt saw a link between history tourism, recreation, and country planning. He hoped that New York City residents, for example, would take weekend pilgrimages to Saratoga, Fort Ticonderoga, and West Point as a means of connecting with their past. All three of these historical sites had marvelous green spaces. Revolutionary War battlefields and burial grounds were sacred, deeply pastoral landscapes. Roosevelt thought these sanctified historic park sites needed to be managed with a long-term preservationist ethic. Historic Saratoga Springs, he argued, would collapse if its famous spa waters (thought to ease dyspepsia, jaundice, rheumatism, and gout) were ever contaminated. For the tourist economy of upstate New York to prosper, the region's assets—historical and natural—needed stringent preservationist standards. Millions of Americans would be able to ponder the Battle of Saratoga better if the greensward transported visitors back to the glorious days of 1776. In Roosevelt's mind, historical preservation in New York was tied in with woodlands ecology.[38]

Meanwhile, in 1911, Congress had swung around and passed the Weeks Act, a carefully constructed federal conservation law that authorized interstate compacts for water and forest conservation. The act encouraged the federal acquisition of land in the eastern United States to protect forests and watersheds.[39] Recognizing that forests were not an inexhaustible resource, the Weeks Act created the eastern section of what would eventually be the National Forest System. In the East, over the course of American history, most land had been disbursed by the government to private citizens, and this legislation provided funds to buy large expanses of canopy—particularly in New Hampshire and Vermont's White Mountains—as national forests. The federal government eventually acquired more than twenty million acres of eastern forests under the authority of the Weeks Act.[40] These acquisitions led to the establishment of forty-one national forests, including White Mountain (New Hampshire and Maine); Green Mountain (Vermont); Finger Lakes (New York); Allegheny (Pennsylvania); and George Washington and Jefferson (Virginia).

In the West (unlike the East), 300 million acres (about one-sixth of America) that were "vacant" or "unappropriated" were owned by the U.S. government. Nobody was sure what to do with the acreage. What was the federal government to do with the cactus lands of the Sonoran Desert, the Painted Hills of eastern Oregon, the West Texas plains, the rolling Palouse country of eastern Washington, or the badlands of North Dakota? Much of the land could be used for mining or grazing, but to progressive conservationists, short-term blind profit was not an appropriate priority for a long-range government land-management ethic.

Roosevelt grew very interested in western topography. Almost as if he were collecting stamps, he learned about the types of western flats—creosote, gumbo, antelope, tidal—that needed to heal. When Roosevelt visited eastern Kansas he saw fertile farmlands that made the Great Plains the "breadbasket of America." But the western counties of that open-range state was suffering from chronic drought and nearly bone-dry reservoirs. Roosevelt knew there was no magic that could make these deserts bloom without irrigation projects and dependable forests to produce dependable quantities of rainfall. A single mature tree issued 1,000 liters of water daily into the atmosphere.[41] Cattle herds, he understood, had made the open-range ecosystem worse and worse. Sheepmen and cattlemen clashed over water rights and grass on the public lands.

As a New York state legislator, FDR didn't have ready answers for the problems in the West. His conservation concerns were focused on establishing national forests in the East where the waterfalls cascaded and the hardwoods were glorious. But the word he used—like a magic wand for the West—was *reclamation*, by which he meant that plenty of public domain land would be irrigable to grow food for a few million people. That meant dams. Roosevelt knew as an amateur silviculturist that the success of hydroelectric dams depended on federal protection of the forests at the headwaters of rivers. In New York, without the Adirondack Preserve to help regulate the flow of the Hudson and to keep the waters fresh, there could have been flooding. To Roosevelt the three major tasks for the federal government in managing the public domain were wisely using and protecting forests (increasing the national forest acreage); building public dams (irrigating arid and semiarid land); and establishing recreation areas (national and state parks to attract tourists). Unlike Theodore Roosevelt, young FDR didn't find cowboys romantic figures. They were, as a whole, prone to abuse the land by overgrazing cattle.

III

In 1912 Franklin Roosevelt's interest in conservation, which had been focused on New York, broadened; he aligned himself with groups like the Camp-Fire Club of America and the National Audubon Society. It was a presidential election year, with Woodrow Wilson winning the Democratic nomination and speaking out on conservation—a term usually circulated only in Republican circles. Roosevelt always felt there was more to be done on behalf of forest protection in the American East. "The destruction of the forests," he lamented, "is proceeding today unchecked on many private holdings." He was right to worry. State oversight was desperately needed because deforestation on private acreage too often led to soil erosion. "It is an extraordinary thing to me," FDR wrote to Dexter Blagden, who was a Wall Street broker and a fellow Harvard man, "that people who are financially interested should not be able to see more than six inches in front of their noses."[42]

Senator Roosevelt promoted a slate of bills in 1912—five concerned with forestry and four with hunting and fishing regulations. His signature piece of legislation was written in cooperation with the Camp-Fire Club of America. The Roosevelt-Jones Conservation Bill aimed to ban the felling of trees, even on private lands, below a certain girth.[43] Gifford Pinchot, an active Camp-Fire Club member, surveyed the immense Adirondack wilderness and wrote a powerful pro-preservation report, which echoed some of the concerns that had been raised in the Roosevelt-Jones Conservation Bill—most notably, that large trees on private property were in need of protection to ensure continuous seed supplies. The Roosevelt-Jones Conservation Act, backed by Pinchot, also made it illegal for timber companies to clear-cut even on private lands. The act further mandated the hiring of additional fire wardens and forest rangers and the building of new fire lookout stations.

Roosevelt shakes hands with a prospective voter in Dutchess County during his run for the state senate in 1910. After a brief career as a lawyer in New York City, he returned home and allowed himself to be recruited into an uphill campaign, as a Democrat running in the Republican stronghold of the Hudson Valley. His career was launched when he beat the predictions and won.

As Senator Franklin Roosevelt was gearing up his reelection campaign in 1912, Theodore Roosevelt was running again for president, this time as a third-party candidate against incumbent Republican William Howard Taft and Democrat Woodrow Wilson. Even though FDR was a stout supporter of Wilson in 1912, he was firmly in TR's camp. He happily adopted his illustrious cousin's clique of outdoor-life friends—forester Gifford Pinchot, scout extraordinaire Buffalo Jones, naturalist John Burroughs, editor Robert Underwood Johnson, and zoologist William Temple Hornaday among them.

After overseeing a day of tree planting at Springwood, Roosevelt went on a whirlwind trip to Jamaica and Panama that April. The largely Caribbean jaunt provided an added bonus of witnessing the Panama Canal's construction firsthand. "I had forgotten how magnificent the cliffs and defiles are, covered with great creepers and clinging crooked trunks of trees," Roosevelt wrote about Panama. "At Rio Cobre we decided to go on and see the Natural Bridge, a very curious formation where the stream has broken through a cliff."[44]

On returning home, FDR rendezvoused with Eleanor in New Orleans. After a scrumptious Cajun meal, they stole off on a train through the Louisiana bayous to the Rio Grande Valley and the Guadalupe Mountains before slipping into Silver City, New Mexico. A close friend of FDR—Robert Munro Ferguson, a former Rough Rider—was fighting tuberculosis and hoped the desert air of Cat Cañon, near Silver City, would revitalize his ailing lungs. At that time, many medium-sized towns in the arid West had a facility offering relief from respiratory illness, and Ferguson owned the high-end convalescent clinic in Silver City.[45] His wife, Isabella, an old friend of Eleanor's, knew that Franklin, the tree aficionado, would enjoy seeing the hills dotted with "scrubby dwarf evergreens."[46]

In 1912, TR's Progressive Party, nicknamed the Bull Moose Party, was drawing thirty-eight-year-old Harold Ickes into the national debate over conservation. Ickes was born on March 15, 1874, in Hollidaysburg, Pennsylvania, a picturesque landscape of rolling hills, bosky dells, pristine brooks, and abundant wildlife just south of Altoona. "We were of the soil," he recalled, "and proud of it."[47] But when Ickes was sixteen years old, the family moved to Chicago, which struck him as an alien steel-and-soot jungle; he hungered for the fresh air of the Pennsylvania outdoors.

Determined to lift himself out of hard-bitten poverty, Ickes worked his way through the University of Chicago, graduating in 1897. The more he read, the more reform-minded he became. A distrust of "concentrated wealth" came naturally to him, and he put his faith instead in such reformers as Jane Addams of Hull-House and John Muir of the Sierra Club. Ickes started writing for the *Chicago Record* and later wrote for the *Chicago Tribune*. In Theodore Roosevelt he found a leader to trust and joined the Progressive Party. One of the greatest moments in Ickes's life was hearing TR accept the Bull Moose nomination at the party's convention in Chicago. All across America a new breed of conservation-minded Democrats and Republicans were rushing to join the Progressive fight, just as Ickes had done, with intensifying resolve. The party's supporters included such illustrious (or soon to be illustrious) names as Gifford Pinchot, William Allen White, Alfred Landon, George W. Norris, Robert M. La Follette, Frank Knox, Henry A. Wallace, Felix Frankfurter, Norman Thomas, Francis Biddle, and Dean Acheson.[48]

During the Wilson years, Ickes developed a reputation in Chicago as a champion of civil rights for African Americans and Native Americans. Because Ickes hated the detrimental role of special interests in politics, he made himself a nuisance and sassed the rich. Ickes believed that environmental conservation, in the spirit of Muir, was a righteous and noble calling. A two-week horseback trip he took through Glacier National Park turned him into an ardent advocate of wilderness protection. "I love nature," Ickes said. "I love it in practically every form—flowers, birds, wild animals, running streams, gem-like lakes, and towering, snow-clad mountains."[49]

Another Bull Mooser and Camp-Fire Club impresario was William Temple Hornaday, the first curator of the New York Zoological Society (the Bronx Zoo). Born in 1854 in Avon, Indiana, Hornaday had drawn a link between Darwin's theory of natural selection and the need to save certain animal species.[50] His 1913 book *Our Vanishing Wild Life* launched the modern endangered species movement. He also fought fiercely against the clear-cutting of forests, gouging mountains for coal and gold, and treating the Hudson River like an open sewer.

Roosevelt would soon fall under the profound influence of Ickes and Hornaday, but in 1912, his primary contact in conservation was Pinchot. In 1910 Pinchot published *The Fight for Conservation*, which defined the burgeoning field of natural resource management in layman's terms.[51] FDR regularly conferred with him about how to protect forests from the grave perils of fire, unsustainable cuts, insects, and disease. Cognizant that lumber companies such as Reynolds Brothers Mill and Logging were heedlessly annihilating the upper

Hudson forests for profit, Roosevelt hoped the second-growth techniques advocated by Pinchot and others could be implemented in upstate New York's already heavily cut-over lands (and that tougher forestry laws would be enacted in Albany). Roosevelt's overarching concern was the International Paper Company, which, in monopolistic fashion, had linked twenty sawmills in New York and New England into a giant corporation.[52]

To fight the big boys, Roosevelt brought in the conservation giant of America. He invited Pinchot to speak to the legislature in Albany on the damaging effects unregulated laissez-faire economics had on global woodlands ecology.[53]

CHAPTER FOUR

"WISE USE"

I

Roosevelt (*left*) chatted with Gifford Pinchot at a governors' conference in 1931. By then, the two conservationists had known each other for more than twenty-five years. They were serving as governors of New York and Pennsylvania, respectively.

Over the summer of 1911, Gifford Pinchot, at the behest of the Camp-Fire Club of America, conducted an in-depth scientific study of the Adirondack forests in twelve counties.[1] He was accompanied by William T. Hornaday and Overton Price (editor of *Conservation* and author of *The Land We Live In: The Book of Conservation*). After inspecting 3.3 million acres—of which the state owned 1.5 million, the rest belonging to timber companies, private associations, individuals, and clubs—Pinchot recommended increasing the number of fire wardens, forest rangers, and fire lookout stations. Taking aim at private landowners, Pinchot called for state control and regulation of their properties. Most dramatically, he called for a constitutional amendment allowing the management of the public Adirondack forests in uncompromised adherence to the Yale University doctrine of scientific forestry.[2]

Senator Franklin Roosevelt, in full agreement with the Camp-Fire Club, invited Pinchot to address the Joint Legislative Committee on Forest, Fish, and Game in Albany about the coming Adirondack report. Even though Pinchot had been dismissed as chief forester by President Taft in 1910, his unshakable belief in "wise use" of forest resources still cast a mighty spell on Progressives.[3] "Mr. Pinchot was waging a rather lonely fight then," Eleanor Roosevelt later recalled, "and few people paid much attention to his warnings."[4] The term *environment*, in fact, was hardly used at all in the early twentieth century. But Pinchot, owing to his deep learning, implicitly understood the concept of the "web of life." Although the conservation of forests and water was his bread-and-butter issue, he was expanding his ecological expertise to include soils and wildlife.[5]

Until that day in Albany, FDR was merely an admirer of Pinchot—not an actual disciple. But Roosevelt experienced an epiphany when he watched Pinchot—meticulously dressed in a handsome suit, watch-chain across his vest pocket—tell the New York legislators about his Camp-Fire Club survey in the Adirondacks the previous summer.[6] "Forestry in New York is flourishing everywhere except in the woods," Pinchot said. "It is time to make it flourish there."[7]

Striding around for theatrical effect, Pinchot displayed on a screen two lantern slides of the same north China landscape from drastically different eras. The first was an ancient painting (circa 1510) bursting with the vibrant greenery of trees. The other was a photo, taken around 1900, of exactly the same vista, reduced to an ecological dead zone.[8] Pinchot, a devout Christian who attended church every Sunday, dramatized his history lesson with a warning in the style of Revelation: deforestation was the devil's work. "One need not be an alarmist," Roosevelt said of Pinchot's bravura performance, "to foresee that, without intelligent conservation measures, long before half a millennium passes some such contrasting pictures might be possible in our own United States."[9]

Roosevelt, decades later, told Yale University students that Pinchot's 1912 lecture in Albany, especially the contrasting pictures from China, had affected him like smelling salts. Pinchot had convinced him that deforestation had ruined Mayan, Mediterranean, and Chinese civilizations. "I discovered immediately that one of the problems before us was the denudation of the Adirondacks," Roosevelt explained. "Timber had been cut there without rhyme or reason or thought and many of the upper slopes were being washed away until only the bare rock appeared. . . . Well, that picture sold conservation and forestry to the Legislature of the State of New York."[10] And, as a result of Pinchot's plea, FDR was enabled to "get through the first important legislation for conservation."[11]

That reform legislation was the Roosevelt-Jones Bill, but it was controversial. The very notion of state control of lumbering on private lands (Section 88 of Roosevelt-Jones) struck some legislators as grotesque government overreach, more like Karl Marx than like the founding fathers. A huge debate over conservation was under way, with FDR fanning the flame. At a People's Forum in Troy, New York, in March 1912—not long after Pinchot's talk—Roosevelt, promoting the Roosevelt-Jones Bill, chastised greedy New Yorkers who, "for the sake of lining their own pockets during their own lifetime," plundered the Adirondacks. "They care not what happens after they are gone," he lamented, "and I will go further and say that they care not what happens even to their neighbors, to the community as a whole, during their own lifetime. They will argue that even though they do exhaust all the natural resources, the inventiveness of man, and the progress of civilization, . . . they will supply a substitute when the crisis comes. When the crisis came to that prosperous province of China the progress of civilization and inventiveness of man did not find a substitute. Why will we assume that we can do it while the Chinese failed?"[12]

The fight over the Roosevelt-Jones Conservation Bill was fierce. At issue was how much "police power" New York State would have to regulate activities on privately held lands. Roosevelt saw the bill as a codification of all laws pertaining to lands and forests. Opponents rose to fight it on grounds of government overreach. Members of the Empire Forest Products Association alleged that Roosevelt, a mid-Hudson dandy, aimed to get reelected to the state senate by preaching his "tree sermon." Robert Parker, a major Adirondack landowner with a "mountain man" attitude, was trotted out to testify before the Forest, Fish, and Game Committee. Parker chided Roosevelt for trying to cripple free market capitalism through overregulation of privately owned forests. FDR's name had become inextricably linked to Gifford Pinchot's in timber and mining industry circles. Disappointed that wealthy men like Parker had no long-term land ethic, Roosevelt knew that conservation education had to take root with rural Americans. "Mr. Parker is without a doubt," Roosevelt wrote to Dexter Blagden, "a gentleman of the highest standing, but from all the evidence that the Committee has before it the [timber] concern which he happens to be connected with has done about as much as any other to destroy the Adirondacks without giving back a quarter of what they have taken. This, of course, is entirely between ourselves, but the same old fight is going on up here between the people who see that the Adirondacks are being denuded of trees and water power and those who, in the early days, when grants of timber and water were given for a song, succeeded in getting for nothing what they would have to pay well for today."[13]

After rounds of negotiation, a compromise deal was finally struck. Clear-cutting on private lands, to a large degree, became regulated by the state. Further environmental safeguards were put in place by New York's legislature, including a new forest fire prevention strategy and the implementation of scientific forestry principles by private sector logging operations. Only Roosevelt's no-cut rule for large trees was discarded in order for the bill to get passed. The Roosevelt-Jones Conservation Act was a pioneering and trendsetting piece of Progressive-era legislation that further safeguarded the Adirondacks.[14] "I have taken the conservation of our natural resources," he said, "as the first lesson that points to the necessity for seeking community freedom."[15]

Franklin Roosevelt was on a roll. Identifying with the Dutchess County farmers who planted, plowed, cleared, irrigated, and harvested—they were his core constituency, after all—Roosevelt eloquently championed using public lands for outdoors recreation and public health purposes; at the redbrick river town of Troy, he rallied his listeners to the struggle for "liberty of the community" rather than "liberty of the individual."[16] Driving all over upstate counties, inspecting the foundations of farmhouses and tall silos and huge barns and cow stables, Roosevelt was sometimes misidentified as a state agricultural inspector. Although cynics asserted that Roosevelt praised farm life largely for political gain, there was more to it than that; Roosevelt, at heart, was truly a Jeffersonian agrarian in outlook and conviction. Even if he had the luxury of delegating the more brutal chores at Springwood to hired help and experienced the sunup-to-sundown pressures of farm life from a certain aristocratic remove, he had still grown up in a hay-strewn world. Republicans later accused Roosevelt of being a "traitor to his class," attacking business interests, a populist in the mold of William Jennings Bryan who aimed at easing the burden of debt for farmers at the expense of bankers. But Roosevelt interpreted this intended put-down as a badge of honor.

For his reelection campaign for the New York state senate in 1912, FDR wrapped himself in the flag of Wilson's "new freedom" movement; he emphasized Democratic issues such as education reform, sewage treatment, and taxation. In the spirit of the Roosevelt-Jones Bill, he argued that in modern industrial society actions taken by private individuals could harm the health of an essential public resource like the Adirondacks, proudly calling his conservation philosophy the "liberty of the community."[17] Championing public commons came easily to Roosevelt. Historian Arthur M. Schlesinger Jr., in *The Crisis of the Old Order*, postulated that TR's "power" in 1912 emanated from "what he did" while Wilson's came from "what he held in reserve."[18] It seemed that FDR was a hybrid of both progressive leaders. While FDR was intellectually more in sync with TR (especially in conservation), cultivating a more "reserved" legislative persona was the safer road to follow in the largely Tammany-controlled Democratic Party.

That summer, just as the 1912 election campaign started sizzling, both Franklin and Eleanor Roosevelt contracted typhoid fever (probably after brushing their teeth with contaminated water en route from Campobello).[19] With two rivals—a Republican and a Bull Moose—jockeying for his seat, FDR could hardly afford to be sidelined. However, bedridden, he had no choice but to convalesce in his Manhattan apartment during the early fall. FDR turned to Louis Howe, his clever campaign manager, to pick up the slack; Howe, a former newsman, launched an effective mail campaign.[20] "I found myself a campaign manager with a candidate who could not lift a finger to win votes for himself," Howe recalled. "The doctors wouldn't even let him speak to me over the telephone."[21]

Come November 5, 1912, Wilson trounced Theodore Roosevelt and Taft to become America's twenty-eighth president. FDR was pleased that Theodore had bested Taft, winning 88 electoral votes to finish second. In the state senate race, FDR himself won handily, by twenty-six thousand votes. It was a stunning political achievement on Howe's part, and until he died in 1936 he would remain Roosevelt's indispensable adviser.

Just how beloved Roosevelt had become with New York farmers became self-evident when he was named chairperson of the New York State Senate's Agricultural Committee. But as fate would have it, after the election, Wilson invited FDR to the resort town of Sea Girt, New Jersey, to discuss employment in the new administration. Roosevelt came away from the Atlantic beachside meeting thinking the position of assistant secretary of the navy—held by TR from 1897 to 1898—might soon be his. He undertook a crash course, reading Alfred Thayer Mahan's *The Influence of Sea Power upon History*, among other works. The official offer came on the eve of Wilson's inauguration, in the lobby of Washington's Willard Hotel. Josephus Daniels of North Carolina, the incoming secretary of the navy, approached FDR and asked, "How would you like to serve as assistant secretary of the navy?" "How would I like it?" Roosevelt replied. "I'd like it bully well."

Once again following TR's career path, Franklin prepared to move from Albany to Washington for duty at the Navy Department. The outcry from New York conservationists on losing FDR to the Wilson administration was heartfelt. The movement needed him in Albany. "I am not in the slightest danger of losing my interest in nature," Roosevelt reassured Ottomar H. Van Norden, who was the president of the Long Island Game Protective Association and a prominent leader in the Camp-Fire Club, "either in conservation or in any other of the big issues of our state." Just before resigning from the state senate, Roosevelt lobbied for stronger protection laws for upland game and migratory birds. As Van Norden noted, Roosevelt had done "splendid work" on behalf of the forests, rivers, and wildlife of New York.[22]

Roosevelt also oversaw the publication and distribution of *Woodlot Forestry* (Bulletin 9) by forester Robert Rosenbluth. This pamphlet, funded by the New York Conservation Commission, became FDR's manual for the transformation of Springwood into a productive tree farm. Rosenbluth, the state director of forest investigations, complained that "the farms and country estates of New York" had been "treated with *mistreatment*!"[23] He charged that even progressive farmers like Roosevelt irresponsibly culled their acreage. Farmers could no longer afford to think solely in terms of the end product: board feet, ties, poles, posts, and firewood. They needed to consider planting woodlots to act as shelterbelts and bird sanctuaries, to generate fresh water, and to combat soil erosion.

Implicit in *Woodlot Forestry* was a theme that later became manifest in New Deal programs, including the Civilian Conservation Corps, Shelterbelt Project, and Soil Conservation Service. Government had an obligation to teach farmers things like how to grow mixed hardwoods (oak, chestnut, beech, birch, hickory, and maple) on good land and white pine and black ash in swampy areas.[24] Putting an amalgamation of Pinchot's and Rosenbluth's teachings to practical use, Roosevelt constructed a checkerboard of wood roads, or "fire lanes," at Springwood. This endeavor proved valuable in preventing forest fires from rushing uncontrolled through the trees and in chopping winter-stripped trees for surplus firewood. Decades later, when FDR took Winston Churchill and King George VI down those unpaved roads at Springwood, the British leaders mistakenly thought their host wanted to show off his trees. But Roosevelt's *real* pride on these sightseeing tours was his system of Rosenbluth-inspired "wood roads."[25]

During Roosevelt's tenure in the Navy Department—from March 18, 1913, to August 26, 1920—naval strategy necessarily trumped forestry concerns. Yet, he nevertheless remained active in the conservation movement during his off-duty hours at home, at 1733 N Street.[26] Refusing to abandon New York State entirely, Roosevelt worked tirelessly from Washington to cast the town of Plattsburgh as a vibrant recreational hub in the Adirondacks–Lake Champlain region and helped search for sites in Dutchess, Putnam, and Columbia counties for potential state parks.[27] Forging an alliance with A. S. Houghton of the Camp-Fire Club of America, Roosevelt continued to oppose the overhunting of birds and advocated bird protection measures.[28]

Roosevelt's official duties in Washington began on St. Patrick's Day of 1913, his eighth wedding anniversary. Boxes of nautical paraphernalia—including model ships, elegant prints, and antiquarian books—were moved from Springwood to adorn his office at the Navy Department. Although the American fleet was the third-largest in the world (after the British Royal Navy and the imperial German navy), the Navy Department's annual budget was only around $60 million—and the department didn't even rate its own building. Much of FDR's work dealt with procurement, supply, and compensating civil personnel. It was crucial work, but tedious. Fortunately his assistant at the department was Louis Howe, who proved to be as indispensable as ever.[29]

Secretary Daniels, the former publisher of the Raleigh *News and Observer*, took a shine to the young Roosevelt. Daniels was not nautically inclined and knew practically nothing about maritime geography, while FDR was a gifted and experienced sailor. An odd pair, they worked fairly well together despite rocky moments. Unfortunately, Roosevelt couldn't resist taking advantage of the perks his new position offered; he had the audacity to arrange for the battleship *North Dakota* to cruise to Campobello for Independence Day in 1913 so it could fire a seventeen-gun salute just for his children. Such "flummery," as Daniels called the ceremonial side of the navy, was always present in FDR's wheelhouse.[30] When there was a war scare with Japan in 1913, FDR sought counsel from TR. "We shall be in an unpardonable position if we permit ourselves to be caught with our fleet separated," TR advised him. "There ought not be a battleship or any formidable fighting craft in the Pacific unless our entire fleet is in the Pacific."[31]

Washington, D.C., was a confining place for a gregarious person like Roosevelt. Despite his occasional inspection trips for the Navy Department, he was usually consumed by a daunting pile of paperwork in his office. He yearned for seagoing escapes from his bureaucratic position. When he heard that Gifford Pinchot was cruising to the Dry Tortugas islands off Florida, he burned with envy. Roosevelt pledged that someday, after he got a firm grasp on his Navy Department workload, he'd sail to the Tortugas himself to catch snapper, eat crab, and tour Fort Jefferson, an unfinished brick fort seventy miles from Key West that had been used as a military prison during the Civil War.[32] In the meantime, in the fall of 1913, FDR embarked on an official seven-week cruise of the Mediterranean aboard the *Dolphin*, a yacht on loan to the Navy Department.[33] The top priority for the secretary and his assistant after World War I began in Europe in 1914 was the creation of a reserve component for the U.S. Navy. (Daniels and Roosevelt were ultimately successful in this goal; legislation was passed in March 1915 making the Naval Reserve a reality.)

II

President Wilson had chosen Franklin K. Lane of California, an ardent outdoors enthusiast, to be his secretary of the interior. At first it seemed like an inspired choice. Raised around the San Francisco Bay Area, Lane entered the University of California at Berkeley but dropped out to work as the New York correspondent of the *San Francisco Chronicle*. After moving home to run his own newspaper, he'd hike on Mount Tamalpais and fish in the Russian River. In 1898 he became the city attorney of San Francisco, casting himself as an anticorruption crusader. Lane had been selected to head the Department of the Interior by President Wilson because of his understanding of western water rights and timber issues; the president also admired Lane because he was one of the few California Democrats who fought rail tycoons and the conservative Hearst newspaper syndicate. And Lane had distinguished himself in California's conservation circles by becoming the first president of the Save-the-Redwoods League.[34] "He loves the forest," one of Lane's friends explained, "living every moment that a busy life could spare in the shadows of the trees."[35]

Some preservationists, however, had problems with Wilson's choice. In particular, the Sierra Club, cofounded in 1892 by John Muir to protect the natural features of the Sierra Nevada, took exception to the appointment.[36] As San Francisco's city attorney, Lane had helped launch a campaign to build a dam on the Tuolumne River in the Hetch Hetchy Valley of Yosemite National Park. In the aftermath of the devastating San Francisco earthquake and fires of 1906, Lane wanted to guarantee that the rebuilt city had access to a reliable supply of fresh water.[37] The proposal to build a dam in Hetch Hetchy to accomplish this goal faced a firestorm of protest, and the controversy dogged Lane's footsteps when he moved to Washington, D.C.

In 1913, when Congressman John E. Raker of California introduced legislation to authorize the Hetch Hetchy dam, Lane cheered him on. The ensuing battle was the most bitterly contested environmental showdown yet seen in American politics. The Sierra Club launched a campaign to save the gorgeous valley. Americans outside the Bay Area were largely opposed to building the dam and reservoir in the national park.[38] Ruining a sacred part of Yosemite, to Muir, would be no less a crime than firebombing the Sistine Chapel. "Dam Hetch Hetchy!" Muir exclaimed. "As well dam for water-tanks the people's cathedrals and churches, for no holier temple has ever been consecrated by the heart of man."[39] The *Congressional Record* printed 380 pages of heated debate on the issue, which culminated with the fifth and final version of Raker's bill.

That October, Robert Underwood Johnson, former editor of *Century* and close friend of Muir (and ambassador to Italy for a short time), wrote FDR a six-point letter explaining why damming the

Tuolumne River was a "backward step in conservation," an "invasion of the national park system for commercial purposes."[40] Johnson fervently believed that America would be defined in later centuries not only by what engineers had built but also by what it had refused to destroy. Because he had persuaded TR to hold a conservation conference at the White House in 1908, Johnson hoped to persuade FDR to join his side in the Hetch Hetchy debate.[41] Johnson told Roosevelt it made no sense to pull water from an arid part of California while ignoring the Sacramento River Valley, which had a "superlative abundance of good water."[42] As a lobbyist, he knew that convincing a Navy Department official, particularly one with the surname Roosevelt, would surely be a boon to the anti-dam cause.

However, Johnson harbored an uncomfortable suspicion that Pinchot, who was in favor of the dam, had tempered Franklin Roosevelt's preservationist convictions with a "use it or lose it" mentality. There was truth in this. But more to the point, Roosevelt was a firm believer in public waterpower projects. Regardless of his personal beliefs, though, he wasn't going to defy President Wilson over a dam project on the far-off Sierras. He opted to dodge the controversy by simply remaining silent. "I am very glad to hear from you about the Hetch Hetchy bill and it will stir me into going over the matter more carefully," Roosevelt wrote to Johnson. "I had already noticed that the great majority of the important papers have opposed the bill, but, on the other hand, many prominent people, such as Gifford Pinchot, who have studied the matter, seem to have come to the conclusion that the bill in its present form is very different from the original and have given their support to it. I can assure you that I shall study the whole question with great care and do what I can to see that justice is done. I feel sure that when the bill comes to the president he will give it careful study before he passes it on."[43]

Johnson replied quickly, warning FDR that Pinchot had made a "mistake in judgment" and that he believed the Wilson administration was wrong to let the bill come up for a vote in Congress.[44] Johnson charged that the White House, by not opposing the bill, had become complicit in the destruction of Yosemite National Park.[45] If

Yosemite couldn't be protected, Johnson wrote, it didn't bode well for other national parks and monuments.[46]

Assistant Secretary of the Navy Roosevelt spent time at a shooting range in Maryland in 1917. Taking aim at left is Secretary of the Interior Franklin Lane. Roosevelt had handled guns since his boyhood, but had little enthusiasm for hunting animals as an adult. He only occasionally participated in hunts for ducks or opossums.

What Johnson most likely didn't know was that FDR had become close friends with Secretary of the Interior Lane. Both were new to executive branch politics, they shared a jocular enthusiasm for cocktail hour, pranks, and the civil engineering "miracles" of hydroelectric power—in that order. In terms of national resource policy, "reclamation" and "irrigation" were favorite words of both men. They heartily approved of the attitude of chief engineer William Mulholland, who watched water flow from the newly built Los Angeles Aqueduct for the first time and declared, "There it is. Take it."[47]

On December 19, 1913, Congress approved the Raker Act. It was a devastating blow to the Sierra Club, and John Muir's heart was broken (he died only a few months later). Johnson was disappointed in FDR but he realized that it was difficult for any Navy Department official to publicly cross swords with the commander in chief. In spite of the Hetch Hetchy issue, Johnson and FDR's friendship solidified. They found common ground in protecting the eastern forests of Virginia, North Carolina, and Tennessee. (In later years FDR would rationalize his silence on Hetch Hetchy in 1913 as a show of support for municipal ownership of public power utilities.)[48]

Roosevelt's fraternal bond with Secretary Lane also grew weekly. As a joke, they founded the "Franklin Club"—an informal group of high-ranking Wilsonian Democrats who had in common a first name and an anti–Wall Street mind-set. The other prominent member was Secretary of Agriculture Franklin Houston. The three Franklins regularly played poker and pinochle over cocktails.[49] Horace Albright, an employee at the Department of the Interior who worked closely with Lane and was in a position to know about the Franklin Club, once said that his primary recollection of FDR during those years was of his skirt-chasing. "He always escorted a beautiful woman, not his wife," Albright recalled later in life. "She was never really identified. Seventy years later I read an article in *American Heritage* about the great love of his life, Lucy [Mercer] Rutherfurd. And sure enough, there was a picture of this lovely lady of long ago."[50]

Despite their friendship, Roosevelt was mistaken in thinking that Lane was a genuine nature preservationist. Lane had a commercial developer's mentality, and in his view, his department's main function was to make money by leasing public lands to drilling, mining, and timbering interests. Overall, he felt that nature should be conquered by American engineering might. "A wilderness," Lane wrote, "no matter how impressive and beautiful, does not satisfy this soul of mine (if I have that kind of thing). It is a challenge to man. It says, 'Master me! Put me to use! Make me something more than I am!' "[51]

III

In the summer of 1914, Franklin and Eleanor—like most people—were transfixed by the events unfolding in Europe. Archduke Franz Ferdinand of Austria, heir to the Austro-Hungarian throne, was assassinated in Sarajevo that June. His assassination inflamed a long-simmering situation, and by the end of July, when Germany declared war on Russia, the World War I was under way. The turmoil in Europe accelerated the pace of Roosevelt's work at the Navy Department. Daniels dispatched him for an inspection tour of the Pacific Northwest. This was the beginning of his lifelong love for the state of Washington. His three great passions—forestry, ships, and ornithology—all converged in Puget Sound.

With the war beginning, ensuring the strength of the two-ocean navy took precedence over all else, including the power of the Weeks Act to create more national forests in the American East. "[I] cannot devote as much time to forestry matters as I would like," Roosevelt admitted. "However, I am able to go ahead with my own planting in Dutchess County, and succeed in interesting a good many people in that locality."[52]

Roosevelt's primary conservation goal during the war was to promote reforestation as a key means of curtailing floods in the Ohio and Mississippi valleys. A devastating flood in southern Ohio in March 1913, right after Wilson was sworn in, had made an indelible impression on Roosevelt, who, with Secretary Daniels traveling abroad, was left in charge of the Navy Department's relief efforts. More than 460 people died in the disaster, and there were indications that the Ohio River had risen twenty-one feet in a single day. "Suddenly called upon by the President to make all arrangements for sending surgeons, attendants, supplies, etc. out to the flood district in Ohio," Franklin wrote to Eleanor. "I had a hectic time getting the machinery going."[53] In Roosevelt's estimation, the violent flood in Ohio was caused, in part, by deforestation—a word not often found in Navy Department memos. George Perkins Marsh, however, had helped propagate this very theory in his groundbreaking *Man and Nature*.[54] Another influential voice was that of Colonel Curtis Townsend, president of the Mississippi River Commission and District Engineer of the Army Corps of Engineers, based in St. Louis. In his speech "Flood Control of the Mississippi River" (1913), Townsend advocated for dependence on levees, dismissing reforestation, parallel channels, and other means of tempering water levels.[55]

Roosevelt tried to become a national voice in the reforestation movement. Traveling to New Orleans and Pensacola in conjunction with Navy Department work, he spoke with civil engineers about reforestation and confining rivers into single straight channels spread across floodplains. He hoped to have some effect on Townsend. "Personally," Roosevelt wrote to Townsend, "I cannot help feeling that as the country develops more attention is going to be paid to reforestation everywhere and that with increased knowledge of the value of trees, not only to themselves but to agriculture in general, especially in hilly districts where streams originate, floods will be decreased to a certain extent, probably a comparatively small percentage by the absorption of rainfall at the source."[56]

While living in Washington, D.C., in 1915, the Roosevelts took the opportunity to remodel Springwood. Sara and Franklin planned the renovation to accommodate the growing brood: Franklin and Eleanor's fourth child, Franklin Jr., was born in August 1914, and Eleanor was soon expecting again. (John, the fifth child, was born in March 1916.) Franklin referred to them collectively as "the chicks" in letters to his wife. The west side of the estate's main house, which offered sweeping Hudson River views, was barely changed.[57] Two fieldstone wings, however, were added to the residence; Franklin gained an office, as well as a spectacular library (which eventually held around fourteen thousand volumes; more than two thousand naval paintings, prints, and lithographs; and 1.2 million stamps). Sara Delano Roosevelt displayed his fine collection of Dutchess County birds in glass cabinets in the front hall.[58] The old porch was torn out in favor of a fieldstone terrace with a balustrade and columned portico that stretched around the entrance of the house. Everything Franklin's uncle Frederic Delano had taught him about Downing-style landscape architecture was integrated to at least some degree at Springwood.

During World War I, anyone from upstate New York whom FDR met in official Washington was embraced as his "neighbor" and "friend"—no two words were more emblematic of his style of politicking. He eagerly gossiped about upstate happenings. In the coming decades, he would downplay his town house in Manhattan's

East Sixties and continue to present himself as a proud Hudson River Valley tree farmer. When the New York State Forestry Association had the temerity to print its letterhead with Manhattan listed as FDR's primary place of residence, he squawked. "I wish you would change my address on the Association's letter paper to Hyde Park, Dutchess County, New York," he wrote, "as I never have been and hope I never will be a resident of New York City."[59]

Another way that Roosevelt stayed involved with New York affairs from official Washington was through the Boy Scouts of America (BSA). When he was in Manhattan, Roosevelt made a point of stopping by the BSA's national office. FDR believed that elements of the BSA program—tree identification, bird and animal identification, woodcraft, astronomy, hiking rules, the ability to complete long walks, starting campfires, preventing forest fires, tracking animals, swimming, and lifesaving techniques—involved the same outdoors values he'd internalized as a boy. Learning to be land stewards and forest caretakers created a sense of virtue and public service in the youngsters.[60]

Much has been written about how FDR ably used his navy years as a stepping-stone to higher office, but his attempt to rush the process in wartime 1914 is often overlooked. Impetuously, FDR decided to run for a seat in the U.S. Senate that year, but Wilson didn't back his action and Roosevelt lost the primary to the Tammany Hall candidate. Secretary Daniels didn't hold his assistant's ambition against him. The next year, he gave Roosevelt the honor of attending the Panama-Pacific International Exposition in California as the Navy Department's representative. On the cross-country train ride, Roosevelt's companion was Vice President Thomas Marshall, who simply did not romanticize the natural world. When FDR marveled at seeing the Front Range of the Rockies, Marshall puffed on a cigar and merely remarked, "I never did like scenery."

IV

The establishment of the National Park Service (NPS) was secured in March 1915 at a conference at the University of California in Berkeley, led by two UC alumni: Stephen T. Mather (class of 1887) and Horace Albright (class of 1912). Mather was tall, athletic, and blessed with prodigious energy. A borax company executive, he could make even scorching Death Valley—where borax was mined—seem like paradise. Knowing the NPS was strapped for cash, Lane hired Mather, a millionaire, as its director at a token salary of $1 per year. Even though Lane hoped that Mather would reconcile the competing interests of the NPS, Mather's sympathies lay with the conservationists—he was an early member of the Sierra Club. In a fortunate turn of events for the American conservation movement, Mather had the foresight to employ Albright as his personal assistant. Albright was only twenty-three. He would go on to serve as superintendent of Yellowstone National Park from 1919 to 1929 and then as NPS director from 1929 to 1933.

The tragedy of Hetch Hetchy had brought these smart conservation leaders together in Berkeley for a symposium on how best to protect America's scenic wonders. The abuse of Yosemite National Park had galvanized the best and the brightest. Brought into the effort at Berkeley was Gilbert Grosvenor, the editor of the National Geographic Society magazine. During the summer of 1915 the "Mather mountain party" hiked two hundred miles of the Sierra to publicize the need for a meaningful National Park Service.[61] The conference helped to establish the University of California as the unofficial partner of those overseeing the national parks, in the same way that Yale Forestry School became a think tank for the National Forest Service. The activities in and around Berkeley combined with Mather's unceasing campaign to build a coalition around the country—and notably, in the national capital. At the same time, a grassroots movement was in motion, led by the Sierra Club, the American Civic Association, and the General Federation of Women's Clubs. And their efforts were rewarded with a bill forming a new government agency to monitor and maintain parks on the federal level.

On August 25, 1916, President Wilson signed the National Park Service Organic Act into law. Mather was to head the new agency, with Horace Albright assisting him. Franklin Roosevelt was delighted that President Wilson—following in the footsteps of Republican presidents Theodore Roosevelt and William Howard Taft—had taken this tremendous step forward for the conservation cause. "There is nothing so American as our national parks," Roosevelt averred years later. "The scenery and wild life are native and the fundamental idea behind the parks is native. It is, in brief, that the country belongs to the people; that what it is and what it is in the process of making is for the enrichment of the lives of all of us. Thus the parks stand as the outward symbol of this great human principle."[62]

The outline of the purposes of the National Park Service (NPS) was, in FDR's opinion, the most important part of the legislation. "To conserve the scenery and the natural and historic objects and the wild life therein and to provide for the enjoyment of the same in such a manner and by such means as will leave them unimpaired for future generations."[63] Indeed, Roosevelt was glad that the Department of the Interior was in the business of both providing public recreation opportunities *and* preserving nature. He worried that if proper camping facilities, scenic roads, restrooms, and hiking trails weren't built at all national parks, then their pristine features would eventually be destroyed by tourists.[64] This was just one way in which the difficulties of the NPS's dual mission—to prevent ecological injury to parks while simultaneously promoting tourism—manifested itself.

The trio of leaders at the Department of the Interior seemed ideal. Lane was the bureaucrat with Wilson's ear, Mather the high conceptualizer, and Albright the workhorse who selected small towns to serve as "gateways" to national parks. They had a bevy of challenges. In 1916 the Department of the Interior—responsible for fourteen national parks and twenty-one national monuments—had no proper protocol from which to manage. The Army, for example, detailed troops to Yellowstone and Sequoia to police against hunting, grazing, timbering, and vandalism. Franklin Roosevelt was concerned that the new NPS had yet to establish a foothold in the American East. The timber lobby had effectively fought against protection for such sites as Virginia's Shenandoah Valley and New Hampshire's White Mountains. No national park would be designated in the East until 1919, when Lafayette National Park was established in Maine. (It was enlarged and renamed in 1929 to become Acadia National Park.)

In early 1917 Assistant Secretary Roosevelt undertook an inspection tour of Hispaniola in connection with the occupation of Haiti and Santo Domingo by the U.S. Marine Corps. There is some evidence that he wanted to open a chain of grocery stores in Port-au-Prince as a business proposition. At one point Roosevelt decided to play a practical joke on a reporter assigned to cover the inspection tour in Haiti. It was bird-faking—lying about seeing a rare avian species—of the highest order. FDR wrote about his close encounter with the nonexistent "Haitian Shrink Bird" in his (unpublished) 1917 trip report: "As you know, there are various species of birds believed to be extinct—the Dodo and the Great Auk. The same thing was true of the great Haitian Shrink Bird. The last specimens were seen seventy or eighty years ago. This rare bird has the curious quality of shrinking away to nothing when shot."[65]

In March 1917, while boarding the USS *Hancock* to return home from Hispaniola, Roosevelt was informed that Germany had issued a warning: any American vessels remaining in the waters off western Europe after twenty-four hours would be fired on "without further notice." Roosevelt hustled his way out of the Caribbean tropics back to Washington. A few days later, Woodrow Wilson told Congress that diplomatic relations with Germany had been severed. On April 2 the president called Congress into a special session to request a declaration of war against Germany. "Everyone wanted to attend this historic address and it was with greatest difficulty that Franklin got me a seat," Eleanor Roosevelt recalled in *This Is My Story*. "I went and listened breathlessly."[66]

Once the United States entered World War I, Roosevelt wasn't able to contribute meaningfully to the conservation movement. In addition to his increased desk duties, he became a leading proponent of a 230-mile minefield between the Orkney Islands and Norway in order to bottle up the German U-boat fleet in the North Sea. His optimistic sense of America's ultimate triumph was contagious and led to regular, glowing press coverage of his efforts. He also became very knowledgeable about how Wilson reshaped the federal government by creating numerous emergency agencies to seize the economy by the scruff of the neck by boosting war production, restraining home consumption, and managing inflation.[67] Roosevelt regretted that he was never able to serve in combat during the war. But over the summer of 1918 he crossed the Atlantic on the destroyer *Dyer*, convoying troopships to Europe. His diary of the trip included elegant entries about "the good old ocean" and his experiences in the Azores.[68]

When World War I ended on November 11, 1918, newspapers began quantifying the carnage. Over 16.5 million people, approximately one-third of whom were civilians, had perished in the war. Another 21 million had been wounded. While the Europeans bore the brunt of the war's devastation, 116,000 Americans had died "over there."[69] Memorials honoring fallen doughboys sprang up in town squares across America. Three such remembrances were erected on the National Mall in Washington. Trees were planted as living memorials to the war dead. One idea that FDR and Frederic Delano supported never came to fruition—a "National Capital Forest." Notes taken at the June 6, 1919, meeting of the Commission of Fine Arts (CFA) suggest the vision that would have made greater Washington into a far greener place. According to the plan, a large tract near the city would be filled with trees to honor the war dead. The report promised that the National Capital Forest would be "an admirable and suitable memorial to those who have suffered and died in this great war of democracy against tyranny and to the great Service rendered to humanity by our Navy, our Armies, and the Nation and that reclamation of the lands under consideration and creating on them a great pleasure and demonstration forest and park would be [of] the greatest value [to] the Capital and the Nation and to future considerations."[70]

The CFA noted that the topography of suitable land around the District of Columbia was similar to that of the Argonne Forest, Belleau Wood, and Verdun. At the CFA discussion were representatives of the U.S. Forest Service and National Park Service. But without a forceful personality such as Theodore Roosevelt, who died in January 1919, to lead the project, the dream of the National Capital Forest faded away.

V

After the war's end, FDR became the most admired Democrat in New York aside from the popular governor, Alfred "Al" Smith. People could not resist his talent for projecting optimism. When pressure mounted for him to declare his candidacy for governor or senator over the 1919 holiday season—thereby getting a leg up on the competition for 1920—Roosevelt demurred. "Being early on the job is sometimes wise and sometimes not," Roosevelt wrote to a booster. "I sometimes think we consider too much the good luck of the early bird, and not the bad luck of the early worm."[71] With the progressive tide still strong, there were draft movements in 1919 in New York for "Roosevelt for U.S. senator" (if Al Smith chose to run for a second term as the state's governor) and "Roosevelt for governor" (if Smith elected instead to seek a U.S. Senate seat).

Smith, who would go on to serve four terms as governor of New York, was born on the East Side of Manhattan to Irish Catholic parents. He had a hardscrabble upbringing. He didn't attend high school or college, instead developing his intuitions about people by observing their behavior at the Fulton fish market. Though beloved by Tammany, Smith was largely untainted by charges of corruption. A champion of labor, immigrants, and the downtrodden, the progressive Smith took up the cause of the "forgotten man" while serving in the New York state assembly. His rise was a godsend to those who felt economically marginalized, kicked to the curb of life. From his perch at the Navy Department in Washington, FDR kept a watchful eye on Smith, who was the competition—another progressive New York Democrat building a national reputation.

Years later, an embattled Herbert Hoover quipped that Franklin Roosevelt was a chameleon on plaid, changing his colors depending on the hour and audience. This unflattering characterization rang true in 1920 when Roosevelt became the Democratic vice presidential nominee on the ticket with Governor James M. Cox of Ohio. On August 9, Roosevelt officially accepted the nomination from Hyde Park, delivering a seventeen-page speech from his porch at Springwood (it was not yet customary for candidates to appear at the party's national convention, let alone accept the nomination during the event). In his address—his national political debut—he espoused some of his ideas about conservation through wise use. Noticeably, there was no talk of state parks or wildlife preservation or outdoor recreation. It was a boilerplate speech that Theodore Roosevelt, a wilderness romantic, probably wouldn't have given (but Pinchot might have). "So with regard to the further development of our natural resources we offer a constructive and definite objective," Roosevelt said. "We begin to *appreciate* that as a nation we have been wasteful of our opportunities."[72]

FDR resigned from the Navy Department on August 6 to campaign full-time. Thereafter, it didn't take long for conservationists to feel betrayed by him. Whenever Roosevelt was in New England or upstate New York, he spoke like an apostle of George Perkins Marsh. However, in the American West, he lampooned Republicans for making the public domain off-limits (instead of opening it for settlement and development). *Harper's Weekly* quoted Roosevelt spouting forth in Montana, saying that "what is needed is development rather than conservation." Governor Thomas Riggs of the Alaska Territory was so elated that FDR had apparently joined forces with the extraction industries that he sent Roosevelt a telegram on August 16: "Bring out in Seattle speech that conservationists still want to bottle-up Alaska and you will influence Alaskans now residing in Washington to our side. All West rabid against Pinchotism."[73]

What Roosevelt keenly understood was that for Cox to defeat the Republican nominee, Warren G. Harding of Ohio—a long-shot victory at best—the Democrats had to curry favor with voters in the West, many of whom were dependent on "big timber" or "big oil." A suspicion grew among conservationists that FDR wasn't a chip off TR's block. They had good cause for doubt. FDR refashioned himself in the West as a pro-extraction candidate and an advocate of public hydropower during the course of the campaign.[74]

Franklin Roosevelt understood that conservation in the arid West focused on issues of water access. TR had given the American West—and the nation—both a policy and the authority under which public control of waterpower became operational: the Newlands Reclamation Act, the Rivers and Harbors Act, and the Inland Waterways Commission. Therefore, to FDR, western conservation wasn't just protecting national parks. It also meant pushing for expensive public works projects such as the Los Angeles Aqueduct, which had drawn water from the eastern side of the Sierra Nevada two hundred miles across deserts and mountains to southern California. While FDR was campaigning for the vice presidency, Congress passed the Federal Water Power Act of 1920, which established the Federal Power Commission and gave it extensive authority over waterways and the building and the use of waterpower projects. If FDR seemed like a hypocrite in 1920, it was because he saw two very different sets of needs for the East (forest preservation) and the West (securing water access to growing cities like Los Angeles and Phoenix).

That year, during a journey around the Columbia River Basin in Washington State, FDR concluded that the future of the American West would depend on the manipulation of nature by engineering marvels. Huge dams would be required to trap water in reservoirs for use during drought seasons and to redirect it via aqueducts into cities. Washington and Oregon, he believed, could become a showcase for hydroelectric power and irrigation districts.

After speaking in 1920 in Fargo, Sioux Falls, Tacoma, Portland, Bakersfield, Salt Lake City, and Pueblo, Colorado, Roosevelt learned that some Americans in the West stubbornly maintained their belief that rain would always fall. Few in states such as Utah or New Mexico wanted to accept that a desert was a desert and that too much population growth in arid areas was unsustainable. Only the greening of valleys by public water projects seized the imagination of westerners. Roosevelt, hungry for votes, presented the West as a Garden of Eden waiting to bloom with the help of scientific experts and civil engineers.

Even more disconcerting to die-hard conservationists was Roosevelt's full support for the Mineral Leasing Act (which allowed mining on federal lands for nominal rental fees) and the Water Power Act (which authorized federal hydroelectric projects). It's fair to charge that in 1920 FDR turned his back on conservation—at least in the West—for the sake of political expediency. It seemed that all of his new allies were engineers, miners, and industrial foresters. FDR didn't utter anything on the campaign trail that contemporary historians could term "green" or "sustainable." Decades later Roosevelt explained away this type of inconsistency as "juggling," or never letting "my right hand know what my left hand does."[75] More simply put, he was a strategist who put his political future ahead of all else in those days. "He had, not a personality, but a ring of personalities," historian Arthur M. Schlesinger Jr. astutely wrote in *The Coming of the New Deal*, "each one dissolving on approach, always revealing still another beneath."[76]

Overall, Franklin Roosevelt was an asset for James Cox in the West, delivering a string of electrifying speeches and exuding an air of disciplined charisma. But Roosevelt's equivocation about conservation during the 1920 campaign was derided by old Bull Moosers. Theodore Roosevelt Jr.—a decorated war hero and Eleanor's first cousin—shadowed FDR throughout the West in order to rebut FDR's pro-extraction message. He considered FDR not only a political enemy but also an embarrassment to the Roosevelt name. Speaking in Sheridan, Wyoming—not far from where TR established America's first national monument at Devils Tower in 1906—Theodore Jr., the late president's son, told the *New York World* that the squire of Hyde Park, when it came to conservation, did not have "the brand of our family."[77]

Political observers suggested that one reason FDR had been added to Cox's ticket was the appeal that the handsome, personable young man might have with women, who had the right to vote as of 1920, after the Nineteenth Amendment was ratified. The number of potential voters suddenly doubled. FDR courted female audiences and cultivated a strong following in chapters of the Garden Club of America, whose members were mostly women interested in horticulture and other aspects of civic improvement. Pouring on the charm, FDR spoke to various Garden Clubs in Maryland, New York, New Jersey, Virginia, and Pennsylvania, each filled with women conversant in the beautification ideas of Olmsted and Downing that he himself held dear.

The intrepid Roosevelt ended up delivering one thousand speeches during the three-month campaign. That year, as the Forest Service pondered creating the designation "primitive areas" to save primeval wilderness tracts, Roosevelt pushed hard for public dams.[78] He concluded the 1920 campaign with a stem-winder at New York's Madison Square Garden. Exhausted, he then retired to Hyde Park to wait for the election results. In the end, the Harding-Coolidge ticket triumphed over Cox-Roosevelt by seven million votes. Joseph P. Tumulty, Wilson's private secretary, called it "not a landslide" but "an earthquake." The Progressive movement was beaten and in retreat. The remaining question, which was not particularly pressing in the aftermath of the defeat, was what direction the post-Wilson Democratic Party would take—and who would step forward to lead it. To that end, the only sunny news at Springwood was the pundits' consensus that Cox—not FDR—was the problem with the Democratic ticket.

Chalking it all up as a learning experience, Roosevelt traveled to Louisiana around Thanksgiving for a binge of fishing and hunting. A little time in the outdoors, he told Eleanor, would redouble his spirits before he returned to practicing law. Accompanied by his brother-in-law, Hall Roosevelt, then an engineer for General Electric, FDR purchased ammunition and hunting attire appropriate for a bayou setting.[79] At first, he stayed at the Hotel Grunewald in New Orleans, rendezvousing with M. L. Alexander, the Louisiana conservation commissioner, who acted as his hunting guide for a week. They were soon joined by Edward A. McIlhenny, who said that there were "millions of birds" along the Sabine River and Lake Arthur. The four outdoorsmen lived on a Sabine River houseboat for days, bagging mallards, wood ducks, and gadwalls in both Louisiana and East Texas.[80]

Although Roosevelt occasionally shot waterfowl in Dutchess

County after losing the 1920 election, the Louisiana trip was the last time he seriously hunted until 1932. Fishing with Harvard friends and birding with Maunsell Crosby became his primary forms of outdoor recreation. Once back in snowy Hyde Park, FDR followed his father's tradition of sending Christmas trees grown at Springwood as presents to friends. His only complaint during that holiday season, it seemed, was that Theodore Roosevelt Jr.—a sharp thorn in his side during the 1920 campaign—had been named Warren Harding's assistant secretary of the navy.[81] Though FDR was happy personally, his public career had stalled, and he was uncertain about his next move.

Sara Roosevelt wrote in her diary that Franklin was "rather relieved to not be elected Vice President."[82] It's hard to believe that was true, but clinging to his Dutchess County "tree farmer" identity even more fiercely that year, FDR sold a hundred white and red oak for $8 each. Two years later, in Poughkeepsie, he sold $1,400 worth of timber to be used as railroad ties.[83] FDR returned to private life at the new law firm of Emmet, Marvin & Roosevelt; he also took a part-time position running the New York office of the Fidelity Deposit Company of Maryland, the third-largest surety bonding firm in the country. While dividing his time between New York City and Hyde Park, FDR also became a national megaphone for the state park movement. There was a big opening for his advocacy: only New York, Indiana, Wisconsin, and California had organized a streamlined state park organization at that time.[84]

Fascinated by social planning in rural areas, Roosevelt recognized that commodity prices were tumbling and land was suddenly quite affordable in upstate New York. In his mind, that presented an opportunity to expand the state park system. To FDR, the problem with outdoors recreation in the state was that citizens would readily rally to save specific places—e.g., Niagara Falls and Lake George—because they were so spectacular. But that was, in effect, a painfully incremental approach to building a world-class state park system. Instead, Roosevelt thought that New York state needed to evolve a more comprehensive system. With the growth of cities, the automobile revolution, and the rise of a middle class that had leisure time, Americans needed outdoor getaways to maintain healthy minds and fit bodies. The strain of urban existence with its hectic pace in business and social activities made escape into nature necessary for many Americans, and essential for their sanity. For easily accessible escapes, the answer may just have been state parks. Most were smaller than national parks, but there were two big exceptions: Adirondack Park in New York and Anza-Borrego Desert Park in southern California.

Beginning in 1921, Roosevelt scouted for unspoiled New York woodlands, fast-flowing creeks, beaches, lakes, mountain peaks, and scenic vistas. He then urged the New York legislature to acquire such areas for recreational activities, the preservation of wildlife, and the study of natural history. He especially wanted public parks and preserves situated within a twenty- to fifty-mile drive of urban centers. In electoral defeat, Roosevelt had rediscovered his conservationist footing, only now with state parks instead of reforestation as his preferred mantra.

CHAPTER FIVE

"NOTHING LIKE MOTHER NATURE"

I

A hot spell gripped New York City in mid-July 1921 as Franklin Roosevelt made his way to greet five thousand Boy Scouts assembled for a festival held at the Grant's Tomb green space. The sweltering humidity, in an era before air-conditioning, had people woozy. The stench of garbage, sewage, and decay that hung in the city streets was almost unbearable. Roosevelt loathed the way New York City had turned the Hudson and East rivers into cesspools of sewage discharge, heavy metals, and soil runoff. Riverside Park, near Columbia University, could have been made into another well-planned Central Park; instead, the four-mile stretch of riverfront had the low-rent look of a dump or landfill.[1]

Since the inception of the Boy Scouts in 1910, Roosevelt had been engaged as both leader and father.[2] His eldest son, James, was a Scout and he planned on getting Elliott, Franklin Jr., and John involved in time. The young, city-bred Scouts with whom Roosevelt mingled at Grant's Tomb had spent months perfecting woodcraft feats for the festival.[3] Following a parade, Roosevelt boarded the *Pocantico*, a steam vessel owned by advertising mogul Baron Collier that was anchored in the Hudson. Self-assured and always hungering for a financial deal, Collier had amassed 1.2 million acres in southwestern Florida, which made him the largest landowner in that state. The building of the Tamiami Trail, which connected Tampa to Miami across the Everglades, was a project Collier intended to complete.

A select group of hardworking Scouts had been rewarded with a free trip up the mile-wide Hudson River with Roosevelt, Collier, and a retinue of other wealthy benefactors. Their destination was Bear Mountain State Park, forty miles north of Manhattan: An additional

2,100 Boy Scouts had already set up eighteen campsites within the state park, offering respite from the heat, industrial angst, and automobile traffic for all in attendance.[4]

With an elevation of 1,301 feet, the granite Bear Mountain was the highest point between Manhattan and West Point. The summit offered a sweeping vista of the surrounding Hudson Highlands on the western shore of the river. It belonged to the public thanks to the philanthropy of Mary Averell Harriman, widow of railroad tycoon E. H. Harriman. The Harrimans were ardent conservationists and supporters of public recreation. Before his death in 1909, the railroad baron had taken naturalists John Muir and John Burroughs on a legendary "scientific" cruise to Alaska.[5] Franklin Moon, a forestry professor who wrote several seminal textbooks, served as warden of the Highlands of Hudson Forest Reservation, one of the tracts that eventually made up Bear Mountain State Park. Professor Moon was convinced that the scenic grandeur warranted national park status.

Bear Mountain and the adjacent Harriman State Park were essentially the first modern state parks in America. In 1912, Major William A. Welch, whom the *New York Times* considered the true leader of New York's state park movement, was hired as chief engineer for both parks. While FDR was in the state senate and at the Navy Department, both Bear Mountain and Harriman had been vastly improved by work crews under Welch's direction. The laborers dammed streams to make lakes and built picnic pavilions, campgrounds, boat docks, icehouses, and a vast network of trails and roads; sixteen permanent campsites were also erected, for the summer, around three handsome man-made lakes. "Then crowds of New Yorkers came," historian James Tobin writes, "and by 1920 the park's boosters were calling it 'the greatest playground in the world' with more visitors annually than all the national parks combined."[6] During World War I, Bear Mountain drew one million visitors each year.

When the *Pocantico* docked at Bear Mountain, Roosevelt was in an effervescent mood. Buses drove the Roosevelt-Collier group from the dock over mountain roads to the camp where the other Boy Scouts had convened. The campsite, surrounded by ancient stands of oak and hemlock, was reminiscent of the forests around Hyde Park. That afternoon, the Scouts demonstrated firefighting, compass reading, tent pitching, and outdoor cooking. Although saddened that the hard-luck kids had been raised mostly on city streets, Roosevelt was enthusiastic and spoke with them about learning the joys of nature along the Appalachian Trail.[7] In an eventful day of speeches, parades, and a fried chicken feast, Roosevelt "led his guests around the camp, hiking through campsites, talking about their scouting activities, and showing off the numerous teepee tents resting on built-up wooden platforms."[8]

Two of Roosevelt's most cherished ideas for societal improvement—the Boy Scouts and the state park movement—converged at the Bear Mountain jamboree. But there was also a political motivation for Roosevelt's participation: after eight years in Washington, he was recasting himself as a spokesperson for New York conservationism. Therefore, his personal involvement in the Boy Scouts and the state park movement was politically advantageous. "What do the boys learn?" Roosevelt asked in an article he wrote for the *New York Times*, extolling the Scouts. "Under the guidance of voluntary leaders, they become acquainted with the subjects into which the outdoors is scientifically divided. They learn the rudiments of botany, biology, forestry, and astronomy. They learn about mapping, surveying, signaling, living in the open, fire-making without the aid of matches, compass reading, first aid, tower and bridge building, and a great many other things as well."[9]

From a distance Bear Mountain was idyllic, but on a more minute scale there was a hazard. In the fall of 1920, New York's Public Health Council dispatched water-quality expert Earl Devendorf, a civil engineer, to inspect the sanitary conditions at both Bear Mountain and Harriman state parks. His findings were startling. The drinking water at the parks was compromised by human waste. The pit privies were in an "insanitary condition," and the newly installed chemical toilets weren't flushing properly. Devendorf reported that the well water and groundwater had a high probability of contamination because so many potentially dangerous "carriers" were present. Almost all of the water samples Devendorf collected contained specimens of coliform bacteria.[10] Evidence suggests that when Roosevelt went for a swim at Bear Mountain, he contracted the poliovirus that would soon fell him. The lake had been contaminated by human waste.[11]

After serving as toastmaster that evening, Roosevelt returned to New York City, bringing along with him the dangerous virus quickly multiplying in his bloodstream. He then left for his summer vacation, with the fresh sea air to replace the city smoke. A couple of days later Roosevelt arrived at Campobello, planning a full slate of outdoors activities, including a three-day fishing trip up the Saint Croix River. After lunch on August 10, with the help of his eldest son, James, FDR readied the *Vireo*—his twenty-one-foot boat, named after a favorite songbird. With Eleanor and two of his children as companions, he set a course for a whole day of blue-water sailing.

After only ninety minutes on the Bay of Passamaquoddy, the Roosevelts spotted smoke rising from a wooded islet. Franklin, a pyrophobe since childhood, anchored the *Vireo* off the island and instructed his family to join him in extinguishing the blaze, flailing out the rising flames with boughs.[12] It was rigorous work. Smoke burned their eyes, and sparks swirled around like fireflies. Years later, Anna Roosevelt recalled watching a towering spruce crackle with fire, followed by the "awful roar of the flames as they quickly enveloped the whole tree."[13]

The Roosevelts left the islet covered in ash, grime, and dust. FDR guided the *Vireo* back home. Once ashore, he decided that they should walk to Lake Glen Severn for a cleansing swim instead of taking baths at the cottage. Gleefully, his family leaped into the frigid waters, swam across the pond, and then dried off with towels. Franklin, however, felt listless. Neither the laps nor the beating sun invigorated him. Flushed and despondent, he had to accept the fact that "the glows" (his term for the rewards of outdoor recreation) weren't coming easily that day.[14]

Back at the cottage Franklin, exhausted, put on a sweater and tried to warm up. But shivers soon took hold. He retired early, turning out the light without reading. After tossing and turning in a delirious state that night, he woke to find he could barely move. He was also running a dangerously high temperature. Eleanor rushed in a doctor from nearby Maine as quickly as possible for an emergency diagnosis. The preliminary determination was that Franklin had a very severe summer cold. The possibility that he had contracted polio while swimming with the Boy Scouts at Bear Mountain, however, was raised.[15]

With Eleanor and Louis Howe, who was vacationing at the cottage, unable to massage any kind of feeling into his body, the situation was dire. Within days, FDR regained the use of his upper body but nothing else. His legs were paralyzed. Eleanor was stoic throughout the entire ordeal, calming Franklin's fears, squeezing his hands, scrambling to find the best local physicians to diagnose his ailment. In a moment fraught with emotion, she helped her husband onto a canvas stretcher; his legs were completely useless. Out of public sight, clutching Duffy, his beloved Scottish terrier, he was hoisted through the open window of a train in Maine, looking utterly helpless. Uncle Frederic Delano, taking charge of the crisis, helped rush his nephew to Presbyterian Hospital in New York City. It was there that FDR was officially diagnosed with polio and told that he was unlikely to walk ever again. "No playacting, no bluster, no promises to be good if given just one more chance," Geoffrey C. Ward wrote of FDR's diagnosis, "could make this obstacle go away."[16]

The strain of the summer of 1921 didn't end once Roosevelt returned to the family's town house. His normal life came to a halt, as he worked full-time on restoring strength to his legs, with extremely limited results. Refusing to be a weakling, Roosevelt did chin-ups, worked out on parallel bars, and lifted weights. Determined to have a productive life, he told his friends and family that he was in training. In the spring of 1922, FDR, fighting depression, moved to Springwood. His progress was measured in feet and pain. Struggling to regain feeling in his legs, he would strap fourteen pounds of metal to his legs most days and set out alone, trying by sheer grit to drag them down a tree-covered lane. When not adhering to his own physical therapy regimen, he read, bird-watched, collected stamps, took long country drives, and supervised tree plantings at the property. Slowly his shattered psyche mended. With Louis Howe he launched, in the Hudson River, a model sailboat they built and designed together.[17] His enthusiasm for nature actually seemed to grow after he contracted polio. With time to spare, watching the Hudson thaw and the migratory birds return was a salve for his low morale. "The only thing that stands out in my mind of evidence of how he suffered when he finally knew he would never walk again was the fact that I never heard him mention golf from the day he was taken ill until the end of his life," Eleanor Roosevelt recalled. "That game epitomized to him the ability to be out of doors and to enjoy the use of his body."[18]

Sara Delano Roosevelt was terribly concerned about Franklin. Her son's friends were no longer stopping by Springwood to chat with him about crop rotation, international affairs, retail politics, or forest conservation. She tried cheering him up but needed the help of others. Howe persuaded Eleanor, always shy in public, to become more vocal in the Democratic Party so that the Roosevelt name wouldn't be forgotten. But when she traveled to make political speeches and official appearances at "big labor" events, Eleanor left Franklin without a sympathetic companion for days at a time. In the spring of 1922 Sara invited a distant cousin, Margaret "Daisy" Suckley (pronounced Sook-lee) of neighboring Rhinebeck, for tea with Franklin. Suckley chatted easily and was content to watch Franklin perform exercises aimed at regaining the use of his legs. A deep and enduring friendship developed between them.

Suckley was born on December 20, 1891, at her family's estate, Wilderstein. While Rhinebeck was Daisy's home, she spent much of her youth in New York City and Switzerland. After attending Bryn Mawr College for two years, she dropped out and returned home to care for her ailing mother. During World War I she sold war bonds door-to-door in Rhinebeck. A string of tragedies then befell Suckley, beginning when her oldest brother, Henry, was killed in a Red Cross ambulance in Greece. Her father died not long after that, of a heart attack. Sara Delano Roosevelt was wise to bring Daisy and Franklin together, as they were both lonely in their own ways.[19] It wasn't until 1933, however, that Franklin and Daisy became intimate friends, collaborating to build a Dutchess County dream home at a spot they called "Our Hill."[20] It would be the site where FDR built Top Cottage.

II

The 1920s were a decade of materialistic excess and the mass production of consumer products, from canned soup to ready-to-wear clothing. Thousands of Americans moved to cities to work in factories, establishing a new middle class with disposable income. America's net wealth almost doubled between 1920 and 1929. It was an era of flappers and jazz, Prohibition and speakeasies—none of which FDR found attractive. As industrialization swept America, Roosevelt turned to rustic writers like Gilbert White and John Burroughs who celebrated tidy farms, harvest moons, uncharted forests, and county fairs. He wasn't alone: thousands of Americans worried that the country's natural resources were being squandered, that wilderness was disappearing, that Jefferson's agrarian dream was gone.

While Roosevelt tried to confront the problem by joining the state park movement, a group of midwesterners led by an advertising executive, Will Dilg, founded the Izaak Walton League in Chicago. Its conservationist mission could be distilled into three W's: woods, water, and wildlife. Starting with only fifty-four founding members in 1922, the league expanded to a deep bench of 100,000 members within three years. What caught FDR's attention was its pioneering work, through its Pollution Bureau, toward restoring America's streams. Roosevelt worked behind the scenes with Dilg to expose the fact that 7.5 million people in New York were dumping untreated sewage into interstate waterways.[21]

By the fall of 1922, FDR felt well enough to resume work at the Fidelity and Deposit Company. Determined to stay connected to the great outdoors in spite of his paralyzed legs, he agreed to become president of the Boy Scout Foundation of Greater New York City, a leadership role held until 1937. He worked closely with George

Dupont Pratt—the son of an oil magnate. Pratt sponsored the Boy Scouts, was a big-game hunter, and served as conservation commissioner of New York from 1915 to 1921. With him, Roosevelt promised to "do everything possible" to better the "understanding of nature by these city-bred boys."[22]

Roosevelt's great hope was that Boy Scouts would begin learning the dual arts of forest conservation and wildlife management. To this end, he started Franklin D. Roosevelt Conservation Camps for Boy Scouts in Palisades Interstate Park. Each camp was attended by sixty boys aged fifteen or older. He was particularly committed to helping children from poor urban families who were unable to find work during the summers. Four skills were taught at these camps: how to plant trees, cut firebreaks, control wildfires, and protect wildlife. The program was so successful that in the late 1920s FDR led an expansion at a new 10,600-acre compound in Sullivan County, New York.[23] FDR was so excited by the conservation work at Bear Mountain, Harriman, and Palisades that he proposed forming a syndicate to buy even more upstate woodlands. The acreage, an hour north of New York City, would then be designated as a large public forest managed with "private interests" in mind. FDR, in this regard, was more in line with wealthy business magnates like the Rockefellers and Harrimans, who at least professed to understand European forestry traditions, than with Harding Republicans who were tied to "big timber" interests in New York.

Another manifestation of Roosevelt's recreation philosophy was his becoming a founding board member of the Adirondack Mountain Club, known as the ADK Club.[24] This mountaineering fraternity promoted the use of the State Forest Preserve in the Adirondack Park according to the "forever wild" provisions of the New York state constitution.[25] While FDR could no longer hope to scale an Adirondack peak, he could still enjoy views of unbroken forests from rough-hewn-log lodges and read *High Spots*, the ADK newsletter. And when in 1922 Governor Al Smith sought legislative approval for *A State Plan for New York*, a comprehensive statewide system of parks and parkways. Roosevelt enthusiastically lent his support. He adamantly believed in connecting New York City residents to upstate forests by means of the automobile.

FDR kept a close eye on the 1922 gubernatorial election in Pennsylvania. The bright, willful Gifford Pinchot was running for the office, and had adopted conservation reform and the regulation of financial practices as his bread-and-butter issues.[26] With his boundless capacity for moral indignation, Pinchot had recently become the president of the National Coast Anti-Pollution League, an advocacy group determined to stop the dumping of oil into harbors. Supported by farmers, conservationists, and former suffragists, Pinchot ran a campaign against private utility companies like Pennsylvania Power and Light and General Electric. His own ancestry and outlook caused Pinchot to scold these companies for monopolizing and exploiting natural resources in Pennsylvania. When the Republican Pinchot was assailed as a "multimillionaire with socialistic ideas" and a "communistic conservationist" who would "Sovietize" Pennsylvania, the same fiery darts could have been thrown at Franklin Roosevelt. Pinchot's spirited campaign captured Roosevelt's attention. That November, holed up in Hyde Park, he was elated to hear that Pinchot had beaten Democrat John A. McSparran.

Despite his return to work, albeit on a limited basis, and his many honorary posts, FDR spent most of 1922 to 1924 waging a battle against his paralysis. Learning to walk again or, barring that, recovering some limited use of his legs was his priority. Neither goal was ever really achieved, but FDR did develop muscular shoulders and forearms, thanks to his intense weight lifting. He followed a daily exercise routine that involved toning his shoulders and torso on parallel bars. He strapped fourteen-pound steel braces onto his legs, under his pants. Demonstrating infinite patience and persistence, he evolved a method of walking, with crutches or help or both. Once Roosevelt became mobile, he took on the equally difficult task of being relevant again.

During the mid-1920s, Roosevelt tried mightily to persuade

Pratt, who had become president of the American Forestry Association, to join forces with him in raising funds and developing a system of public and privately owned forests—like those in Germany and Austria. Roosevelt wanted to transform the mid-Hudson region into the Black Forest of America: large, dividend-paying woodlands that could sustain "wise management" timber production for years to come. "In this country no such forest exists," Roosevelt wrote to Pratt. "The federal government and various states own large forest areas. Much planting has been done but most of it has been for the purpose of reclamation of barren or denuded lands. As far as I know, no forest in this country is run on a strictly business basis."[27]

Pratt flatly rejected Roosevelt's idea of community-based forestry as unrealistic. If the lumber industry steered clear of the concept of German-style forests, then investors wouldn't be likely to "take it up." It was easy for Pratt to brush FDR aside in 1922; many New York financiers assumed that, because of his polio, FDR wasn't a major political player anymore. "I do not want to appear like a wet blanket," Pratt wrote to Roosevelt in early December, "but I do think you would have difficulty in finding people who would be willing to go into a proposition to raise $500,000 with no prospect of a dividend for at least 25 years."[28]

III

In early February 1923 Roosevelt, feeling hemmed in, decided that the blue-green waters off Florida would fortify his health. "I am going to Florida to let nature take its course," he wrote to his doctor; "[there's] nothing like Mother Nature anyway."[29] Referring to the fresh air and sunshine as "therapy," he lived happily on the rented houseboat *Weona II* and sent family members letters that were joyous in tone. The fact that so much of Florida was still "wild and tropical" appealed mightily to FDR.[30] Eleanor spent a few days with him exploring the Everglades, but ultimately judged the swamp "eerie and menacing."[31] In a small johnboat, in the company of friends, Roosevelt casually traveled around the fringes of the storied swamp, casting into sloughs so densely covered by mangroves that they were nearly impenetrable.

That very year, Florida began abolishing all state income and inheritance taxes. As a result, land speculation was an extremely lucrative business in America's so-called New Eden. A couple of Roosevelt's Groton and Harvard friends had already invested heavily in Florida real estate. Towns like Fort Lauderdale and Miami were booming and in need of hired hands for construction projects. But Roosevelt also learned about the impoverished conditions in much of hurricane-plagued Florida. Tenant farmers in Dade County lived in lean-tos without electricity or running water. Hookworm, pellagra, typhoid fever, and malaria were common. Soil depletion was crippling the state's agricultural sector.[32]

The passengers of the *Larooco* celebrate the landing of a trophy fish in 1924. Roosevelt is seated at right. The *Larooco* was a secondhand houseboat, owned by Roosevelt and his friend, John Lawrence. Sailing out of Miami (or "My-am-eye," as FDR spelled it in his journal), the *Larooco* not only gave Roosevelt a place to relax with a measure of privacy, it introduced him firsthand to Florida's varied coastal habitats.

With John S. Lawrence, a friend from his Harvard days, Roosevelt purchased the *Larooco* (a portmanteau for Lawrence, Roosevelt, and

Company) and planned—along with Maunsell Crosby of Rhinebeck—a grand trip to Florida for early 1924.[33] Not long after his birthday in January, Roosevelt found himself in Jacksonville, ready to explore a multitude of Florida towns. "Went fishing in the inlet," Roosevelt recorded in the *Larooco* log on the first day with Lawrence and Crosby. "Caught one sea trout. [Crosby] identified 33 different species of birds, including a very large flock of black skimmers. Also a flock of Greater Snow Goose."[34]

Liberated from the New York cold and in the easy company of friends, Roosevelt thrived in the Florida sunshine. All his troubles seemed to dissolve when he was surrounded by dolphins and shorebirds. Roosevelt and Crosby, now a professional ornithologist, compiled an AOU checklist of the various Florida birds they encountered; it was just like old times, but it was an unfair competition because Franklin, unable to walk, was usually confined to using his

With the *Larooco* moored in the background, Roosevelt (*center*) and friends Maunsell Crosby (*left*) and Sir Oswald Mosley (*right*) played statues in shallow water off Florida in 1926. Crosby, an influential ornithologist, lived near the Roosevelts in Dutchess County and was a longtime adviser to FDR. Sir Oswald, a British politician and socialite, was lively company, but he lost his standing with Roosevelt and many others when he founded the British Union of Fascists in 1932.

binoculars from the *Larooco*'s deck.[35] On most days he, Lawrence, and Crosby goofed around in their bathing suits, went for lazy swims, discussed flora and fauna, picnicked on sandy beaches, and tried to catch perch on a hardline for their supper. "I know [the recreation] is doing the legs good, and though I have worn the braces hardly at all, I get lots of exercise crawling around, and I know the muscles are better than ever before," Roosevelt wrote to Sara from Miami. "Maunsell has been a delightful companion and we have any number of tastes in common from birds and forestry to collecting stamps!"[36]

On the trip, Roosevelt learned a lot about the Everglades firsthand, sometimes nosing the *Larooco* into its waters. The expedition saw so much marine wilderness—so many coral reefs, tangled mangrove swamps, pods of dolphins, and herds of manatees—that Roosevelt couldn't help being affected. To him and Crosby, the Everglades were a wilderness that needed to be properly cared for by scientific experts. Biological studies were being made only spottily in Florida. Roosevelt, who had interacted with many local farmers, worried that the drainage, drying, and oxidation of the Lake Okeechobee–Everglades ecosystem had led to the rapid depletion of peat soils. Stephen Mather, director of the National Park Service, would soon propose that the Everglades be considered for national park status. Roosevelt, having finally experienced the swamp's majesty firsthand, was strongly in favor of the idea. But Roosevelt also backed Collier's successful effort to finish the Tamiami Trail. This was an example of the trade-offs—preservation versus development—for which FDR became famous as president: ambitious infrastructure construction followed by the creation of a new park.

Convinced, along with many Floridians, that protection of the Everglades should be a priority, FDR collaborated with Robert Sterling Yard, executive secretary of the National Parks Association. Yard was one of the most respected authorities on national parks. A journalist who became an ardent supporter of Theodore Roosevelt's conservation crusade, Yard acted as the unofficial publicist for the national parks movement—for instance, as the editor of *Century* magazine during the early Wilson years, Yard promoted John Muir's work on behalf of Yosemite National Park.[37] Yard's 1919 *Book of the National Parks* became very popular with members of the half dozen outdoors organizations to which Roosevelt belonged. Yard told Senator Duncan Fletcher of Florida, a Democrat, that FDR had become "one of our valued coworkers" in the crusade to make the Everglades a national park. Yard encouraged Roosevelt to lobby against sugar growers of Dade County, home to one section of the slow-flowing wetlands. "If we wait until these lands show commercial values," Yard wrote to FDR, "we'll never get our park."[38]

For birders like Roosevelt and Yard, 1924 was a banner year. In June, Congress at long last, authorized the appropriation of $1.5 million for the acquisition of bottomlands along the upper Mississippi River to establish what would become a seminal migratory bird refuge. Roosevelt, still a card-carrying AOU member, became interested in the idea of the U.S. Department of Agriculture's Biological Survey's purchasing habitat in Florida to save the great white heron (great egret, *Ardea alba*) and the roseate spoonbill (*Platalea ajaja*). Heartsick from reports that some of America's famed marshes—Montezuma (New York); Lower Klamath Lake (California); Malheur Lake (Oregon); Kankakee (Indiana); Horicon (Wisconsin); and the Souris river bottoms (North Dakota)—were shrinking away, Roosevelt dreamed of a national refuge system from Washington to Florida and from Louisiana to the Canadian border to dramatically increase the dwindling population of waterfowl.

In October 1924, working as an AOU advance man, Roosevelt was plotting to take the *Larooco* on more Florida cruises during the coming winters. "The *Larooco*'s engines are going in," Roosevelt wrote to Crosby, "and I shall join her about February 1st and am keen to have you with me."[39] Crosby, who was compiling a guidebook titled *Birds of Dutchess County*, signed up for the adventure, hoping to identify suitable habitat for federal purchase.

"I wish that Pearson [president of National Audubon] would get busy and accomplish something along the line of creating a bird refuge, national, state, or by private enterprise all through the southern end of Florida," Roosevelt wrote to Crosby on October 13, 1924. "To do so now is practicable; to wait another 10 years would be to lose the whole project. The protected area should include practically all the land south of a line drawn from Florida City on the East Coast to the Everglades or Chokoloskee on the West Coast. This would include all of the so-called Shark River, or Ten Thousand Islands Country. I have been there now for 2 years and there is no question that it is an ideal reservation for the protection of an enormous amount of bird life."[40]

After his first Florida outdoors adventure in 1923, Roosevelt wrote out his own conservation manifesto. Deforestation and "game hogs" hell-bent on slaughtering an unsustainable number of forest animals were at the top of his list of concerns. He hoped to help "charismatic" species like the white-tailed deer (*Odocoileus virginianus*) and North American river otter (*Lontra canadensis*) rebound. His wish list was the following:

> (1) The creation of additional game refuges on public lands and certain areas owned by private individuals or associations, and along certain portions of our seacoast and streams.
> (2) The gradual elimination of "special" laws affecting open season in individual localities, i.e., making the existing federal laws more general in their enforcement throughout the nation.
> (3) Educating the public that songbirds, rodents, etc. are not game.
> (4) The further elimination of the hunter.
> (5) A standardization of the license system and an improvement of morale among so-called sportsmen.[41]

Encouraged by Crosby, who would soon be in Central and South America to collect specimens of wild birds for the American Museum of Natural History, Roosevelt stepped up his conservation efforts and participation in nonprofits. He became an active fund-raiser for the American Museum of Natural History (to further honor Theodore Roosevelt) and oversaw the enlargement of the Roosevelt Wild Life Forest Experiment Station at Syracuse University.[42] He pushed to have California's Sequoia National Park renamed Theodore Roosevelt National Park, as well as for its boundaries to be enlarged to include pristine Kings Canyon.

The magnificent sequoias reminded Roosevelt of the gnarled, weathered, though beautiful Balmville tree near Algonac. Roosevelt believed the sequoias, the grandest of living things, king of conifers, were priceless. California's other "giants"—the colossal sugar and yellow pines, Douglas spruce, and silver firs—should be likewise cherished as heirlooms. But preserving the noble forest trees in California in the age of Harding and Coolidge was an uphill battle. "The Roosevelt sequoia project looks discouraging," Yard wrote to Roosevelt in 1924. "We could have got everything six or seven years ago, but now I'm afraid that the water-power interests have made up their minds that the canyons shall never enter a completely conserved reservation."[43]

After a few years of fighting, FDR and Yard lost the Sierra battle. On April 3, 1926, Congress decided not to change the name of Sequoia National Park, nor to enlarge it to include Kings Canyon. Later, as president, FDR would revisit the preservation of Kings Canyon and the majestic trackless forests of the High Sierra.[44]

IV

The year 1924 was important for Franklin and Eleanor Roosevelt. Governor Al Smith of New York, who was running to be the Democratic nominee for president, asked FDR to deliver his nominating speech that June at Madison Square Garden. It was a high-risk venture for Smith. What if Roosevelt could not stand without assistance at the rostrum or fainted from exertion? But Roosevelt knew that hopping onto Smith's bandwagon was the best way to reinsert himself into American politics. Spurred on by Louis Howe, FDR tirelessly rehearsed his walk to the dais. Howe, seeing that FDR was straining from the effort, decided that he would walk in on one crutch and lean on his son James with his other arm. This made Franklin seem more human and less mechanical. "I was afraid, and I know he was too," James Roosevelt later wrote. "As we walked—struggled really—down the aisle to the rear of the platform, he leaned heavily on my arm, gripping me so hard it hurt."[45]

What happened that day at Madison Square Garden has become the stuff of American legend. Reporters swooned over his bravery. There are stories about Roosevelt being swathed in an unusual light—like a halo—during his speech. He spoke with poetic flourishes lifted from Wordsworth, calling Al Smith "the Happy Warrior." When Roosevelt finished his speech, he received a thunderous seventy-three-minute ovation. The speech thrust FDR back onto the front page of the *New York Times*. Here was a stricken man, unable to walk, exuding vibrancy and moral strength.[46] Al Smith joked that the squire of Hyde Park had stolen the show at Madison Square Garden. Speculation grew that FDR would be running for governor or for the U.S. Senate in a few years. James Roosevelt poignantly called his father's oration an "hour or so stolen from his illness."[47]

Late that summer, after the confetti was swept away, Franklin again escaped into the natural world of Springwood to continue his convalescence. Meanwhile, he had suggested to Eleanor that she (and her new friends Marion Dickerman and Nancy Cook) build a retreat on 180 acres he owned just two miles from Springwood along Fall Kill Creek. Recognizing that Eleanor needed a place to escape from the domineering Sara Delano Roosevelt from time to time, he drafted a lease for them, offering the women lifetime use of the property. He also assumed the role of general contractor for the project. Roosevelt visited John Burroughs's log cabin, known as Slabsides, on the west bank of the Hudson, for inspiration but deemed its aesthetic too rustic for Dutchess County. He commissioned architect Henry Toombs of Poughkeepsie to design a fieldstone Dutch colonial house for his wife's home away from home. Roosevelt wanted it to blend in with the natural surroundings.

Named Val-Kill, the home became more than just a quaint holiday sanctuary. Marion and Nancy, who were activists and educators, moved into the cottage immediately, making it their permanent residence until 1941. Eleanor joined them for weekends and holidays during the summers. In 1926 the three women oversaw the construction of a larger building on the site to house their experimental business, Val-Kill Industries. They were distraught by the exodus of rural New Yorkers to the cities in search of jobs, and believed that if these farmworkers learned manufacturing skills to supplement their agricultural knowledge, then they would have an alternative source of income if farming became unprofitable. For a decade, men and women were employed making replicas of Early American furniture, pewter pieces, and weavings at Val-Kill. Although it produced finely crafted products, the furniture factory folded in 1936, like so many others a casualty of the Great Depression.

V

On October 3, 1924, just a month after construction began on Val-Kill, FDR discovered the town of Warm Springs, then a dilapidated resort, in Georgia. Investment banker George Foster Peabody of Saratoga Springs, New York, had grown up in Georgia and it was he who introduced FDR to Warm Springs. When FDR first arrived, he immediately felt at home amid the sprawling pine forests, clean air, and thermal pools. Eleanor had escorted him there, set him up in his own cottage, and then left. Those first nights in woodsy Georgia, a restless Franklin listened to squirrels make a nocturnal racket and felt oddly tranquil. Shortly after dawn, he got up to swim; it became a daily habit. Never before had he experienced such magical water. "We are here safely and I think Eleanor has written you this morning," Franklin wrote to his mother. "I spent over an hour in the pool this a.m. and it is really wonderful and will I think do great good, though the Dr. says it takes three weeks to show the effects."[48]

Whether the eighty-eight-degree water at Warm Springs healed Roosevelt in any medically tangible way is debatable. But, after a few arduous swims, he was indeed able to stand in four feet of water—a feat unrealized in New York or Florida. "The legs are really improving a great deal," he wrote to Eleanor. "The walking and general exercising in the water is fine and I have worked out some special exercises also. This is really a discovery of a place and there is no doubt that I've got to do it some more."[49]

The Warm Springs cure catapulted FDR back to his childhood, when James Roosevelt had brought his family to the spa in Bad Nauheim, Germany. And he had heard stories from Louis Howe about similar therapeutic treatments at Saratoga Springs. Roosevelt embraced holistic ideas of healing. Unhappy in New York City hospitals—his father had died in one—he believed that nature afforded a cure for him. But something else also drove Roosevelt to adopt Warm Springs. The scenic promontories, blue hills, inviting water, and the scent of the high forests enhanced his overall pattern of thought. "Some day you must see that spot," FDR wrote to Margaret Suckley from Warm Springs later in life. "You would like the great pines and red earth—but it's very different and can never take the place of *our* River."[50]

On motor trips around Pine Mountain—a nearby ridge with a peak exceeding 1,300 feet—he inspected pear orchards, pine groves, rivulets, and neglected Georgia farms. It was a relief for FDR to be far away from the pitch and whine of New York politics, to get the newspaper two or three days late, and to enjoy the rural provincialism of Meriwether County. Discovering the Georgia hills had been providential for him and for others, as he tirelessly changed the village of Warm Springs into a thriving haven for other victims of paralysis.

Encouraged by the effects of the hydrotherapy there, Roosevelt arranged to have the main pool to himself for two hours each day. Before long, though, he found that he enjoyed swimming with others in need of therapy. There are home movies of him at Warm Springs, happily exercising in the therapy pool. "He swims, dives, uses the swinging rings, and horizontal bar over the water, and finally crawls out on the concrete pier for a sun bath that lasts another hour," the *Atlanta Journal* revealed in a special *Sunday Magazine* profile. "Then he dresses, has lunch, rests a bit on a delightfully shady porch, and spends the afternoon driving over the surrounding country, in which he is intensely interested."[51]

Much as when he was in New York, Roosevelt drove himself around the country roads of Georgia for hours on end, using a hand-controlled car. He presented himself as a farmer. Often when he saw someone working in a field, he'd pull up for an informal chat about crops, wildlife, and the weather. On a parcel of pastureland on Pine Mountain, he experimented with agricultural alternatives to planting cotton. Rejecting peanuts and soybeans, he decided that the depleted Meriwether County soil was best suited for raising cattle. Reading about animal husbandry, he purchased purebred bulls and bred them with scrub cattle belonging to local farmers. His operation was working proof that beef cattle could be reared successfully on otherwise exhausted lands.[52] Perhaps because, to be a viable national political candidate, he needed southern Democratic voters, he never rocked the boat about Jim Crow segregation when he was in Georgia. Instead he acted like a social worker, peering into tar-paper shacks and around broken-down farming villages.

Roosevelt jubilantly took a swim in one of the pools at Warm Springs, Georgia, in 1930. Six years earlier, he had discovered the town's waters, always a soothing 88 degrees, and developed the area as a spa—not for the wealthy, but for anyone suffering from paralysis.

Just as soothing to Roosevelt as the thermal waters was the southern hospitality he received in Warm Springs. After Eleanor left, locals looked in on FDR regularly to make sure he wasn't in need of eggs, kindling, or other sundries. Later in life, reflecting on his early visits to Warm Springs, Roosevelt realized his attachment to the place stemmed more from the generosity of the people he met there than from the soothing waters of the pools.[53]

Another aspect of Warm Springs that attracted Roosevelt was the town's warm-water fish hatchery. Established in 1899 to replenish stocks of striped bass, sturgeon, robust redhorse, and paddlefish in the Southeast and on the Atlantic coast, the Warm Springs National Fish Hatchery was just a few miles away from Roosevelt's cottage. There were about forty ponds at the hatchery. Roosevelt's beloved sturgeon—both the Atlantic and the lake sturgeon—spawned at the station. Their eggs were allowed to hatch and then the fry were introduced into the upper Tennessee River (in Tennessee), and the Coosa River (in Georgia). This hatchery was exactly the kind of laboratory Robert Barnwell Roosevelt had spent his life promoting after the Civil War.

From his initial trip to Warm Springs in 1924 until his election as governor of New York in 1928, FDR made thirteen separate visits to the therapeutic waters.[54] After Franklin purchased his house there in 1927, it was administered by the Warm Springs Foundation, a nonprofit that he established (and funded in its first years). Biographers usually note what a shrewd political move it was for FDR to make rural Georgia his second home. His choice was indeed politically convenient. The South was Democratic territory, but with a pre–Civil War conservative streak. Many southerners distrusted Yankees—even Yankee Democrats. Roosevelt, however, was able to create a true bond with people he was happy to call his "neighbors" and "friends." This is not to say that Roosevelt put down roots in Warm Springs as part of a Machiavellian campaign plan. FDR was first and foremost an obsessive rural improver. He felt needed, as well as needful, when he spent time in Warm Springs. "Distinct progress has been made in regard to the fishing lake," Roosevelt wrote to his friend Herman Swift, a lawyer from Columbus, Georgia, in October 1926 about improvements to Warm Springs. "Can't you run down here any day that suits you so we can talk things over? I may go to Atlanta the 19th or 20th to address the Appalachian Trail Association, but the rest of the time will be here."[55]

VI

During the Harding-Coolidge years, the state park movement took hold in America. In 1921, when Roosevelt contracted polio, only nineteen states had even a primitive state park system. Four years later, forty-eight states had park systems.[56] New York was in the forefront, with the legislature soon appropriating $1 million for the development of a first-class state park system. In 1927, George Foster Peabody donated to the state a pristine tract of Prospect Mountain, which overlooked Lake George. It became a "recreation destination," operated by the state. Roosevelt fought for the permanent protection of other such unique sites.[57]

There were a number of reasons why FDR considered the state

park movement essential. The automobile was making Americans more mobile than ever. Taking a drive took on a whole new meaning in the early 1920s. A tank of gasoline could easily take a family to the nearest picnic area or fishing hole, and the family road vacation was fast becoming an American tradition. No longer were places like Yosemite and the Grand Canyon primarily for the elite; nature was being democratized. In New York State, there was a boom in outdoor recreation opportunities.[58]

Yet the state of New York, headquarters of many great railroad companies, lagged behind in the construction of paved roads. The leading advocate of the new automobile recreation movement was Robert Moses, executive secretary of the New York State Association. If Franklin Roosevelt had an archenemy in the 1920s, it was Moses, the master builder. Six years younger than Roosevelt, Moses grew up in New Haven, Connecticut, just a few blocks away from Yale University. His parents were German Jews who fervently believed education was the surest path to successful assimilation. Brilliant, hardworking, and blessed with the ability to cut easily through red tape, Moses excelled first at Yale and then at Oxford in England. After earning a PhD in political science from Columbia University, he plunged headfirst into the reform politics of New York City.

Spurred on by the impulse to "do good," Robert Moses hitched his wagon to Alfred Smith's star and became a key adviser to Smith, the state's leading Democrat. Once Smith, a hero to his fellow Catholics and "wets" during Prohibition, was elected to the governorship in 1922, Moses started his lifelong crusade to turn New York City into the greatest metropolis in the world. Like Roosevelt, Smith was intensely interested in large-scale public works. Both of them believed that state governments and the federal government functioned best when centralized, and both promoted parks. But there were geographic and practical sticking points between them. Moses, thought to be the smartest bill-drafter in Albany, wanted to develop parks and roads on Long Island first rather than in the Hudson River Valley or further upstate.

By contrast, Roosevelt wanted to preserve Country Living throughout New York State as a counterbalance to unchecked urbanization. He genuinely believed that the upstate farmers were better-adjusted citizens than the lucre-driven citizenry of New York City. "If Moses knew Long Island as few men knew it," historian Robert Caro ventured in *The Power Broker*, "Roosevelt could, in the days when he could walk, say the same thing about Dutchess County and the three other counties—Putnam, Columbia, and Rensselaer—whose gently rolling hills made with Dutchess a continuous soft green border, broken only by the patchwork of cultivated fields, all along the east bank of the Hudson from Westchester to Albany."[59]

Moses, by contrast, thought that New York City residents—the ambitious mob who built the Ritz Tower, Chrysler Building, and Empire State Building in the 1920s and 1930s—were the best America had to offer.[60] In his mind, "country living" was old-fashioned claptrap served up by the Crowell-Collier publishing empire in magazines like *Collier's Weekly*, *Woman's Home Companion*, and *Country Home*. Moses's pet state park projects—such as Jones Beach on Long Island—would be a respite for city dwellers in desperate need of relaxation. Moses wanted to build parkways as a means to transport New York City residents conveniently and quickly to the parks and beaches. Moses, more pragmatic than FDR, placed a higher premium on efficient transportation than on beauty when it came to parkways. Unlike Moses, FDR thought that a balanced "regional city"—such as Poughkeepsie—which included villages and hinterland, was the American ideal, *not* a metropolis like New York City or Buffalo. In spite of their downstate-upstate rivalry, Roosevelt did admire Moses's quest to open the beaches of Long Island to average citizens. Coastal areas, Moses believed, shouldn't belong only to the wealthy. Roosevelt agreed.

Because FDR so loved the Hudson River Valley, Governor Al Smith appointed him the first chairman of the Taconic State Park (TSP) Commission in 1925.[61] A restless Roosevelt drove all over the mid-Hudson region scouting for natural areas worthy of "state park" designation. One place Roosevelt fell in love with was Lake Charlotte, just a short drive from Springwood. It took him a few years, but he eventually persuaded the Livingston family to donate the lake (and uplands) to the state of New York in 1929. Livingston, however, set one condition: the name of the lake and park would have to be changed to Lake Taghkanic. At Roosevelt's urging, the state of New York stocked the lake with brown bullhead, white perch, rock bass, and chain pickerel.

While working with the TSP Commission, Roosevelt promoted the idea of a Taconic State Parkway. Roosevelt's intention was to present the three upstate New York counties to the traveling public with their natural features preserved.[62] He took up the task of routing the Taconic Parkway with zeal. No detail was beneath his notice, and the project afforded him the opportunity to enjoy the natural world. On his scouting trips, Roosevelt determined locations for exits, overlooks, rest stops, and recreation areas. Villages along the proposed parkway—including Chatham, Millwood, and Hopewell Junction—would economically benefit from the increased automobile traffic.

As Country Living planner, Roosevelt hoped that someday New York would have between 200 and 250 state parks, with bridle trails, stocked lakes, sanitation facilities, tent sites, and handsome rustic cabins, thereby offering affordable outdoors recreation opportunities to everybody—just as Bear Mountain did. His park philosophy was based on the altered contemporary American landscape, including urbanization, demographic shifts, increased leisure time, and the proliferation of inexpensive cars. As cities grew in population, automobile congestion would become unbearable. Escape from the mayhem could be found on day trips or weekend trips to one of the state's eleven park regions: Allegheny, Niagara, Genesee, Finger Lakes, Central, Thousand Island, Saratoga–Capital District, Taconic, Palisades, New York City, and Long Island. Recreational facilities would take advantage of scenic and underused lands; historic locations that FDR fervently championed included Lake George Battleground Park (the location of a skirmish between colonial troops and an allied group of French soldiers and Native Americans in 1755), and Sackets Harbor Battlefield (the center of American naval and military activity following the War of 1812). Championing regional vernacular, insisting on recycled stone from old buildings in recreational structures, Roosevelt was an avatar of the rustic movement. For these structures he preferred "cozy places back from the highway . . . far from a neighbor."[63]

On July 21, 1928, Roosevelt hosted the State Parks Council at Springwood. High on his agenda was the opening of new campgrounds in Westchester, Rensselaer, Putnam, Dutchess, and Columbia counties.[64] Land within a day's drive of the metropolis was going to get more expensive. To Roosevelt, this meant that the state park movement had to stop foot-dragging before land prices, particularly in the mid-Hudson, hit prohibitively high levels. Affluent New Yorkers, he believed, needed to snap up large parcels of woodlands to deed to the state for perpetual protection (as the Rockefellers and Harrimans had done). In this enterprising spirit, Roosevelt also suggested that the State Parks Council buy large tracts of land—or solicit their donation—in southern Columbia County and northern Dutchess County, in order to establish a sizable wildlife preserve. "I am personally familiar with both sections, having hunted and collected birds all through these counties," Roosevelt wrote to Charles Adams, the director of the New York State Museum in Albany, "and southern Columbia County has far more natural wild life than Putnam."[65] The response was favorable.

VII

Franklin Roosevelt was in Warm Springs, Georgia, on October 2, 1928, when Al Smith, the Democratic nominee for president, reached him by telephone. He wanted FDR to run for the New York governorship, but Roosevelt was reluctant to do so. His goal for the late 1920s had been to continue his physical rehabilitation, often in Warm Springs, and strategize with Howe about a possible gubernatorial run in 1930 or 1932. But Smith laid it on thick, refusing to take no for an answer. Roosevelt finally agreed to seek the office. "I'll be back in Warm Springs—win, lose, or draw—two days after the election," Roosevelt insisted. "My health has greatly improved since first coming to Georgia and I intend to take every possible advantage of the benefits I obtain here, regardless of everything else."[66]

Once FDR arrived back in New York, he preached the gospel of state parks, soil conservation, public utilities, and scientific forestry and took a stand against corruption. While campaigning for the governorship that October, Roosevelt specifically referred to the previous year's devastating Mississippi River flood. All of his warnings about deforestation—warnings that had begun in 1911—had been tragically borne out in the Mississippi Delta. Levees had failed in 120 places along the Mississippi, flooding more than 165 million acres. Six hundred thousand people were left homeless. At least 246 people died. Many more were simply listed as missing. The 1927 flood, in Roosevelt's mind, was a wake-up call for all Americans to take reforestation seriously. Roosevelt insisted that the Army Corps of Engineers needed a comprehensive national plan to improve levees, replant forests, and construct reservoirs to divert floodwaters, but he also thought some kind of state "tree corps" was needed to help prevent flooding in New York's Mohawk and Black River valleys.

When speaking to farmers in New York state, the gubernatorial candidate sounded, overall, like an authority on agriculture, offering tips about how to scrape out a living from soil-depleted land. Wherever Roosevelt traveled in upstate New York, he struck up conversations about the weather and crop cultivation. With iron determination to heal the land, Roosevelt sounded almost biblical, speaking of rains, floods, pestilence, and human error that all had the power to decrease the land's crucial topsoil. No one could doubt his heartfelt commitment to rural development and his own accomplishments at Springwood. "If you run into troubles," he was fond of saying, "bring them to me; my shoulders are broad."[67]

Roosevelt was a loyal reader of the *American Agriculturist*, published weekly by Henry Morgenthau Jr. Born in 1891 to a respected Jewish family in New York City, Morgenthau grew up preferring the countryside to the city. His father, Henry Morgenthau Sr.—or Uncle Henry, as FDR affectionately called him—was a deep-pockets Democrat who had an affinity for Al Smith. In 1910, after contracting typhoid fever, Morgenthau Jr. was sent to West Texas by his father to recuperate. Amazed by the beauty of the Davis Mountains, he fell in love with such rivers as the Rio Grande and Pecos. His letters home intimated that rural life was a higher form of existence than life in the city. "Although I have only been here two days," Morgenthau wrote, "I begin to feel at home. This country and the life is wonderful to me, who has always lived in the City."[68] In 1913, soon after graduating from Cornell University, where he had majored in architecture and agriculture, the owlish Morgenthau Jr. bought a farm ten miles from Springwood that he called Fishkill Farms. He skillfully grew tomatoes and peppers, kept a chicken coop and an apple orchard, and maintained pastureland on which his horses could roam.

FDR first met Morgenthau in 1915 at a luncheon in Hyde Park. Their wives, Elinor Morgenthau and Eleanor Roosevelt, took a shine to each other, becoming almost like sisters. Keeping his eye on Dutchess County affairs from the Navy Department, Roosevelt urged Morgenthau to run for sheriff as a Democrat; the answer was a firm no. Morgenthau and FDR nevertheless grew close, bonding over Democratic politics, Hudson River folkways, rural life, and the landscape architecture of the Olmsteds. As historian Kenneth Davis wrote, Roosevelt and Morgenthau both had "a profound commitment to natural-resource conservation."[69] A myth circulated in Hopewell Junction that Morgenthau was *profitably* growing corn, hay, and trees. In truth, the farm was at best breaking even. His *American Agriculturist*, however, offered information on advanced farming techniques and about such topics as soil conservation, normal versus abnormal erosion, tree planting, and terracing.[70]

Morgenthau once asked FDR, whose opinions he highly valued, to judge a contest to select the most "outstanding farmer" in the state of New York. Roosevelt accepted but raised an unusual question. "May I ask whether forestry has any place in the questionnaire which is sent to farmers?" Roosevelt asked Morgenthau. "It seems to me that in view of the fact that the farm wood lot is or should be a very important producing portion of the average farm area, this phase should be considered."[71]

When FDR was chairman of the TSP Commission, Morgenthau became his loyal sidekick on inspection and mapping trips. Sometimes, they picnicked or did a little fishing on these excursions. Morgenthau was one of the few people other than relatives and staffers whom Roosevelt would allow to lift him out of cars or help him with any aspects of his handicap. Just as important, Morgenthau had a swimming pool at Fishkill Farms, and Roosevelt routinely sneaked over from nearby Springwood to take a dip. Morgenthau's farm had mature maple trees and ancient oaks that Roosevelt greatly admired. The two would sometimes spend hours reveling in the joys of country living or debating the prices of produce.[72]

What seized the attention of both Roosevelt and Morgenthau in 1928 was a report of the Joint Committee on Recreational Survey of Federal Lands that called for large natural areas near big cities to be converted and protected as "green" recreation zones. The report complained that America's existing state parks were located too far away from population centers. What Roosevelt and Morgenthau took from the report was the "urgent need" for state parks with lots of good water and camping facilities located within a fifty-mile radius of cities. If the federal government could purchase lands considered "submarginal" for agriculture, have relief workers make them green oases, then turn them over to state and local park systems, a true beautification of America would occur.[73]

Henry Morgenthau Jr. (*right*) with Roosevelt at a conference in Ithaca, New York, in 1931. The two men, residents of Dutchess County, shared an interest in responsible farming and land management. As governor, FDR appointed Morgenthau to the state's Conservation Commission.

Furthermore, new national parks such as Glacier and the Grand Canyon had become icons of the American West. Roosevelt recognized that the Appalachian Trail offered easterners a landmark conservation project of their own making. Excitement over the "super trail" had caused a movement to designate Shenandoah and the Great Smoky Mountains as national parks for the East to gift to the country.[74]

VIII

On November 6, 1928, Roosevelt defeated his Republican opponent by a slim 25,000-vote margin. All of his outreach to upstate rural districts had paid off. Without farmers, riverkeepers, and conservation-minded voters pushing Roosevelt's candidacy forward, he probably wouldn't have been elected governor.[75]

On the stump in rural districts, Roosevelt's most effective speech had been aimed at folks trying to produce a few dollars from worn-out farms.[76] Most of the farmers' problems stemmed from overproduction, which led to lower crop prices. He claimed that Calvin Coolidge and Herbert Hoover, who'd resigned as secretary of commerce to be the Republican nominee to succeed the president, either didn't care or didn't have enough information to ameliorate the situation. When FDR spoke to New York farmers, it became clear how different he was from either Republican president. FDR was a rare Democratic success story in an election year when the Republicans won big. Al Smith, attacked by anti-Catholic bigots in the South and Midwest, ended up losing to Hoover by a wide margin.

Hoover was born in 1874 to a Quaker family in rural Iowa. Orphaned at age nine, he worked hard in school and developed a keen interest in geology. After earning a degree at Stanford University, he spent his early adulthood traveling the world, working as a mining engineer and looking for mineral-rich lands. Hoover's wife, Lou Henry, was the only female geology student at Stanford at the time and spoke eight languages. By 1914, Hoover was a millionaire with mining investments all over the globe.

Like FDR, Hoover spent World War I working for the federal government in Washington. As head of the U.S. Food Administration, Hoover was trusted by President Wilson to run numerous relief efforts during the war. To FDR's chagrin, Wilson also selected Hoover to serve on the American delegation to the Versailles Peace Conference. Throughout the 1920s, Hoover's star rose in the Republican Party. His tenure as secretary of commerce during the Harding and

Coolidge administrations helped him win the Republican nomination for president in 1928. He ran on a platform that frowned on farm subsidies and TR-style national forests and supported Prohibition and lower taxes. Although Hoover was an enthusiastic fisherman with strong connections to the Izaak Walton League, he was personally in favor of allowing companies to self-regulate. As Hoover later explained in his book *The Ordeal of Woodrow Wilson*, Americans in the 1920s "were tired of the war, the economic controls, the debt, and the huge taxes they paid during and after the war."[77] Hoover had won the White House, but the Progressive conservation movement was still alive in Albany with FDR in the governor's mansion.

It's clear in hindsight that FDR's 1928 gubernatorial victory allowed him to become the unofficial head of the national Democratic Party. Al Smith's star had been eclipsed. Smith had bolstered Roosevelt's presidential chances by helping to bring urban, blue-collar, and Catholic voters into his coalition. Roosevelt's win, however, eventually caused Smith more grief than Herbert Hoover's had. A feud was brewing between FDR and Smith over a very human conflict between two good friends. Smith thought he would control the new governor . . . and FDR was adamant that he had no need of advice from the old governor.

After taking office, Roosevelt regularly read reports from the Great Plains, where settlers had mistakenly tried to replace grass with crops more beneficial to their economic aspirations. These farmers and townspeople soon discovered that although the vast grasslands were productive in wet years, they were also subject to serious drought and bitter winters. The prairie earth was becoming dry as ashes. Eleanor Roosevelt correctly predicted that Franklin would make woodlands ecology and soil conservation the linchpins of his governorship immediately on assuming office.[78] While other Democrats moped, Roosevelt, who knew Hoover from his World War I years in Washington, anticipated the president's weakness as a leader: unlike Jefferson or TR, Hoover could focus on only a single big idea at a time with deep conviction. FDR believed that a successful U.S. president needed "versatility of mind" to "take up one subject after another during the day and find itself equally at home in all of them."[79]

CHAPTER SIX

"A TWICE-BORN MAN"

I

The first thing Roosevelt did at the governor's mansion in Albany was hang a U.S. Geological Survey map delineating the varied topography of New York state. As governor Roosevelt planned to develop an intimacy with every village from Niagara Falls to Long Island, the Wallkill River Valley to the Catskills, the Thousand Islands to Finger Lakes. Roosevelt knew that residents of the Empire State who didn't live in New York City identified mightily with their hometown; the plowed fields, deep woods, mountain streams, and oyster harbors were a source of local pride. FDR's modus operandi was to encourage rural and small-town citizens to measure tree diameters, learn soil types, protect drinking-water sources, and help wildlife prosper. He would prove a genius at making conservation a positive exercise of self-worth and skill, not simply a warning that abstinence and caution were needed.

When, on January 1, 1929, Roosevelt delivered his inaugural address before a huge mass of people, he claimed that scientific forestry, public hydropower, land rehabilitation, and pollution control were ways to truly honor the "gift of God."[1] These weren't quite the fighting words of Theodore Roosevelt, but they indicated that the astute management of natural resources would be a major focus of FDR's administration. The heart of Roosevelt's land policy was anchored around crop restriction, by means of the retirement of marginal land. Instead of farmers abandoning tired soil, allowing erosion by water and wind to carry it away, the state of New York would purchase the marginal land, plant millions of seedlings, and establish well-maintained state forests. To procure the money needed for the ambitious program, Roosevelt wanted the legislature to issue bonds.[2]

Just a few weeks later, Governor Roosevelt spoke before the New York State Forestry Association, challenging citizens to plant thirty million trees on abandoned farmland the state would acquire—a clarion call by any standard.[3] Reforestation would retain moisture in the soil, regulate the flow of streams, and set as an insurance policy against flooding and drought. "My own personal feeling," Roosevelt told the annual meeting, "is that we ought, in going into the question, to take a leaf out of the notebook of European experience and get larger forest areas at work so that the state would not be impeded by multiplicity of detail and an awkward load."[4]

Governor Roosevelt also proposed an amendment to the state constitution that would establish state-run tree nurseries—like the one he frequented in Saratoga Springs—in every New York county. Saplings would be handed out free of charge to farmers. The amendment was never ratified, but Roosevelt revitalized the state's tree seedling program. He maintained that it wasn't "a charity, but a financial investment in the future."[5] Under FDR's leadership, New York's nurseries soon distributed more than forty million trees—40 percent of *all* the trees planted in America from 1929 to 1931.[6] Working in partnership with the New York State College of Forestry at Syracuse University (since renamed the SUNY College of Environmental Science and Forestry), Roosevelt admonished that trees were a long-term investment; they didn't offer immediate economic gain. "Of course, one thing that we have to face in this whole proposition," Roosevelt said, "is that we people with grey hair who start in to plant trees now will be under the ground a good many years before those trees are grown to maturity or to marketable size."[7]

His paralysis prevented Governor Roosevelt from hiking, but he regularly studied the natural world of New York through his car windshield, the next best thing. When asked how he acquired encyclopedic knowledge of the state's creeks and woodlands, battlefields and historic buildings, Roosevelt had a ready answer: "You fellows with two good legs spend your spare time playing golf, or shooting ducks and such things, while I had to get all my exercise out of a book."[8] And by driving around the countryside, he should have added, studying scenic landscapes.

Although the southernmost section of the Taconic Parkway wasn't completed until 1931, excellent reviews trickled in once Roosevelt took office in 1929. To Roosevelt's delight, urban architect Lewis Mumford, a Dutchess County resident, normally a critic of highways, described the Taconic as a masterly combination of modern engineering and conservation, designed for high-speed travel.[9] Mumford was awed by how engineers designed the parkway connecting the northern suburbs of New York City to the Hudson Valley without perpetrating "brutal assaults against the landscape."[10] As Roosevelt envisioned it, the Taconic would be a 110-mile postcard of handsome woodlands, haystacks, gardens, bird reservations, cultivated farmlands, picnic areas, mountainscapes, and expansive views.[11] Although the Taconic Parkway wasn't completed until 1963, the finished parkway followed the very route FDR laid out in 1929.

Over the summer of 1929 Governor Roosevelt, living with Eleanor and the children in the governor's mansion in Albany, took the barge *Inspector* on a multi-stop voyage from Albany to Buffalo. Nostalgic for the past, they traveled on Lake Ontario and the Saint Lawrence River before taking the Hudson to the Champlain Canal. In the time-honored tradition of politicians showcasing their family values, all five Roosevelt children went with their father on the excursion. FDR designed the *Inspector* trip to meet constituents, investigate forests, promote recreation, and scout for new state park sites. He went pole-fishing from the barge for brook trout and carp, and he wrote to a former colleague of the Navy Department that his vessel made him laugh whenever he stopped to "compare it with the old Navy days."[12] But the journey was illustrative of the intrepid Roosevelt's penchant for learning by seeing things firsthand.

Roosevelt found himself in a scrap with the Association for the Protection of the Adirondacks in early 1929. Assemblyman Fred Porter of Essex County, where Lake Placid is located, introduced a bill to construct a bobsled run on state lands in the forest preserve. The bill passed the legislature and Roosevelt signed it into law. His perception blurred the line between preservation and recreation. "In it

[Adirondack Park] are approximately two million acres of State-owned land constituting the perpetual forest preserve for the protection of the mountain water sheds and the regulating of the stream flow out of that great area and also to protect it as a great recreation ground for all the people of the State," Roosevelt said. "In truth, as a recreation ground, enjoyment of it is not by any means limited to the people of this State. Thousands come to it from all parts of the world, for it is one of the world's great nature playgrounds and health resorts—larger indeed than the great Yellowstone Park itself."[13] But in the case of the bobsled run, the Association for the Protection of the Adirondacks screamed foul: the course would entail the removal of 2,500 trees and the law stated, clear as day, that "no cutting" was allowed. This was a worrisome precedent. Innocently, Roosevelt thought the bobsled course would help Lake Placid become a possible site for the Winter Olympics. The case went to New York's highest court, which upheld the lower court's rejection of Porter's law. The bobsled run was built on private land—and the forest remained "forever wild."[14] Roosevelt—and the rest of the state—learned that a forest preserve didn't need golf courses, baseball diamonds, or bobsled courses after all.[15]

The governor likewise got on the wrong side of the preservationist crowd by supporting a proposed highway from Wilmington, New York, to the top of Whiteface Mountain. To the purists, mountain roads, by definition, were unwelcome violations of the Adirondack wilderness. The construction work alone was construed as an act of violence against nature. While Governor Roosevelt had a measure of sympathy for that view, his main conservation goal was democratic in spirit: to allow all citizens, rich and poor, equal recreational access to the best of New York's treasured landscapes. Again the ecological sanctity of Adirondack Park was at stake and Roosevelt sided with the developers. Catching it on both fronts, the governor decided to explain himself by giving a speech, high in the Adirondacks on conservation as recreation.[16]

On September 11, 1929, Governor Roosevelt braved heavy rain to drive into the heart of Adirondack Park to attend a groundbreaking ceremony for Whiteface Mountain Veterans Memorial Highway. By turning the road, slightly more than eight miles long, into a World

War I memorial, he muted some criticism. This Adirondack project, like the Taconic Parkway, had been a dream of FDR's since the early 1920s. Roosevelt bragged that, when completed, the Whiteface would be the only eastern road that snaked up a major mountain peak (though a similar one was being built on Mount Washington in New Hampshire). From the crest of Whiteface Mountain (elevation 4,871 feet), thirty to forty Adirondack hamlets could easily be seen on a clear day. Even the Green Mountains of Vermont could sometimes be savored from the top vantage point. The only hairpin road Roosevelt thought comparable in engineering ingenuity was the acclaimed nineteen-mile highway up Pikes Peak in Colorado Springs.

But the rain proved problematic that September. At the last moment, roads impassable, the event's organizers switched the location of the formal ceremony from the peak of Whiteface to the valley village of Wilmington Notch. The very same two-lane scenic mountain road that Roosevelt wanted paved had been washed out, proving his point. Standing with assistance, shovel in hand, Roosevelt turned the first sod, praised veterans of the Great War, and then returned to Albany to put on dry clothes.[17]

Governor Roosevelt had only virtuous intentions in supporting the Whiteface highway, built to increase tourism in the Lake Placid area. The road was a godsend for out-of-towners and locals alike. Easy automobile access up Whiteface allowed citizens, even those who were handicapped, the opportunity to soak up a breathtaking panorama. The highway surpassed even some of the roadside lookouts Roosevelt enjoyed while traveling in Europe. When Roosevelt next returned to Whiteface Mountain, on September 14, 1935, he was president. During the dedication, Roosevelt made a rare public reference to his handicap. "I wish very much," he told the crowd, "that it were possible for me to walk up the few remaining feet to the actual top of the mountain."[18]

As a proud amateur ornithologist, Governor Roosevelt was pleased when Congress passed the Migratory Bird Conservation Act, sometimes referred to as the Norbeck-Anderson Act, in February 1929. The law directed the federal government to acquire land and establish a national system of wildlife sanctuaries for waterfowl and other migratory birds. FDR applauded the provision that gave the federal government the right "to lessen the dangers threatening migratory game birds from drainage and other causes, by the acquisition of areas of lands and of water to furnish in perpetuity reservations for the adequate protection of such birds." Roosevelt noted that Norbeck-Anderson authorized appropriations over a period of ten years for the purchase, development, and maintenance of migratory bird refuges.[19]

The new law offered a mechanism for establishing a network of refuges not unlike the national forest system, which TR and Pinchot had made a reality in 1905, and the national park system, which Woodrow Wilson and Franklin Lane crafted in 1916. Unfortunately, President Hoover thought incrementally when it came to wildlife protection. While the Hoover administration did add bird refuges in Florida, California, North Carolina, and Nebraska to the national portfolio, Roosevelt bemoaned that it was done in a piecemeal, disorganized fashion.

Pet projects like the Taconic Parkway and Whiteface Mountain Veterans Memorial Highway, and the Norbeck-Anderson Act, took a lower position in Roosevelt's agenda when, on October 29, 1929, a day remembered as "black Tuesday," the stock market collapsed, and the American economy rapidly unspooled. Some twenty-seven thousand businesses collapsed, including 1,372 banks that took $3 billion in deposits down the tubes with them.[20] Widespread homelessness and hunger became a national curse. Over one fifth of the nation was soon unemployed. Squatters' camps lined the Hudson River in Manhattan. In Central Park and open spaces in other cities, indigent people erected settlements of makeshift lean-tos or shanties called Hoovervilles. Roosevelt was distraught by the squalor.[21]

As the decade drew to a close, Governor Roosevelt's progressive instincts intensified. Proactive with regard to the needs of the poor, he pushed hard for the advancement of union rights, old-age pensions funded by employers' and employees' contributions, and an eight-hour workday for government personnel. Under Roosevelt's leadership, New York state was first to provide meaningful relief to the unemployed. FDR's liberal philosophy, based on activist government intervention, suddenly served as a beacon of hope for millions of down-and-out citizens in his state whose pro–big business stance of the 1920s shifted to pro–government intervention in the 1930s.

What differentiated Roosevelt from other eastern state governors was his pronounced concern for destitute farmers who had been fleeced by bankers and financiers at the time of the crash. The American Friends Service Committee, a Quaker organization in which Eleanor would soon become active, reported that about 90 percent of children in rural Kentucky and Virginia were malnourished and without medical care.[22] Credit, the "lubricant of agricultural capitalism," was no longer forthcoming. Beginning in 1929, Roosevelt delivered half-hour radio addresses, forerunners of his presidential "fireside chats." Hoover, by contrast, refused to use the radio to rally farmers to change their habitual ways of planting and plowing. "The spread of government," Hoover trenchantly remarked, "destroys initiative and thus destroys character."[23]

By contrast, Governor Roosevelt identified with American farmers who mistakenly planted more and more crops on their land, which inevitably pushed the exhausted soil beyond its capabilities. When the *Mitchell* (South Dakota) *Republican* mocked him as a city slicker intent on telling midwesterners how to farm their land, FDR fired back a rebuttal: "By the way I am not, as you say, an 'urban leader.' For I was born and brought up to have always made my home on a farm in Dutchess County."[24] Once again Roosevelt advised American farmers to plant fewer crops in order to boost prices, and to plant trees—part of God's design—because of their abundance of practical uses. Roosevelt was in line with serious agronomists: the overused land needed rest and nourishment.[25]

Throughout his governorship, Roosevelt fervently promoted the development of hydroelectric power for the upstate counties along the Saint Lawrence River. Since 1921, when he dreamed of using the fierce tides of Passamaquoddy Bay, near Campobello, for cheap public power, Roosevelt had sought to bring hydraulic engineers into the main thrust of American politics. As governor, he fought for the Power Authority Act in 1931 in order to bring state money and leadership to hydrogeneration projects, including one planned for the

Saint Lawrence River. Roosevelt also called for private utility companies to dramatically lower rates for consumers. Using Albany as his megaphone to speak to a national audience, he called for more public waterpower projects like the Saint Lawrence and Columbia rivers, Muscle Shoals, and Boulder Dam. Not only should the U.S. government build great dams, FDR maintained, but private utilities also had to be hyperregulated, or "attacks on other liberties will follow."[26]

Refusing to be deskbound in an election year, Governor Roosevelt traveled around New York throughout 1930, taking the pulse of the people. With upthrusted head, with a confident look in his eyes, belittling the malefactors of wealth, FDR often reminded Democrats of a rich man's William Jennings Bryan. Talking about God came easily for him. At town halls and community forums, he stuck to talking points about helping farmers survive the economic downturn through tax relief. While socialists shouted for a social revolution and hard-line conservatives clamored for more laissez-faire capitalism, Roosevelt espoused a slate of economic policies, but he never forgot his emphasis on reforestation, pollution control, soil conservation, waterpower, and crop restoration as the best solutions to combat the Depression in rural areas.[27] And, on a national level, he pushed for the U.S. Forest Service and National Park Service to protect old-growth forests before the declining economy forced timber concerns into overdrive. "I am doing everything possible to help the saving of the Yosemite and other trees," FDR wrote from Albany to Nicholas Roosevelt, a distant cousin who was a member of the *New York Times* editorial board, "and am writing at once to [Senator Robert] Wagner and [Senator Royal] Copeland as you suggest."[28]

Roosevelt religiously read the monthly magazine *American Forests*, edited by his conservationist ally Ovid Butler. Very few of its stories were, strictly speaking, about scientific forestry. There were articles about pronghorn antelope (*Antilocapra americana*), nature preserves in California, and John James Audubon's artist studio in New York; how-to designs for building a mountain cabin; and triumphalist celebrations of the Oregon Trail. There remains no better way to understand Roosevelt's land ethic and historical preservation instincts than by reading copies of this Depression-era magazine, published in

Washington, D.C., by the American Forestry Association. Though it didn't have the benefit of peer review, its thoughtful viewpoint had a huge impact on Roosevelt's understanding of modern conservation and historical preservation.

II

Governor Roosevelt asked Irving Isenberg, a graduate of the New York State College of Forestry, to draft a forest management plan for all of the Roosevelt family's holdings in Dutchess County. FDR had come to believe that the Kromelbooge Woods, on his Hyde Park property between the Big House and the river, were virgin forest. Isenberg concurred with FDR's assessment, deeming the hemlock stands to be old-growth. "The stand should remain untouched," Roosevelt wrote to Isenberg. "Do not remove even the dead trees. Do not build new roads. Thus it will be preserved just as nature has treated it."[29]

Furthermore, in 1930 Governor Roosevelt hired Nelson C. Brown to manage his old-growth Springwood forests. Impressed by Brown's expertise, as reflected in his books, *Elements of Forestry* (1914), *Forest Products and Their Manufacture* (1919), and *America's Lumber Industry* (1923), Roosevelt came to rely on Brown for counsel about lumber production and pine, fir, arborvitae, and other conifers. Roosevelt also used Brown as his conduit to the American Green Cross, the New York State Conservation Commission, and the National Committee on Wood Utilization.[30] "He told me of his interest in taking care of his native woods," Brown recalled of their conversation, "and in planting trees in some of the worn-out old pastures and fields that had once been cultivated."[31] Brown, a New Jersey native three years younger than FDR, was inspired by Pinchot's utilitarian conservation. Born in 1885 in South Orange, New Jersey, he graduated from Yale University with a degree in forestry in 1908. For the next few years, he worked for the U.S. Forest Service in the Pacific Northwest and the Deep South, hungry for hands-on silviculture experience.

During World War I, Brown was a professor at the New York State College of Forestry in Syracuse and then worked briefly at the Federal Trade Commission in Europe. He became the top forest products adviser to the American Expeditionary Force. In August 1921 he was reappointed to NYSCF to teach forest utilization. When Roosevelt's friend Franklin Moon was appointed dean of the school, Brown became his loyal second in command and served as acting dean for three brief stints. He also oversaw FDR's trees. "The Governor's most impressive and stately stand of timber . . . is the white and red oak forest lying to the east of the [Albany] Post Road," Brown wrote in *American Forests*. "By judicious and careful cutting, the beauty and capital growing stock have been preserved. It has yielded valuable products and is today a living example of successful American forest management. . . . The most impressive plantation is one of white pine—now fifteen years old. This has been thinned and pruned by the most acceptable forestry methods. It is very similar to the American white pine stands in the Rhine Valley or the Weymouth pine plantations as they are called on the British Islands."[32]

While this was not obvious to the press, Roosevelt was bringing together a brain trust of agronomists and forestry experts (both professionals and amateurs) in his inner circle. FDR, in particular, was smitten with his gentleman farmer friend Henry Morgenthau Jr., who had hired unemployed men to chop firewood, collect kindling, and plant trees on his rolling thousand-acre estate in East Fishkill, New York. He asked Morgenthau about how many of these day laborers it would take to transform a one-hundred-acre plot of worn-out land into a profitable vegetable garden or woodlot. Recognizing that Morgenthau understood how forestry could supplement a farmer's earnings, Roosevelt asked him to chair the New York State Agricultural Advisory Committee. Together they discussed whether New York could hire thousands of unemployed workers to reforest abandoned farms on a wide scale. In Roosevelt's and Morgenthau's combined thinking about unemployment and deforestation lay the seeds of the New Deal's Civilian Conservation Corps.

Eleanor Roosevelt believed her husband thought of Morgenthau as a "younger brother," part of his official family.[33] Both men loved dairymen's cooperatives and the local Grange, state parks, and county fairs. FDR also appointed Morgenthau commissioner of the Conservation Department overseeing New York's state parks, rivers, forests, seashores, and lakes. Pastoral nationalism ran in both men's blood. A year or so later, their friendship had strengthened immeasurably, thanks to Morgenthau's sharp intellect and willingness to experiment. Improving rural life was their shared public policy passion. "Henry always goes about his work with a real feeling of consecration," Morgenthau's wife Elinor wrote to FDR, "but the fact that he is working under you and for you, fills him with . . . enthusiasm. . . . The part which pleases me most is that while you are moving on in your work . . . it gives Henry a chance to grow."[34]

A favorite topic of conversation between Roosevelt and Morgenthau was Dutchess County's crops. They formed a partnership called Squashco to see if gourds could become a new cash crop. "I have written Moses Smith on my farm to get four or five acres ready for squash to seed," Governor Roosevelt wrote to Morgenthau. "I told him to plow the land now and harrow it twice before the seed is put in about July 1st, and to put about six or eight loads of manure to the acre, harrowing it in."[35]

Roosevelt and Morgenthau shared a philosophy of "country living" and a determination to teach hard-luck urban youngsters about the wonders of New York's natural world. Street kids from Hell's Kitchen and the Bronx and Harlem suddenly found themselves in the Catskills and Hudson Highlands, along Lake Erie and the Finger Lakes, digging and planting and irrigating in a kind of work-relief Boy Scouts program. "We took the gas house gang, the bad boys who were loafing on the streets and getting into trouble, and we put them on the 4 a.m. train that ran up to the Bear Mountain area where they worked all day," Morgenthau proudly recalled. "Then because there was no housing for them we took them back at night. FDR was much interested in this conservation of human resources, as in all conservation work."[36]

Long before President Lyndon Johnson signed the Highway Beautification Act of 1965 in hopes of removing billboards from scenic areas, Governor Roosevelt, the tireless automobile traveler, had led the way in New York. In March 1930 he asked the New York state legislature to have the Department of Public Works team up with the Conservation Department to carry out a herculean tree-planting effort in each of the state's ten highway districts. "These plantings would be in part of an experimental nature but primarily for the purpose of demonstration to the people of the state that the highways could and should be more sightly," Roosevelt explained. "An increasing large body of public opinion recognizes the beauty of tree-lined highways as well as their economic value. If the state itself sets the example, even in a small way, I am certain that communities and individuals will follow it in a large way. Perhaps, too, a greater realization of beauty by those who use our highways may lead us some day to the elimination of those excrescences on the landscape known as advertising signs."[37]

Roosevelt used all available media—newspaper, radio, pamphlets—to arouse public consciousness regarding conservation. The June 1930 issue of the journal *Country Home*, for example, had an article called "A Debt We Owe," written by Governor Roosevelt, about the importance of forest resources. He was deeply concerned that Americans consumed five times more timber than was being planted, and he feared that if habits weren't changed ecological disaster was imminent. Roosevelt admitted that a "certain amount of sentiment" clouded his unwavering love of the scenic Hudson's glorious hills and bird-rich woodlands, but he wanted to put aesthetic considerations aside in favor of an economic argument for large-scale reforestation. "As a people we need wood for innumerable purposes, from ball bats and rocking horses to shingles, print paper and artificial silk," Roosevelt wrote. "For the conservation of our soil resources we need the forests to break the force of rainfall to delay the melting of snows, to sponge up the moisture that would otherwise pour down the slopes and grades, carrying with it invaluable fertility and creating floods that destroy. Much of the water that falls in forested land never needs to be carried away, for it is said that one average white oak tree will give off by evaporation one hundred and fifty gallons of water on a hot day."[38]

During the spring and summer of 1930, a severe drought affected twenty-seven states.[39] Scant rain fell in the eastern United States.

Only the Pacific Northwest escaped the ongoing drought between 1930 and 1936. Groundwater in the eastern two-thirds of the nation disappeared. Crops withered. Seventeen million people were directly affected by the drought while Roosevelt was governor. Along the Mississippi River—scene of the great flood of 1927—water was so low that barges were grounded. In some counties in Arkansas, the temperature remained over one hundred degrees for forty-three days. With no moisture on forest floors, wildfires raged throughout Missouri's Ozarks. Many trees that managed to survive the drought of 1930 looked skeletal and malnourished. Florida was not spared; the dry spell lowered its water tables, and there was a threat that huge amounts of seawater would enter the municipal wells of Miami, Fort Lauderdale, and other coastal cities.[40]

Starting in 1931, the "persistent center" of the drought, as the environmental historian Donald Worster has explained, shifted from the East to the Great Plains, turning large swaths of Montana and the Dakotas into a region "as arid as the Sonoran Desert."[41] Banks foreclosed on thousands of family farms across the affected areas. The Hoover administration seemed to be paralyzed, barely shrugging in response, and believing the cyclical economy and Mother Nature would both eventually right themselves. By and large, the White House stuck to the laissez-faire economic view that the market would eventually correct itself. But how could "the market" fix conditions in states dying from the "great plow-up" in the 1930s? Hoover offered only piecemeal grants, loans, and relief programs—nothing substantial enough to make a difference in the human suffering throughout those twenty-seven states in ecological crisis.

In approaching conservation issues, Hoover wanted to transfer some public lands back to the states instead of increasing federal control over land and wildlife management—as Roosevelt desired. While Hoover deserved credit for establishing Arches National Monument (Utah) and signing the legislation that created Carlsbad Caverns National Park (New Mexico), he scoffed at progressive conservationists' notion that preserving America's treasured landscapes would also act as an economic salve. "On the whole, Hoover's attempts to set conservation policy on an entirely new course did not succeed," historian Kendrick Clements observed in *Hoover, Conservation, and Consumerism*. "It was almost as if there were two governments in the field of conservation—one, led by Hoover, trying to divest Washington of responsibility and preaching volunteerism and localism; and another, led by established federal agencies, quietly carrying on and even expanding traditional programs."[42]

Throughout the early 1930s, Governor Roosevelt bemoaned Hoover's failure to push forward on the report of the Southern Appalachian National Park Commission to establish Shenandoah, Great Smoky Mountains, and Mammoth Cave as new national parks. Hoover seemed interested in Shenandoah only because he had a fishing camp along the Rapidan River there. The fact that Hoover wanted to slash the annual budget of the Department of the Interior from $311 million to $58 million by fiscal year 1934 incensed Roosevelt. If anything, its budget and that of the Biological Survey (then under USDA control), needed to be increased.

A major accomplishment of Roosevelt's second term as governor was an innovative program that placed unemployed men on farms to work. On August 28, 1931, at a special session of the state legislature, Roosevelt recommended the creation of the Temporary Emergency Relief Administration (TERA). It had an array of facets, designed to put money in the hands of New Yorkers who were destitute. As often as possible, that involved work of some kind. Modeled after FDR's effective Boy Scout conservation camp initiative, one TERA program provided work relief to young men who planted trees throughout New York. The first year, TERA would have $20 million in funds, and it wasn't enough. Major tax increases followed, but TERA staved off the worst of the Depression for many New Yorkers. With the formation of TERA, the governor established in New York the first comprehensive state relief program. Isador Straus, president of the R. H. Macy department store, was appointed chair of the new agency.[43] Frances Perkins, the state labor commissioner, a social welfare activist from Massachusetts with degrees from Mount Holyoke College and Columbia University, had recommended to the governor that public works projects like TERA would be the "greatest source of hope for the future."[44] FDR also put men to work harnessing the state's water resources, which he felt nature had "supplied us through a gift of God" to produce electricity.[45]

Roosevelt entrusted the daily operation of TERA to the feisty Harry Hopkins, whom he considered thereafter to be one of his closest advisers. Hopkins was born in Sioux City, Iowa, on August 17, 1890. Deeply interested in social work, on graduating from Grinnell College, he started working in the slums of New York City's Lower East Side. Skinny and frequently ill, he instinctively pulled for the underdog. A true-blue reformer, Hopkins helped found the American Association of Social Workers in the early 1920s. Under Hopkins's direction, TERA functioned as a disciplined program that helped the state's poor to survive the Depression. A marvelous tactician, Hopkins had the mathematical and technical skills that Roosevelt lacked. Unafraid of donnybrooks, Hopkins defended progressive liberalism throughout the 1930s and 1940s and, along with Morgenthau, promoted FDR's conviction that government spending would lift America out of the Depression and into economic prosperity.

Most New Yorkers admired the optimistic Governor Roosevelt. Whether or not they were aware of the extent of his paralysis—and most were not, due to FDR's efforts to be as mobile as possible—people naturally perceived the governor's story as a noble fight against adversity. As a politician, Roosevelt was unique, empathizing with everyone he spoke with, feeling others' joys and woes profoundly. Certainly none of his detractors dared intimate that his polio affected his daily performance or impaired his ability to tackle his workload. Always in motion, he lived up to his pledge to visit New York villages that had never hosted a governor before. "His severest test was the 'polio,' " his uncle Frederic Delano wrote, "and to my mind, that is what really made him what he is—a twice-born man."[46]

In 1930, when he ran for reelection, Roosevelt did not just beat his Republican opponent; he trounced him, winning by 725,107 votes.

The victory positioned FDR as the party's presidential front-runner in 1932.[47]

Judge Irving Lehman (*center*) administered the oath of office to Roosevelt, after his reelection to the governorship in 1930. Looking on, at left, is the judge's brother, Lieutenant Governor Herbert Lehman. At right is Eleanor Roosevelt and behind her, Sara. FDR's strong, progressive governorship included models for the New Deal in many areas, including farmland reclamation, reforestation, and parkland for the public.

A tragedy cast a pall over the beginning of Governor Roosevelt's second term. On February 12, 1931, Roosevelt received news that Maunsell Crosby, only forty-four years old, had died following a botched appendectomy. One of the nation's best ornithologists, who discovered new species in Central and South America, Crosby was perhaps the friend with whom Roosevelt was most relaxed. During the mid-1920s he had been Franklin's legs in Florida, the friend who lifted him to enjoy beach picnics and long swims. In coming years Roosevelt would have a mural panel painted at the Rhinebeck post office showing the planting of locust seeds at Grasmere as a tribute to Crosby.[48] But the fact that his dear friend was gone made Roosevelt feel that he himself was living on borrowed time.[49]

That spring of 1931, Governor Roosevelt asked Conservation

Commissioner Morgenthau to scout for soil-depleted land and clear-cut acreage across New York for the state to purchase. The Great Depression had driven real estate prices down to pathetically low levels. Buying otherwise worthless land not only put cash in the pockets of distressed farmers but also gave Roosevelt and Morgenthau's youth corps tracts on which to plant millions of seedlings. What was unacceptable to Roosevelt and Morgenthau was that annually 250,000 acres of poor crop land in New York were abandoned.[50] It was Roosevelt's idea that the state would restore forests and, in some cases, even create new state parks and camping areas—the Springwood model writ large. To pay for this plan, Roosevelt lobbied for an amendment to the state constitution that would make $19 million available for reforestation over an eleven-year period. The amendment's goal—supported vigorously by the Adirondack Mountain Club, of which Roosevelt was a board member—was for the state to buy abandoned farmland, plant trees on it, and appoint scientific experts to manage its productivity as woodlands.

The Hewitt Reforestation Amendment (named for Senator Charles Hewitt, the Republican who introduced it) was cheered by most conservation groups in New York. If the Hewitt Amendment passed, the state would be able to buy property for reforestation outside the so-called blue lines that marked off Adirondack Park and Catskill Park. The difference between the amendment and previous laws promoting reforestation was that it provided for the eventual harvesting of the trees, in order to pay for further land purchases and planting. The Hewitt Amendment was, in effect, Roosevelt's homage to the German state forests he'd so admired all his life. Gifford Pinchot, who was elected to a second (nonconsecutive) term as governor of Pennsylvania in 1930, backed FDR's action with gusto from Harrisburg and Milford. Morgenthau, in a series of columns in *American Agriculturalist*, lobbied frantically for upstate backing of the amendment.

What Governor Roosevelt was proposing had precedent. In the 1890s, the silviculturist Charles Bessey was determined to replant a treeless area of northwestern Nebraska to serve as an experimental tree nursery. The Department of Agriculture acted on this idea, eventually overseeing the largest human-planted forest in North America. In 1902 Theodore Roosevelt established two national forests: Dismal River and Niobrara River (they were consolidated in 1908 to form Nebraska National Forest). "In a great many states farmers may obtain desirable seedling trees for a nominal cost from state nurseries," FDR wrote in an article in the *Country Home*. "Nebraska, for instance, the home of Arbor Day, charges a cent apiece. During 1928 Nebraska distributed 682,000 trees to 2,600 farmers. Many of these were for windbreaks, a most useful purpose in the Plains country."[51] Governor Roosevelt reasoned that if the USDA did it in the sand hills of Nebraska, then Henry Morgenthau could do the same in deforested upstate New York counties.[52]

Throughout 1931 Roosevelt contended that large-scale reforestation efforts in New York would help the soil retain moisture and thereby protect land against future droughts and floods. "Heretofore our conservation policy has been merely to preserve as much as possible of the existing forests," Governor Roosevelt declared on the radio. "Our new policy goes a step further. It will not only preserve the existing forests, but create new ones."[53]

To help pass the amendment, Roosevelt brought Pinchot into the fight—the idea being that Pinchot would pull in Republican votes. But the effort by Hewitt, Roosevelt, and Pinchot faced an unexpected stumbling block, former governor Al Smith.[54] As governor, Roosevelt had learned from his predecessor's example that conservation could also be good politics. But the men had their differences as they presented voters with starkly opposing views on the Hewitt Amendment. Smith—then chairman of the New York Fish, Game, and Forest League, a consortium of hundreds of sporting clubs—turned outdoors enthusiasts against the amendment, claiming that it would "carve up the great potential that was the Adirondack Park." He argued that by allowing logging operations on forestlands just outside the blue lines, the amendment would kill forever the chances of expanding the park. Conceding the point, FDR nevertheless insisted that the amendment would prohibit logging within the park's boundaries and put people to work planting renewable forests; he saw the Hewitt Amendment as a win for everyone.[55] Smith, however, was perhaps the purer preservationist. FDR considered the Catskills Park the best model.

On Election Day 1931 the Hewitt Reforestation Amendment was ratified by a vote ratio of 3 to 2, thanks to Roosevelt, who had mobilized voters to follow the commerce-conservation partnership embodied in Pinchotism.[56] "What a queer thing that was for Al to fight so bitterly on No. 3," Roosevelt wrote to a friend shortly after Election Day. "I cannot help remembering the fact that while he was Governor I agreed with almost all the policies he recommended but I was against one or two during those eight years. However for the sake of party solidarity I kept my mouth shut."[57]

Whether or not the passage of the Hewitt Amendment was the birth of New Deal conservation, the *New York Times* interpreted it as the start of FDR's 1932 campaign for the presidency. He was catapulted overnight to the position of "leading Democratic aspirant" while Smith was downgraded to the status of "the loser in the presidential sweepstakes."[58] Dodging questions about his prospective candidacy, Roosevelt was nonetheless considered the voice of struggling farmers everywhere. The *Times* ran an article about how Democrats from all over America called to congratulate him. Governor George H. Dern of Utah spoke for many when he wrote to FDR that "this is my first opportunity to say to you after the election, Hurrah for Trees!"[59] The *Sentinel*—a little newspaper in Shenandoah, Iowa—was prescient when it wrote that FDR's "advocacy of reforestation, the planting of trees in the waste places of New York" might make a grand national policy.[60]

In 1931 New York's Bank of the United States lost $200 million; this loss resulted in the largest single bank failure in the nation's history up to that point. In the face of the economic crisis, state funds were strained and the land purchases got off to a very slow start in 1932. In response, Governor Roosevelt tackled tree planting on a county-by-county basis and encouraged forestry experimentation. Tasking the Cornell Department of Agriculture to oversee Tompkins County as a pilot project, Roosevelt argued that soil conservation studies could lead to the "efficient planning of farm-to-market roads, rural electrification, and the scientific allocation of school facilities."[61]

Whenever possible FDR promoted the four major waterpower projects in the country: Boulder Dam, Saint Lawrence River, Muscle Shoals, and Columbia River. As a proud member of the Marine Research Society and Adirondack Mountain Club, he took stands on other national issues in conservation, as well.

Governor Roosevelt started working closely with Senator Harry B. Hawes of Missouri, a fellow Democrat, to bring national attention to the dwindling waterfowl populations west of the Mississippi River. The two could talk for hours about the preferred food of geese and the causes behind the mysterious woodland glow known as fox fire. Born in Covington, Kentucky, in 1869, Hawes grew up with the Ohio River as his backyard. His paternal grandfather was a Confederate governor of Kentucky during the Civil War, and Hawes became a believer in big government. Educated at Washington University School of Law in Saint Louis, he became the preeminent expert in issues pertaining to U.S. statehood and territory requirements at the turn of the twentieth century. He pushed hard for Hawaii to become an American territory, but conversely, he proposed granting the Philippines independence.

Global-minded, and able to see a connection between conservation of natural resources and America's economic future, Hawes was respected throughout the Midwest for organizing the Lakes to the Gulf Deep Waterway Association, whose mission was to construct a network of dams and locks along the Mississippi, Illinois, and Missouri rivers. After serving in the army during World War I, Hawes was elected to Congress from Missouri's Eleventh District. In 1926, he was elected to the U.S. Senate. A proud outdoorsman, capable of identifying most North American birds by their song, he was the mover and shaker behind the Migratory Bird Conservation Commission of 1929. Concerned that ducks and geese were dying off in large numbers, he partnered with Frederic C. Walcott, founder of the American Game Protective and Propagation Association, to establish a special committee in the Senate devoted to wildlife issues.[62] Not long after the crash of 1929, Hawes announced that as of 1933 he would quit the Senate to become a lobbyist for wildlife protection. To start spreading the gospel of recreation in 1930, Hawes published *My Friend, the Black Bass: With Strategy, Mechanics and Fair Play*.[63]

Roosevelt and Hawes were allied in blaming the drought, deforestation, and soil erosion on bad-faith capitalist agriculture. They criticized Great Plains and Southwest cowboys for overgrazing livestock in the range to meet consumer demand for beef. As a result, grass disappeared. Drought, soil erosion, and high winds led to crop losses and rising unemployment. Huge swaths of the Dakotas, Nebraska, Oklahoma, and Texas became deserts where fierce winds sucked up swirling dust clouds of topsoil. Water became scarce. In the plains and throughout the West, with lakes, marshes, and ponds drying up rapidly, migratory game birds—which relied on these waterways for resting, nesting, and feeding—were dying off. Waterfowl populations plummeted to the lowest numbers in recorded history. This loss also stripped away a primary high-protein food source from rural folks scraping out a subsistence living.

It was the nationwide drought that impelled Governor Roosevelt to embrace Hugh Bennett, the "father of soil conservation," who worked at the Bureau of Soils in the USDA. Bennett had scored a major discovery with his observation that in clear-cut lands, soil eroded at an alarming rate. Cheered on by Gifford Pinchot, Bennett sampled soils from all over the world. After visiting Alaska, he suggested that the Forest Service prohibit farming and other activities in the Chugach National Forest there. Regarding the Dust Bowl, he predicted its onset in a number of periodicals, including *North American Review*, *Country Gentleman*, and *Scientific Monthly*.

Always seeking a larger compass for his ideas, in 1928 Bennett was a coauthor of a USDA bulletin that Roosevelt deemed mandatory reading: *Soil Erosion: A National Menace*. "The writer, after 24 years studying the soils of the United States," Bennett stated, "is of the opinion that soil erosion is the biggest problem confronting the farmers of the Nation over a tremendous part of its agricultural lands."[64] FDR himself considered this booklet not only a manual describing how to stop soil erosion but also a moral call to action that drew attention to "the evils of the process of land wastage."

Luckily for America, Roosevelt wasn't the only one jarred into awareness by Bennett's cogent analysis and relentless concentration on soil erosion. Congressman James Buchanan of Texas secured appropriations during the Hoover years to establish soil experiment stations throughout the South. Stations popped up in the cotton fields of central Alabama and peanut plantations of Georgia. George Washington Carver, the "Wizard of Tuskegee," pointed to these experiment stations as one of the best things that ever happened to America's rural poor. But funding for agriculture projects became scarce in the Hoover administration, after the onset of the Great Depression.

Roosevelt had come to learn from his demonstration plots at Hyde Park and Warm Springs that commercial fertilizers in the soil—such as phosphates, nitrates, and potash—had mixed results. Some compounds made the soil too sterile for yearly cultivation. It was the job of Hugh Bennett's Soil Conservation Service to determine the correct formulations to help farmers excel. The SCS also sprayed acreage with different chemicals, in an effort to protect against insects. Farmers applauded the use of pesticides during the New Deal. Only later, during the cold war, were the harmful health consequences of such spraying properly understood by agronomists and biologists.

III

Even when FDR was in the midst of battling the effects of the Great Depression, the daily cocktail hour remained a sacred ritual to him. Nothing controversial was discussed while Roosevelt had the cocktail shaker in his hand, and whoever was in his orbit around twilight generally joined him in two or three cocktails. At one after-work affair, Roosevelt pressed Samuel Rosenman, one of his appointed legal advisers, to refill his glass. Rosenman disliked the taste of alcohol. He barely finished his first drink, and the thought of a second made him queasy. But the governor took over, pouring him a refill. Rosenman, when no one was looking, poured it into a nearby flowerpot.

A week later Governor Roosevelt, in front of friends, called Rosenman out. "You know, Sam," Roosevelt said, his lips twisted into a sarcastic smile, "a peculiar thing has been happening to the plants in the Executive Mansion. Some time ago, the leaves of some of them began to change their color. Whitehead [the steward] got worried about them and asked one of the experts from the Department of Agriculture to come over and take a look at them. The expert said he had never seen such a strange condition before, and would like to take the plants over to his laboratory for analysis of the soil. The report has just come back, and what do you think they found? They found that the soil was filled with a large percentage of alcohol. Whitehead is thoroughly mystified as to where we ever got that kind of soil." Everyone laughed heartily as Rosenman admitted culpability. "Well, Governor," he said, "if you don't want to lose all your plants, you'd better pass me up on seconds."[65]

While America was in crisis FDR's political future looked brighter than ever. Louis Howe, his chief adviser since 1912, thought the time was right for the governor to seize the White House. Howe, who had been raised in the spa town of Saratoga Springs, was perennially sick, and gnome-like in appearance, with pockmarked skin resulting from a freak accident. Therefore, even though he was a genius as a political tactician, his trajectory in New York electoral politics was limited. His appeal was cerebral, not personal. Howe soon quit journalism to run FDR's state senate reelection campaign in 1912; it was a lucky break for Roosevelt.[66] They became inseparable.

Howe believed in 1932 that Roosevelt had sufficiently overcome his polio. But certain rules were to be abided by: Roosevelt would be lifted in public only when *absolutely* necessary; he was not to be photographed in a wheelchair; he had to "walk" at most events without using crutches (he would instead receive help from an aide or one of his sons); and he needed to swim laps regularly.

IV

In January 1932, a week before his fiftieth birthday, Roosevelt declared his candidacy for U.S. president. He assembled an excellent team to help with his campaign: politicians such as senators Cordell Hull of Tennessee and Alben Barkley of Kentucky as well as trusted advisers like Howe, Morgenthau, Hopkins, and New York politico James Farley. Southern Bourbons, Grange farmers, western radicals, and anti-Hoover conservatives were brought into the coalition. Roosevelt's campaign kept conservative Democrats in his camp by staying mum on Jim Crow segregation laws and seldom criticizing business and industry.

By the spring, millions of worried Americans jumped onto FDR's bandwagon. Momentum built daily. In a landmark speech in May 1932 at Atlanta's Oglethorpe University, about eighty miles from Warm Springs, Roosevelt declared that the country needed "bold, persistent experimentation" in order to survive.[67]

Unusually for a New York governor, Roosevelt urged Easterners to start protecting their treasured landscapes. The Great Smoky Mountains, for example, were under consideration for national park status, with its towering two-hundred-foot pine trees and poplars of twenty-five-foot circumference. Roosevelt, an enthusiast for automobile tourism, was excited that he could see more species of trees on a thirty-mile drive through the Smokies than if he drove diagonally across Europe from Calais to Vienna. The University of Tennessee had launched a "tree count" in the Smokies—modeled after the AOU and Audubon Society bird counts—that documented over 550 species of flowering trees, shrubs, and plants within the proposed national park boundaries. And the Appalachian Trail hit its peak elevation (6,643 feet) in the Smokies, at the summit of Clingman's Dome.

What interested Governor Roosevelt politically was that citizens of Tennessee and North Carolina, with the help of their respective state governments, had raised more than $5 million to establish Great Smoky Mountains National Park. John D. Rockefeller Jr., who was eight years Roosevelt's senior and an advocate of forest health, matched this amount. Rockefeller, son of the founder of Standard Oil, likewise helped finance other national parks, such as Shenandoah, Grand Teton, and Acadia (where Rockefeller kept a summer house).[68] To Roosevelt, this tapping of wealthy patrons to invest in helping to protect America's landscapes held great promise for enlarging the National Park Service in the future.

Another natural feature that Roosevelt wanted to save was the Potomac River. Untreated waste from the Washington, D.C., metropolitan area (population 575,000) had made the Potomac unfit for recreational activities. Dead fish floated on the surface. Bacterial contamination forced the closing of the river for swimming from Three Sisters Island to Fort Washington. Fisheries were endangered. Roosevelt vowed, if elected, to help establish a waste treatment plant. (In 1938 Blue Plains was indeed completed and the Potomac slowly began to recover.)[69] Trying to purify the Potomac all the way from its source to its mouth was a lifelong ambition for Roosevelt.

Governor Roosevelt's efforts to preserve the Okefenokee Swamp equaled his advocacy for the Great Smokies and the Potomac River. The Okefenokee (a Seminole word meaning "Land of Trembling Earth") was a sprawling, tangled wetland, covering approximately seven hundred square miles, populated by alligators, bobcats, raccoons, and waterbirds. Most of the Okefenokee was in southern Georgia, though parts of it spread into Florida along the Suwannee River. The slow-moving swamp was a biologically rich region of pure and mixed cypress forests, bogs, swamp islands, black gum and bay forests, live-oak woodlands, and pine savannas that, taken together, defied easy ecological classification.[70] For any devoted forester, the centuries-old bald cypress trees (*Taxodium distichum*) soaring out of the wild watery loneliness on arching roots, some 130 feet tall with gnarled branches draped with shawls of Spanish moss, were something to behold.

Jean and Francis Harper paused at a campsite on Chesser Island in the Okefenokee Swamp, Georgia, in a picture taken May 29, 1930. Francis first visited Okefenokee as a Cornell student in 1912. His wife, Jean, a former tutor for the Roosevelt children, was instrumental in drawing FDR into their lifelong fight to save the magnificent swamp.

Roosevelt grew determined to save the Okefenokee from Georgia's lumbering and tanning businesses after listening to two friends rave about its surreal beauty. Because of aggressive extraction practices, more than 1.9 million board feet of lumber had been harvested from the swamp by the time Roosevelt became governor. No subsequent replanting efforts were made. Dr. Francis Harper of Cornell University was a specialist in vertebrate fauna who had made the Okefenokee watershed his living laboratory. His wife, Jean Sherwood Harper, was a Vassar alumna who tutored Anna and Elliott Roosevelt at Springwood during 1920 and 1921. Even from afar, the Okefenokee sprang to life in Roosevelt's mind as he heard the Harpers describe the calls of cranes "taking wing from the piney woods" of Honey Island and the "demonical guffaws of courting barred owls during winter nights on Floyd Island."[71]

Between 1912 and 1951, Francis Harper, who sometimes visited the Okefenokee in company with a Cornell University survey team, filled thirty-eight volumes with firsthand observations and swamp lore.[72] As a field naturalist, Harper wrote up his biological reconnaissance in numerous articles, monographs, and a fine posthumously published book, *The Okefinokee Album*.[73] "The denser cypress bays are places of deep shade and at times oppressive gloom," he observed, "but there is somber beauty here."[74]

While running for president in 1932, Roosevelt heard grisly reports about an uncontrollable wildfire, set off by the drought, destroying the few remaining Okefenokee pine, gum, and cypress forests that the lumber companies hadn't clear-cut. Foresters considered naturally occurring fire a natural, sometimes desirable event in the Okefenokee. The 1932 blaze, however, was not such a burn. A wicked combination of drought and careless timber practices—in particular, leaving flammable debris behind at job sites—had sparked an untamable inferno.[75] Roosevelt knew Georgia timber outfits were also responsible for damaging the Okefenokee by overworking the vast pine groves in search of turpentine. The fire had already endangered the swamp's struggling populations of Louisiana black bear (*Ursus americanus luteolus*) and ivory-billed woodpecker (*Campephilus principalis*). As president, Roosevelt hoped to rehabilitate the Okefenokee ecosystem and maybe even secure national park designation for it—as he intended to do for the Everglades.

If Roosevelt had a special weapon, it was his fine radio voice. Voters were spellbound by his clear articulation of hope and change. While much has been made of Roosevelt's instantly recognizable tenor, it was his sterling enunciation that made him so successful. There was something about his delivery that made him believable—the ultimate gift for a politician. If anything, Roosevelt's polio helped him hone his genius for communication.[76] He sounded like strength personified on the campaign trail. This didn't hurt when it came to persuading voters that he had a hearty attitude and a natural, easy charm. He called himself "Old Doc Roosevelt," giving the impression that he could make the ailing nation feel better, even if he could not prescribe a cure.

And then there was Roosevelt's infectious smile; Hoover's countenance, by contrast, was typically fixed in a grimace. America desperately needed to be lifted out of its economic turbulence and governmental stupor. While the Republican incumbent clung to the bunker mentality, FDR believed he could thaw a nation, as he put it, "frozen by fatalistic terror."[77]

V

The 1932 Democratic National Convention was held in Chicago from June 27 to July 2. The competition for the nomination came down to FDR, Al Smith, and Speaker of the House John Nance Garner of Texas. Each man represented a different Democratic faction. Roosevelt had the support of farmers, conservationists, western progressives, and women. Smith had strong support among city dwellers, intellectuals, minorities and regionally, the New England states. Garner had a few key senators and California newspaper magnate William Randolph Hearst in his corner. FDR clinched the nomination after Garner, a distant, but potent, third on the first ballot, joined his ranks. Roosevelt, in turn, offered Garner the nomination as vice president. "All you have to do," Garner told Roosevelt at a meeting in Hyde Park, "is stay alive until election day. The people are going to vote against the depression."[78]

Born during the Reconstruction era in Red River County, Texas, Garner was proud of his Confederate lineage. He had served in Congress since 1903. His hometown was Uvalde, where TR had once hunted for javelina. Garner was nicknamed "Cactus Jack" because of his (failed) effort to make the prickly pear the state flower of Texas.[79] Roosevelt initially liked Garner, calling him "Mr. Common Sense." But over the course of two terms the two men were divided on so many policy issues that Garner would essentially be ostracized from FDR's inner circle. Disgruntled, he immortalized himself in the annals of U.S. political history by deeming the vice presidency not worth "a bucket of warm piss."[80]

While Roosevelt wasn't personally close to him, he learned from Garner about the damage in Texas to natural soil resources from unabated cultivation of unproductive farms, which were then frequently abandoned. In West Texas, as well as the semiarid portions of the western Great Plains, wind erosion devastated not only plowed land but also the adjacent overgrazed pastures and rangelands. Garner told of soil particles in the guise of dust and sand blown from cultivated fields, fallow acreage, and overgrazed range. Roosevelt knew that the next U.S. president would have a serious drought on his hands.

It had become clear to the nation that Governor Roosevelt was decent, energetic, kindhearted, and open to trying new remedies for the economic crisis. His public "walking" and standing were a painful balancing act, but many people had no idea he was paralyzed.[81] Full of sunshine, with even mundane chores bringing him bliss, he regularly smashed precedent, believing that only dramatic gestures would grab the public's attention. A perfect example occurred when he learned that, after four hotly contested votes, he would be the Democratic Party's nominee for the presidency. On July 2, Roosevelt flew from Albany to Chicago unannounced to accept the nomination. "You have nominated me and I know it," he told the delegates, "and I am here to thank you for the honor. Let it . . . be symbolic that in so doing I broke traditions." With oratorical verve, he went on to promise "a new deal for all the people."[82] And by persuading Speaker of the House Garner to be his running mate, Roosevelt turned the Democratic Party into a united front against Hoover.

Not only did Roosevelt's speech accepting the Democratic nomination introduce the term "new deal," but it was also a triumph for the conservation movement. He proposed the creation of a national conservation corps to combat unemployment while improving America's protection of its national resources and scenic beauty. That so much of his Chicago convention speech was about conservation would have pleased Theodore Roosevelt. "We know that a very hopeful and immediate means of relief, both for the unemployed and for agriculture, will come," FDR declared, "from a wide plan of the converting of many millions [of acres] of marginal and unused land into timberland through reforestation."[83] At heart this was the Hewitt Reforestation Amendment being applied to America.[84]

"There are tens of millions of acres east of the Mississippi River alone in abandoned farms, in cut-over land, now growing up in worthless brush," Roosevelt went on. "Why, every European Nation has a definite land policy, and has had one for generations. We have none. Having none, we face a future of soil erosion and timber famine. It is clear that economic foresight and immediate employment march hand in hand in the call for the reforestation of these vast areas. In so doing, employment can be given to a million men. That is the kind of public work that is self-sustaining, and therefore capable of being financed by the issuance of bonds which are made secure by the fact that the growth of tremendous crops will provide adequate security for the investment."[85]

A federally funded workforce dedicated to the environment struck critics as more notional than realistic. Ever since the fall of 1929, when Roosevelt asked New York state for additional funds to support conservation projects, he had touted reforestation as the key to putting unemployed men back to work. Others before him had envisioned soldiers working as the caretakers of America's natural resources. Benton MacKaye, champion of the Appalachian Trail idea, had suggested a large conservation corps back in 1917. Ideas about such an entity appeared in newspaper op-eds from time to time.[86] But, in the main, FDR's conservation corps came from an amalgam of influences, the most important being his forestry experiments on his own land (both in Hyde Park and in Warm Springs), his fondness for German forests, the TERA work with Henry Morgenthau, his erudite conversations with Pinchot, his activism with the Boy Scouts of America, and his relationship with British silviculturist Richard St. Barbe Baker (with whom FDR had dined in Albany before the Democratic National Convention in Chicago).[87]

Nevertheless, a Democratic delegate could have been forgiven for thinking that Governor Roosevelt had misspoken when he mentioned a million jobs. Just in case the audience thought the nominee was exaggerating, he reiterated his conservation strategy: "Yes, I have a very definite program for providing employment by that means. I have done it! And I am doing it today in the state of New York. I know that the Democratic Party can do it successfully in the nation. That will put men to work and that is an example of the action that we are going to have."[88]

With millions of Americans out of work and no sign that the country was climbing out of the Depression, economic issues clearly were paramount during the 1932 presidential campaign; nevertheless, it is striking how conservation issues were presented to the voters. The Republicans predictably pounced on Roosevelt for offering ill-conceived blue-sky oratory. If the public actually believed that one million unemployed men could suddenly find work reforesting millions of acres of submarginal land, then President Hoover was doomed to be a one-term president. But if FDR's idea of a forestry corps took hold, then unemployment, erosion, and the nationwide timber famine would all be addressed in short order. The GOP needed to counter Roosevelt. The distinguished forester Charles Lathrop Pack, on behalf of the American Forestry Association, wrote to FDR that his proposal was poppycock. "Shall we, as foresters and conservation leaders," Pack asked, "shut our eyes to facts?"[89] From Pack's perspective, Roosevelt was exaggerating the policy implications of mass reforestation.

The meanest challenge Roosevelt faced was from Secretary of Agriculture Arthur M. Hyde. Deeming the New Deal un-American, he publicly ridiculed FDR's views on forestry, arguing that the number of tree planters the federal government could actually hire in 1933 would be "inconsequential."[90] Hyde asserted that a million men *could* presumably plant "1,000,000,000 trees in a day," but America's nurseries simply didn't "possess 1,000,000,000 seedlings." Therefore, as president, Roosevelt—God forbid—would be able to provide jobs for only 27,900 men; so the candidate's rhetoric was therefore disingenuous and misleading.

The attack by the Hoover campaign didn't work. Roosevelt cleverly dispatched Congressman Marvin Jones of Texas, the Democratic chairman of the House Agriculture Committee, to counter Hyde's arguments. Jones alleged that Hyde failed to understand that "reforestation required labor-intensive soil preparation and erosion and flood control, not just sticking seedlings in the ground."[91] Ovid Butler, executive secretary of the American Forestry Association, defended FDR's bold work-relief plan, noting that a national conservation corps would also protect against fire, insect invasions, and fungi—all the while helping to build roads, trails, and telephone lines.[92]

The Hoover campaign, in turn, enlisted Charles L. Pack to counter Butler's enthusiastic endorsement. In a lengthy manifesto, Pack criticized Roosevelt's reforestation plan as an unrealistic pledge based on "shibboleths and dangerous generalizations." Roosevelt next assigned James O. Hazard, the state forester of Tennessee, to knock Pack down a peg. "I hope," Roosevelt wrote to Hazard, "Mr. Pack will receive such discouragement that he will be induced to abandon his apparent intention to oppose the early adoption of a comprehensive national plan of reforestation as a means of combating unemployment."[93]

An unexpected consequence of Hoover's derision toward the idea of forestry as work-relief was the defection of Republicans like Senators George Norris of Nebraska and Robert M. La Follette Jr. of Wisconsin. Pinchot was particularly angry at Pack, a forester he respected, for acting as Hoover's hit man. It sickened him that Hoover had seemingly become a rubber stamp for the "economic royalists" who were eager to extract every dollar possible from public lands. Pinchot was instructing all of his Republican conservationist friends to vote against Hoover.[94] "Roosevelt believes as I do," Pinchot wrote in an unpublished statement, "that the good of the People comes first. Hoover is and always has been the errand boy of the private utilities."[95]

With the Democratic nomination in hand and Progressive Republicans like Pinchot backing most of his policies, Howe persuaded Roosevelt to travel through the Southwest and West to learn more about the percussive heat waves, which had developed into a permanent drought, and rustle up votes. His first speech was in Topeka on the grim farm conditions.[96] Kansas was plagued by sudden floods, vanishing topsoil, and, on occasion, invasions of locusts. The number of dust storms was steadily increasing throughout the Great Plains and at the Southwest.[97] Great rivers like the Cimarron and Canada were becoming mudflats.[98] Corn and grain refused to sprout. People were desperate to pay their mortgages. Loose dust accumulated on country roads. At some spots dust drifts were as high as haystacks. America's breadbasket was becoming the Sahara. "In a rising sand storm," wrote Margaret Bourke-White, "cattle quickly became blinded. They run around in circles until they fall, and breathe so much dust they die. Autopsies show their lungs caked with dust and mud."[99] Governor Ross S. Sterling of Texas declared a "grave" ecological crisis in his state.[100] Perhaps the only good news from the West about the land was that the drought killed invasive weeds and grasses, giving the pampa and buffalo grasses a chance to gain ground for the first time in generations.

Many struggling Great Plains families just picked up and left but wildlife didn't necessarily have that option. The drought was lethal to prairie chickens, antelope, woodcock, and deer. Severe bag limits were ordered by the Hoover administration, but controlling the number of duck and geese killed was far from an effective solution to the basic problem of birds' lack of suitable nesting cover, food, and protection.

Throughout the southern plains, Roosevelt assured windblown

farmers that the drought of 1931 and 1932 could be combatted with proper soil conservation measures. Farming didn't have to be a speculative venture. Green could replace brown on the land again. Dust storms didn't have to be so frequent. Plagues of grasshoppers and swarms of rabbits could be managed with federal pest control measures. But a new land ethic was needed on western homesteads. If Roosevelt went to the White House he would have the U.S. Department of Agriculture dig wells and establish grazing districts and restore grasslands. "He had an extraordinarily acute power of observation and could judge conditions in any section from the looks of the countryside as he traveled through," Eleanor Roosevelt recalled. "From him I learned how to observe from train windows; he would watch the crops, notice how people dressed, how many cars there were and in what condition, and even look at the washing on the clotheslines."[101]

Governor Roosevelt's campaign was cresting while President Hoover's circled the drain. Through a combination of charisma, indomitable will, and bold policy, FDR persuaded the economically devastated American people to trust him. President Hoover also made the strategic mistake of refusing at least publicly, to acknowledge that in the Great Plains, the Depression was a crisis of the ecology, rather than Wall Street. He chose to stick with the message that "time will heal," as if the perils of soil erosion and deforestation didn't worry him. By contrast, Roosevelt insisted that the government needed to immediately take a dynamic role in managing the natural resources of the public domain, and he thought the number of national forests and wildlife refuges should be doubled. Eleanor Roosevelt explained her husband's conservation philosophy succinctly: "Where land is wastefully used and becomes unprofitable, the people go to waste too. Good land and good people go hand in hand."[102]

As Election Day neared, Hoover grew desperate to counter Roosevelt's New Deal conservationist vision. He seized on an informal letter FDR had written to Lowe Shearon of New York City. The GOP reprinted Roosevelt's reply to Shearon and circulated it far and wide. "I believe in the inherent right of every citizen to employment at a living wage and pledge my support to whatever measures I may deem necessary for inaugurating self-liquidating public works," Roosevelt wrote, "such as utilization of our water resources, flood control and land reclamation, to provide employment for all surplus labor at all times.' "[103]

President Hoover, supported by the National Lumber Manufacturers' Association, mocked the letter at a rally in Detroit. "There can only be one conclusion from this statement," Hoover charged. "It is a hope held out to the 10,000,000 men and women now unemployed that they will be given jobs by the government. It is a promise no government could fulfill. It is utterly wrong to delude suffering men and women with such assurances. . . . There are a score of reasons why this whole plan is fantastic. These 10,000,000 men, nor any appreciable fraction of them, cannot be provided with jobs in this fashion. The only way is by healing the wounds of the economic system to restore them to their normal jobs. . . . But above all I ask you whether or not such frivolous promises and dreams should be held out to suffering unemployed people. Is this the new deal?"[104]

On November 8, 1932, the Democrats scored a landslide victory, with Roosevelt-Garner winning 472 electoral votes to Hoover-Curtis's 59. The Democrats also won large congressional majorities (310–117 in the House, and 60–35 in the Senate). Roosevelt won every Western state—his strongest regional showing. So lopsided was the outcome that the Democrats successfully ran against "Hooverism" for the next decade. When Americans were offered a choice between Roosevelt's new deal and Hoover's old deal, they picked change.

Given the president-elect's clear mandate, Congress planned on allowing FDR wide latitude for jump-starting the woeful economy. With approximately one third of Great Plains farms facing foreclosure, the banking system essentially ceasing to function, and unemployment at 25 percent, Roosevelt had a Herculean task before him.[105] But he was ready. Everybody sensed that his March 4, 1933, inauguration would be a turning point in American history. "If Roosevelt burned down the Capitol," humorist Will Rogers declared, "we would cheer and say, 'Well we at least got a fire started anyhow.' "[106]

PART TWO

NEW DEAL CONSERVATION, 1933–1936

CHAPTER SEVEN

"THEY'VE MADE THE GOOD EARTH BETTER"

I

Having won in a landslide, Franklin Roosevelt quickly brought conservationists of all stripes into the New Deal fold. On January 8, Governor Pinchot, a moderate Republican, drove from Pennsylvania to New York at the president-elect's behest to strategize about how best to implement a national forestry strategy. Pinchot had been focused on deforestation in the Appalachians since 1930, so his expertise was rusty with regard to public lands in the drought-plagued western Great Plains; still, he agreed to help FDR address the ecological crisis. The following day, Pinchot enlisted Robert Marshall—one of America's greatest foresters and wilderness activists—to help launch the New Deal conservation movement.

Bob Marshall, born on January 2, 1901, was from a prominent family of New York social progressives. His father, Louis, was a top-drawer constitutional lawyer who, as president of the American Jewish Committee from 1912 to 1929, fought against anti-Semitism in all of its ugly forms. Louis helped establish the New York State College of Forestry in Syracuse. All three of his sons followed his lead in protecting the natural world. As an impressionable youth who summered at the family's estate on Lower Saranac Lake, Bob devoured books about the Maine woods of Thoreau, the Colorado River of Powell, and the upper Missouri of Catlin. Adopting the preservationist precepts of Muir and TR, the young Marshall was determined to fight for government protection of what remained of the American wilderness. Exploring the Adirondacks—with backpack, a down sleeping bag, rainproof clothing, and leather boots—became his fixation. As a teenager Marshall protested against the urban-industrial sprawl of New York City and joined the cult of the "primitive," mourning the disappearing western frontier. "My ideology," he once said, "was definitely formed on a Lewis and Clark pattern."[1]

Bob Marshall, pictured at home in a forest, circa 1935. Marshall, an independently wealthy Manhattanite, was a passionate naturalist. With Roosevelt among his admirers, he was appointed head of the Forestry Division of the Bureau of Indian Affairs in 1933. Two years later, Marshall helped found the Wilderness Society, underwriting it in its fledgling years.

Marshall earned a BS degree at the State College of Forestry at Syracuse, a master's degree in forestry from Harvard, and a doctorate from Johns Hopkins. His first book, *High Peaks of the Adirondacks*, was published in 1922. After working for the U.S. Forest Service at the Northern Rocky Mountain Experiment Station in Missoula, Montana, he moved to Washington, D.C., in 1931. Camping in the wilderness was for Marshall, who had spent fifteen months in Arctic Alaska, an American birthright.[2] "There is just one hope of repulsing the tyrannical ambition of Civilization to conquer every niche on the whole earth," Marshall wrote in the *Scientific Monthly* in February 1930. "That hope is the organization of a spirited people who will fight for the freedom of the wilderness."[3] In 1929 and 1930 Marshall wrote other impassioned articles, including "Forest Devastation Must Stop" (*The Nation*), and "A Proposed Remedy for Our Forestry Illness," in (*Journal of Forestry*)—must-reads for progressive conservationists.[4]

In 1932 Marshall, working for the Forest Service, was asked to contribute several sections to a monumental report prepared at the request of the U.S. Senate: *A National Plan for American Forestry: Report on Senate Resolution 175*, known as the Copeland Report.[5] Marshall was responsible for the recreation portion of the appraisal; he recommended putting 10 percent of all U.S. forestlands into recreational zones, ranging from big parks to wilderness areas to roadside campsites.[6] The report, completed at the beginning of 1933, would stretch to 1,677 pages. Pinchot and Roosevelt wanted a sneak preview; they also wanted it boiled down to a manageable memo. Pinchot tasked Marshall with the latter. "He limited me to six double-spaced pages," Marshall recalled. "I stressed two things: a huge public [land] acquisition program; and the use of the unemployed in an immense way for fire protection, fire-proofing, improvement cuttings, planting, erosion control, improvements (roads, trails, fire towers, etc.), and recreation developments."[7]

The Copeland Report specifically recommended that the U.S. government purchase approximately 240 million acres of privately owned woodlands—an extremely ambitious emergency conservation measure regarded as imperative by the authors.[8] The seeds of the New Deal's Emergency Conservation Work—soon known as the Civilian Conservation Corps (CCC)—are also, in part, suggested in the Copeland Report. "Private forestry in America, as a solution of the problem, is no longer a hope," Pinchot reminded the president-elect in late January. "Neither the crutch of subsidy nor the whip of regulation can restore it. The solution of the private forest problem lies chiefly in large scale public acquisition of private lands."[9]

The idea of a CCC readily solidified in Roosevelt's mind. As governor of New York, Roosevelt had proved the value of a similar, though less ambitious, reforestation corps. On three occasions during the 1932 presidential campaign—the acceptance speech in Chicago, a press event in Atlanta, and a give-and-take in Boston about work-relief—Roosevelt had called for a national conservation corps.[10] If the primary selling point to Congress was work relief, the long-term vision was nothing less than to heal the wounded American earth.

Pinchot wanted the U.S. government to purchase clear-cut forestlands for bottom-dollar prices to be replanted and then turned into federal forests. The time was ripe for action. Because the Great Depression had lowered property values, abandoned farmland was inexpensive. Moreover, lumber companies were financially stressed; purchasing their tracts would save them—and their land. The Forest Service could purchase acreage from all sources very cheaply. "By utilizing the unemployed, highly necessary and productive improvements could be made in the forests thus acquired at substantially no cost to the public," Pinchot explained. "The major fields of work include planting, thinning, release cuttings, the removal of highly inflammable shags and windfalls, a large scale attack on serious insect epidemics, the control of erosion, the construction of roads, trails, and telephone lines, and the development of campsites and other recreational facilities. All this would supply large numbers of men with highly useful work."[11]

Roosevelt met with Pinchot again on February 1 to discuss the CCC. Inevitably Congress would need to appropriate a large amount of money to create such an agency. Roosevelt was undaunted, however. Owing to his definitive victory and the urgent economic problems that had precipitated the Great Depression, he had the prerogative to spend federal money as needed. He felt that, with a two-year grant of emergency powers tacitly granted by Congress, he could immediately hire many unemployed men for Pinchot's scientifically managed forest, as well as for his own special concern with creating a more recreation-friendly atmosphere in state and national parks.[12] If CCC labor could become operational during his first months in the White House, Roosevelt could, at last, prioritize "localized recreational opportunities" in all forty-eight states. It was his dream to initiate meaningful state park systems throughout America.

Wintering that February in south Florida, bone-fishing in Biscayne Bay, Roosevelt lived on philanthropist Vincent Astor's 264-foot yacht *Nourmahal*. Wearing a linen Palm Beach suit, he discussed policy imperatives with Morgenthau and Hopkins, and considered various potential New Deal unemployment relief initiatives while basking in the Florida sunshine. Buoyed by the victory in November, feeling oddly carefree, Roosevelt was downright jovial. The problems of the Great Depression seemed far away. "And I didn't even open a briefcase," he bragged.[13]

Roosevelt's vacation, however, turned grim on February 15 when the president-elect spoke before some twenty thousand American Legionnaires at Miami's palm-tree-lined Bayfront Park.[14] Sitting on the back of a Buick convertible, FDR delivered a friendly talk of less than one minute describing his twelve-day holiday. Afterward, still in the car and surrounded by reporters, he spotted the mayor of Chicago, Anton Cermak, and good-naturedly waved him over for a chat. Just as they finished talking, a deranged Italian-born anarchist, Giuseppe Zangara, sprayed the gathering with gunfire. Plagued by severe stomach problems, Zangara believed that killing a "capitalist" would alleviate his chronic pain, so he had been aiming for Roosevelt, but he missed. One of the bullets struck Mayor Cermak, who slumped to the ground.[15] Millions of Americans credited FDR's escape to "divine intervention."[16]

Bystanders and police officers apprehended Zangara while FDR ordered the Secret Service to lift Cermak into the car and race him to the nearest hospital. On the ride to the hospital, Roosevelt held Cermak's hand and told him to remain calm. At Jackson Memorial Hospital in Miami, short of breath, Cermak bravely whispered to Roosevelt, "I'm glad it was me, instead of you." The mayor lived for a few weeks, long enough to hear FDR's famous inaugural address on the radio. But, afflicted with pneumonia and gangrene, he died on March 6. Roosevelt's grace under pressure in Miami, and his compassion for the wounded Cermak, instilled in the American people an unshakable confidence that the right leader was headed to the White House.[17]

II

In late February 1933, another Chicagoan entered—and transformed—Roosevelt's life. Harold Ickes, an enthusiastic supporter of Theodore Roosevelt's run for the presidency as a Bull Moose in 1912, had long been active in Chicago politics. As a Republican crossing the aisle, Ickes had helped FDR win the presidential election by wrangling the midwestern Independent Republican vote for the Democratic standard-bearer (to the dismay of Ickes's wife Anna, who was running for the state legislature as a Republican that same year).[18] As a reward, in early 1933, the Republican Ickes was being promoted by New Dealers as a serious contender for the job of secretary of the interior.

The Department of the Interior, formed in 1849, had faced more than its share of scandals through the years, largely because it oversaw vulnerable groups, including Native Americans and indigenous Hawaiians, as well as the vast resources in land and minerals owned by the federal government. The most recent disgrace, as of 1933, was the Teapot Dome scandal of the early 1920s involving bribes for oil leases on public lands in Wyoming by Harding administration officials. A former secretary of the interior, Albert Fall, went to jail for complicity in that scheme. Roosevelt needed someone to clean up Interior and its image. He was looking to replace Hoover's outgoing secretary of the interior, Ray F. Wilbur, with a progressive dynamo. To show that the New Deal was a bipartisan effort, he and his advisers had designated the appointment for a Republican. Roosevelt had already offered the position to two well-liked U.S. senators—Hiram Johnson of California and Bronson Cutting of New Mexico—but they had rejected it.[19] Johnson did, however, tout a progressive Republican politician active in local politics in Chicago: Harold Ickes.

While Ickes's name wasn't a household word in 1933, conservationists in the know cheered the appointment. Ickes, nicknamed "Honest Harold" by progressives, was a dependable crusader for the middle class and the poor, like his Chicago-based hero, social activist Jane Addams. Rotund, bespectacled, a bullish writer and scholar, Ickes considered the wise management of America's natural resources the number one policy objective for a president. Emulating the great John Muir, Ickes had earned a celebrated reputation in the Middle West for lashing out against entrenched interests and exploiters of nature, fearlessly taking on Chicago bullies and power brokers such as Samuel Insull and Mayor William "Big Bill" Thompson. Rapacious industrialists who ran roughshod over Illinois's scenic landscapes often met Ickes's acid-tongued ire.[20] "In my own city of Chicago we generously handed over to the railroads miles of the wonderful shore-line of Lake Michigan," Ickes said in an essay published in the *American Civic Annual*. "For more than a generation now, the people of Chicago have been taxing themselves for millions upon millions of dollars to recapture their shore-line. The total cost to Chicago for its great generosity, without taking into account those aesthetic values which cannot be measured in money, has already run into the hundreds of millions of dollars, with additional hundreds of millions to come before the shoreline can be completely reclaimed."[21]

Secretary of the Interior Harold Ickes *(right)* met with delegates from the Confederated Tribes of the Flathead Reservation in western Montana in 1935. The three tribes received the first constitution ratified under the Indian Reorganization Act of 1934, which ceded cultural and economic autonomy back to Native Americans.

Ickes and his wife, Anna, had also been longtime crusaders for Native American rights. Harold had founded the Indian Rights Association of Chicago, and Anna served on the General Federation of Women's Clubs' Indian Welfare Committee in the early 1920s.[22] Because Anna had asthma, they'd escape the intolerable humidity of Chicago's summer for an adobe home near the Navajo reservation at Gallup, New Mexico. In 1933, Anna, who had learned Navajo, published *Mesa Land*, about her love of the Four Corners region. Both Harold and Anna knew that the United States' history of stripping lands from Native American tribes was deplorable; by 1933, these tribes had lost two thirds of the 138 million acres allotted to them by government treaties in 1871; the remaining fifty million acres were designated as "reservation" lands, most located in the West.[23] (Once in Roosevelt's cabinet, Ickes urged the president to hire John Collier, a dyed-in-the-wool progressive reformer, as commissioner of Indian affairs in 1933, and Bob Marshall as chief forester of the Bureau of Indian Affairs.)[24]

That February, FDR, a genius as a talent scout, invited Ickes to his posh double town house on East Sixty-Fifth Street in Manhattan, which served as his pre-White House headquarters, to discuss the failing economy with a small cadre of policy wonks. During the meeting the Roosevelt kept looking at Ickes, sizing him up as if buying a horse or a house, trying to read his body language. Journalist Walter Lippmann homed in on the paradoxical essence of Ickes by characterizing him as "violently virtuous" and "angrily unbigoted."[25] Intuiting that Ickes's hard-nosed, even caustic, exterior was a facade for an inner sincerity, when the meeting ended Roosevelt requested a private word with Ickes in his library. Roosevelt told Ickes that regarding the stewardship of public lands they were indeed brothers in arms. Both men were particularly enthusiastic about having the National Park Service help build multiple-unit state park systems. Ickes took the job offer. Many of Roosevelt's "country beautiful" ideas, as manifested in national and state parks, would be made realities through the hard work of the policy pugilist Ickes. He and FDR shared a fervor for saving what Muir called "pieces of extraordinary nature for the national trust."[26]

III

On March 2, 1933, Franklin and Eleanor left New York City for Washington, D.C. Holed up in room 776 of the Mayflower Hotel, the president-elect tweaked and rehearsed his inaugural address.[27] With the stock market down 89 percent from 1929 and a quarter of American adults (nearly sixteen million citizens) out of work, many questioned whether capitalism could survive. Too many good people had hit rock bottom, roaming by highways and railroads looking for a square meal, squatting in shantytowns that mushroomed in urban slums. Diminished expectations were threatening to obliterate the American dream. How could words move a nation in utter despair? While conservation wasn't a central part of Roosevelt's inaugural speech per se, the CCC work-relief program was already planted firmly in his mind as part of his overall job recovery program. The moment was propitious for the renewal of America. Countrymen knew that in many regions—particularly the South, Midwest, and Great Plains—the abused land needed to be repaired for the sake of the future.

Expectations were high on March 4, when Roosevelt, standing at a lectern in front of the Capitol, placed his hand on his family's Bible and took the Oath of Office. The sun peeked through, as if signaling Roosevelt's message to the country. Movie star Lillian Gish, seeing him in person that day, said that FDR glowed brightly, as if he had "been dipped in phosphorous."[28] There was an almost palpable current of anticipation zipping through the land, where millions of Americans were listening to the president over the radio. On that red-letter day, over six hundred stations broadcast the moment of national uplift.[29]

Roosevelt's first inaugural address was a heartfelt clarion call for New Deal reform and marked a rare moment in history when the hopes and aspirations of a world power depended on a single leader. Roosevelt's radical decision to declare a four-day national bank holiday proved to be the headline news from the historic day. Great attention was also given to his call for a special session of Congress within four days. "Whatever laws the President thinks he may need to end the Depression," Senator Burton K. Wheeler of Montana, an early backer of FDR, predicted that day, "Congress will jump through a hoop to put them through."[30]

During his first moments as the thirty-second president of the United States, Roosevelt announced, "This great Nation will endure as it has endured, will revive and will prosper. So, first of all, let me assert my firm belief that the only thing we have to fear is fear itself."[31] The memorable phrase about "fear itself" became almost synonymous with FDR and would be his reelection slogan in 1936. Historians have celebrated Roosevelt's first inaugural address for its brave and rousing rhetoric. The concept of living without fear wasn't new; famous writers from Cicero to Shakespeare to Daniel Defoe had all contemplated it. But such unhinged confidence had never been so operatically expressed by an American politician. The inauguration of 1933 turned out to be one of the three most important in American history, along with Washington's first inaugural in 1789 and Lincoln's in 1861. When asked years later about FDR's greatest public moment, Eleanor Roosevelt didn't cite the Social Security Act of 1935, the military mobilization after Pearl Harbor, or the logistic genius of the Normandy invasion; she chose his "fear itself" oration.[32]

Between March 9 and June 16, Roosevelt proposed and Congress passed fifteen major emergency acts, which, taken together, would deeply and permanently alter America's social, political, and environmental complexion. With the full force of Congress behind him, Roosevelt prescribed minimum wages and prices, told farmers which crops to plant, regulated Wall Street, and underwrote credit for homeowners. What became known as the "Hundred Days" set a new benchmark for governmental productivity. Even Prohibition was repealed. Congress, under pressure from the White House to act quickly, passed the Emergency Banking Act of 1933, which stabilized the banking industry and restored the country's faith in the financial system. That law was followed in May 1933 by the Emergency Farm Mortgage Act, allocating $200 million to refinance mortgages so that farmers could avoid foreclosure. The following month saw the passage of the Farm Credit Act of 1933, which established local banks and created local credit associations. Humorist Will Rogers famously quipped, "Congress doesn't pass legislation any more, they just wave at the bills as they go by."[33]

During those first "Hundred Days" Congress also passed many new legislative measures inspired by FDR, to provide job relief for the unemployed and jump-start the economy. These measures—known as the New Deal's "alphabet soup" programs—included the Federal Emergency Relief Administration (FERA); the Agricultural Adjustment Act (AAA); the National Industrial Recovery Act (NIRA);

and the Public Works Administration (PWA). Few programs in 1933 would shine brighter than the conservation-based Civilian Conservation Corps (CCC), Soil Conservation Service (SCS), and Tennessee Valley Authority (TVA). The CCC probably best captured the public's imagination as the showcase of the New Deal, along with the more grown-up and grandiose, Works Progress Administration (WPA). Roosevelt knew that large-scale dams and scenic highways would take years to complete. But employing 250,000 young men to cut trails, plant trees, dig archaeological sites, and bring ecological integrity to public lands was immediately effective can-do-ism. That spring half a dozen governors cabled to FDR, wanting first-wave CCC camps established in their states. Roosevelt knew he was panning gold when even Senator Ellison Smith of South Carolina, an early and vocal skeptic, called the Emergency Conservation Work Act a "marvelous piece of legislation."[34]

Just days after his inauguration, Roosevelt instructed Nelson Brown to plant thousands of trees on the Springwood subsidiary, Creek Road Farm, in Hyde Park. Since 1930 Roosevelt and Brown had developed a first-rate forest demonstration plot at Hyde Park. With proper cultivation and plenty of fertilizer, Roosevelt would showcase a model tree farm, twice the size of the nearby Vanderbilt estate holdings. "I wish you would make a note of having a careful inspection made of the swamp area planted last year," Roosevelt wrote to Brown. "The permanent tree crop consisted of tulip poplars and black walnuts and these were interspersed with, I think, red cedar and larch. This planting should be filled out to replace trees that have died. During the winter I had all the sprouts cut off from the stumps of the old trees that had been cut."[35]

That FDR, with the crushing weight of the Great Depression on his shoulders, found time for planting trees at Springwood in March 1933 was tangible proof of his long-held conservationist convictions. "Forests, like people, must be constantly productive," Roosevelt told the *Forestry News Digest*. "The problems of the future of both are interlocked. American forestry efforts must be consolidated, and advanced."[36] To that end he wanted to use forests to ease the crisis at hand. "If you have not got accurate figures as to the number of men who could be permanently employed per hundred acres of national forests, state forests, or private forests," Roosevelt wrote to Ferdinand A. Silcox, the head of the Forest Service, "would it be a good idea to send someone to Europe to get us the data? . . . They have been at it for hundreds of years."[37]

IV

On March 14, Roosevelt issued a memorandum for the secretaries of war, the interior, and agriculture, making good on his campaign promise. "I am asking you to constitute yourselves an informal committee of the Cabinet to coordinate the plans for the proposed Civilian Conservation Corps. These plans include the necessity of checking up on all kinds of suggestions that are coming in relating to public works of various kinds. I suggest that the Secretary of the Interior act as a kind of clearing house to digest the suggestions and to discuss them with the other three members for this informal committee."[38]

People soon wondered where FDR came by the idea of the CCC. The obvious answer is that in 1932, after FDR had established conservation-reforestation programs in New York, other states, such as California and Pennsylvania, had done the same. It just made practical common sense. A rumor circulated, and still has academic currency, that the concept of the CCC emanated from the eminent Harvard philosopher, William James's pacifist essay "The Moral Equivalent of War," delivered at Stanford University in 1906 and published in 1910 in *McClure's* magazine. Dr. James believed that instead of military service, a "conscription of the whole youthful population" to form an "army enlisted against Nature" would cause young men to get the "childishness knocked out of them," so that they would "tread the earth more proudly."[39] Roosevelt brushed that rumor aside, pointing out that he had not studied with James, nor did he remember carefully reading the celebrated essay. Roosevelt had once seen James walking in the streets of Cambridge, and marveled at his famous beard—but that was all. Yet FDR did publicly declare at a Harvard reunion that William James's mind was a "sword" in the "service of American freedom."[40] And despite his dismissal, Roosevelt's personal papers at the presidential library in Hyde Park include a typescript excerpt from the *McClure's* essay.[41] Whatever the case, the spirit of the CCC, including many of its organizational details, was entirely concocted by Roosevelt.

FDR convened the first meeting of the quartet of cabinet members that March. The team was composed of George Dern, from War; Henry A. Wallace, from Agriculture; Harold Ickes, from Interior; and Frances Perkins, from Labor. During the meeting Roosevelt nonchalantly sketched, on a scrap of paper, a flowchart of the CCC chain of command. As Perkins later explained, Roosevelt "put the dynamite" under his cabinet members and let them "fumble for their own methods."[42] The president envisioned Interior and Agriculture selecting project locations and dispatching recruits to staff the camps. Lewis Douglas, the director of the Office of Management and Budget, would provide the financial resources for improving public lands infrastructure. The U.S. Army's judge advocate general, Blanton C. Winship, would offer legal services. Roosevelt envisioned three types of camps: forestry (concentrated in national forest sites); soil (dedicated to combating erosion and implementing other soil conservation measures); and recreational (focused on developing parks and other scenic areas).

From the get-go Ickes was the New Deal's notorious taskmaster, with the impatience of a drill sergeant. In a symbolic gesture, Ickes ordered the doors to the Interior headquarters locked every morning at 9:01. Showing up late for work, by even a few minutes, meant instant dismissal. Throughout Roosevelt's first term, Ickes promoted state and national parks with pluck and vigor, rebuffing right-wing senators who claimed the CCC was a Bolshevist threat to democracy.[43] The New Deal's sense of striking fast on behalf of conservation appealed mightily to Ickes. "The pace couldn't have suited Ickes any better," biographer Jeanne Nienaber Clarke wrote, "for he had been a compulsive worker throughout his life."[44]

Frances Perkins—the first female cabinet officer in U.S. history and one of only two cabinet secretaries to work for the entirety of Roosevelt's tenure in the White House (the other being Ickes)—was tasked with coordinating the recruitment and selection of able-bodied CCC enrollees. Initially a quarter of a million unemployed "boys" or juniors between ages eighteen and twenty-three (later expanded to twenty-eight) were sought. The pool was later widened to include twenty-five thousand veterans of World War I who had fallen on hard times; twenty-five thousand "Local Experienced Men" (LEM) who worked as project leaders in the junior camps; ten thousand Native Americans, who would be assigned to improve reservations; and five thousand residents of the territories of Alaska, Hawaii, Puerto Rico, and the Virgin Islands.[45] Perkins worried that Roosevelt was biting off more than he could chew. However, as the trees got planted, she soon became a believer.

What made the CCC more than just a dazzling work-relief program was the professional expertise the LEM brought to land reclamation. Skilled young physicians, architects, biologists, teachers, climatologists, and naturalists learned about conservation in a tangible, hands-on way. If not for the Great Depression, these workers would have found themselves engaged in upwardly mobile jobs. But by a twist of fate, as many of their diaries and letters home make clear, these LEM were indoctrinated in New Deal land stewardship principles. Later in life, after World War II, many became environmental warriors, challenging developers who polluted aquifers, and unregulated factories that befouled the air.

Having developed a working model for the CCC, Roosevelt delivered his plea to launch it with the passage of the Emergency Conservation Work (ECW) Act, which would provide the authority to create, by statute, a "tree army" to provide employment (plus vocational training) and conserve and develop "the natural resources of the United States."[46] He sent his bill to Congress on March 21. As instructed by Perkins and other cabinet officers, Roosevelt made it *very* clear that reforestation projects wouldn't interfere with "normal employment."[47]

I propose to create a Civilian Conservation Corps to be used in simple work, not interfering with normal employment, and confining itself to forestry, the preventing of soil erosion, flood control, and similar projects. . . .

More important, however, than the material gains, will be the moral and spiritual values of such work. The overwhelming majority of un-

employed Americans, who are now walking the streets and receiving private or public relief, would infinitely prefer to work. We can take a vast army of these unemployed out into healthful surroundings. We can eliminate to some extent at least the threat that enforced idleness brings to spiritual and moral stability.[48]

Roosevelt did a marvelous job of selling the CCC, answering congressmen's questions forcefully but politely. At six press conferences he invoked public works and the CCC. The biggest obstacle was Congressman William Connery Jr. of Massachusetts, the chairman of the House Labor Committee, whose constituents lived in the mill towns of Lawrence and Lowell. To Connery the CCC was "sweatshop work," a tricky way to undercut labor unions. Insisting that the CCC wasn't based on conscription but voluntary enrollment, Roosevelt sent Perkins to testify before Congress. She disarmed Connery and older skeptics with her wit and ardor.[48] On March 28, the Senate passed the CCC bill with great enthusiasm for the program. What stuck in the craw of many Republicans, however, was the unease that Roosevelt had quickly amassed too much rubber-stamp clout. Senator L. J. Dickinson of Iowa predicted, "We will rue the day when we put so much power into one man's hands."[49]

The opposition was more visceral in the House. A dissenting voice was Congressman Oscar S. De Priest of Illinois, the only African American member of the House, who objected, to no avail, that the proposed CCC was racially segregated. After a round of debate, the Seventy-Third Congress passed S. 598, Public Law No. 5 on March 31, creating the CCC as a temporary emergency work-relief program.[50] Roosevelt's unstated hope was that what the *New Republic* called his "tree army" would eventually become a permanent agency. On April 5 Roosevelt signed Executive Order 6101, allowing him to appoint the CCC director and the advisory council composed of representatives of War, Agriculture, Interior, and Labor.[51]

Roosevelt hired AFL labor leader Robert Fechner as the first director of the CCC.[53] Originally from Chattanooga, Tennessee, Fechner, born in poverty, left school at age sixteen and moved to Georgia to become a "candy butcher" on trains. Mild-mannered and collaborative by nature, Fechner earned his way through an apprenticeship to become a trained machinist, working in mines, smelters, and harbor projects in the United States and throughout Mexico and Central America.[54] Modeling himself on labor leader Samuel Gompers, Fechner eventually became vice president of the International Association of Machinists, and he persuaded its members to vote for Roosevelt in 1932.[55] A proponent of the nine-hour workday in 1901 and the eight-hour workday in 1915, Fechner was an intrepid labor reformer. Although he was a southerner, Boston became his home base. Because of his sterling reputation for fairness—as well as his union background—he proved an inspired choice.

William Green, president of the American Federation of Labor (AFL), complained vigorously to the *New York Times* that the CCC reforestation program caused "grave apprehension in the hearts and minds of labor."[56] Green testified before Congress that the U.S. Army's involvement in the CCC would "militarize labor."[57] Convinced that the CCC would steal work away (at bargain-basement prices) and depress wages, Green charged that the CCC "smacked of fascism, of Hitlerism, or a form of Sovietism."[58] By appointing the likable Fechner, a pro at making things run smoothly, the president cleverly assured Congress that Green was wrong, that an experienced labor leader held the keys to the entire CCC enterprise. Indeed, the shrewd hiring of Fechner mollified detractors of the work-relief program in organized labor circles.[59]

By mid-April, the program was coming to life. According to Roosevelt's estimation, an eleven-man CCC crew could, weather permitting, plant five thousand to six thousand trees per day.[60] Surrounded by maps of America, the president studied rivers and streams, deserts and forestlands. "I want," Roosevelt declared, "to personally check the location and scope of the camps."[61]

Roosevelt's "tree army" became a legend from the start, and he became a forester in chief hero to many conservation groups. The Izaak Walton League, a Chicago-based nonprofit, honored the nation's new leader by unanimously electing him its "honorary national president." That was just an opening salvo. The Society of American Forestry soon awarded FDR the first Schlich Memorial Medal for conceiving the CCC.[62] Roosevelt proudly assumed the mantle of natural resource protector, which TR and Pinchot had once worn. As historian David M. Kennedy noted, the public quickly understood that Roosevelt had a "lover's passion" for trees.[63]

All over America, CCC tent cities popped up like Boy Scout camps; they were soon replaced with rustic barracks. To address the ecological concerns registered in the Copeland Report, Roosevelt allotted the initial fifty CCC camps to the Forest Service (over time, 50 percent of all CCC work was done in national forests). Each company unit of two hundred CCC "boys" (unmarried) was a temporary village in itself. All sorts of bylaws, pledges, and rules of engagement were announced. The "CCCers" received three full meals each day, and were issued olive drab uniforms, which included pants, a shirt, gloves, two pairs of underwear, a canteen, and a pair of heavy steel-toed boots. In time, Roosevelt and Fechner changed the uniform to a spruce green color for the coat, pants, overseas cap, and mackinaw (the shirt remained olive). "The issuance of a new uniform distinctive from other governmental services will improve the appearance of the corps," Fechner noted. "It will also aid in building up and maintaining high morale in the camps."[64]

Unlike the army, there were no guard houses, drills, saluting, or court-martials. But morale was important from 6 a.m. reveille to taps at 10 p.m. At just $30 per month, these young men weren't going to get rich: $22 to $25 of their pay was mandated to be sent home to their families; what remained was spent at the canteen, on haircuts and snacks, or at the local nickelodeon. Topnotchers were able to boost their salary by becoming technicians. A gag circulated among new enrollees—"Another day, another dollar; a million days, I'll be a millionaire." While they uttered it with a degree of sarcasm, the recruits were at least looking at the glass as half full.[65]

Uniformed enrollees started working at a breakneck pace to plant millions of trees, restore grass, build check dams, practice rodent control, and kill invasive or destructive animals, prevent wildfires, and teach bankrupt farmers how to form soil conservation districts. No sooner did Congress pass the Emergency Conservation Work Act than Roosevelt prepared to do battle with the "three horsemen" of environmental destruction: fire, insects, and disease.[66] Particularly concerned about California's forests—which drought and arid conditions had made hyper-vulnerable to fire—Roosevelt instructed CCC crews to cut a six-hundred-mile Ponderosa Way firebreak along the base of the Sierra Nevada in California, the longest such protective barrier in the nation.[67] The CCC "boys" were also dispatched to do immediate battle in the drought-ravaged Great Plains and the soil-stricken American South. Even Puerto Rico had CCC camps, employing 2,400 men.[68]

Roosevelt never meant the CCC to be a panacea for the systemic woes of the Great Depression, but it did save a vast number of young men from homelessness or, even worse, hopelessness. As Ray Smith of CCC Company 991 in California put it, his work-relief mates had "the mark of shattered ambitions and blasted hopes written in their faces," and the "fruitless tramping of the city streets showing in every stride."[69] Roosevelt viewed his "boys" not merely as temporary relief workers, but as makers of a permanent, greener new America. Bursting with optimism, he believed the work-relief experience would transform the young recruits intellectually as well as physically. Teamwork and citizenship and conservation would all be learned in the CCC. Many kinds of Americans—Slavic, Jewish, Italian, and Irish among them—found themselves working as a band of brothers, saying, "We're all Americans," with a newfound sense of patriotic unity.[70]

V

Only thirty-seven days after Roosevelt's stirring inaugural speech in May, the first CCC enrollee—Henry Rich of Virginia—was dispatched to Camp Roosevelt near Luray, Virginia. Camp Roosevelt, located in the 649,500-acre George Washington National Forest, was the first camp to open. Six additional CCC camps soon followed in Virginia's Shenandoah National Park, employing nearly a thousand men to thin overcrowded stands, remove dead chestnut trees, plant saplings, install water systems, build overlooks, and lay stone walls along Skyline Drive.[71] There were approximately 120 native species of trees in the southern Appalachians, and the CCC boys were tasked

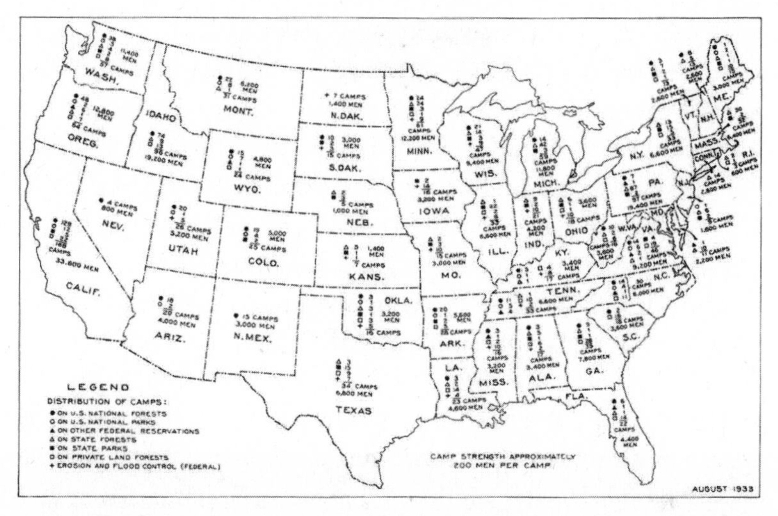

A map showing the distribution of Civilian Conservation Corps camps across the United States, as of August 1933. After just five months' of existence, the CCC had reached its enrollment quota of 300,000 young men. That month, FDR generated a movable celebration of the CCC by visiting two of the first CCC camps, located in the Shenandoah Mountains of Virginia.

with preserving them, especially the hardwoods.[72] An ecological rebound was under way. "Without them," historian John A. Conners wrote, "it would have taken many years longer and cost far more to implement the conservation measures and construct the public facilities that made Shenandoah National Park a reality."[73]

These CCCers, as historian Neil M. Maher wrote in *Nature's New Deal*, "literally built" the recreational infrastructure of Shenandoah National Park "from the ground up" by constructing interpretive centers, campgrounds, picnic areas, hiking paths, and roadways.[74] (The WPA also made an important contribution.) In 1926 Congress had authorized the establishment of the national park, a choice parcel of over 180,000 acres in the beautiful Blue Ridge Mountains. But there was a hitch. The commonwealth of Virginia had the arduous task of acquiring remaining land deeds and planning the park. Recalcitrant landowners, legal disputes, land claims, and squatters were more than Congress had bargained for. Furthermore, the NPS wouldn't "accept title" to the park if people were still living in it. Within a year the NPS evicted three thousand to four thousand people living within the park's boundaries. "Thus began the removal policy," historian Dennis E. Simmons wrote, "a highly emotional issue that left scars which have not entirely healed to this day."[75] Not until 1935—when Ickes accepted a fully executed deed from the commonwealth of Virginia conveying 174,429 acres, largely vacated by previous residents—would the national park be officially opened.

Between 1933 and 1938, owing to New Deal care, state park acreage in America increased by 70 percent. When FDR became president, Virginia had only two state parks. Determined to rectify the recreation crisis, Roosevelt sent 107,000 CCC enrollees into the state, and within three years, six more parks would open: Douthat, Westmoreland, Hungry Mother, Fairy Stone, Staunton River, and Seashore (now First Landing).[76] The CCC in essence created Virginia's statewide park system.[77] Restoration money and manpower were dispatched to Colonial National Monument in Virginia to ward off wind and tidal erosion of the historic York River and the banks of Jamestown Island on the James River. Two Civil War battlefields of Virginia—Fredericksburg and Spotsylvania—were likewise revamped to attract tourists hungry to understand America's past.[78]

Roosevelt insisted that all new state park and national park recreational structures be built in a rustic Americana look that borrowed from the Tennessee log cabin, the Navajo adobe, the Maine saltbox, the Cape Cod half-timbered cottage, and the arched stone bridges of the Lexington-Concord era. Andrew Jackson Downing's emphasis on the perfect harmony between architecture and nature was Roosevelt's goal. Architect H. H. Richardson and landscaper Frederick Law Olmsted shared the president's predilection for using natural materials. But it was the Adirondack school of architecture with its "camp beautiful" ideal that the CCC crews most imitated. The plans, specifications, and philosophical tenets presented by the NPS in its 1935 publication *Park Structures and Facilities* would set the basic style of construction. In stating the ideals of a rustic camp, Albert H. Good, architectural consultant for the NPS, wrote, "Successfully handled, it is a style which, through the use of native materials in proper scale and through the avoidance of severely straight lines and oversophistication, gives the feeling of having been executed by pioneer craftsmen with limited hand tools. It thus achieves sympathy with natural surroundings and with the past."[79]

Although the CCC actually started in Virginia, it was the trans-Mississippi West that the "boys" quickly occupied like an invading army. As Roosevelt saw it, the American West was about a half dozen subregions, all with distinct conservation needs. Shortgrass plains, alpine mountains, geyser basins, plateaus, mesas, canyons, cliffs, badlands, sinks, and saguaro deserts were all components of western ecology that Roosevelt thought deserved attention. In the early and mid-1930s perhaps the most notable CCC infrastructure work in the West was in Colorado, a state ideally suited for a youth corps. Only five states exceeded Colorado's native forest acreage. Meanwhile, unemployment was at 25 percent. So in the summer of 1933, twenty-nine CCC camps were established. At Rocky Mountain National Park, established in 1915, the CCC built roads at a dizzying elevation of 9,200 feet. The CCC boys set about improving Trail Ridge Road, over the Continental Divide, ten miles of which were at an elevation above eleven thousand feet.[80] FDR—who loved the Taconic Parkway—hoped to make the breathtaking route northwest of Denver a similarly evocative roadway.

Many CCC recruits lived in the gateway town of Estes Park and rode red tourist buses (called "woodpeckers" by locals) up to the construction sites. It took six CCC companies, working on a dozen mountain peaks, to help turn FDR's "Top of the World" road into a forty-eight-mile reality. "A few months ago I was broke," Charles Battell Loomis wrote in *Liberty* magazine in 1934. "At this writing I am sitting on top of the world. Almost literally so, because National Park No. 1 CCC Camp near Estes Park . . . is 9,000 feet up. Instead of holding down a park bench or pounding the pavements looking for work, today I have work, plenty of good food, and a view of the sort that people pay money to see."[81]

Working at such high altitudes was physically daunting. Shortness of breath, altitude sickness, dehydration, and nosebleeds were commonplace. An infestation of the mountain pine beetle was devastating the national park's stands of ponderosa and pine. CCC manpower helped peel the bark off 11,194 trees to expose *Dendroctonus ponderosae* larvae to the elements.[82] These CCCers also cut more than a hundred miles of hiking trails, stocked lakes and rivulets with 1.5 million trout, laid telephone lines across mountains, and transformed an old lodge into a museum. CCC manpower made Rocky Mountain National Park a tourist destination comparable to Yellowstone or Yosemite.

In the state of New York, enlistment in the CCC began on April 7 and 8 with 1,800 young unemployed men, all carrying welfare agency certificates, showing up at the Army Building at 39 Whitehall Street in lower Manhattan. Cheers and renditions of "Happy Days Are Here Again" were heard. From the Wall Street area, these initial New York City recruits were bused to Fort Slocum in Westchester County, Fort Hamilton in Brooklyn, and a segregated African American CCC camp at Fort Dix in New Jersey.[83]

Roosevelt ordered that half of New York's sixty-six CCC camps were to be based in state parks. With over $134 million in New Deal appropriations for New York alone, Roosevelt wanted its CCCers to immediately battle tree diseases and gypsy moths.[84] At Bear Mountain State Park a half dozen artificial lakes and a network of hiking trails were built. The president offered suggestions for the beautification of several Adirondacks locales: Fort Ticonderoga; Tahawus in Essex County; and Bolton and Lake George. All of these camps built new lookout stations to watch for wildfires and illegal timbering.[85] Putting Pinchotism into action, four innovative New York CCC camps were erected on private land with the cooperation of the owners.[86] Throughout the Adirondacks there were numerous old dams, once installed for logging purposes but now collapsed, leaving flats overrun with stumps and bog vegetation. The CCC, by improving these old dams or building new ones, created nine new lakes. One CCC dam on the west branch of the Sacandaga River was erected to protect a trout stream from the heavy run of northern pike (*Esox lucius*) which formerly ascended this river from the Sacandaga Reservoir; an apron was designed to discourage the jumping of these fish.

No one was happier about the boldly interventionist CCC than Pinchot. Rolling up his shirtsleeves, he helped organize CCC camps in Pennsylvania's Poconos, Penn's Woods, and Alleghenies. In a very real way, the CCC was an extension of Pinchot's Pennsylvania program for "human conservation through safeguarding nature." Because of his influence, Pennsylvania soon boasted the second-highest number of CCC camps of any state, trailing only California. Protected forest, amid the hustle and human destiny around Philadelphia and Pittsburgh, were Pinchot's priorities. Federally funded historical restoration projects took place at Fort Necessity and Valley Forge. Thirty-seven new fire observation towers were erected in state parks.[87] Ickes, an ardent supporter of the NAACP, dispatched an African American CCC company (led by black military officers) to landscape and renovate Gettysburg National Military Park with the hope that the experience would foster pride in the unit. The company enhanced Adams County, Pennsylvania, for tourists interested in the Civil War.[88]

Under Pinchot's watchful eye, CCC Company S-73, assigned to the Allegheny Mountains, did brilliant conservation work, planting thousands of trees, excavating recreational lakes, and building Parker Dam in Clearfield County. Man-made lakes were created at Greenwood Furnace and Parker Dam, and family cabins were built at Moshannon and Clear Creek. "As I see it there is no single domestic step that can be taken that will mean so much to the future of the United States as this one," Pinchot wrote to the president, "and at the same time none that will meet with such universal approval."[89]

Pinchot lauded Roosevelt's CCC plan to reduce unemployment, but hoped that around one quarter of the enrollees would come directly from the forested regions and rural communities where the new reforestation camps would be situated. Rural recruitment, Pinchot reasoned, would ensure that America's public forests were tended by young men with local affinities. He refused to believe, for example, that a jobless man from Philadelphia was better suited to cut trails in Cook Forest State Park in western Pennsylvania than a youngster from nearby Oil City or Clarion. "I hardly need to tell you how much I am delighted with the President's reforestation program," Pinchot wrote to Howe at the White House. "There is just one question that I would like to raise, if I may. I understand enrollment of the men will be largely from the cities, and I am rather fearful that working city men only in the forests of Pennsylvania will result in millions of acres of forest being burned over through dissatisfaction of our native population, many thousands of whom are out of work."[90] The administration made sure 55 percent of the CCC enrollees were from rural areas.

What Roosevelt hoped to do by employing youths, whether rural or urban, was to reduce juvenile delinquency in cities. The *New Republic* went so far as to editorialize that the CCC was Roosevelt's way to "prevent the nation's male youth from becoming semi-criminal hitchhikers."[91] Education was a key component of the camps.[92] Once the young men were officially enrolled, they would take classes in "Forestry," "Soil Conservation," and "Conservation of Natural Resources."[93] CCCers were further required only to do calisthenics, polish their shoes, brush their teeth clean, and maintain a sense of humor. For all the emphasis on land stewardship ethics, however, thousands of CCCers never actually engaged in "green" tasks. "I worked as a carpenter," Ray Condor recalled, speaking for many. "And a CAT operator, and I [did] welding up here in the shop. I ran the drag line to haul the cement to [the] tank. . . . I helped slope. We had to shovel and load the rocks up on the switchbacks and dynamite rocks. That was most of the work."[94]

Chores and responsibilities ranged from the heroic (building the biggest log cabin in Minnesota) to the mundane (picking up roadkill in Tennessee) to the public health–oriented (eradicating rabbit ticks in Minnesota). Each CCC camp provided basic medical services, including inoculations against typhoid fever and smallpox. A "Camp Life Reader and Workbook" was assigned to the "boys" early on to determine literacy.[95] "We didn't really have training," Floyd Fowler, a recruit from Silver City, Nevada, recalled, "We didn't really *require* a lot of training. We just followed instructions and they had men that knew what they were doing along with us and they just set the example and we just followed them. There wasn't very much technical knowledge involved."[96]

From inauguration day forward, Roosevelt ably projected the image of an openhearted liberal who cared mightily about the downtrodden, struggling families, and the homeless. Increasing the size and scope of the federal government to alleviate suffering blindsided the GOP opposition. The CCC was part of this expansion. The public response was so favorable to the CCC that on October 1, 1933, Roosevelt instituted a second period of enrollment. Three months later, 300,000 CCCers were serving America. In 1935, Congress renewed the program, allowing participation to be over 350,000.[97]

If Roosevelt had any worries about the procedural aspects of the CCC it was the extraordinary power the War Department had in processing the agency's applicants, training enrollees, organizing camp locations, and then transporting them to far-flung camps across the country. After all, in Germany, Chancellor Adolf Hitler had initiated a fascistic military program of his own (*Hitlerjugend*—"Hitler Youth"), which engaged young men, who were allegedly "rotting helplessly" on the streets, to work on community forestry and land rehabilitation. Hitler Youth were also trained in military activities, with aggressive behavior strongly encouraged. Not wanting his "boys" to be compared to Hitler's social engineering schemes, Roosevelt urged reporters not to refer to CCC enlistment centers as "cantonments" and encouraged the public to think of the CCC's work sites as "camps"—like Boy Scout camps. He repeatedly stressed that the crucial functions were overseen by the Department of the Interior and the Department of Agriculture.[98] The War Department only helped train enrollees and handled logistic concerns at the camps. And unlike U.S. military personnel, CCCers could initially sign up for only six-month stints. Later they could re-up for a total of eighteen months, but after that time expired they had to leave the CCC for six months before they could reenlist.[99]

VI

In Missouri, Roosevelt had former Senator Harry B. Hawes helping him in the field. The main CCC office for the state was in Saint Louis. Notices about the CCC were posted in shopwindows throughout downtown, on the Mississippi River docks of LaCleide's Landing, and in the rough-and-tumble tenement neighborhoods of Columbus Square. Applicants were rejected if they suffered from a serious illness or had a criminal record. "No matter to what camp you are sent you'll probably have to learn to use a pick and an ax, to dig with a shovel," a CCC brochure stated. "And pull one end of a crosscut saw and build fences, to move rocks and do all sorts of hard labor. And, count on it, sometimes it will get monotonous."[100]

When CCC acceptance letters arrived by mail or telegram or even word of mouth in Missouri, whoops and celebrations usually occurred. Overnight a young man went from being destitute to being a breadwinner. Each Saint Louis enrollee was instructed to first report to the U.S. Army post at either Jefferson Barracks in Lemay, Missouri, or Fort Leavenworth, Kansas. While there is no proper documentation of how precisely the CCC office in Saint Louis selected the first twenty-five thousand men from Missouri, a distinctive pattern emerged. Most Missouri CCCers were skinny as a rail, Caucasian, averaging an eighth-grade education, and lacking meaningful work experience. After passing a strict physical evaluation and receiving vaccinations, they were clay ready to be molded.

The Ozark Mountains, with their sublime natural beauty, were the first priority. That spring, three state parks—Sam A. Baker, Meramec, and Roaring River—were selected as prototype sites. Although the Ozarks' vast pineries, grassy barrens, and prodigious burr oaks had been heavily damaged by heedless farming and timbering, the New Deal was going to bring ecological glory back to southern Missouri.

Within a year, over four thousand CCCers, directed by the National Park Service, fanned out in twenty-two CCC camps in fifteen Missouri state parks, the majority in the Ozarks. A total of 342 examples of "rustic architecture" erected in these state parks by the CCC have been listed in the National Register of Historic Places as of 2015—an astounding testimonial to the craftsmanship of the CCCers. Aesthetically, the stone bridge at Bennett Spring, Missouri, may be the most perfect masonry work the CCC produced.[101]

Tainting this fine record of achievement in Missouri, however, was the institutionalization of racial prejudice. It had many Confederate sympathizers during the Civil War and bigotry against African Americans was still potent in 1933. Although Roosevelt had originally considered integrating the CCC, the program wasn't sold to Congress as a civil rights crusade. Nor did he want to offend his Democratic political base in the South—which had been instrumental in his election—by attacking Jim Crow. Early on the CCC created separate companies for African American enrollees; 250,000 blacks enrolled in 150 "all-Negro" CCC companies throughout the nation from 1933 to 1942. The president's uninspired "separate but equal" principle regarding the CCC infuriated civil rights groups. When NAACP leader Thomas Griffith complained about the lack of integration in the CCC, Tennessean Robert Fechner shot back that segregation was not, in fact, discrimination.

To compensate for such institutionalized racism, the Roosevelt administration taught African American history in the camps, with Booker T. Washington and Frederick Douglass honored as sustainable heroes. African American supervisors were placed in the chain of command on Roosevelt's orders. "In the CCC camps," Roosevelt wrote to Fechner, "where the boys are colored in the Park Service work, please try to put in colored foremen, not of course, in technical work but in the ordinary manual work."[102]

Three all-black CCC companies in Missouri worked in the loamy soil and pine forests with dedication. In Missouri, CCC Company 1743 left behind a stunning legacy of rustic stone architecture. Indeed, Company 1743, based near Stoutville and nicknamed the "Thunderbirds," became legendary for its work in Tom Sawyer State Park. But, for the most part, these African American companies were assigned to the more menial duties, such as digging ditches and planting trees, and did not build the handsome stone bridges and picnic areas that were the CCC's high-profile projects in the state.

Each Missouri camp had a company commander, a project superintendent, and an educational adviser. There were also chaplains, doctors, silviculturists, agronomists, and engineers. At bugle call, the enrollees made their beds and scrubbed the barracks under the watchful eye of an army foreman. Because the CCC had certain practices in common with the U.S. Army, it isn't surprising that many military leaders, after initial skepticism, appreciated the CCC. Reserve officers were often in charge of a camp's transportation needs, day-to-day management, and operational regulations. Part of the attraction of the CCC for many men was the seductive promise of three meals a day. Once the orange juice was downed and plates of eggs and sausage were consumed, the shovel-ax brigade climbed into the pickup trucks that drove them to work sites. At some sites CCCers drove giant bulldozers and concrete mixers and wielded hydraulic rock-busters and electric saws. It's been said that the CCC recruits in Missouri were more likely to stink of gasoline than smell of pine.

Although the Missouri CCC camps varied in layout, depending on topography, each camp was equipped with barracks, a mess hall, bathhouses, latrines, quarters for the officers, and a water storage tank. Photographs of Missouri camps show long, narrow barracks with a row of bunks on either side. A woodstove was usually installed in the center of the barracks, and men would fight for the privilege of sleeping nearest to the stove during the bleak winter months. As if in a Jack London story, the pecking order at CCC camps was sometimes determined by boxing matches. Conversely, what went on behind the barracks, away from the eyes of the camp directors—like fisticuffs and cursing—doesn't often appear in letters to mom and dad. Boxing, authorized or not, wasn't the only popular sport at the Missouri camps. Intercamp leagues were formed for baseball, softball, and basketball.

Not long after the CCC was established, the mushrooming camps each launched their own newspapers or mimeographed newsletters to chronicle daily life. Collectively, these publications represent a trove of Great Depression journalism, regional lore, and conservation accomplishment. Roosevelt asked Melvin Ryder—who had served on the staff of *Stars and Stripes* during World War I—to publish from Washington a nationally distributed weekly newspaper, *Happy Days*, which would feature propagandistic pieces on life in the CCC camps, sporting events, entertainment, and developments in conservation education.[103] *Happy Days* sometimes printed entries for the best CCC motto. Many were funny, such as "They Came, They Saw(ed), They Conquered," or bittersweet: "Farewell to Alms."[104] The *New York Daily News* approvingly quoted from *Happy Days* that the CCC motto was "They've Made the Good Earth Better."[105]

CHAPTER EIGHT

"HE DID NOT WAIT TO ASK QUESTIONS, BUT SIMPLY SAID THAT IT SHOULD BE DONE"

I

Shenandoah National Park is an area of lofty peaks, continuous ridges, peaceful valleys, gentle forests, and far-reaching vistas. In the full flush of springtime, all 180,000 acres—from Front Royal to the southern boundary at Jarman Gap—lived up to their reputation as the ethereal backbone of the famous Blue Ridge Mountains. Although Roosevelt had been in Washington for only a month, he wanted a short vacation, a day trip, in Virginia's radiant greenery. And Shenandoah was the ideal place in the mid-Atlantic to meander on the kind of drawn-out country drive Roosevelt loved.

Another reason Roosevelt wanted to explore Shenandoah National Park that April of 1933 was to inspect Herbert Hoover's fishing camp along Rapidan River. Toying with the idea of a presidential retreat, where he could escape from the heat, Roosevelt wanted to find out whether Hoover's camp fitted the bill.[1]

Back in 1931, Hoover had started building the ninety-seven-mile Skyline Drive, the only public road that ran through Shenandoah National Park, a long, narrow, north-to-south scenic road through the Appalachians of Virginia. In addition to visiting the Rapidan camp, Roosevelt would also inspect Skyline Drive, meet with work-relief crews, and enjoy a few scenic Virginia overviews.[2]

At the last minute, Roosevelt asked Horace Albright, the prime advocate of Shenandoah National Park in the Hoover years, to accompany him on the much-anticipated drive. Flattered by the invitation, Albright quickly accepted. Most NPS directors got to meet the president for only stray minutes at an occasional ceremony. Albright, by contrast, would have uninterrupted hours with FDR early in his administration enjoying together the Virginia roadsides blooming with wildflowers.

Albright, who had succeeded Stephen Mather in January 1929, had been at the acme of the NPS since it was established in 1916. Genial, reliable, and good with numbers, he got along well with the succession of secretaries of the interior—Franklin K. Lane, Albert B. Fall, Herbert Work, and Ray Wilbur—before Ickes. Following the "Mather tradition" of selling the NPS to the American people as the surest way to win congressional appropriations, Albright developed into a master publicist for wild places. Roosevelt knew that by asking him to take a country drive on April 9 he'd learn how NPS units were faring in the Great Depression.[3]

Albright, taking advantage of his seniority, also had a hidden agenda: he planned to lobby the president to transfer America's military-historical areas and national monuments (those in the Department of Agriculture and the War Department), to the NPS bureau within the Department of the Interior. And he wanted Roosevelt to establish a protocol of uniform administration for the national parks and to create a Branch of Education and Research (staffed with naturalists and historians). On a smaller scale he wanted Roosevelt to authorize the Colonial National Monument to manage historic Jamestown, Williamsburg, and Yorktown, in Virginia.[4] Doors had closed on Albright's grand idea for reorganization in the 1920s—though Hoover had been sympathetic. With FDR in the White House, *anything* was possible.

Roosevelt was excited to be playing hooky from Washington, hungry for good-natured banter and fun, as he climbed into the open-roof touring car. It was a beautiful day. Once the motorcade arrived at Hoover's fishing camp, in the Blue Ridge Mountains, a picnic lunch was served, at which there was "much informal conversation and laughter among members of the party."[5] But Roosevelt quickly determined that Hoover's fishing camp wasn't for him. After lunch, FDR told Albright he wanted to cruise the central portion of the Skyline Drive, set to open in September. The fact that 101 miles of the Appalachian Trail ran parallel to Skyline Drive, impressed Roosevelt, as did the beauty of flowering azaleas and mountain laurels. "From the Hoover camp, we were driven up an old wagon road to the partially finished highway, later to be named the Skyline Drive, in what was to be the Shenandoah National Park," Albright recalled. "The President first asked questions about the establishment of the Park Service and its policies relative to new national parks in the East, among them Shenandoah, and about the new road over which we were traveling."[6]

Experiencing the Skyline Drive was exhilarating for FDR. The parkway's macadamized and hand-built rock embankments, guard walls, and easy gradients; a seven-hundred-foot tunnel; the wide sweeping curves; the pull-offs for panoramic views; and the absence of billboards and tacky commercial enterprises were worthy of Switzerland or Austria.[7] No construction detail around Thornton Gap was too small to escape Roosevelt's eagle eye. Even though rock retaining walls were being built, he instructed Albright that the speed limit on the federal road shouldn't be any higher than forty-five miles per hour.

Although Albright was a strong-minded, even dictatorial NPS administrator, one-on-one he was very personable, an unquenchable conversationalist. Delighting in Albright's Civil War stories about Bull Run and Appomattox Court House, Roosevelt pushed back his return to the White House. Mustering his courage, Albright raised the transfer to the National Park Service of *all* sixty-four of the historical sites then administered by the War and Agriculture Departments. If Albright had his way, Interior would absorb such heritage sites as the Statue of Liberty (New York), Kings Mountain (the Carolinas), Gettysburg (Pennsylvania), Vicksburg (Mississippi), Shiloh (Tennessee), Fort McHenry (Maryland), Abraham Lincoln's birthplace (Kentucky), and Chalmette Monument Grounds (Louisiana).[8]

Roosevelt was very enthusiastic.[9] "He did not wait to ask questions but simply said that it should be done," Albright recalled, "and told me to take up the plan with his office and find out where to submit our papers at the proper time."[10] Under the expansive authority granted him by Congress, Roosevelt quickly moved to consolidate all federal park business under the jurisdiction of the National Park Service. The time was opportune for fast action. All sixty-four national sites could be suddenly transferred to NPS by executive order.

The Forest Service was angry about Roosevelt's abrupt embrace of the NPS. But Roosevelt had long thought that a single system of federal parklands was needed: genuinely national in scope and embracing historic as well as natural places. That he chose the NPS over the Forest Service made sense. As of 1933, the Park Service was already managing twenty archaeological and historic places including Bandelier (New Mexico), Scott's Bluff (Nebraska), and Siskia (Alaskan Territory).[11] Although the big wilderness parks would be the heart and soul of the National Park Service, historic sites were taking on new importance.[12]

Never before or since had an American president made so many pilgrimages to national and state historic sites. Wherever the president traveled in 1933 and 1934, his staffers assured him that historical sites would be added to his itinerary: he visited James Monroe's office in Fredericksburg, Virginia; dedicated a national monument to Samuel Gompers of the AFL in Washington, D.C.; delivered a Memorial Day address at the Gettysburg Battlefield in Pennsylvania; journeyed to Harrodsburg, Kentucky, to commemorate the Revolutionary War hero George Rogers Clark; bowed his head at James Polk's grave in Tennessee; explored the "Lost Colony" of Roanoke in North Carolina; toured the sixteenth-century Castillo, San Felipe del Moro citadel in Puerto Rico; inspected Blue Beard's castle in the Virgin Islands; and dedicated Duke of Gloucester Street in Colonial Williamsburg.

Enthralled by these secular shrines, Roosevelt set into bureaucratic motion what became the Historic Sites Act of 1935. This act made the conservation of historical sites a National Park Service obligation (something that had been only implied in the Antiquities Act of 1906). Section 462 of the Historic Sites Act also granted the NPS a wide range of enhanced powers, including the authority to oversee preservation work and the legal rights to survey and select sites and buildings of national significance (this provision eventually evolved into the National

Historic Landmarks Program).[13] To Roosevelt, the preservation of historic places was fundamental for building a deep sense of citizenship.

Eleanor Roosevelt once remarked that her husband "never liked to dwell on the past" and "always wanted to go forward."[14] That was certainly his personal outlook. But as a national planner, FDR fervently believed that National Historic Sites and state parks (with interpretive nature museums) instilled in citizens a sense of pride in, and ownership of, the United States. As a corollary, Roosevelt instructed that all the New Deal's state park and national park structures draw on indigenous local styles and materials. Local historians and landscape architects clearly never had a stauncher ally than Roosevelt, who ordered the completion of such dormant projects as Frederick Law Olmsted's Aquatic Park in San Francisco and Metropolitan Park District in Cleveland.[15] Albright was stunned by FDR's "intense interest in American history and his memory of men and events."[16]

One historic spot that Roosevelt wanted to bring under immediate NPS control that spring of 1933 was the Saratoga Battlefield in Stillwater, New York. On the drive around northern Virginia he brought the matter up with Albright. Without missing a beat, Albright informed Roosevelt that a bill had been introduced in Congress in 1930 but hadn't made it out of committee. After a few moments of reflection, Roosevelt said, "Suppose you do something tomorrow about this?"[17] To Roosevelt's consternation, Harding, Coolidge, and Hoover hadn't acted to protect Revolutionary War sites like Saratoga Battlefield from encroaching commercial development. It bothered him that, if it weren't for the Society for the Preservation of Virginia Antiquities and the Mount Vernon Ladies Association, numerous Revolutionary War–era historic landmarks in Virginia would have been lost to commercial development. It was the work of such private associations that had preserved Williamsburg, Kenmore, the Monroe Law Office at Fredericksburg, Gadsby's Tavern, Pohick, Christ Church, Mount Vernon, and Hanover Court House. The nonprofit preservationists had done their part. Now, under Roosevelt's leadership, the federal government would accept long-term responsibility.

That April of 1933, the president put the reorganization of America's heritage sites into motion at the White House. After further reflection, he decided that all the public federal buildings—not just parks and monuments—located inside the District of Columbia should be transferred to Interior. On June 10, 1933, he signed Executive Order 6166, consolidating all national parks, national monuments, and batttlefields into the National Park Service. Transferred to Interior from the Office of Public Buildings and Parks were the White House, the Washington Monument, and the Lincoln Memorial. FDR simultaneously abolished the Arlington Memorial Bridge Commission and the Public Buildings Commission, among similar overly specialized government offices.[18] Under FDR's prodding, the NPS soon assumed one more task: the Historic American Buildings Survey, to protect from the wrecking ball the nation's most precious architectural gems.[19]

When EO 6166 went into effect on August 10, the modern-day National Park Service (NPS) was born. "It effectively made the Park Service a very strong agency with such a distinctive and independent field of service as to end its possible eligibility for merger," Albright wrote, "or consolidation with another bureau."[20]

Once Roosevelt signed the NPS consolidation documents, the forty-three-year-old Albright, on July 17, tendered his resignation. After four years as director of the National Park Service he wanted to make money in the private sector. He joined the boards of both the U.S. Potash Corporation and the U.S. Borax and Chemical Corporation. On Albright's recommendation, the president chose Arno B. Cammerer, from Arapahoe, Nebraska, to be the new NPS director.[21]

Cammerer was a Washington "cave dweller," having worked since 1919 in various offices at the Treasury and at Interior. Ickes had preferred Newton B. Drury of the Save-the-Redwoods League, believing that the NPS needed "outside blood" from California. But in the end Cammerer—an associate director of the NPS—won out by virtue of seniority. Ickes soon developed an antagonistic relationship with "Cam," considering him an ingratiating and downright annoying man, lacking imagination, and afraid of special interests. In his *Secret Diary*, Ickes carped about Cammerer's habit of "vigorously chewing gum in an openmouthed manner."[22]

As Roosevelt hoped, Executive Order 6166 brought the National

Park Service into urban areas for the first time. In Greater Washington, NPS park rangers now ran Ford's Theatre, the Curtis-Lee Mansion, and Rock Creek Park. Before the grand reorganization of June 10, 1933, the NPS had sixty-seven units; after Roosevelt signed the transfer documents, it boasted 137. Roosevelt hoped all Americans would visit far-flung heritage sites run by Interior. And they could. After all, when TR was president there were only nine thousand cars registered in the United States. By 1933, despite the Great Depression, that number had risen to twenty-three million. And the consolidation of parks and monuments under the NPS helped centralize conservation efforts. With the NPS serving as the supervisory agency for infrastructure improvement (roads, restrooms, campgrounds) in federal, state, and local parks, Roosevelt had turned the bureau into something extraordinary.

Roosevelt's predilection for historical preservation emanated from his uncle Frederic Delano, who helped him craft the executive orders for the reorganization. During the early New Deal, many people in the Eastern Establishment thought Delano—clean shaven, well dressed, mannerly, and refined—was the best channel to get an idea about policy directly to FDR's attention. Delano was a personification of the old boy network of the Northeast; serving as chairman of the National Capital Park and Planning Commission, he was also a powerful Washington insider. Many senators, cabinet officers, and Supreme Court justices learned the value of a quiet word with Delano. As the Washington *Evening Star* noted, Delano's gravitas was never built on claiming "personal credit for anything."[23] Wary of charges of nepotism, Delano, a cohesive force in the early New Deal, didn't leave much of a paper trail. But it is doubtful that EO 6166 could have happened without his sage and persistent advocacy.[24]

Delano's experience in city and regional planning was significant. As chairman of the Committee on the Basis of a Sound Land Policy, he had engineered an influential report on how America should manage its land. What made him unique was the importance he placed on beautification in the building of modern America. His Jeffersonian agrarianism had a tinge of the romantic, and as a reformer, Delano wanted nothing less than for all American roads to have a scenic quality. Even in front of the Ford Rouge factory in Dearborn and the United Steel furnaces in Pittsburgh, Delano wanted to see islands of well-maintained trees specially planted to "green" industrial zones. In urban areas Delano insisted on water treatment plants, daily garbage pickups, sanitation laws, and some added appeal for every sidewalk. "Few Americans had had more impressive experience in city and regional planning than Delano," historian Arthur Schlesinger Jr. attested. "Chicago, New York, and Washington all bore his mark in their programs for urban development."[25]

Delano encouraged his nephew to build the Blue Ridge Parkway—an extension of the Skyline Drive that would connect Shenandoah National Park with the Great Smokies. Money for the project was found through the National Industrial Recovery Act. With Congress's authorization, the Blue Ridge Parkway was entrusted to the National Park Service and Bureau of Public Roads. A strategy was formed to develop the Natchez Trace Parkway (to run along an ancient Indian trail connecting Nashville, Tennessee, to Natchez, Mississippi).[26] Meanwhile Eleanor Roosevelt became the first woman to drive up Whiteface Memorial Highway in the Adirondacks, the road that her husband had promoted as governor.[27] "We are definitely in an era of building," Roosevelt said of the early New Deal, "the best kind of building—the building of great public projects for the benefit of the public and with the definite objective of building human happiness."[28]

II

Some in the War Department felt great anguish after hearing of the President's abrupt decision of June to transfer military cemeteries—including Arlington National Cemetery—out of their jurisdiction to Interior. But George Dern, the secretary of war, wasn't among them. Born in Dodge County, Nebraska, Dern moved to Utah in 1894 to make money in mining. By coinventing the Hoyt-Dern ore-roasting process (for tracking silver from low-grade ores), he made a fortune. As governor of Utah from 1925 to 1933, Dern had been lukewarm about the Interior Department. When Albright tried to establish a Kolob National Park in 1931—near Zion and Bryce Canyon—Dern shot down the proposal.[29] When Roosevelt was governor of New York, he met Dern at a governors' conference, and they hit it off immediately. Both were ardent Wilsonian Democrats with an abiding love of the great American outdoors. Dern was, as the *New York Times* put it, "especially active" in all things related to conservation.[30] Just a few weeks before the 1932 election, Roosevelt had spent two days consulting with Dern in Salt Lake City.[31]

Roosevelt's choice of Dern to head the War Department was unexpected. But Roosevelt knew that nobody in the leadership of the Democratic Party was more inclined to promote his CCC project than Dern. Outdoors recreation was a promising new route for Utah's economic prosperity, the troubled state had a whopping 36 percent unemployment rate. Dern, a miner and the most pacifist secretary of war in American history, understood that the Colorado Plateau needed roads, and Roosevelt was the man to build then. When asked how a secretary of war could advocate a tree-planting corps, Dern spoke about the many amazing experiences U.S. Army reserve officers would have through the CCC. "They have had to learn to govern men by leadership, explanation, and diplomacy rather than discipline," Dern said. "That knowledge is priceless to the American Army."[32]

On August 22, 1933, Roosevelt signed a presidential proclamation under the authority of the Antiquities Act of 1906 to save a 6,155-acre site of rugged Utah wilderness as Cedar Breaks National Monument; the 6,711 acres were taken from Dixie National Forest.[33] The president's action set an important New Deal environmental precedent: pulling "natural wonders" away from the Forest Service (a component of the Department of Agriculture) for Interior to manage. Ever since the Utah Parks Company (a part of the Union Pacific Railroad) built a handsome lodge at high-altitude Cedar Breaks—a half-mile-deep natural amphitheater carved by wind, water uplift, and erosion—it had drawn tourists. Iron oxides in the limestone made the cliffs shine with continually changing tints as new angles of the sun's rays were reflected.

With great determination, Roosevelt further expanded the portfolio of the National Park Service in southern Utah. Here he had the help of Ephraim Portman Pectol, who was a bishop in the Church of

Latter Day Saints, a scrupulous businessman in the town of Torrey, a supporter of the New Deal, and a hiking enthusiast. Pectol had been enchanted with a hundred-mile stretch of gorgeous canyons, ridges, buttes, and monoliths in Wayne County. The official name of this sixty-five-million-year-old steep depression in the earth's crust was Waterpocket Fold, but in the 1920s Pectol and other local boosters dubbed it Wayne Wonderland.[34]

Pectol was elected to Utah's legislature in 1933. With FDR in the White House, he was not inclined to wait patiently for Congress to approve a national park—a process that often took ten years. Instead, he lobbied the White House directly, extolling the rugged beauty of the area and enclosing photos of the canyonlands. Roosevelt wanted Wayne Wonderland to be brought into the NPS. A number of CCC reports had agreed that this largely inaccessible part of wild Utah was nearly as breathtaking in terms of scenery as the famous Zion National Park. Ickes asked Pectol to take Interior officials on an inspection tour of the wilderness. These officials quickly determined that the area would make an ideal national monument.

All of Pectol's boosterism finally seemed to be paying off, as a federal border survey team was dispatched to Utah to draw up boundaries for the prospective monument. However, Ickes didn't love the name "Wayne Wonderland." Two words describing this hauntingly beautiful part of Utah jumped out at him: *reef* (for the ridges of sharply jutting rock that interrupted the land) and *capitol* (for the white-domed peaks of sandstone that dominated the reef's high horizon).[35] Ickes, perhaps unilaterally, directed lawyers at Interior to draft an executive action for FDR to sign establishing Capitol Reef National Monument. Unlike Cedar Breaks, which became a national monument quickly, Capitol Reef languished for four years. But on August 2, 1937, Roosevelt signed a presidential proclamation creating Capitol Reef National Monument (it became a national park in 1971).[36] And that same year, Roosevelt signed Proclamation 2221 establishing Zion National Monument: 49,150 pristine acres in the Kolob region of Washington County, Utah.[37] (It was incorporated into Zion National Park in 1956.)[38]

Utah was getting so much work-relief help from Roosevelt that a field representative from the Federal Emergency Relief Administration dubbed it "the prize 'gimme' State of the Union."[39] An astounding 116 CCC camps were run by the Forest Service in Utah. Two visionary Roosevelt administration experiments were also set up in Utah—the Widstoe Project (located in the Sevier River drainage near the Escalante Mountains) and the Central Utah Project (later to become the Intermountain Station's Benmore Experimental Range)—aimed to irrigate arid valleys to help communities thrive.[40] Because Utah became a "red state" in the twenty-first century, reliably Republican, the fact that FDR did more to help Utah prosper than all other presidents combined has been largely overlooked.

III

Luckily, Southern Utah University in Cedar City had the foresight to establish a CCC oral history program for southeastern Utah. Collectively, these New Deal oral histories illuminated how aimless young Utah men, ages eighteen to twenty-three, including many Mormons from impoverished families, worked in the canyonlands. "On the trail we saw Echo Canyon, a deep narrow gulley cut in the rock by stream," Belden W. Lewis wrote in an unpublished diary of his CCC experiences in Zion. "From observation point we could see the Great White Thrones to the south. We could look down on top of it. The rest was a God-damned trail and a Son-of-a-bitch climb. The scenery was swell, though."[41]

To bolster Utah's agriculture, CCC projects included improving grazing conditions on rangelands, conserving water, controlling rodents and predatory animals, and constructing hundreds of miles of fences and guardrails (many along large water diversion dams). Roosevelt's tree army built service roads, sewers, water and electrical systems, bridges, footpaths, restrooms, gas stations, and new campgrounds.[42] Seemingly overnight, CCC trucks loaded with gravel could be seen chugging around Zion and Bryce Canyon national parks, determined to turn the dirt roads into tourist-friendly ribbons of asphalt. (The Union Pacific Railroad, which had built a lodge in Zion, didn't like the New Deal's promotion of automobile tourism one bit.) Enrollee Tony Melessa was already working for the CCC in

Ohio when unconfirmed rumors spread about assignments opening up in Utah. "I heard there was a troop train going to Utah, to the West," he recalled. "Everyone back there said, 'Go west, young man. Go west.' "[43]

Many CCCers working in the Utah wilderness became lifelong ecology enthusiasts because of their experiences in the Zion backcountry. Hiking in the cliffs and mountains, while tiring, offered them a sense of the serene immensity of the Colorado Plateau. While members of the U.S. Army were known to carve their initials into trees and rocks, this type of defilement was forbidden in the CCC. "I never marked a tree in my whole life," one CCC worker in Zion, Quince Alvey, recalled. "I didn't put my name on anything because I also adhere to the old saying 'Fools' names and fools' faces are always found in public places.' "[44]

William C. Dalton, born in 1916, grew up in a left-wing Wobbly (Industrial Workers of the World) household in Salt Lake City, where his father instilled anti-Mormon sentiments in him. When the Great Depression hit, the Daltons lost their home and found themselves running low on both food and money. So William went to work for the CCC in 1934 to help feed his two younger brothers. "My father might have been headed for college and a career in law if his family had kept their money," his daughter Kathleen Dalton, author of *Theodore Roosevelt: A Strenuous Life*, recalled, "but in the CCC he learned surveying, drainage, and building culverts, and he gained a variety of construction skills he used for the rest of his life."[45]

William Dalton did a lot of tree planting, well digging, trail cutting, and survey crew work in the canyonland ecosystems of Bryce Canyon, Cedar Breaks, and Zion. It was tough work in the boiling sun. He slashed his leg in a construction accident, was struck by lightning in one of the makeshift shower stalls, and was hit by a wedge when quarrying stone (one of his eyes was permanently damaged). All the deprivations and mishaps were worth that $30 per month. By 1938 Dalton had enough civil engineering skill that he was hired to work on major infrastructure projects like the Alcan Highway in Alberta and the Panama Canal Zone Isthmian Highway. Just as FDR had foreseen, once World War II broke out Dalton—like numerous other former CCCers—served in the Seabees, building bridges, storage tanks for gas, Quonset huts, landing strips, and other types of combat infrastructure in both theaters. "The CCC was the basis for his family's semi-upward mobility," Kathleen Dalton said. "The CCC saved his family's lives, really."[46]

Secretary of Agriculture Henry A. Wallace wasn't keen on Roosevelt's trying to turn Utah (and other western states) into National Park Service fiefdoms. In fact, Executive Order 6166 had put Wallace in an uncomfortable position.[47] It stripped the Agriculture Department's portfolio, wholesale. Albright, refusing to be embroiled in an interagency fight between Wallace and Ickes, had resigned.[48] Instead Wallace aimed his ire over EO 6166 directly at Ickes.

Henry Wallace, circa 1939. Wallace, a wealthy Iowan, served as FDR's secretary of agriculture from 1933 to 1940 and was one of the most effective of all New Dealers. The plight of the American farmer turned around with his help, and FDR rewarded him with the second spot on the 1940 presidential ticket.

Henry Wallace was born in Adair County, Iowa, on October 8, 1888. His grandfather was a beloved Presbyterian minister and founder of *Wallaces' Farmer*, an agronomist magazine whose slogan was "Good Farming, Clear Thinking, and Right Living."[49] His father had served as secretary of agriculture under presidents Harding and Coolidge. The eldest of six, Henry studied animal husbandry at Iowa State College in Ames, graduating in 1910. He served on the editorial staff of *Wallaces' Farmer* from 1910 to 1924. Intelligent, and proud of his farming heritage, Wallace mocked Iowa officials who favored corn that *looked* handsome over corn that *grew* properly. Wallace's Hi-Bred Corn Company raised hybrid, high-yield strains that he had personally developed; they made him a millionaire.

While Wallace was a Republican, he had grown disappointed with the way the GOP leadership prioritized big business over the nation's farmers. This discontent led him to vote for Al Smith for president in 1928 and to support FDR in 1932. This was a happy moment for FDR, as he was a great admirer of Wallace, the genius Iowa corn-grower whose ideas about elevating farming interests mirrored his own beliefs and those of Henry Morgenthau. The president appointed Morgenthau to serve as the chairman of the Federal Farm Board and Wallace as secretary of agriculture to implement an unprecedented New Deal farming strategy. It included price supports, production adjustments, insurance for crops, resettlement programs for impoverished farmers, soil conservation, a tenant purchase initiative, farm credit, rural electrification, and food distribution. Many of the innovative partnerships Wallace built between farmers and the government were based on conservation.

Like Roosevelt, Wallace hoped the New Deal's conservationist agricultural projects could "leave something that contributes toward giving life meaning, joy, and beauty for generations to come."[50] During the "Hundred Days," however, Wallace was a leading skeptic in the administration with regard to the CCC. Unlike Ickes and Dern, Wallace thought it outrageous that Roosevelt would pay men a measly dollar a day for hard labor. It also seemed to Wallace an impossibility for 250,000 formerly idle men to engage in meaningful work.[51] And Wallace was frustrated with FDR's method of appropriations: his avoidance of lump-sum budget payments and his shifting of funds from agency to agency.

While the New Deal generated optimism in rural America, much of

Europe remained paralyzed by the Depression. In several nations, democracy was at a crossroads. Totalitarianism and racism had taken on new dimensions. In German cities, such as Berlin and Munich, Nazis burned books in public bonfires. Japan was in the middle of a campaign of ruthless aggression against the Chinese, especially in Manchuria. Militaristic fascism was rising throughout Europe and Asia. Anxiety had replaced the hope offered by the League of Nations. But Roosevelt focused on America's sick economy. Rehabilitating abused land and saving ecosystems was prioritized over foreign policy.

"The plain truth is that Americans as people," Hugh Bennett would write, "have never learned to love the land and to regard it as an enduring resource. They have seen it only as a field for exploitation and a source of immediate financial return."[52] Bennett had been a passionate crusader for the very soil that constituted American lands, but his speeches, articles, and studies meant little in Washington until Franklin Roosevelt arrived there as president. Determined to revive exhausted American land, Roosevelt launched the Soil Erosion Service (SES) in September 1933, with Bennett as director, and with $5 million provided under the authority of the National Industrial Recovery Act (NIRA). The agency would be part of the Department of the Interior.

Bennett, a North Carolinian, raised in the cotton fields, had worked for the federal government in soil research since the days of Theodore Roosevelt's presidency. "Big Hugh," as he was called, preached in a scientific style—which FDR enthusiastically embraced—against the denudation of soils.[53] Armed with reams of data, Bennett asked struggling farmers to reduce crop production on highly eroded land. In coming years, Bennett became FDR's guru on all things related to soil, particularly on establishing a planned and permanent federal agricultural system.[54] "He could talk cotton in the South, wheat in Kansas, oranges in California," historian Tim Egan wrote of Bennett. "He loved nothing more than digging with his hands in the earth that was the greatest of American endowments."[55]

Given a free hand by the Department of the Interior, Bennett ordered the directors of soil experiment stations, twenty-three all across America, to tailor a set of conservation methods that could be easily taught to local farmers and field-tested on their farms.[56] Millions of acres of wonderful farmland had been destroyed by primitive farming practices. The Soil Erosion Service (SES) provided equipment, seeds, and advice to farmers. Employees of the CCC, CWA, and PWA were lent to Bennett's agency to help with demonstration projects. Bennett's philosophy was that there was no single best method of soil conservation; it required a recipe of interlocking techniques. A typical team consisted of a soil scientist, a forester, and a field biologist working closely with local agricultural interests and landowners. Appreciative farmers soon deluged the Department of the Interior with requests for Hugh Bennett to open more agricultural field stations in their backyards. Many scientists, however, thought Bennett's evangelism was rather overwrought.[57] The problem in the Great Plains, these critics argued, was weather, not farming abuse. "Millions of acres of our land are ruined, other millions of acres already have been harmed," Bennett wrote in *Soil Conservation*. "And not mere soil is going down the slopes, down the rivers, down the wastes of the oceans. Opportunity, security, the chance for a man to make a living from the land—these are going too. It is to preserve them—to sustain a rewarding rural life as a bulwark of this nation, that we must defend the soil."[58]

As the SES grew in stature, Henry Wallace took notice and began insisting that its work be overseen by the Department of Agriculture. Yet another interagency feud soon erupted between Ickes and Wallace. Overnight the status of SES changed: having started as a temporary first-year New Deal agency, one of Roosevelt's many ideas, it became a permanent influence on land management. By mid-1935, when SES changed its name to the Soil Conservation Service (SCS), it was transferred to the Department of Agriculture, becoming the biggest division there. The SCS oversaw 147 demonstration projects, forty-eight tree nurseries, 454 CCC camps, and numerous federally paid workers in the fight to stop soil erosion.[59] Always on the move, Bennett, a showman, mingled with farmers, dined with college presidents, chatted with county agents—whatever it took to save the land from sick soil. For the land to be reborn, Bennett argued, farmers needed to return to the years before "the plow broke the prairie."[60]

Like a football coach recruiting star talent, the charismatic Bennett cast wide, enlisting the best agronomists to halt erosion, restore soil, and improve the rural environment for farmers and wildlife alike after fifteen decades of abuse. Commissioner of Indian Affairs John Collier asked Bennett to study the soil of the Navajo reservation in Arizona in mid-1933. Tribal leaders soon adopted Bennett's comprehensive recommendations for soil conservation. This boded well for the New Deal's Native American policies and conservation. "[Bennett] sees the matter *steadily and whole*," Collier wrote to Ickes, "and is not an engineering fanatic nor reseeding and ecological fanatic nor an animal husbandry fanatic."[61] Great men are usually honored in marble or bronze, but the earth within America's borders was the lasting memorial for Hugh Bennett, the "Father of Soil Conservation."[62]

IV

On May 18, 1933, Congress, at Roosevelt's urging, passed the legislation establishing the Tennessee Valley Authority (TVA) to address a wide range of waterpower needs in Alabama, Kentucky, Mississippi, and Tennessee, and small sections of other southern states. Illiterate croppers, tenant farmers, day laborers, and backwoods squatters were ravaged by the Depression. Seeing that the neglected Tennessee Valley region had desperate needs, Roosevelt brought a massive solution: the TVA, a thoroughly considered approach to a region in crisis. The population overall was poor, unhealthy, and tired—and therefore it was no coincidence that the land was likewise ruined. The TVA, an enormous hydroelectric project, included intensive and extensive agricultural programs, habitat restoration, and educational efforts aimed at people who were often isolated from mainstream avenues of information. In addition, the TVA provided inexpensive electricity.

A headline in the *New York Times* noted, correctly, that Roosevelt saw the Tennessee Valley as the "laboratory" for all of America.[63] The first meeting of the TVA board was held in June 1933. A few months later, a colossal dam project, under the able leadership of David Lilienthal ("Mr. TVA"), was under construction. Within three years of the TVA's creation, two new dams—Norris and Wheeler, each with a capacity of over 200,000 horsepower—were operational and making the Tennessee River "continuously navigable." Meanwhile, the CCC set up camp in the Tennessee Valley to help fight chronic erosion along the Clinch River. Roosevelt aimed to heal the abused southern landscape. All of it was managed through a new type of entity: the public corporation. Altogether, the TVA was an unusual, almost socialist entrance of the U.S. government into the private sector.[64]

As a bold experiment, promising improvements in river navigation, flood control, and electrical power,[65] the TVA was primarily opposed by private utility companies and their political surrogates.[66] FDR selected David Lilienthal to oversee the TVA, but first, New Dealers had to get it out of Congress.

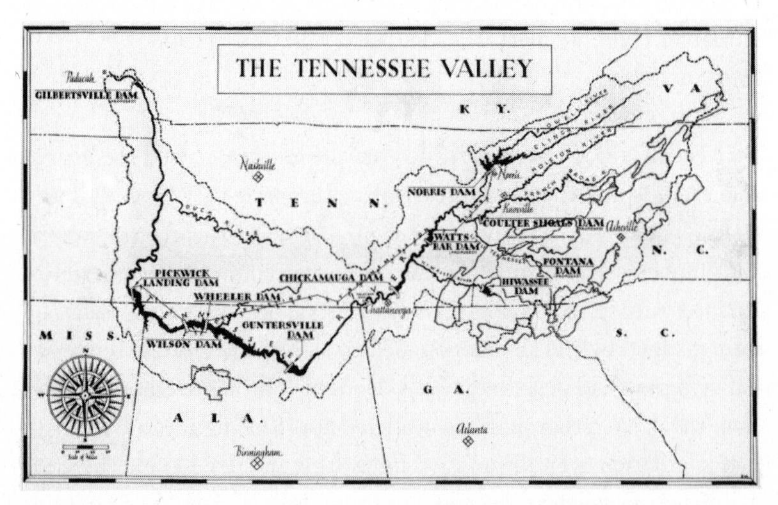

A map of the Tennessee Valley Authority reflects its scope across six southern states. Established in May 1933, at the outset of the New Deal, the TVA addressed the stifling poverty of the region through power generation and reclamation of abused lands, among other programs. The TVA was also a Roosevelt invention in government itself, as the first public corporation.

Roosevelt had a shrewd political ally in Senator George Norris of Nebraska (for whom the first TVA dam would be named). Roosevelt considered the seventy-one-year-old Norris, the noble agrarian Republican from the heartland, the bravest senator of his lifetime. Norris thought the 1902 Reclamation Act—which brought irrigation projects to western states—was the greatest thing Theodore Roosevelt ever accomplished.[67] Having backed the construction of Boulder (Hoover) Dam of Nevada, Senator Norris threw his full support behind the TVA. FDR and Norris shared an unwavering belief that ensuring the basic welfare of the American public was the great duty of the federal government. Roosevelt came to believe that the TVA, which Norris had helped create "with body and soul," was the legislator's gift to American betterment.[68]

The TVA eventually flooded 153,000 acres of land in the Tennessee Valley region, creating huge reservoirs for outdoor recreation while providing cheap electricity to the rural poor. By 1934 more than nine thousand workers were building not just Norris Dam but Wheeler Dam in Alabama. One of FDR's TVA commissioners said, bragging without hyperbole, that thirty-five Boulder Dams could have been constructed out of the materials allotted to the TVA site.[69] The work went far beyond pouring concrete, however. As an auxiliary to the TVA, the Civilian Conservation Corps planted three million trees and 2.6 million square yards of brush to keep soil from washing away.

Roosevelt, like Norris, was very bullish on the theory behind TR's Reclamation Service (renamed the Bureau of Reclamation in 1923). But he thought it had been poorly managed under the preceding Republican administrations, which favored private utilities over federal irrigation projects. One notable exception was Boulder (or Hoover) Dam, which the Republicans initiated. Built in the Black Canyon of the Colorado River on the border between Arizona and Nevada, Hoover Dam, at over seven hundred feet tall, would dwarf the Aswan Low Dam on the Nile in Egypt. Hoover Dam changed the torrent-like river forever; more than 1.7 million horsepower of hydroelectricity could be produced, allowing cities like Los Angeles, San Diego, and Las Vegas to grow. As a side effect of the dam, beautiful Lake Mead, the world's largest man-made reservoir (110 miles long), was formed—holding enough water to flood all of Pennsylvania. This was hydroelectric power on an unprecedented scale. While Herbert Hoover deserved credit (or contempt, if you're an environmentalist) for damming the wild Colorado River, it was FDR who fast-tracked this engineering project during his first term as president.

Roosevelt approved the construction of other multipurpose dams in the West, including the Shasta on California's Sacramento River and Montana's Fort Peck on the upper Missouri River.[70] From an environmental perspective, however, dams withheld water and disturbed river ecology; they eliminated highly variable flows, which a healthy native ecosystem required.

V

On August 12, 1933, Roosevelt invited AFL president William Green—a critic of the CCC—on a picnic in Shenandoah National Park's Big Meadows, a grassy tableland high on a bucolic ridge. It was quite an outing. Roosevelt left from Washington; traveled to Harrisonburg, Virginia; and met up with Senator Harry F. Byrd Sr. and Representative A. Willis Robertson—both Democrats from Virginia—and members of his cabinet, including Ickes, Wallace, and Dern. After inspecting a few CCC camps and meeting with "the boys," the visitors participated in the huge picnic at Big Meadows. A series of group photo opportunities were staged, but they could not capture the festive atmosphere that actually prevailed that day.

Franklin Roosevelt *(right)* and Harold Ickes were served lunch by a CCC enrollee during their spotlight visit to the Big Meadows camp in Virginia's Shenandoah Valley in August 1933.

Roosevelt and other dignitaries ate from the regulation mess kit, comprising a plate, a tin cup, and silverware. Reporters never saw Roosevelt so happy, eating fried beefsteak, smoking cigarettes, laughing with CCCers the ages of his sons. Everybody was excited about how the CCC was building Virginia's state park system from scratch. "It was a thoroughly democratic function," the *Page News and Courier* wrote, "for eight of the rank-and-file members of the camp were honored with a seat at the President's table, which was set on the blue-grass on the edge of the camp while the remainder of the visitors and camp personnel were eating in the mess hall."[71]

Reporters were astonished at how loose and relaxed the president was with the CCC boys of Company 350 in Skyland, Virginia, discussing baseball and how Dutch elm fungus could be controlled by burning infested trees. "I wish I could spend a couple of months here myself," he told the men at the Big Meadow picnic. "The only difference between us is that I am told you men have put on an average of 12 pounds each," Roosevelt joked. "I am trying to lose 12 pounds." Visiting Camp Nira in Shenandoah National Park, where CCC company 1316 was based, Roosevelt watched a play depicting the defeat of "Old Man Depression." Two actors—one labeled CCC, the other labeled NIRA—chased away the demons of the Great Depression to the enthusiastic approval of the president.[72]

There were many comical moments at Big Meadows that afternoon. Roosevelt got a huge kick out of listening to a U.S. Army general boast that his work-relief boys could compete with Ivy Leaguers. "He told how someone was teaching trigonometry to one of the boys, how another was learning French," Ickes recorded in his diary, "and he concluded with this gem: 'There won't none of these boys leave these camps illiterate.' "[73] After the Shenandoah outing William Green, president of the AFL-CIO, did an about-face. He became a booster for the CCC, admitting that he "could not help but view the whole project in a most sympathetic way." He told the president that picnicking in Shenandoah was "one of the most pleasing experiences" of his life.[74]

VI

As America's landscape planner, Roosevelt paid attention to municipalities that had no proper park system, especially Oklahoma City and Tulsa. If Olmsted gets credit for designing Central Park, then

Roosevelt certainly deserves recognition for helping to establish a system of New Deal beaux arts parks in Dallas, as well as the River Walk canal system in San Antonio. Places that many elite easterners thought of as no-man's-land, such as Texas's Permian Basin and Nevada's Great Basin, were—for the president—worth beautifying. "I love maps," Roosevelt declared. "I have a map mind and I can explain things."[75]

Recognizing that many in Oklahoma lived in rural poverty, especially in the drought-stricken region in the western part of the state, Roosevelt wanted forested eastern Oklahoma to be refashioned into a giant recreation hub, where people could fish, hike, and picnic. It would be called the Chickasaw National Recreation Area (NRA).

Early in 1933 he tasked the NPS to help grow and administer Oklahoma's state park system. On unproductive land the CCC built dams that formed lakes, planted trees and wildflowers, and quarried stone. While Texans during the New Deal bragged that they had "a state park every hundred miles," Oklahomans adopted the New Deal motto "a state park wherever nature smiles." Roosevelt's CCC "boys" did truly impressive recreation infrastructure work in what became the Chickasaw NRA. This green oasis—in the shadow of the Arbuckle Mountains in south-central Oklahoma—offered a reprieve from the stultifying heat of Oklahoma City and Tulsa during the summer months. It had a lovely assemblage of CCC-built mineral spring pavilions, examples of "government rustic" architecture.

By taking an "eroded resort landscape" and transforming it into a "holistic place of great beauty," Roosevelt made Chickasaw NRA (which in 1976 incorporated Platt National Park) into southeastern Oklahoma's very own version of Warm Springs. The gateway town of Sulphur boomed as tourists discovered the outdoor health and recreational benefits of the numerous mineral springs.[76] None of Sulphur's residents wanted the beloved Company 808 to close shop. When the CCC left Sulphur in 1942, one historian recalled that a sense of permanent loss fell on the community. "This New Deal legacy is still enjoyed in the twenty-first century by millions of Oklahomans who swim, boat, camp, and hike in the parks across the state," Suzanne H. Schrems explained.[77] By bringing artists, craftsmen, and conservationists into southeastern Oklahoma, Roosevelt showed he cared deeply about the Sooner State.[78]

VII

By the summer of 1933 a grassroots movement to protect 1.5 million Joshua trees (*Yucca brevifolia*) in southern California caught FDR's attention. Centuries ago Native Americans recognized the spiky Joshua trees for their useful properties: strong leaves were worked into baskets, and the flower buds and raw seeds made healthful additions to the diet. Joshuas bristled with strange dagger-shaped leaves and had upraised branches gnarled by wind and heat. When explorer John C. Frémont first laid eyes on Joshua trees in the 1840s, he derided them as "the most repulsive tree in the vegetable kingdom."[79] Mormon pioneers of the 1850s, however, trekking across the Mojave Desert, thought the somber trees resembled the Old Testament prophet Joshua, arms uplifted, beckoning the faithful to the promised land of California.[80] The Mormons' name stuck.

With a discontinuous distribution in California, Nevada, Arizona, and Utah, these strange trees grew at elevations between three thousand and five thousand feet. Very few Angelenos thought these spine-studded yuccas were worth saving; but the International Deserts Conservation League (IDCL) of Los Angeles did.[81] The founder of the league was Pasadena socialite Minerva Hamilton Hoyt, who, in the early 1930s, devised a far-reaching, unprecedented conservation proposal to designate a Joshua Tree National Monument (not far from Palm Springs) to protect *Yucca brevifolia* so that the federal government paid attention to the importance of these trees.[82]

Hoyt, a popular leader with the Garden Clubs of America, was also the founder of the IDCL, serving as the group's president until her death in 1945.[83]

Minerva Hoyt was a genteel daughter of the American South and an ardent New Dealer. Her parents owned a thriving plantation in Mississippi, and her father, Joel George Hamilton, had served in the Mississippi state senate and was a delegate at the 1872 Democratic National Convention. Minerva Hamilton was a debutante and attended Ward's Seminary in Nashville before enrolling at music conservatories in Cincinnati and Boston. She married Dr. A. Sherman Hoyt of New York City—a physician and financier—and they moved to Denver, New York City, and Baltimore before establishing a permanent home in South Pasadena. The elaborate Hoyt mansion on Buena Vista Street was admired by the *Los Angeles Times* for its five acres of manicured gardens and showcase ponds filled with alligators.[84] As a leading socialite in the Valley Hunt Club, Minerva organized the annual Rose Parade on New Year's Day and also served as president of the local chapter of the Boys and Girls Aid Society.

Hoyt's life changed, however, in the 1910s when she visited Palm Springs, a tiny but growing California resort community frequented by such Hollywood luminaries as Theda Bara and Cecil B. DeMille.[85] During the 1920s, the population of Los Angeles doubled, reaching two million. To help beautify the oceanside city, Los Angeles commercial developers stripped the Palm Springs area of Joshua trees, fan palms, and cacti of all kinds for ornamental planting. Another modern problem was that the proliferation of motorcycles and automobiles in southern California was turning the land of the Joshua trees into a motorized playground without the benefit of speed restrictions. The entire Mojave was treated as a lawless wasteland. To many Californians, H. L. Mencken's definition of nature as "a place to throw beer cans on Sunday" was especially applicable to the land of Joshua trees.[86]

Minerva Hoyt was photographed in the desert of the American West, probably in the mid-1930s. Born in Mississippi, Mrs. Hoyt was living the pleasant life of a socialite in Pasadena in the 1910s, when she discovered the overlooked beauty of the California desert. Finding an ally in Ickes, she was instrumental in the establishment of the Joshua Tree National Monument (now Park) in 1936.

Not to Minerva Hamilton Hoyt. "This desert [the Mojave]," she said, "with its elusive beauty . . . possessed me and I constantly wished that I might find some way to preserve its natural beauty."[87] During the springtime Hoyt would backpack in California's deserts, choosing blossoms from among seventy different species of wildflowers for her vasculum. Among the mundane and strange varieties of plants in the desert, the exotic Joshuas always stood apart. Disdainful of the swelling tide of cars swallowing up Los Angeles, insulted by the noise, she founded the IDCL solely to protect arid landscapes such as Antelope Valley, Coachella Valley, Death Valley, the Amargosa Desert, and the Sonoran Desert.[88] In 1931, Hoyt joined an expedition into Mexico to study rare desert plants and animals; this expedition earned her an honorary doctorate and the title Professor Extraordinary of Botany from the National University of Mexico. At the ceremony President Pascual Ortiz Rubio of Mexico praised Hoyt as the "Apostle of the Cacti."

During the 1932 Summer Olympics in Los Angeles, Hoyt had made a point of calling the Joshua trees "sentinels" of the desert.[89] "I fear," she told the British press, "we cut our trees down in rows sometimes forgetting that town and country must have a personality, so to speak, and tradition born of time."[90]

Hoyt's dream of Joshua Tree National Monument became more realistic with Franklin Roosevelt in the White House. Aware of her preservationist work with the IDCL, Ickes encouraged Hoyt to visit him in Washington. In person, Hoyt recommended to Ickes that a large *federal* park or monument be established in the California desert between Palm Springs and Twentynine Palms. But Hoyt's Joshua Tree project was strewn with bureaucratic roadblocks. The Southern Pacific Railroad, for example, had claims that no local politician could afford to challenge.

In late June 1933, Ickes arranged a private audience for Hoyt with the president. Roosevelt was impressed to learn that Hoyt had introduced the Joshua tree to the public in botanical exhibitions in New York, Boston, and London. Taking one sharp, practiced glance at the desert preservationist, Roosevelt endorsed the movement to protect the Joshua tree forests. When Hoyt warned about potential obstructionism from the Southern Pacific Railroad, homesteaders, miners, and private ranchers, Roosevelt brushed her concerns aside. Lawyers at Interior, he told Hoyt, would take care of any obstacles. "It is of great satisfaction to me to know that both the President and Ickes want the Monument," an elated Hoyt soon wrote to a National Park Service official. "Secretary Ickes said to me after I had seen the President, 'The President is for it, I am for it.' And now that the withdrawal is technically on its way, can you not devise some clear plan by which the area can be protected as soon as possible as it *now* stands in all its natural beauty?"[91]

A major part of the appeal of Joshua Tree National Monument for Roosevelt was the protection its designation would afford numerous species of wildlife, especially desert bighorn sheep (*Ovis canadensis nelsoni*). On October 25, 1933, three months after Hoyt's visit to the White House, President Roosevelt set aside 1.1 million acres of federal land in Riverside and San Bernardino counties for the protection of the Joshua tree; this was the first crucial step toward the eventual designation of Joshua Tree National Monument.[92] (By contrast, Cedar Breaks National Monument in Utah was only 6,711 acres.) But Hoyt didn't celebrate too quickly. She knew that Interior still needed to rectify legalities before the executive order was officially signed. For all of the quickness of the New Deal, withdrawal of public acreage was, almost invariably, a time-consuming affair.[93]

Around Thanksgiving of 1933, W. A. Simpson, the president of the Los Angeles Chamber of Commerce, wrote to Ickes, asking him to name the Joshua Tree monument after the Apostle of Cacti. "I agree with you that Mrs. Minerva Hoyt is entitled to much credit for her conservation work," Ickes replied, "but it is the established policy of the Department of the Interior to refrain from naming national parks and monuments after people. Our leading conservationists are of the spirit that it is far more fitting to choose a name that bears a direct relation to the area's natural features or early history."[94] Ickes also cautioned both Roosevelt and Hoyt that finalizing all the necessary land deeds for Joshua Tree National Monument would take time—not until August 16, 1936, was it established.

CHAPTER NINE

"ROOSEVELT IS MY SHEPHERD"

I

Frederic Delano, circa 1905. Delano was FDR's uncle, and Sara's brother. A railroad executive and financier, he developed influential opinions about land use and especially the value of long-range, public planning. He was regarded by Roosevelt as an unbiased adviser.

In mid-June 1933, President Roosevelt visited the U.S. Naval Academy in Annapolis and Groton School before traveling to Fairhaven, Massachusetts, to enjoy the Delano family's "Homestead." Uncle Frederic Delano, FDR's confidential adviser on all things related to conservation and preservation, met him at the town fishing port with a kiss on the head. The president wanted his uncle to be the new chairman of the National Planning Board (which was re-formed with alternate names three times, including its final iteration in 1939 as the National Resources Planning Board). Delano, then seventy, agreed to start in July. The board and its staff made or commissioned reports on matters of economic capacity, highlighting potential efficiencies.

One arm, the Land Planning Committee, made a study of the nation's needs, circa 1934, with projections to 1960. "What was remarkable in the report dealing with land requirements," wrote Marion Clawson, an Agriculture Department official during the New Deal, "was its extended treatment of the need for forestland, recreation land, land for wildlife purposes, and land for other purposes."[1] The report advised that the government purchase 75 million acres in farmland, 244 million acres in timberland, and 114 million acres for recreation/conservation.[2] FDR would act on this advice.

In addition to Delano and several cabinet members, including Perkins, Ickes, and Wallace, the board was composed of private citizens. The vice chair was Charles Merriam, of the political science department at the University of Chicago. Another recruit was Charles W. Eliot II, a town planner originally from Boston whose most recent accomplishment had been a parkway granting automobile access to George Washington's Mount Vernon estate. In coming years, Delano would staff the NRPB with academic and bureaucratic experts adept at multiuse water planning, woodlands ecology, sewage treatment, and population problems.[3] Such future star American economists as Milton Friedman, John Kenneth Galbraith, Wassily Leontief, and Paul Samuelson would cut their teeth in Delano's planning shop.[4]

Careful to avoid charges of nepotism, Delano operated quietly, never using his relationship with FDR to gain power for the board. In fact, some observers felt that another chairperson would have been more pushy with FDR, and therefore more effective. But Delano did have influence. Much of FDR's sense of "thinking big" about conservation was learned from Delano, who long sought a "new bill of rights" for America that would guarantee adequate food, clothing, shelter, and medical care, and the "right to rest, recreation, and adventure," for all citizens.[5] Whenever Delano discussed historic preservation or state parks issues, the president would invariably smile, his eyes twinkling, nodding knowingly, "Yes, yes."

After chatting with Uncle Frederic at Fairhaven, the president took the forty-five-foot schooner *Amberjack II* on an eleven-day cruise along the Atlantic coast to Campobello Island. After the frenetic pace of his "Hundred Days," Roosevelt craved the tang of sea air, the ebb and flow of the tides, the splendid isolation. His son James joined him on the Atlantic voyage. It was Roosevelt's first visit to Campobello since the onset of polio in 1921. After a couple of months in the White House, the president realized that periodically escaping from the confines of official Washington by sailing was essential to maintaining good health. There was plenty of precedent for these extended getaways. Thomas Jefferson, for one, hid regularly in the blue mist of the Virginia hills when he needed solitude. Theodore Roosevelt kept a little rustic cabin—"Pine Knot"—on the woody outskirts of Charlottesville. And Herbert Hoover, the gifted fly fisherman, often retreated to his 164-acre rustic camp along the Rapidan River.[6]

What made FDR different from other U.S. presidents was how much time he spent at sea. When pundits weren't paying attention during the "Hundred Days," he transferred the *Sequoia* from the Commerce Department to the U.S. Navy to become his official presidential yacht. Roosevelt loved the name *Sequoia* because it linked his passion for forestry with his love of seafaring. When an irate midwesterner complained that it was wrong for FDR to have a yacht when millions were unemployed, Louis Howe wrote back that the *Sequoia* would give the president "a source of relaxation and a haven of quiet where he might thoroughly go over problems of National import without interruptions."[7]

In coming years, Roosevelt switched from the *Sequoia* to the larger, safer *Potomac*. On either vessel he would flee Washington, D.C., on Fridays and cruise the Potomac to Chesapeake Bay for fishing. Whenever a longer offshore ocean voyage was in order, Roosevelt had his choice from the U.S. naval fleet. Cruisers named after American cities—*Houston*, *Indianapolis*, *Philadelphia*, *Tuscaloosa*, and *Augusta*—became his favorites.[8] Navy admirals were amazed at the commander in chief's first-rate navigation skills, especially his ability to avoid lockage into a single course in open water, and his knowledge of where sandbars, reefs, and other underwater dangers lay.[9]

On June 16, en route to Campobello, Roosevelt, under the authority of the National Industrial Recovery Act, approved the Public Works Administration (PWA) to "encourage national industrial recovery" by means of large-scale public works projects.[10] Harold Ickes was asked to run it. With a first-year budget of $3.3 billion, the PWA organized and funded approximately twenty-six thousand large-scale state-sponsored capital investment projects. Bridges, trestles, pipelines, dams, water and sewage treatment systems, stadiums, tunnels, and low-income housing—to name just a few—were all PWA infrastructure projects. Over nine thousand highways and streets would be built by PWA work-relief crews between 1933 and 1943.

Within a few years the handiwork of the PWA could be seen in eight hundred health care structures, six hundred city halls and courthouses, 350 airports, and fifty housing projects. Signature American structures such as the Conservatory Garden and Zoo in New York's Central Park, Chicago's Outer Drive Bridge, and San Francisco's Bay Bridge were constructed by PWA, as were thousands of public schools. From an engineering perspective, however, no bridge or road matched the PWA's colossal dams, such as Grand Coulee (Washington), Bonneville (Oregon), and Fort Peck (Montana). Yet Ickes rejected numerous other dam projects—for example, in the Oklahoma panhandle in November 1933—because he thought them a waste of money. Instead, in the case of Oklahoma, the Roosevelt administration saw federal grassland parks. "We'll have to move [the people] out of there," Ickes said, "and turn the land back to the public domain."[11] Over forty thousand western Oklahomans complained bitterly to the White House about the snub. As the *Boise City News* wrote, Ickes was "entirely ignorant" of the panhandle's virtues.[12]

Many PWA projects were, by definition, the antithesis of conservation-mindedness. Because Secretary of War Dern was a staunch backer of the CCC, Ickes unfortunately didn't challenge his order to the Army Corps of Engineers, under the aegis of the PWA, to dredge the Mississippi and Missouri rivers. He probably couldn't have stopped it anyway; the transformation of the rivers into deepwater shipping channels had been a fervent dream in the region for thirty years. But Ickes, determined to uphold an environmental ethos, did direct PWA work-relief crews to improve dozens of state park systems by reforestation, restoration, and reproduction of historic landmarks, and by enhancing pristine wilderness areas, not overrunning them with roadways. Shrewdly, Ickes placed the brilliant young landscape architect Conrad Wirth in charge of the NPS's state park improvement program. Also brought into the fold was workhorse Herb Evison of the National Conference on State Parks (NCSP), who, in the early New Deal, helped establish 239 CCC camps in state parks across thirty-two states.[13]

Although this is often overlooked by scholars, Roosevelt put America's state parks at the forefront of the New Deal in 1933. To the delight of members of the NCSP, the president no longer made conservationists go around state capitals with tin cups looking for alms. The New Deal poured money into states like Texas, California, and Florida that were willing to establish meaningful recreation systems. By 1935 there would be a CCC force of over ninety thousand men working in the state parks and national parks of America, championing Roosevelt's recreation philosophy for the twentieth century.[14]

Roosevelt purposefully rewarded Iowa for cutting-edge wildlife management programs already in place. "Give Iowa," he instructed both Ickes and Wallace, "all it wants."[15] This wasn't because Roosevelt was smitten with Iowa's loess hills or the Mississippi River bluffs near Dubuque. Instead, Roosevelt, like any shrewd politician, wanted to receive credit for helping Iowa develop a showcase state park system. After all, it had been the epochal 1921 meeting at Des Moines, with twenty-five states represented, which gave birth to the National Conference on State Parks. Many of the smartest conservationists, Roosevelt knew of—Aldo Leopold, Ira B. Gabrielson, William Temple Hornaday, and Irving Brant—had Iowa roots. Because Iowa had pioneered standardized conservation initiatives, positive political results would come sooner there than in other states.[16]

Encouraged by the availability of CCC labor and funds, eleven states acquired their first parks because of Roosevelt's commitment to the cause. In a masterstroke, Roosevelt, by having the National Park Service oversee the development of the state park movement, ushered in one of the most successful programs in U.S. environmental history.[17] The architectural style known as "government rustic" was applied to build picnic areas, log and adobe lodges, cabin clusters, and nature interpretive centers. As historian Phoebe Cutler wrote in *The Public Landscape of the New Deal*, "a pattern of rough-cut stone, crude timbers, and wooded demesne made it simple to identify almost all of Roosevelt's state parks."[18] This "pioneer design" of stone, wood, and other natural materials might be called the Downing-Olmstead-Delano-Roosevelt recreational aesthetic look.

Despite the uniform look of these parks, every Depression-era state park was different in presentation, depending on the local artists, craftsmen, engineers, and conservationists. Some—like Oak Openings in northwestern Ohio—were simple picnic areas with bird-watching platforms. In California, a number of state parks—such as Armstrong Woods and Humboldt-Redwoods—were destinations as impressive as national parks, with idyllic campsites that rivaled any in the world. Some state parks—like Daingerfield in East Texas—were nothing more than a little man-made lake; in neighboring Louisiana, at Chicot State Park, by contrast, CCC crews excavated a two-thousand-acre water hole. Oklahoma used the New Deal work-relief expertise to build six elaborate state parks with gorgeous cabins and fish-stocked lakes. The interiors of some Roosevelt-era state park lodges were lovingly detailed. Bastrop State Park near Austin, Texas, surrounded by Lost Pines Forest, for example, had a lodge adorned with a cowboy bas-relief, a suspended wagon wheel, and elegant handcrafted furniture that gave the structure a regional swagger. For those Americans looking for a more solitary state park experience, there were little overnight cabins built in Idaho, Pennsylvania, and New York.[19]

Roosevelt's state park aesthetic extended to national parks as well. In 1933, the PWA made financial allotments to Yosemite National Park, which Roosevelt considered the "showcase for national park values." Most of the outstanding buildings in Yosemite—like the Sentinel and Wawona hotels—were Adirondack-style rustic lodges built using local Sierra Nevada stone and rock materials. Yosemite became a benchmark for eastern national park projects, such as Great Smoky Mountain and Shenandoah.[20] As when FDR led the Taconic

Parkway Commission in the 1920s, he believed that the artistic CCC workmanship and building materials themselves embodied American values worth preserving.[21]

A wonderful example of New Deal rustic architecture can still be observed in Sequoia National Park. CCC labor and PWA expertise built model patrol cabins, ranger stations, and lookout towers. Local granite and thick redwood logs gave the wooden structures in Sequoia much-needed protection from fierce winter snowstorms. The CCC and PWA continued to design and build national park structures there until 1942. An informed, modern-day visitor could easily discern which national park edifices the Roosevelt administration was responsible for building—usually the most organic, yet sturdiest; the simplest, yet most remarkable.

While heading north to Campobello that June, enjoying the Maine coast, Roosevelt wrote to Wallace about the devastated southern forests. Congress, as part of the July 1933 National Industrial Recovery Act, with a portion of its $3.3 billion budget earmarked to purchase forestland for conservation work.[22] "I am inclined to believe that a certain portion of the $20,000,000 should be spent in areas not exclusively within the forty-two existing national forest purchase areas," Roosevelt wrote to Wallace. "My reason for this is that as a matter of public policy the wider the distribution of federally owned and developed forests the wider will be the public interest and education in regard to the importance of organized forestry. For example, in Florida, Georgia, Alabama, Mississippi, and Louisiana there are practically no mature forests and the inhabitants of those states take little or no interest in forestry."[23]

What concerned Roosevelt was that the longleaf pine (*Pinus palustris*) forests of the southeastern United States were disappearing at an alarming rate. Owing to greed and mismanagement, 98 percent of longleaf pines had been lost to cut-and-run loggers. There were scattered stands along the Gulf Coast, the Piedmont, and the southern foothills of the Appalachian Mountains that Roosevelt wanted saved from industrial turpentiners and paper companies. To Roosevelt, the destruction of the longleaf pine forests of the South constituted a dastardly crime.[24]

Roosevelt campaigned for president in southern Georgia on October 23, 1932, by chatting about forestry with a farmer he met along the way. FDR was on his way for a one-day visit to Warm Springs, in between two official appearances in Atlanta, where hundreds of thousands turned out to greet him. The *Atlanta Journal-Constitution* called him "a modern Moses, who is to lead a darkened America out of a wilderness of depression."

Roosevelt pushed Wallace's USDA to start educating southerners living in the land of longleaf pine about the perils of reckless timbering. At the very least, a scientific replanting strategy for longleaf pine was necessary.[25] More knowledge was also needed in relation to other types of farming. If the planting of rice, sugar, cotton, and tobacco continued without regard for crop rotation and soil renewal, the South would become an agricultural dead zone. From his time with farmers at Warm Springs, Roosevelt knew that many rural southerners misunderstood the precise soil composition needed to grow robust orchards, vineyards, and pineries. They were too haphazard about planting protocol. Dumping out seed from a gunnysack while praying for rain was unscientific and inefficient.

II

When President Roosevelt finally arrived at Campobello, he was greeted enthusiastically by Mainers and New Brunswickers alike. "It seems to me that memory is a very wonderful thing, because this morning when we were beginning to come out of the fog off Quoddy Head, the boys on the lookout in the bow called out 'land ahead,' " Roosevelt told the crowd. "Memory kept me going full speed ahead because I knew the place was Lubec Narrows. . . . I was thinking also, as I came through the Narrows and saw the line of fishing boats and the people on the wharves, both here and at Welch Pool and also in Eastpond, that this reception here is probably the finest example of friendship between Nations—permanent friendship between Nations—that we can possibly have."[26]

Roosevelt's Campobello holiday perhaps marked the pivot to his less celebrated "second hundred days," during which restoring wildlife populations—particularly migratory waterfowl—also became a New Deal priority.

Once settled at his Campobello cottage, doing a little bird-watching and fishing, Roosevelt wrote an open letter that appeared in the July 8, 1933, edition of *Happy Days*, the national CCC newspaper: "I welcome the opportunity to extend, through the medium of the columns of Happy Days, a greeting to the men who constitute the Civilian Conservation Corps," Roosevelt wrote, congratulating his cabinet secretaries—Ickes, Dern, Perkins, and Wallace—for the expediency with which they brought his pet New Deal program to life. "It is my belief," Roosevelt asserted, "that what is being accomplished will conserve our natural resources, create future national wealth and prove of moral and spiritual value not only to those of you who are taking part, but to the rest of the country as well."[27]

An ingenious aspect of the CCC during its second phase was Roosevelt's decision to enroll veterans of the Spanish American War and World War I. Back in July 1932, a "Bonus Army" of forty-three thousand veterans had marched on Washington demanding back pay for their retirement bonuses, setting up camp on the lawn of the Capitol. The Hoover administration had these veterans forcibly evicted; their demands were summarily dismissed. Two veterans were shot dead in an altercation. On May 10, 1933, a second Bonus Army congregated at Fort Hunt, Virginia. Others bivouacked at Mount Vernon, George Washington's estate.

The Roosevelts, determined not to repeat Hoover's mistake, sprang to action. Not only did Eleanor Roosevelt visit the veterans on a soggy field, sharing coffee and singing folk songs; she soon persuaded her husband to allow the veterans to join the CCC. The *Washington Post* reported on May 20, "Half of Bonus Army Ready to Work in Woods."[28] After extensive negotiations, the Bonus Army broke up, with around 1,800 men headed to reforestation training at Langley Field and Fort Humphries, not far from Bonus Army encampments in Virginia. The rest of the protesters were shipped home at government expense. Once trained, these veterans were sent to CCC camps in New England.[29] By bringing the Bonus Army into the CCC, and by paying veterans early bonuses, Roosevelt became a hero to them and others who had formerly felt forgotten.

Another idea, coming from the same impulse to remember the downtrodden, was providing "quality outdoor recreation facilities" at the "lowest cost for the benefit of people of lower and middle incomes."[30] Millions of Americans simply didn't have the wherewithal—money, transportation, or initiative—to visit far-flung state parks, much less national ones. Roosevelt wanted such people to be able to visit nature or relax outdoors in an easy daytrip. The effort was spearheaded by Conrad Wirth, assistant director of the NPS, as well as director of the State Park Emergency Conservation Program. Using $5 million from the FERA billions, the new plan called for the purchase of dilapidated farmland within fifty miles of an urban area. The tracts also had to encompass between two and ten thousand acres. They did not necessarily have to be gorgeous; the New Dealers would make them attractive.

Once Roosevelt green-lighted purchases at four hundred sites around the country, he ordered the NPS, CCC, and PWA to reforest the blighted areas. Campsites with the capacity to accommodate individuals or large civic groups were built by the CCC and WPA. Water-based recreation (either natural or man-made) was given top priority. Roosevelt thought that no state park should be without a swimming pool or lake. Once rehabilitated by the CCC, these landscapes could be given to state park systems.[31]

By 1934, Roosevelt had approved forty-six areas in twenty-four states for what he called Recreational Demonstration Areas (RDAs). These lands allowed Roosevelt to help alleviate overcrowding in America's national parks. The RDAs were to be within fifty miles of an urban center. At many RDAs day-use picnic grounds and bathing beaches were established for urbanites to get a quick dose of nature. Of all the states, Pennsylvania had the most RDAs, five: Raccoon Creek (near Pittsburgh); Laurel Hill (near Johnstown); Blue Knob (near Altoona); Hickory Run (near Scranton and Wilkes-Barre); and French Creek (between Philadelphia and Reading).[32]

Roosevelt pushed for (and helped plan and design) Maryland's Catoctin Mountain Park RDA. The idea was to help Washington, D.C.-Baltimore residents have a convenient destination for outdoors getaways. FDR had been impressed with the beautiful mountain topography of northern Maryland, and sickened at the sight of blighted hillsides that had been recklessly timbered. The park was built exclusively on worn-out uneconomical farmlands, and the president hoped to prove the potential of eroded lands plus the possibility for recreational opportunities.[33] He ordered the U.S. government to purchase thousands of acres and declare them an RDA. Work-relief crews laid out Roosevelt's dream RDA park—Catoctin Mountain Park—to his minute specifications, with group camping facilities (fire rings, water systems, restrooms) like those at Bear Mountain State Park. Rustic cabins, hiking trails, scenic lookouts, filling station areas, stone amphitheaters, and picnic centers were erected, with Maryland's natural heritage kept in mind. Picturesque snags, six per acre, were left to please the eye. The CCC also built Appalachian Trail shelters west of the park. Other popular RDAs that Roosevelt's crews created include Custer (South Dakota), and Mendocino (California).

Taking advice from Roosevelt, Ickes bought Head Waters, an estate in the town of Olney, Maryland, not far from Catoctin Mountain Park. And during World War II, in 1942, Roosevelt claimed one of these Catoctin Mountain cabin camp centers, built in 1939, as his own retreat away from Washington's hot, humid summers. Originally dubbed "Shangri-La," the president's retreat was renamed Camp David in 1953 by President Eisenhower (after his grandson).[34]

Roosevelt also established the greenbelt town of Greenbelt, Maryland, with funds from the Resettlement Administration (later reorganized as the Farm Security Administration). The administration acquired two thousand acres of Maryland countryside a dozen miles from downtown Washington, D.C. The area was protected on the north by an additional five thousand acres procured for a Beltsville National Agricultural Research Center, to be built as a model suburban community. Around the developed 225-acre Greenbelt commercial district was a "permanent protective belt of fields and forests." Here was Roosevelt's ideal "garden city" sprung to life, a planned community where nature received care as surely as the townspeople did.[35]

Another spirited CCC outfit that wrote to FDR about its successes and high jinks was Camp Cabeza de Vaca (Company 843) in Magdalena, New Mexico.[36] Perched high in the Magdalena Mountains, the camp was named after the Spanish explorer who led a group of the first Europeans to see the Mississippi River. Cabeza de Vaca had written exaggerated tales of his travels in the fantastical New World, about the resplendent (and fictional) "Seven Cities of Gold." Now in 1933, under the humorous leadership of Edward Smith, an aspiring writer, Camp Cabeza de Vaca started publishing its own newspaper—the *Woodpecker*—filled with satirical squibs.[37] The CCC enrollees all took a facetious oath, printed in the *Woodpecker* that summer, to worship Roosevelt as if he were a biblical patriarch:

Roosevelt is my shepherd, I shall not want;
He maketh me lie down on straw mattresses, he leadeth me inside a mess hall, he restoreth to me a job.
He leadeth me in the paths of reforestation for his country's sake.
Yea, though I walk through the valley of the shadow of poison oak and ivy,
I will fear no evil, for he is for me.
His blankets and uniforms, they comfort me.
He prepareth a saw and ax before me in the presence of my

commanding officer, my shoes runneth over.
Surely beans and employment will follow me all the days of Roosevelt's administration and I shall dwell in a tent forever.[38]

III

During FDR's "second hundred days," with the CCC thriving and banking reforms in place, the New Deal turned to wildlife protection. Shortly after the 1932 election, Roosevelt had arranged a duck hunt in Dutchess County. Struggling with his heavy leg braces, he had hidden himself in a Hudson River blind to wait for waterfowl to arrive in the cold sunshine. The blind was built of local vegetation and blended well with the placid marshland so as not to frighten waterfowl. Decoys were placed around the structure by the small hunting party to lure a flock within the range of FDR's shotgun. The president-elect was more used to fishing—in saltwater, following the menhaden toward trophy fish—but now he listened for the discordant, ethereal sound of ducks that had inspired hunters for millennia. Instead, five Canada geese (*Branta canadensis*)—black-beaked, with brown bodies and long necks—headed directly toward the blind. He took aim and fired. "I hit the leading goose, swung left to try to get another with my left barrel," he wrote to Senator Harry Hawes of Missouri, "and at that moment the first goose hit me a glancing blow on my right shoulder."[39]

Hawes had served in Congress since 1926, but chose not to seek reelection in 1932 and retired from the U.S. Senate early in 1933 to become a lobbyist for wildlife protection. Enthusiastic about Roosevelt's conservation reforms, Hawes, author of *Fish and Game: Now or Never*, goaded Congress throughout the 1930s into prioritizing the restoration of essential wildfowl nesting areas in the North, and vast marsh areas along the age-old flyways and wintering resorts in the South. Unencumbered by the pressures of reelection, he offered Roosevelt counsel on ways to rescue North American wildlife from the vicious ravages of the Dust Bowl and from overhunting. For the first time since Theodore Roosevelt, two powerful politicians—FDR and Harry Hawes—had adopted nonvoting North American wildlife as a constituency.[40] "To fight and conserve our big outdoors and its wildlife," Hawes wrote, "is a patriotic duty. Increasing its area is an achievement for health and better citizenship."[41]

To secure $14.5 million in emergency congressional appropriations so that the Biological Survey could purchase waterfowl habitat throughout America, Roosevelt and Hawes sought the active support of sportsmen's clubs of the same vintage as the Izaak Walton League; their members tended to be Republican. FDR and Hawes also looked to the American Legion, the Boy Scouts and Girl Scouts, the 4-H Clubs, land-grant colleges, bait-casting clubs, railroad companies, and even automobile clubs. Believing that bird-watching could bring considerable tourist dollars to the Atlantic Flyway states, Roosevelt ordered all federal refuges in the East to begin ecological restoration work to attract avians over the summer of 1933.[42] This wasn't easy to do. During the Hoover years, Congress had slashed the Biological Survey's budget to bare bones. Refusing to be stymied, Roosevelt, like a street hustler playing three-card monte, transferred top-salaried Biological Survey employees to the CCC's emergency financial sheet. Blaming human encroachment for at least part of the precipitous decline of North American waterfowl populations, Roosevelt proposed that the preservation of grasslands and wetlands offered many environmental benefits, including aquifer recharge, flood control, fisheries, water quality protection, soil maintenance, and carbon storage.[43]

In 1903, Theodore Roosevelt, who was then the president, established America's first federal bird reservation in Florida.[44] At that time, there were 120 million waterfowl in North America. By 1933, when FDR was in the White House, that number had shrunk to 30 million. The legal precedent FDR invoked in 1933 was the Migratory Bird Conservation Act of 1929, authorizing the federal government to establish inviolate sanctuaries for waterfowl and other migratory birds. This national system was in its embryonic stages when the crash of 1929 froze funding for the habitat purchase program. Appropriations shriveled. Understandably, Hoover was forced to put American citizens ahead of North American wildlife, and the waterfowl crisis kept getting worse. Owing to the drought, great duck hatcheries—like Lake Malheur and Klamath Basin of Oregon—were, as one biologist put it, "dry as a bone." The burden was on the New Dealers to halt this trend.[45]

Roosevelt downplayed the launch of his visionary waterfowl restoration program in 1933. After all, Congress wouldn't easily direct funds to help Chesapeake Bay's threatened wood ducks (*Aix sponsa*) when 90 percent of Virginia's schoolchildren were ill clad and malnourished.[46] Nevertheless, throughout the year Roosevelt badgered Biological Survey director Paul Redington, an unimaginative holdover from the Hoover administration, to order emergency measures for the preservation of North America's wildfowl. The president had even established CCC camps at three key Biological Survey refuges along the Atlantic Flyway—Saint Marks (Florida), Swanquarter (North Carolina), and Blackwater (Maryland)—to start his waterfowl rescue plan by means of feed stations.[47]

The Biological Survey hired university-trained biologists to collaborate with the CCC to band birds, build freshwater impoundments, cut trails, and erect fire towers.[48] Saint Marks, the most celebrated of the New Deal congeries, teemed with glorious waterfowl, including red-throated loons (*Gavia stellata*), yellow-crowned night herons (*Nyctanassa violacea*), and the anhinga (or "snakebird," *Anhinga anhinga*). Blackwater, known as the "Everglades of the North," hosted upwards of thirty-five thousand geese and fifteen thousand ducks, which used the brackish Maryland marsh and forested wetlands during the winter migration.[49]

Roosevelt shrewdly put wildlife restoration under the banner of retiring submarginal land and relieving drought, thus procuring for wildlife a modest portion of the federal moneys that were being spent to alleviate human distress. Under Roosevelt's "New Deal for Wildlife" program, the iron gates at these federal wildlife refuges were unlocked to encourage public and educational recreational use. Bird-watching was also encouraged. Informational leaflets (including lists of the species that lived within refuge boundaries) were disseminated free to the public. At Swanquarter, a stopover on the Atlantic Flyway, hunting guides were replaced with "guide yourself" trail maps. Federal manuals were printed on how to stabilize the level of shallow, freshwater areas in which to foster the growth of aquatic vegetation that would furnish food for birds. "It is difficult to overrate the economic importance of ducks," one USDA pamphlet read, "and undoubtedly their esthetic and recreational worth is fully as great."[50]

Roosevelt's wildlife restoration program had for its first-term objectives the redemption of breeding stocks (remnants of the great pre-Columbian flocks and herds) and the federal rehabilitation and restoration of available lands that were suitable for the proliferation of wildlife. Roosevelt wanted to withdraw these lands generally classified as submarginal; areas that, under cultivation, had never produced crops comparable in value to the fish, game, furbearers, songbirds, and insectivorous birds that inhabited them before they were wholly or partially destroyed by rashly attempted agricultural operations and drainage projects.

On these lands, Roosevelt wanted to erect green, living, organic sanctuaries to protect the fertile areas against soil erosion and against damage by drought and flood. Under a dozen or more New Deal land administration agencies, CCC boys worked to help wildlife rebound in all forty-eight states and the territories. The CCCers built fish hatcheries and wildlife shelters; seeded and planted food and cover; developed springs, lakes, ponds, and streams; and carried on a variety of other wildlife protection activities. Roosevelt had the CCC cooperate in a quantitative way with the U.S. Bureau of Fisheries, in developing, among others, York Pond fish hatchery in the White Mountains National Forest, which generated nearly twenty million eggs annually.[51] With native wild trout disappearing, Roosevelt hoped that nonnative hatchery fish, reared at state and federal fisheries, could make the difference.[52]

Encouraged by the CCC's revamping of Saint Marks, Swanquarter, and Blackwater, Thomas H. Beck, pro–New Deal editor in chief of *Collier's Weekly*, wrote to his friend the president that August about the chronic depletion of North American waterfowl. Beck was the head of the More Game Birds in America Foundation (a respected nonprofit founded in 1930 with financial backing from financier J. P. Morgan; it was later renamed Ducks Unlimited).[53]

A Connecticut duck hunter, Beck now had devised a waterfowl

rescue scheme for the president's consideration. Among his recommendations were an international agency to help waterfowl prosper; wild ducks and geese breeding facilities (for eventual release); raising private-sector revenue from sportsmen's clubs; restoring traditional breeding grounds in the upper Midwest; establishing a string of new federal refuges along the main flyways; and purchasing habitat for refuges in such winter migratory states as Georgia, Florida, Louisiana, and Texas.[54] Roosevelt eventually incorporated all of Beck's reforms into the Biological Survey's overall migratory bird recovery strategy, except the idea of breeding wildfowl in captivity.

When FDR was a boy, there was an almost unbroken series of Great Plains nesting areas for ducks stretching from the sand-hill lakes of Nebraska north to the Canadian line and as far west as Montana. It was prosperous-looking country. The upper Midwest and the Great Lakes region had once been rich with flocks of ducks, but no longer. The president hoped that Beck, along with Hawes, could become USDA's conduit to influential sportsmen's clubs to bring the great flocks back to their historical size. "I think it is very important to keep the good will of the fish and game clubs and associations," Roosevelt wrote to Wallace from Hyde Park as a cover note to Beck's memo, "and the chief point is the necessity of giving them a chance to be heard before promulgating orders changing the dates of open seasons."[55]

Wallace was irritated at Beck for sending his restoration plan directly to the White House instead of going through proper USDA bureaucratic channels. For all of his New Deal progressivism, Wallace was thoroughly old school when it came to respecting the chain of command. Furthermore, with unemployment at 25 percent, tougher restrictions on open seasons and bag limits seemed premature and elitist to Wallace. His concerns were immediate and practical—he ordered six million pigs slaughtered that September to stabilize and regulate prices—and he thought that waterfowl habitat purchases and breeding factories should have a low priority.

Nevertheless, Wallace dutifully passed Beck's letter to Rexford G. Tugwell, his assistant secretary. Tugwell, an agricultural economist who had taught at Columbia University, murmured that Beck was a well-meaning eccentric, naive about budgets and national priorities. There simply weren't any USDA funds available to create dozens of new federal refuges in places like South Carolina, North Dakota, and Texas. While sympathetic to the president's enthusiasm for wildlife conservation, Tugwell sided with Wallace and didn't want to be at loggerheads with the farm bloc in Congress over wood ducks and trumpeter swans. Tugwell, in fact, reported back to Roosevelt that Beck's recommendations weren't feasible from a budgetary standpoint. Quite simply, struggling farmers had to be supported first, far ahead of wild ducks.[56]

Tugwell, prone to overanalyzing situations, always needed empirical answers. He saw Roosevelt as an enigma, a juggler with "an extraordinary ability to leave things in flux and to prevent their taking concrete and final shape before the time is ripe."[57] He was perplexed as to whether Roosevelt really thought waterfowl habitat, state parks, and RDAs should have a high priority in a nation crippled by an economic collapse, or whether the president was merely pandering to old friends and conservationists. In truth, Roosevelt, unlike Tugwell, didn't consider Beck a peripheral player. His *Collier's Weekly* was a truly influential magazine. And Roosevelt, the tree farmer and stamp collector, wasn't one to hold a passionate hobby like waterfowling against an entrepreneur who fought for New Deal progressive principles. As it turned out, the president shared Beck's fundamental view that waterfowl, by way of the government's habitat purchase, could be instrumental in stabilizing water levels for the farmers in all non-desert regions, especially the West.[58]

IV

Another waterfowl area that President Roosevelt was deeply committed to saving in the early New Deal was the Okefenokee Swamp of Georgia. Dr. Francis Harper and Jean Harper had lost no time in plotting to inspire their friend from Hyde Park to help them rescue the Okefenokee. Throughout the Okefenokee for two generations, there had been the morning shouts of the timber bosses, "Daylight in the swamp!" signaling it was time to cut millions of feet of virgin cypress, yellow pine, bay, and gum.[59] Generations of loggers had hewed what was called the "Big Clearing" of Georgia. Now with "dear Franklin" in the White House, the Okefenokee had some hope of being rescued permanently. "There is a matter that needs your immediate attention—the preservation of the Okefenokee Swamp," Jean Harper wrote to Roosevelt in late 1933. "For twenty odd years the naturalists and nature lovers have been working for the preservation of this marvelous wilderness; unique in its nature not only in the country but the world. The character of its fauna, its flora, and its human life is unsurpassed."[60] A sympathetic Roosevelt responded gracefully to Jean Harper: "I too would hate to see the Okefenokee destroyed."[61]

Georgia was Roosevelt's demonstration plot in the American South. If Georgia could be saved from ecological ruin, he believed, so, too, could Alabama, Mississippi, South Carolina, and the rest. As president, Roosevelt made sixteen trips to his "Little White House" in Warm Springs in the rust-clay hills of Georgia, where he practiced scientific forestry, raised cattle, and helped create a meaningful state park system.[62] From his study at the Little White House, Roosevelt designed trails and cottages around Georgia's 1,293-foot Pine Mountain State Park to be built by the local CCC. Georgia's state parks grew from five hundred acres to five thousand acres by 1935. Park revenues increased by over 300 percent in those years.[63] And the longleaf pine ecosystems started slowly regenerating under his New Deal leadership.

Running the nearby CCC's District F out of Fort Screven, Georgia, was General George C. Marshall of Virginia. Roosevelt marveled at how, under Marshall's leadership, the CCC boys brought water systems to Georgia towns such as Hinesville and Homerville. Marshall, FDR came to believe, was a born leader, a genius logistician and surveyor, and the best U.S. Army general of his generation.[64] Marshall was vocal about the "inestimable benefit" of the CCC to the armed forces' "noncombatant needs." Having city dwellers work hard at home in the great American outdoors, dressed in army garb, waking at dawn, learning all about their homeland—this was public service at its finest. Not only did CCC enrollees plant trees and help wildlife prosper, but they also cleared land for rifle ranges and improved the grounds of military bases. While recruits didn't receive actual military training, they did learn the intrinsic value of good citizenship.[65] Always sympathetic to the men, Marshall saw to it that the army provided a part-time doctor and dentist for every CCC camp.[66]

That many southern Democrats thought of FDR as one of their own was no trivial feat. Always calling Georgia his "other" state or referring to himself as an "adopted" son of Dixie, Roosevelt assumed the posture of a true native. The way Georgians had welcomed him, arms wide open, in 1924 had left an indelible impression on him. Two years after arriving in Warm Springs to receive treatment for his polio, Roosevelt had invested two thirds of his fortune in rehabilitating the run-down spa town.[67] Cruising around Meriwether County in his specially built roadster, top down, waving to locals, Roosevelt was in his element. In a speech in 1927 at the Biltmore Hotel in Atlanta, Roosevelt touted Warm Springs as his "garden spot" where the finest food in the land—Brunswick stew, County Captain (a curried chicken dish), fried chicken, corn bread, hush puppies, turnip greens, black-eyed peas, and crackling bread—was served in generous portions. So at home was FDR in Warm Springs, away from the imperious Sara and the serious-minded Eleanor, that he claimed all Georgians as his "kinfolk."[68]

All Roosevelt could do in 1933 to repay rural Georgians for their kindness was to wean them from cotton as a single crop. By planting loblolly pine and trying his hand at raising cattle in Warm Springs, he hoped to convince Meriwether County farmers that exhausted soil was a curse. Wherever Roosevelt traveled, from Okefenokee Swamp to Pine Mountain to Flint River basin, he bragged about Georgia's amazing national heritage. "Some day you must see that spot," Roosevelt told his distant cousin Daisy Suckley, about Meriwether County. "You would like the great pines and red earth."[69] Enraptured by Warm Springs, Roosevelt made a tradition of spending Thanksgiving there as president. "Now let me tell you something cheerful," he wrote to a friend from the Little White House. "This Southland has a smile on its face. Ten-cent cotton has stopped foreclosures, saved banks, and started people definitively on the upgrade.

That means all the way from Virginia to Texas. Sears-Roebuck sales in Georgia are 110 percent above 1932. . . . I am having a grand rest and am catching up on much needed sleep."[70]

By the mid-1930s, there were fifty-eight CCC camps operating in Georgia under the authority of the U.S. Forest Service. Roosevelt inspected the two camps tasked with rehabilitating central Georgia.[71] Exhausted national forests in the region—Cherokee (Tennessee), the Nantahala (North Carolina), and the Chattahoochee (Georgia)—were overrun with CCC boys climbing trees to snatch pinecones to provide the Forest Service with seeds for new forests in other areas. Using tarpaulins they gathered nuts from hardwoods, built staircases to freshwater springs, cut trails, and erected dozens of forest towers. With the devastating Okefenokee wildfires of 1932 on everybody's mind, Roosevelt swore to Georgians, "Never again!" Fire towers were erected all over the swamp.

The CCC was responsible for three other outstanding national forests in Georgia (with miles of trails); two national battlefields (Kennesaw and Chickamauga-Chattanooga); Ocmulgee National Monument; Fort Pulaski National Monument; and sections of Okefenokee Swamp.[72] "If the C.C.C. does nothing more than impress upon us the love for nature, it will be a success," one ecologist was quoted as saying in the 1935 booklet *We Can Take It*. "When we better realize and understand nature, the world will be a better place to live in, and war will be but a dream."[73]

FDR credited Warm Springs with redoubling his commitment to one of his most far-reaching ideas. Continuing to balance the need for beneficial infrastructure with nature preservation, he habitually guaranteed that every hydropower project his administration launched in the South had a corresponding "conservation gift." Working with engineer Morris L. Cooke during his first term, FDR created the Rural Electrification Administration (REA) to provide farmers with cheaper power.[74] The idea for the REA came to him unexpectedly one afternoon, he sometimes said, while he was daydreaming on the porch in pine-scented Warm Springs. At other times, Roosevelt claimed to have thought of it when he saw his first electric bill for the Little White House there. In truth, he had already been a proponent of rural electrification during his time as governor of New York; nonetheless, his time in Georgia no doubt kept him mindful of the problem.

Even though Roosevelt led the campaign to protect Georgia's forests, much of what he accomplished in the name of applied science in the Deep South was, in hindsight, environmental folly. Ecosystems could survive drought, lightning bolts, bug infestations, hurricanes, disease, earthquakes, and fluctuating water levels, but not the hyperindustrialization offered by the Corps of Engineers. In an understandable effort to start a mechanized agricultural revolution, fertile southern lands were segregated from the Mississippi River by New Deal land improvers. Many southerners, untouched by the ecological thinking of George Perkins Marsh and John Muir, contaminated their beloved land, denuding landscapes and poisoning waterways.

Not that Roosevelt was wrong to hope that his crop diversification and his waterpower projects would lift the southern poor out of their economic straits. Like the needy tenant farmers portrayed in the photographic essay *Let Us Now Praise Famous Men* (1941), the "swampers" whom the Harpers were chronicling for an American folklore project were the kind of southern farmers Roosevelt cared deeply about. Trying to assist rural folks who were struggling, the USDA exterminated hordes of cotton lice, worms, and boll weevils by spraying pesticides throughout the South. The unintended result was the contamination of such rivers as the Tennessee, the Cumberland, the Sunflower, and the Atchafalaya. This, in turn, compromised the health of Americans residing in nearby farm communities. The waterways of the Deep South—which had already been degraded by fertilizers, industrial sludge, dredging, chemicals, and sewage—were, ecologically speaking, made worse by the New Deal.

The Harpers were relentless in their efforts on behalf of the Okefenokee. There was, however, a powerful lobby of Georgians anxious to restart construction on the Suwannee Canal, which had been abandoned in the 1920s; this new waterway would use existing rivers and a channel through the Okefenokee to give ships an east-west route from the Atlantic Ocean to the Gulf of Mexico. It would require draining significant parts of the Okefenokee. By the 1930s, new land-moving vehicles had been invented that could cut through the swamp with relative ease. Roosevelt, feeling pressured by the existence of technology that could finish the canal, and aware that doing so would lead to a slew of jobs for unemployed Georgians, ordered a preliminary survey of the Okefenokee.

Other obstacles loomed in the path toward preserving it. The White House learned that private sector money had been raised to complete the Suwannee Canal through the swamp. If the trans-Florida shipping canal were to be operable, it would scar the swamp ecosystem beyond recognition. No one knew which option the president would support: the canal, with its economic benefits (not just short-term jobs, but also long-term ones, resulting from efficient transportation) or the federal wildlife sanctuary that the Harpers were pushing for. "You well know what this [canal] would mean to the beauty of the area and to the wildlife," Jean Harper reminded FDR near the end of 1933. "The destruction that would thus be brought on is unthinkable. Our hope lies in you, to stop the project before it goes farther, and spend the money in the purchase of the swamp for the reservation, where beauty and scientific interest may be preserved for all time. The Okefinokee is regarded by naturalists all over the country as one of the very finest of all our natural areas, and I sincerely hope you will not bring disappointment and bitterness to them by permitting its destruction."[75] Roosevelt responded to Harper, "I think there is much more chance of a ship canal going the southern way than through Georgia. I hope all goes well with you and the family."[76]

A few months later, Francis Harper pressed his case in an article, "The Okefinokee Wilderness," in *National Geographic*. Harper warned that, owing to drainage projects throughout the Deep South, the Okefenokee's ivory-billed woodpecker would join the passenger pigeon as an extinct species.[77] These woodpeckers needed large stretches of hardwood forests to survive, but about 90 percent of the swamp's 400,000 acres had already been logged.[78] While the Biological Survey—with the help of the More Game Birds in America Foundation—had initiated efforts to increase the waterfowl populations of North America, there was no similar conservation effort being made in 1933 and 1934 for owls, hawks, eagles, or woodpeckers.

In a victory for the preservationists, the White House soon publicly abandoned the plan for the Sewannee ship canal. The only remaining question was what kind of site the Okefenokee would become. Logging companies were still encroaching on the heart of the swamp. Quick executive action was needed to save it, but Roosevelt stalled. An impatient Jean Harper prodded him relentlessly: "Every day's delay means more lumbering carried on and more of the swamp lost forever in its primeval state."[79]

Jean Harper's intensified appeal helped spur Roosevelt to act. The president told his staff that he wanted the Okefenokee saved as either a national monument (which would fall under the Department of the Interior) or a national wildlife refuge (under the Department of Agriculture). To ruin the swamp would be the equivalent of clear-cutting the trees in Sequoia National Park for railroad ties. Roosevelt needed a few senators—including both of Georgia's, John S. Cohen and Richard Russell Jr.—to give his plan for an Okefenokee park southern political cover. Resistance from timber and shipping interests didn't worry Roosevelt, but he always preferred to have the local politicians on his side when saving wild places. Ultimately, Roosevelt had no appetite for a protracted fight with a hostile Congress over the Okefenokee. If push came to shove, he would apply the Antiquities Act of 1906; he told Jean Harper that he was "entirely willing to have it made a national monument."[80]

Because Roosevelt was getting razzed by farmers for his penchant for wildlife preserves like the Okefenokee, he decided, as a spoof, to establish the mythical Marvin McIntyre Memorial Possum Reserve. According to Roosevelt, the "magnificent Reserve," named after his appointments secretary, would "do so much to prevent the extinction of that glorious symbol of our freedom—the American possum."[81]

One New Deal project Roosevelt certainly wasn't joking about was the huge new headquarters of the Department of the Interior in Washington, D.C. In late 1933, Roosevelt, influenced by Ickes, approved what became the New Deal's signature architectural statement there. The old Interior Building between E and F streets was a shambles. Space for employees was at such a premium that Interior was leasing offices in fifteen different buildings. From the start, Roosevelt, Delano, and Ickes agreed that classical Greek columns weren't architecturally appropriate for the building that would house such government departments as the National Park Service, Bureau of Reclamation, and Bureau of Indian Affairs. At Delano's suggestion the architect Waddy Butler Wood, a well-known Washington society figure, received the commission with a mandate of "utility and economy" of style. In coming years regionalist painter John Steuart Curry installed a series of commissioned murals for the new building, including *The Homestead* (honoring the Homestead Acts), and *The Oklahoma Land Rush* (1889).

A new, two-block Interior headquarters was soon under construction southwest of the White House, extending from C to E streets. It boasted three miles of hallways, central air-conditioning, a basketball court, escalators, a radio station, and even an ice cream shop. The *Washington Daily News* noted that the real "father" of the building was Harold Ickes. Every free moment Ickes had, from late 1933 until the day when the edifice officially opened in 1936, was dedicated to making the new Interior building a monument for the ages.[82]

CHAPTER TEN

"THE YEAR OF THE NATIONAL PARK"

I

Looking out of train and automobile windows, seeing one rural hamlet after another, smelling the air scented by agriculture and nature was President Roosevelt's idea of outdoor recreation. A keen observer of nuance, he picked up on the subtle regional differences between grain silos and wooden barns, pine trees, and wildflowers. "There was no more excellent custom," Roosevelt would say, "than getting acquainted with the United States."[1]

Hoping to encourage automobile tourism in America's natural wonderlands and antiquities sites, FDR declared 1934 the Year of the National Park.[2] That March, as part of a public relations effort, he hosted a Grand Canyon National Park slide show at the White House.[3] One hundred guests oohed and aahed over pictures of what John Burroughs once memorably called the "divine abyss" of northern Arizona.[4] As a forest conservationist, Roosevelt was concerned that all around the Grand Canyon ponderosa pine were being logged, causing erosion, threatening wildlife diversity, and marring the tourist experience.

At the White House dinner, the journalist Robert Sterling Yard, who was a cofounder of the National Parks Association, spoke encouragingly about a new national park that FDR would soon be opening in the Great Smoky Mountains (North Carolina and Tennessee).[5] Roosevelt considered his longtime friend the most effective publicist for protecting America's parklands.[6] What perplexed Roosevelt was that Yard, a purist with regard to preservation, was opposed to the proposed Everglades National Park. Yard argued that there were already man-made structures in the swamp. He was also against the enlargement of Grand Teton National Park, because the town of Jackson Hole had buildings.[7]

Roosevelt, breaking with Yard, envisioned Everglades National Park as 2,500 square miles of sawgrass and shallow freshwater. Twice the size of Rhode Island, it would encompass much of Dade, Monroe, and Collier counties. The idea of establishing the first national park preserve to protect plants and wildlife rather than gorgeous scenery was garnering considerable support on Capitol Hill. Roosevelt was ahead of his own Department of the Interior in this matter. "I understand that you have already expressed interest in the bill now pending before Congress for the creation of a national park in the Everglades of Florida," Ickes wrote to Roosevelt a few days after the slide show. "If you would be willing to make Speaker [Henry] Rainey and other leaders in both the House of Representatives and the Senate aware of your interest, it would be very much appreciated by the advocates of the bill, which include this Department."[8]

The efforts by the Roosevelt administration to preserve the Florida Everglades were largely influenced by FDR's "map mind"—his knowledge of the subtropical geography of southeast Florida. Any population increase in Miami, he knew, would inevitably put pressure on the Everglades, which had already been compromised by the encroachment of real estate developers.[9] That spring, Roosevelt alerted congressional Democrats that he supported a bill introduced by Senator Duncan Fletcher of Florida to establish Everglades National Park and was sure it would be a huge tourist draw for south Florida.[10]

Fletcher's bill, however, was a tough sell on Capitol Hill. Many southern Floridians saw the Everglades, with its vast timber and wildlife resources, as a fetid swamp in need of drainage—not a great marsh that was, as Marjory Stoneman Douglas later wrote, south Florida's "rain machine," which created mangrove-lined estuaries.[11] It wasn't just Miami-area real estate developers and Robert Sterling Yard who raised objections to what the Roosevelt administration proposed. The Izaak Walton League, surprisingly, wanted the Everglades acreage proposed by the National Park Service to be cut by two thirds; "Ikes," as the league's members were called, were unhappy with the "no hunting" provision that would come with national park status. And, regardless of the Indian New Deal, many

Seminoles were understandably suspicious about having traditional homelands taken by the untrustworthy federal government.

Around this time, Roosevelt consulted Ernest Coe about getting affluent Floridians involved in protecting the Everglades from commercial and agricultural development. Coe, a Yale-educated landscape architect, had moved to Coconut Grove in 1927, and designed green spaces for hotels and private homes around Miami, but he spent his weekends happily exploring the swamp's "great empire of solitude" and was a cofounder of the Tropical Everglades National Park Association.[12] Traveling from Pensacola to Miami Beach, Coe lobbied for the upper section of Key Largo and part of the Big Cypress to be included within the national park. And he called for a federal ban on any dumping of fertilizers, waste, and chemicals in the Everglades. "It is my hope," Roosevelt told Coe, "that the State of Florida will take the necessary steps to make the State-owned lands within the proposed park area available for the project at an early date."[13]

On March 27, six days after the White House slide show, the president traveled to south Florida for a fishing holiday that would also serve as a promotional tour of the Everglades, Biscayne Bay, the Florida Keys, and the Dry Tortugas.[14] For security reasons, the islets that Roosevelt planned to explore, and his other stopovers, weren't made public. Very little advance information about the expedition was given to the press. Makeshift executive offices were established in Miami's Biltmore Hotel under the expert direction of FDR's traveling secretary Marvin McIntyre.[15] Living aboard Vincent Astor's three-thousand-ton yacht *Nourmahal*—and escorted by the naval destroyer *Ellis* for twelve days—Roosevelt gleefully fished for marlin and barracuda and signed major bills including a widening of the Agricultural Adjustment Act.[16] One of FDR's companions on the trip was Kermit Roosevelt, the most conservation-minded of TR's children, who helped the president discover and collect unusual species of fish and crustaceans. Two of FDR's sons—twenty-six-year-old James and twenty-three-year-old Elliott—became crewmates for the Bahamian segment of the cruise.[17] The president's guests in Florida included Joseph Kennedy, the chairman of the newly created Securities and Exchange Commission; and George Hearst, son of press mogul William Randolph Hearst.[18]

Roosevelt, seated, posed with a 190-pound sailfish that he reeled in while on a fishing boat in the Caribbean. The fish put up a fight. According to the log, "the boat was towed about five miles in heavy seas during the strugle [*sic*]." The Smithsonian Institution accepted the fish, preserved it taxidermically, and displayed it as the largest of its species in the collection of the National Museum of Natural History.

Charles W. Hurd of the *New York Times* reported that in addition to fishing for trophies, Roosevelt was collecting marine specimens for the huge saltwater fish tanks on the *Nourmahal*, including queen triggerfish, blue parrotfish, sandfish, grunts, jacks, snails, and numerous turtles. Just before the cruise ended, the president allowed himself to be photographed in his stained fishing clothes, holding Astor's pet dachshund instead of a prized marlin.[19] When Roosevelt returned to Washington, over two hundred congressmen and thirty senators greeted his train as a band played "Happy Days Are Here Again." The *New York Times* noted that such a welcome was "unprecedented" in American history.[20]

In mid-April, the president invited the landscape architect Frederick Law Olmsted Jr.—a fierce advocate for national parks—to a private meeting in the Oval Office. Olmsted had been dispatched to the Everglades by the National Park Association in 1932 to evaluate an ecosystem so unique that it was the only place in the world that had both alligators and crocodiles. The president read Olmsted's comprehensive report, which offered stories about sixteen species of wading birds in the Everglades. "After dusk, flock after flock came in from their feeding grounds . . . and settled in the thickets close at hand," Olmsted wrote. "It was an unforgettable sight . . . [that] ranks high among the natural spectacles of America and can be perpetuated most effectively by the creation of a National Park in the region."[21]

On May 30, 1934, just six weeks after his meeting with Olmsted, the president happily signed the act authorizing the Everglades National Park, America's first "tropical" park.[22] The law stipulated that the Everglades would be "permanently reserved as a wilderness" and barred any development for the benefit of visitors that would disturb the "unique flora and fauna" and the "essential primitive natural conditions" in the area.[23] Although the parakeets and the ivory-billed woodpeckers had long since been exterminated, the habitat of the white ibis (*Eudocimus albus*), roseate spoonbill (*Platalea ajaja*), and Key deer (*Odocoileus virginianus clavium*) were to be saved in the nick of time. "It was," environmentalist Marjory Stoneman Douglas wrote, "like the small beginning of a new hope."[24]

There was, however, a sticking point in the Everglades deal. The state of Florida would have to donate the land for the Everglades to become a national park—the federal government could not use taxpayer dollars to buy acreage. A contingent of Florida legislators, worried about lost revenue, immediately derided the Everglades law as a federal boondoggle, Roosevelt's "gator nursery." But the Florida Federation of Women's Clubs quickly donated four thousand acres to start the national park. It seemed at times as though the Everglades National Park was snakebitten. The project was slowed down by wildfires in the swamp, by land speculators, and by oil prospectors who believed that there was "black gold" in the Everglades. "I am about to die," park advocate William Sherman Jennings moaned, "waiting until this thing is ready."[25] In the long run, it took the Roosevelt and Truman administrations thirteen years of lawsuits and injunctions, grandfather clauses, and buyouts to procure the required land and determine the final boundaries of the Everglades park.

When Everglades National Park was initially established in 1934, forty-nine thousand CCCers, based in more than twenty camps, were already working throughout Florida.[26] These CCC "boys" had a heavy burden. The soil was naturally sandy; its nutrients were easily depleted by poor farming techniques. Because Florida was one of only four states with an unemployment rate of over 20 percent during the Great Depression, Governor David Sholtz, a Democrat, pleaded with the president to make it a laboratory for New Deal conservation efforts. Roosevelt happily obliged.

An unsung part of FDR's New Deal environmental legacy was his farsighted personal intervention in the formation of the Florida Park Service (FPS), modeled directly after the National Park Service. Assuming the role of planner, Roosevelt ordered the CCC to work in tandem with the FPS to create eleven state parks, including the popular Gold Head Branch, Hillsborough River, Myakka River, Florida Caverns, and Fort Clinch. Unemployed Floridians rushed to join the CCC—if only to gain access to the agency's steady supply of high-quality food. Just reading the menu for CCC Company 262 of Sebring, Florida—home of Highlands Hammock State Park—made the mouth water: breakfasts of eggs, oatmeal, fruit, bread and butter, and cereal; lunches of baked beans, potatoes, pork, beets, salads, sandwiches, sides of beef, and gravy; and dinners of roast chicken, beef stew, cabbage, sweet potatoes, peppers, beef liver, roast veal, mashed potatoes, spaghetti, fish, cauliflower, peaches, salad, and bread and butter. Courtesy of the New Deal, enrollees were being *paid* to eat such hearty fare in exchange for cutting hiking trails, planting trees, and creating botanical gardens.[27]

In an important conservation effort in north Florida, Company 4453C did wonders in helping to save the rare torreya tree (*Torreya taxifolia*). These evergreens were found near the town of Bristol on the bluffs overlooking the Apalachicola River. Colloquially called "stinking cedars" because of their acrid-smelling resin, they had short leaves, a dark green aril, and little yellowish branches; and they were a less endangered species once the CCC helped save the stands in Torreya State Park.[28]

Even though the CCC was a boys' club, with women excluded, one of the supervisors at the Highlands Hammock camp was Clara Thomas. Long a vocal environmental activist, Thomas rose to prominence in the corps thanks to her sheer talent and her enthusiasm for the newly created Florida Park Service. She was determined to save the wildlife around the town of Sebring. Under her guidance, water catchments were instituted for demonstration forests on the banks of Tiger Branch Creek and an herbarium was erected, along with a greenhouse and garden plots containing palms, conifers, and bamboo.[29] These were the bold years of conservation, and thousands of women, like Thomas, inspired by the New Deal and Eleanor Roosevelt, led campaigns to protest the destruction of American wilderness areas.

II

The Year of the National Park brought Roosevelt and Ickes closer together; they got along like old school chums—trading stamps, dabbling in horticulture, mocking Republicans, drinking whiskey, and plotting conservation strategy. To Roosevelt, the square-jawed "Honest Harold," was a profoundly loyal New Dealer.[30] They plotted to save the Everglades, Dry Tortugas, Sonoran Desert, and Cascades. "Roosevelt had a real grasp of Interior Department matters and policies from the very beginning," Ickes recalled. "We understood each other and, generally speaking, we were in accord on policies even though we did differ in some particulars when it came to the organizational setup."[31]

Friction never developed between these two progressives. Roosevelt tolerated Ickes's sudden "Donald Duck" tantrums and periodic resignations (which were rejected) because Ickes had a unique understanding of the intricacies of public lands policy and the ins and outs of the hydropower revolution.[32] Ickes was Roosevelt's environmental conscience, the cabinet member who regularly reminded Roosevelt that sometimes New Deal work-relief projects like Skyline Drive and the TVA had to make provisions sensitive to primitive wilderness areas. "I think we ought to keep as much wilderness area in this country as we can," Ickes said on a weekly NBC radio broadcast promoting national parks. "It is easy to destroy a wilderness; it can be done very quickly, but it takes nature a long time, even if we let nature alone, to restore for our children what we have ruthlessly destroyed."[33]

To commemorate 1934 as the Year of the National Park, Roosevelt and Ickes, both avid philatelists, collaborated on a series of stamps showing America's outdoors sites, including Mirror Lake at Mount Rainier, Great White Throne in Zion, and Two Medicine Lake in Glacier. The president offered a few aesthetic alterations. For the one-cent Yosemite stamp, Roosevelt suggested that the artist "try a flatter arch above Yosemite at the bottom of the stamp." When asked to approve the seven-cent Acadia stamp, he scrawled, "Okay, but put a ship on the horizon."[34]

Thousands of stamp hobbyists from all over the world sent the White House rare or obscure stamps. Under Roosevelt's leadership Postmaster General James A. Farley opened a new philatelic museum in Washington, D.C. In fact, Roosevelt's first two years in the White House, 1933 and 1934, are "fondly remembered by philatelists" as the "pastime's 'Golden Age.' " When FDR was elected president in 1932, stamp collectors in America had numbered two million; by 1938 there were more than nine million.[35] The president hoped his stamp issues would not only promote the national parks but also help youngsters learn about American history. Just five days after his inauguration, in fact, he had asked Farley to issue a Newburgh Peace Stamp. Roosevelt personally chose the die proofs, which, as he instructed, elegantly depicted the stone Dutch farmhouse that served as George Washington's headquarters at the end of the Revolutionary War, with an idyllic Hudson River Valley background.[36]

What most concerned Roosevelt about the NPS in 1934 was that New England was underrepresented. Therefore, he decided to move forward on a proposal from Colonel William J. Wilgus to create in Vermont a Green Mountains National Park, which would also serve to commemorate Revolutionary War hero Ethan Allen.[37] "The continuous reservation area from Maine to Georgia along the line of the

Appalachians is rapidly taking form," Roosevelt wrote to author and conservationist Robert Underwood Johnson. "You will be pleased to know that this year we are buying twenty million dollars worth of additional land—a large part of it being in the Eastern area."[38]

Roosevelt also wrote to Governor Stanley C. Wilson of Vermont about the project in the Green Mountains, suggesting that a parkway could be built in his state (which would provide desperately needed infrastructure jobs). In Johnson's conservationist circle, however, suggesting that the government build a parkway through the Vermont wilderness was a faux pas, and a rift opened up between FDR and the preservationists. When a coalition of Republicans and environmentally concerned liberal Vermonters rejected a proposed bond issue to purchase land for the parkway, the idea of a national park lost steam in the state, and with President Roosevelt as well.[39]

Regardless of the parkway, Roosevelt was bullish on the idea that Vermont could turn ski tourism into big dollars. Cross-country skiing wasn't yet a hugely popular sport in America, but the CCC cut trails for it nevertheless. Downhill skiing areas in the state were built by the CCC at East Corinth, Killington, Shrewsbury Peak, and Stowe. The CCC essentially transformed Vermont into the "ski capital of the East."[40]

Throughout 1934 Roosevelt built the case on Capitol Hill for a national landmarks program, promoting the conservation of the Chesapeake and Ohio Canal, which stretched for 185 miles from Washington, D.C., to Cumberland, Maryland. The canal had been proposed by George Washington in the 1780s, and the first spadeful of dirt had been turned in 1828 by John Quincy Adams. It operated until 1924, but then the Baltimore and Ohio Railroad made the canal "obsolete and profitless."[41] Frederic Delano, head of the National Capital Park and Planning Commission, thought—like FDR—that the lowland between the C&O Canal and the Potomac River should be a European-type towpath for recreation and hiking. At Uncle Frederic's instruction, Roosevelt told Ickes to apply $2 million from the Interior budget to get the C&O canal restoration project operational within five years.[42] He then dispatched the CCC to reconstruct the canal's locks and bridges; the CCCers eventually transformed it into a kind of Appalachian Trail for greater Washington to enjoy.[43] "For large numbers of people it would have the greatest all-around recreational value to be obtained in one unit," Frederic Delano said to the president about the canal, "providing ideal facilities for boating, canoeing, cycling, hiking, picnicking, and even swimming at certain points. It traverses a river valley of the greatest scenic interest and variety, and represents an age and advance in American civilization of permanent historic value."[44]

Roosevelt and Ickes also sought to establish a Quetico-Superior "wilderness sanctuary" in a fourteen-thousand-square-mile region of boreal and temperate forests straddling the Minnesota-Ontario border. In 1909, the Canadian government had set aside more than one million acres northwest of Lake Superior as Quetico Forest Preserve. Around the same time, the USDA established Superior National Forest in Minnesota. Now, FDR was willing to fund the Superior Roadless Primitive Area (which had been designated by the U.S. Forest Service in 1926 to eliminate the private development that had been allowed under old homesteading and mining laws). This part of northern Minnesota became a battleground for the movement to save wilderness areas from human encroachment.

Wilderness protectionist Ernest Oberholtzer of Minnesota led a noble grassroots movement against damming the Rainy Lake Watershed to produce electricity, lumber, and paper. His friendship with Ickes, who vacationed in the area, proved helpful to the protection cause. In June 1934, Roosevelt signed Executive Order 6783, establishing the Quetico-Superior Committee to study the Boundary Waters Canoe Area Wilderness.[45] This became the first step in creating Voyageurs National Park (which was eventually signed into law by President Gerald Ford in 1975). Before the area received national park designation, however, EO 6783 and the Quetico-Superior Committee banned new homesteading, prevented the alteration of water levels, barred the sale of lands, and prohibited logging within four hundred feet of Lake Superior's shore.[46]

While environmental historians haven't given Roosevelt enough credit for intervening favorably in the disputes over the C&O and Quetico-Superior, they do recognize his effort to add Isle Royale—a

207-square-mile island in Lake Superior—to the roster of national parks. On the enthusiastic recommendation of Stephen Mather, Presidents Calvin Coolidge and Herbert Hoover advanced the notion of Michigan's Isle Royale becoming a major midwestern national park. Tourists, it was thought, would come by boat to the wilderness to see the dense forests, crystal lakes, high cliffs, herds of rather tame moose, and four hundred woodland caribou. Swayed by Mather's boosterism, Hoover authorized Isle Royale National Park in 1931 "to conserve a prime example of North Woods Wilderness."[47] The Michigan legislature was responsible for acquiring the privately owned property on the island and then donating it to the federal government.

But in 1934, Isle Royale National Park was in limbo. The state of Michigan was broke and weary of the Depression, and over 50 percent of the island was still owned by copper and timber interests. There were also Michiganers and Minnesotans with second homes on the island who were in no mood to be booted out by the Department of the Interior. Determined to break the impasse, the president allocated $350,000 to buy parcels of land on the island. It bothered him that the National Park Service didn't have major Great Lakes property in its portfolio. Because Lake Superior was the largest freshwater body in the world by surface (five thousand square miles larger than Victoria Nyanza in Africa), Roosevelt thought it should have a major national park. The *New York Times* joined forces with the president in championing the "primeval charms" of Isle Royale, including the hundreds of rocky islets surrounding the main island. "Lonely Isle Royale, pushing its lovely shores up through the green cold waters of Lake Superior off the Michigan mainland here," the *New York Times* wrote, "may soon become the first national park of the Central 'Western States.' "[48]

On June 15, 1934, President Roosevelt brought the Great Smoky Mountains into the portfolio of the National Park Service on a "limited park status." What that meant was that money was still needed before Tennessee and North Carolina could finish the required land purchases. A succession of political squabbles involving the state commissions seriously delayed the completion of the park project.

But FDR had made it clear that the Great Smokies would indeed become a unit of the NPS very soon. To show his administration's investment in the idea of a national park in the Great Smokies, the CCC set up seventeen camps in the area and immediately began structural improvements, including the restoration of three gristmills and the building of two stone visitor centers (one at each state's main gateway to the park).[49] In 1933 and 1934 alone, 4,350 CCCers worked in the Smokies ecosystem.[50] The four-arch Elkmont Bridge they constructed remains one of the CCC's most aesthetically pleasing accomplishments.[51]

Roosevelt recognized that tourists wanted to see live animals in the national parks. This issue had caused a deep rift in the national park movement during the 1920s. Tourists came to see bear-feeding shows in Yosemite and Yellowstone, and Yard was among many who objected to this ("swell swill dumped from the platforms of trucks"). By contrast, Stephen Mather encouraged such attractions in order to boost the number of tourists. Either way, nobody doubted that the star of the Everglades would be the American alligator, and the star of the Great Smokies would be the black bear.[52] Roosevelt leaned more toward Yard than toward Mather in this debate.

The president was opposed to guns and hunting in national parks and monuments. When a disconsolate David Wagstaff of Tuxedo Park, New York, wrote the White House a grim letter about the overhunting of bears in Alaska—a direct result of lax hunting laws—Roosevelt was appalled. "The enclosed speaks for itself," Wagstaff wrote to Roosevelt; "I only hope that something can be done to limit the slaughter of these great bears so that they will not follow the pigeon and the bison through our own lack of interest."[53] Wagstaff lobbied the White House to establish a sanctuary for the bears on Admiralty Island. At that point, bears were considered predatory menaces and therefore weren't protected on public lands in Alaska. Roosevelt forwarded Wagstaff's worrisome letter to Harold Ickes, with a cover note: "This horrifies me as much as it does my friend David Wagstaff. If these bears come under your jurisdiction will you please have the matter checked up? It seems to me that that kind of slaughter ought to be stopped."[54]

Ickes, in addition to running the National Park Service and PWA, administered the General Land Office, the U.S. Geological Survey, and all of the lands in the territories including Alaska (he worked closely with Roosevelt to design a stamp series featuring the American territories). So Alaska—including the protection of bears, harbor seals, northern fur seals, sea otters, harbor porpoises, Dall's porpoises, and Steller's sea lions—was his beat. Therefore, with Roosevelt's blessing, Ickes sent a stern letter to the Alaska Game Commission about the mass slaughter of wildlife. In a stealthy move essentially unnoticed by the East Coast press, FDR had agreed to enlarge Katmai National Monument in southeastern Alaska—established by President Hoover in 1931—by more than 1,609,000 acres. Likewise, starting in 1934, FDR sought to double the size of Glacier Bay National Monument. In the Katmai and the Glacier Bay national monuments, the Department of the Interior worked to protect black and brown bears, known for hunting salmon along the Bartlett River and the Beardslee Islands. "Already there is in Alaska a total of over 6 million acres in parks and monuments under the jurisdiction of the National Park Service where bears are given total protection," Ickes reassured Roosevelt. "The Alaska Game Commission has afforded brown bears additional protection of late and has established several refuges for bears."[55]

Roosevelt was adamant that the New Deal teach Americans how to properly treat wildlife habitats. The CCC boys were taught to respect the habitat of the wildlife in areas where they worked. John B. Adams, working in Utah's Zion National Park, is an example of Roosevelt's idea of how the CCC boys could play a meaningful role as teachers. Raised by foster parents until he was fourteen, Adams then became a hungry hobo on the prowl in rail yards for odd jobs. When he turned nineteen, he abandoned the footloose life and joined CCC Camp PE-222. Outfitted in his jodhpurs, work boots, and CCC batwing pins, Adams became apprenticed to Zion's chief naturalist at its museum. Transfixed by the four life zones of Zion—desert, riparian area, woodland, and coniferous forest—he helped national park visitors properly interpret geographical features like the Checkerboard Mesa (an orange-, brown-, and white-textured mountain peak) and the Three Patriarchs (three peaks of Navajo sandstone).[56]

Believing reptiles should be undisturbed in their ecosystem, Adams studied how snakes and lizards ate, digested their food, and reproduced. Most tourists in Zion were understandably terrified by the rattlesnakes they occasionally encountered on hiking trails. Whenever a rattler was spotted Adams, the amateur herpetologist, rushed in to capture it for study and relocation. Before long, he was holding snake-handling demonstrations for tourists with two species of poisonous snake—the rattlesnake (*Crotalus cerastes*) and the ring-necked snake (*Diadophis punctatus*). "They have fangs, but their poison is about like a bee sting," Adams explained of the ring-necked snake. "Maybe a little worse, but very few people would die from it, no more than a person would from a bee sting. But after a while [I got] used to it, just like picking up a piece of candy or something."[57]

III

On July 2, 1934, President Roosevelt boarded USS *Houston* in Annapolis to begin a long journey to Puerto Rico, the Virgin Islands, and then the Hawaiian Territory to see the Kilauea volcano, on which he had written a childhood essay. The director of Kilauea Volcano Observatory, Dr. Thomas Jaggar, a friend of FDR's from Harvard, had been engaged in advance to serve as the president's guide around the majestic landscape, with its fields of lava.[58] His sons, Franklin Jr. and John, would accompany him on the sun-drenched trip. The U.S. Navy did a marvelous job of transforming the admiral's portside quarters on the *Houston* into a handsome presidential suite. Roosevelt, as was his wont, filled it with books about natural history and detective stories. The kitchen staff was brought over from the presidential yacht *Sequoia* to cook the president three square meals each day.[59]

Roosevelt's plan for the voyage was to make a fast run from Annapolis to Cape Hatteras, North Carolina, where he wanted to spend the Fourth of July with Franklin Jr. and John. Like the Bay of Fundy, the whitecap swells off Cape Hatteras were tricky for ships to navigate; this was a sailor's graveyard of sorts, steeped in nautical lore. The president logged, "The sea was choppy on rounding Hatteras and the destroyers took some punishment."[60] Taking in the cape's bird-filled marshes, Roosevelt understood that Hatteras—like the Everglades, Great Smokies, Isle Royale, the C&O Canal, and Quetico-Superior—was another precious American landscape that belonged in the NPS portfolio.[61]

From the Carolina coastline, the *Houston* cruised southward to Puerto Rico and the U.S. Virgin Islands before visiting Cartagena, Colombia; this was the first time an incumbent U.S. president had traveled to South America. The meeting with President Enrique Olaya Herrara on Colombian soil was an early example of FDR's "good neighbor" policy regarding Latin America.[62] Once the *Houston* was through the Panama Canal, in the waters off Clipperton Island, fishing was the main event.[63]

On July 24 the president got his first glimpse of the Hawaiian Territory on the western side of Hawaii (the Big Island). At the anchorage in Hilo, his cruiser was greeted by a huge crowd of admirers. There were flowers everywhere. A band played "The Star-Spangled Banner" for the arriving commander in chief. Schoolchildren, wearing bright clothes, waved American flags and chanted "F.D.R.," lining the thirty-mile road to the Kilauea volcano.[64] No president had ever visited the Hawaii Territory before, so FDR, leis around his neck, was feted like a hero come home.

Roosevelt, sitting in an open car climbing uphill through Hawaii's vivid scenery, was fulfilling his boyhood dream of seeing Hawaii Volcanoes National Park. (In 1961, Hawaii National Park was divided into two discrete units: Hawaii Volcanoes National Park and Haleakala National Park.) At the park's entrance Roosevelt rendezvoused with "America's volcano man," Dr. Jaggar. Jaggar pointed out Kilauea Iji (a small, dormant crater) to his old school chum and then took him to inspect a blazing green fern forest. Over two hundred CCC enrollees, in uniform, presented themselves to the president and talked to him about the forest improvement work under way there. They told funny tales of working to mitigate the damage wild pigs visited on the lush ecosystem.[65]

President Roosevelt shoveled dirt onto a newly planted tree during a visit to Hilo, Hawaii, on July 28, 1934. As the first sitting president to visit the Hawaiian Islands, he mixed military meetings at the Pacific outpost with a colorful four-day vacation in the company of two of his sons. Just moments before leaving, Roosevelt helped to plant the kukui tree next to the Iolani Palace.

Jaggar, explaining heat changes, lava flows, and the region's weather, led Roosevelt to the lip of Kilauea.[66] In a scripted event, a Hawaiian farmer appeared during Dr. Jaggar's discussion of the volcano's origins with a basket of chelo berries. According to Jaggar, it was an old Hawaiian custom for visitors to throw a few of the berries into the lava floor. Roosevelt relished such ceremonies and gleefully flung a few into the five-hundred-foot abyss. "Moving on again through the steaming crater countryside to the observatory high on a bluff above the pit," the *New York Times* reported, "the President listened to a lecture on bird life in the park and signed the observatory register."[67]

A highlight for Roosevelt was dining in Volcano House, a glass building with jade floors and a lava fireplace that had been built in 1846 on the rim of Kilauea Crater and was praised by such illustrious visitors as Mark Twain, Robert Louis Stevenson, and Louis Pasteur.[68] Another thrill for Roosevelt was fishing off the *Houston* with the giant volcanoes Mauna Loa and Mauna Kea in the distant background. Franklin Jr. hooked a mighty swordfish that, after a spectacular fight, slipped off the hook. Compensating for his son's misfortune, FDR reeled in three skipjack tuna and one kawakawa. When asked about the swordfish that got away, the president joked that it didn't matter—the fish had "spoiled" while jumping into the air.[69]

Once the *Houston* docked in Honolulu, on the island of Oahu, FDR dedicated Moana Park (now called Ala Moana Beach Park), financed by the Federal Emergency Relief Act. At an afternoon rally in Honolulu, more than sixty thousand people came to hear the president speak. It was a touching affair, with floral leis and a luau for which a pig was prepared in a traditional imu (an underground pit oven). Roosevelt also made a trip to Schofield Barracks and Pearl Harbor Naval Station, where he inspected the submarine base's giant machine shops, warehouses, and oil tanks filled with crude from Teapot Dome, Wyoming.[70] Basking in the sunny hospitality and wishing he could stay another week, the president thanked Hawaiians on his departure for "your flowers, your scenery, your hospitality. . . . Aloha from the bottom of my heart."[71]

IV

From Hawaii, the *Houston* cruised back to the continental United States, dropping anchor at Desdemona Sands, three miles from Astoria, Oregon. Sporting his Hawaiian tan, looking fit and happy, Roosevelt was still in a mood to fish. Using a smaller boat, he trolled for salmon along the foggy mouth of the Columbia River. Eleanor Roosevelt and their eldest son, James, both arrived in Astoria from California just in time for a Pacific Northwest salmon bake.[72]

While FDR was in Hawaii, the first lady had spent five days in Yosemite National Park, accompanied by her friend Lorena Hickok, a reporter who wrote about the Dust Bowl in North Dakota. In Washington, Eleanor consistently promoted campground development so that families who couldn't afford to stay in hotels could still camp out in the beauty of the Adirondacks and Sierra Nevada. She wrote in her weekly column for *Women's Home Companion* about the spiritual uplift she herself found in nature. While she warned that women might find their "first day or two a little difficult," she assured them that when they became "accustomed to camp life," they would "look back even on the rainy days, when you had to eat under the flap of your tent and devote your time to reading or writing, a very pleasant experience."[73]

They were escorted around Yosemite by Superintendent C. G. Thompson. Before taking the Tioga Road through Tuolumne Meadows and Tioga Pass, Eleanor Roosevelt met up with Ickes. Together they hiked in Mariposa Grove, where the redwoods were two thousand years old and had been protected since 1864.

Before arriving in Yosemite, the first lady had led a White House Conference on Camps for Unemployed Women. This conference led to the establishment, by FERA, of twenty experimental schools and camps for out-of-work women. The forest camps were run by the National Youth Administration (NYA), the New Deal agency tasked with providing work and education for Americans aged sixteen to twenty-five.[74] After talking with Ickes at Yosemite, Eleanor pushed for ninety female camps to be run by NYA. But the program was defunded in 1937. In the end, only 8,500 women benefited from these camps, whereas 3.5 million men had their lives upgraded by the CCC.[75]

Eleanor Roosevelt was still dissatisfied because the CCC was for men only. The only unemployed women given work through the CCC program were the so-called female campers, who did the housekeeping for the men at various barracks. Whereas men were recruited for a half year of service in the CCC, women could stay for only a few months. Women were also paid far less. While men received $30 per month, women workers were given an "allowance" of 50 cents per week.[76] Only 8,500 women participated in this experimental "female camper" program from 1933 to 1942, compared with the three million men who were recruited for FDR's CCC tree army.

Eleanor Roosevelt continued to push hard for more opportunities for women in the CCC but was brushed off by Robert Fechner, who didn't want any "She She She" camps distracting the all-male outfits.[77] Eleanor persisted and, with the help of Frances Perkins, eventually managed to have several camps for women opened on the outskirts of Elmira, New York.[78] Eleanor herself set up Camp Tera on Lake Tiorati in Bear Mountain State Park, funded by FERA and the Welfare Council of New York City. At Camp Tera women taught sewing but weren't paid.[79]

That the CCC discriminated against women was undeniable. But the rationale was that it had been established as a sort of Boy Scouts for young men—an organization in which they could strengthen their bodies and minds in the midst of the Depression. The image of the typical CCC recruit wasn't very different from the masculine images celebrated in Nazi Germany and Stalinist Russia: muscular men, tan from reforestation work, who were pillars of virility and strength. "The greatest achievement of the CCC," one government bureaucrat said, "has not been the preservation of material things such as forests, timber-lands, etc., but the preservation of American manhood."[80]

With Ickes, Eleanor received great press in California for promoting the Year of the National Park in the shadow of El Capitan.[81] After leaving Yosemite, still ruddy-cheeked from hiking through Yosemite, Eleanor traveled to San Francisco to rest. With Hickok as companion, she crossed the Golden Gate Channel to Sausalito and then headed north to see more redwoods in Sonoma and Humboldt counties.[82]

V

After reuniting with Franklin along the Oregon coast, Eleanor prepared to tour two of the New Deal's biggest public works projects in the West, both part of the Columbia River Basin initiative. If the 1920s had been the decade of skycrapers that served as monuments to American *private* sector can-doism then the New Deal era was characterized by colossal *federal* dams best exemplified by the TVA. At Bonneville (Oregon), in the TVA spirit, a concrete dam was being built 140 miles inland from the Pacific Ocean to provide cheap power to the region.[83] At Grand Coulee (Washington), an even bigger dam was under construction; it would regulate the flow of the upper Columbia River, generate hydroelectric power, and open up a large tract of previously parched land for the benefit of future generations. Bonneville and Grand Coulee dams were Roosevelt's engineering masterpieces.

The Bonneville Project on the Columbia River, forty miles east of Portland, was nearing completion when it was photographed on October 24, 1936. Including dams, a powerhouse, spillway, and the largest single-lift lock then in existence, the project was started in 1933. The plan to harness the Columbia River was fully supported by Roosevelt before and after his first election. It was different from any dam that had been built before, though. The Bonneville Dam was among the first projects affected by the Fish and Wildlife Coordination Act of 1934, which forced hydroelectric construction to reduce the disruption of animal life.

Within four hours of his return to the continental United States, the president spoke to five thousand people in Bonneville, Oregon, about installing new locks on the Columbia River that would allow barges from the Pacific Ocean to travel all the way to the wheat fields of Idaho. Before long, he promised, the Bonneville would produce ninety thousand kilowatts of electricity. There was no such thing "as too much power," he said. "I regard, and have regarded, the Columbia River as one of the greatest assets not only of the Northwest," Roosevelt declared from the palisades of the lower Columbia, "but of the whole United States."[84]

Moments after FDR spoke, the first couple was chauffeured to a special train, parked on a siding at Bonneville, to begin an overnight trip to Ephrata, Washington, to see another colossal dam. No matter how impressive Bonneville Dam was, Grand Coulee, to be completed in 1939, was an even greater engineering marvel. But, as with Bonneville, its construction posed serious environmental problems. Dozens of steam shovels and bulldozers uprooted everything in their paths. In fact, FDR was handed a piece of bad news on that score. Because of the dam's colossal scale, building a fish ladder for spawning salmon was deemed impossible. This meant that salmon were denied access to essential streams. "It is pretty late for us to discover that the salmon ladders at Bonneville may not work," Roosevelt fumed to Ickes after his Pacific Northwest trip. "It seems to me that all persons concerned should get together in a meeting and agree on something so that at least we can say to the country that we have done the best we could. Criticism and objections after the dam is half-built get us nowhere."[85] The thought that Grand Coulee was highly detrimental to Pacific Northwest fisheries prodded Roosevelt to build salmon hatcheries on numerous tributaries of the Columbia River downstream from the dam, including the Methow River, the Entiat River, and Icicle Creek. It was from these hatcheries that FDR's National Hatchery Service would eventually grow.[86]

VI

From Spokane the Roosevelts headed for the "Big Sky" country of northwestern Montana. There they traveled through the vast Kootenai National Forest, which encompassed 2.2 million acres in Montana and 50,384 acres in northern Idaho. When Bob Marshall was roaming the West for the U.S. Forest Service in the 1920s, he was overwhelmed by the raw scenic beauty of the Kootenai River, the Yaak River, and the Purcell Mountains. Thanks to the New Deal, however, the raw wilderness was being encroached on by work-relief crews with construction blueprints and machinery. As Franklin and Eleanor headed to Glacier National Park, they were abuzz about all the great work the CCC was doing in the West. African American CCC companies from Fort Dix, New Jersey, for example, were operating outside of the all-white communities of Libby and Troy, Montana. They heroically fought fires, planted trees, designed recreation areas, laid roads, and even constructed a small airport.

The assigning of African Americans to Montana was a New Deal experiment. Would the white residents in communities like Libby and Troy accept these black CCC workers from New Jersey? These black CCCers assigned to western Montana tried to use athletics to get along with the white population. The Colored Giants baseball team routinely played the all-white teams of Libby and Troy (one local headline read "Colored Giants Scalp Opponents in Sunday's Game.")[87] A "Negro quartet" from Pipe Creek gained fame throughout western Montana and Idaho, singing old-time spirituals. Their concerts, to all-white audiences, were like a church revival. After Governor Ben C. Ross of Idaho heard them perform, he declared that the Negro quartet was "more in demand than [he] had been."[88]

But even with outreach activities like the Colored Giants and the Negro quartet, it was still tragically difficult for these African American CCCers not to feel ostracized, even hated, by the white communities in which they lived.

On August 5, Franklin and Eleanor arrived at the western entrance to Glacier National Park. Boarding a chauffeured 1927 Cadillac touring phaeton they headed along fifty-one-mile Going-to-the-Sun Road, an engineering marvel carved out of the precipitous mountainside. Cutting through the middle of the "Crown of the Continent" top down, the Roosevelts followed the shores of Lake McDonald and Saint Mary Lake and then crossed over the Continental Divide at Logan Pass, soaking in the grand vistas of alpine lakes, thick forests, and jagged mountain peaks that rose thousands of feet above the valley floor. As the roads ascended, Roosevelt went through the west tunnel and saw elegant stone-faced concrete bridges and culverts.

Roosevelt was intrigued by the "international" aspect of the park, situated at the wild heart of North America. In 1932 Glacier and Waterton Lakes National Park in Canada had been joined, establishing the 1,143,272-acre Waterton Lakes–Glacier International Peace Park to protect the jaw-droppingly beautiful wilderness. In four years, open red buses called "jammers" were introduced into the park, so tourists could experience the drive in the fresh air, as the Roosevelts did. The president told his staff that the views at Glacier easily beat the Swiss Alps and Norwegian fjords he had seen with his mother. If the park were located in Germany or France, he felt, enthralled Americans would have crossed the Atlantic Ocean to enjoy the picture-postcard glaciers, flowering alpine meadows, and pristine coniferous forests. The fact that tourists from the East could arrive at Glacier by the Grand Northern Pacific Railroad, disembarking at the Glacier Park Hotel (now Lodge), made this western Montana park less far-flung than it seemed.

On entering the Blackfoot Reservation, adjacent to the park, the president was given a valuable peace pipe, made from Pipestone (Minnesota) clay—a relic from the 1855 treaty between the federal government and the Blackfoot Nation at Judith Basin, Montana.[89] The historic relevance of the Minnesota quarry had to do with its being the source of the soft red stone that Great Plains Indians used in their handsome ceremonial peace pipes, and it was run by the Bureau of Indian Affairs.[90] Three years later, on August 25, 1937, Roosevelt would sign a law making Pipestone a national monument.[91] The Blackfeet were so enamored of Roosevelt that they adopted him into their tribe. The president was given a certificate framed by arrows, which symbolized, as tribal leader Richard Sanderville told him, the Blackfeet method of "showing friendship and gratification."[92]

Realizing that the Great Depression had been particularly brutal to Native Americans, Roosevelt sought to alleviate some of the suffering on reservations and to lessen some of the ethnic tension that had been building for centuries with white Americans. Among these efforts was the Wheeler-Howard (Indian Reorganization) Act of 1934, which established the "CCC Indian Division" (or CCC-ID, a branch of the Indian Emergency Conservation Work project) that employed Native Americans in seventy-two "forest work camps" on Indian reservations in fifteen states in the West and Southwest.[93] That year, 14,400 Native Americans aged seventeen to thirty-five, living at the bottom of the economic ladder, were hired by the CCC in segregated Indian camps. Many of these young Native American men were itinerant, moving from tribal land to tribal land. Over 5,500 Hopi, Navajo, Apache, and Paiute were immediately tasked with fighting pine-blister rust and soil erosion throughout Arizona.[94] In the early New Deal, half the male breadwinners living on the Sioux reservations of South Dakota worked for the CCC-ID. "No previous undertaking in Indian Service," Collier asserted, "has so largely been the Indians' own undertaking."[95]

To help promote solidarity among the Native American CCCers, of which there were twenty-seven thousand, the U.S. government began publishing a periodical, *Indians at Work.* While its primary focus was the CCC Indian Division and related news, it had a mimeograph sheet featuring photographs and drawings of Indian life in general on reservations like the Blackfoot. That the Roosevelt administration was willing to pay to produce *Indians at Work* was remarkable for the time. Articles often criticized various reservation policies, but no one censored the content. When Fechner toured western Indian CCC camps, he was impressed. "I saw some wonderful water conservation work done by them [the Indians], soil erosion, cultural work in the forests, building of fire trails, etc.," Fechner reported, "and their camps compare favorably in every way with those of the white boys."[96]

From the Blackfoot Reservation, the Roosevelts headed south to Two Medicine Chalet, toward the scenic eastern entrance to Glacier National Park. "Today, for the first time in my life, I have seen Glacier Park," Roosevelt told the locals. "Perhaps I can best express to you my thrill and delight by saying that I wish every American, old and young, could have been with me today. The great mountains, the glaciers, the lakes and the trees make me long to stay here for all the rest of the summer."[97]

Franklin *(in car)* and Eleanor *(right)* visited Glacier National Park in Montana on August 5, 1934, only days after seeing the construction at the Bonneville and Grand Coulee Dams. As the first presidential couple to stop at the park, they were officially inducted into the Blackfoot tribe by an elder *(center)*. In a radio address that same day, FDR expressed his "thrill and delight" with the experience of Glacier National Park.

At Glacier National Park, the president reflected on the revolutionary concept of countries having porous borders as wildlife corridors—as at Waterton Lakes–Glacier International Peace Park. This idea came readily to Roosevelt because of Campobello Island, which, while technically in New Brunswick, Canada, operated in unison with Maine on fishing and hunting matters.

Roosevelt's speech at Glacier on August 5, broadcast nationwide on radio, would be historic. Never before had a U.S. president spoken so eloquently about the national park system. "There is nothing so American as our national parks," Roosevelt declared. "The scenery and wildlife are native. The fundamental idea behind the parks is native. It is, in brief, that the country belongs to the people, that it is in the process of making for the enrichment of the lives of all of us. The parks stand as the outward symbol of this great human principle."[98]

The president told stories about how Yellowstone had become the first national park in 1872, and he praised President Ulysses S. Grant for signing the founding document. With Ickes by his side, he mentioned former secretary of the interior Franklin K. Lane for persuading President Wilson to establish the National Park Service in 1916, quoting Lane's three broad principles for national parks: "First, that the national parks must be maintained in absolutely unimpaired form for the use of future generations as well as those of our own time; second, that they are set apart for the use, observation, health and pleasure of the people; and, third, that the national interest must dictate all decisions affecting public or private enterprise in the parks."[99]

Roosevelt reminded his listeners at Glacier that each existing national park had once been caught up in a political maelstrom. "We should remember that the development of our national park system over a period of many years has not been a simple bed of roses," the president said. "As is the case in the long fight for the preservation of national forests and water power and mineral deposits and other national possessions, it has been a long and fierce fight against many private interests which were entrenched in political and economic power. So, too, it has been a constant struggle to continue to protect the public interest, once it was saved from private exploitation at the hands of the selfish few."[100]

Getting a chance to meet the president was a thrill for the CCCers present in Glacier National Park. Bill Briggs wrote about the experience in *Happy Days*. "I saw the President," he boasted. "For eleven minutes he was with us; blue-eyed, genial, smiling . . . his keen eyes flashing over our camp and over us." He added a physical description, more vivid than a movie. "I found him a giant with massive shoulders and powerful arms that belied the steel braces on his legs. His face was ruddy and deeply tanned, his blue eyes flashed vigor and good humor, and his shock of iron-gray hair tossed in the wind. . . . Even the surrounding mountains and green-clad pines must have sensed that a great man was in our midst. Never did they seem so majestic and grand. The air was electric with the sense of a great happening."[101]

VII

After the speech at Glacier National Park, the Roosevelts traveled to Fort Peck Dam, a massive Public Works Administration project along the Missouri River in eastern Montana. Begun in 1933 and completed in 1940, Fort Peck would become the world's largest earthen dam.[102] Its construction led to the creation of Fort Peck Lake, the largest lake in the state and the fifth-largest man-made lake in America. Although the shore of the lake was to be longer than the coastline of California, there was something hideous about marring the unspoiled upper reaches of the scenic Missouri River. While Roosevelt boasted that Fort Peck Lake had a new bird reserve attached to it, in truth, the dredging took a hellacious toll on the landscape. In 1936, photographer Margaret Bourke-White's black-and-white image of the Fort Peck Dam appeared on the cover of the first issue of *Life* magazine. Bourke-White presented the structure as a triumph for the Army Corps of Engineers, off in a lost corner of Montana. Deeming the Fort Peck dam a boon for the nation Roosevelt, sweating prodigously, promised to return to Valley County in a couple of years when the dam was finished.[103]

The president next made the nearly four-hundred-mile journey across the North Dakota plains through the towns of Williston and Minot, arriving at Devils Lake. It jarred Roosevelt to see a once gorgeous lake—four hundred square miles in extent—become largely alkaline. The drought had taken its toll on the ecosystem. A vast "shelterbelt" reforestation effort was needed around Devils Lake. Barbed wire wasn't stopping the wind. Appalled at the devastation from the drought, seeing how badly the land was stripped of vegetation, he promised that the New Deal would resuscitate the soil-sick area.

That spring, Roosevelt had signed the landmark Frazier-Lemke Farm Bankruptcy Act, which placed certain restrictions on banks seeking to repossess farms. The law prohibited banks from foreclosing on the property of bankrupt farmers for five years; it was, to a certain extent, FDR's attempt to empower poor Great Plains farmers in the face of callous banking policies. Even after the Supreme Court ruled the Frazier-Lemke Act unconstitutional in 1935, the president worked to reimagine the landscapes of the Dakotas, to help farmers mitigate the effects of drought.

While FDR was speaking to a crowd in North Dakota, one farmer joked, "You gave us beer"—a reference to the repeal of Prohibition in 1933—"now give us water!"[104] Roosevelt laughed heartily. Lo and behold, only seven hours after Roosevelt departed North Dakota for Minnesota, rain cut a path one hundred miles wide through the region—with the heaviest fall coming in towns he had visited. Roosevelt was hailed as a magical rainmaker by newspapers in Bismarck and Fargo, and even the *New York Times* ran the story.[105]

Roosevelt and Ickes inventoried what needed to be done for the national park system to flourish. Clearly, some locations needed more landscape and tourist facilities. Other areas needed to be left alone in a primitive or wilderness condition, with the possible exception of hiking trails. Yet, for all its heroics, by 1934 the New Deal was being sharply criticized by a number of smart and influential conservationists. Bob Marshall of the Forest Service complained that the NPS was guilty of "putting roads where there was no need for them and destroying wilderness areas."[106] Benton MacKaye, the "father of the Appalachian Trail," declared that Roosevelt's NPS was "a destroyer of the primeval."[107]

Taking notice, Ickes asked Marshall, MacKaye, and Appalachian expert Harvey Broome to report back to him about the impact New Deal road construction projects were having on the Great Smokies. That August, after meeting in Knoxville, Broome's hometown, the trio investigated the new national park, taking careful field notes. "I hiked to Clingmans Dome last Sunday, looking forward to the great joy of undisturbed nature for which this mountain has been famous," Marshall wrote to Ickes. "Walking along the skyline trail, I heard instead the roar of machines on the newly constructed road just below me and saw the huge scars which this new highway is making on the mountain. Clingmans Dome and the primitive were simply ruined."[108]

Ickes, in fact, sympathized with efforts to reevaluate the NPS's series of new road projects. It wasn't hard to understand that "primitive" areas and "wilderness" began where the road ended. But Roosevelt, the politician-conservationist was not offended by roads, well designed and constructed, if they brought more people into his parks. With the Great Smokies and Everglades established in 1934, and his own speeches in Hawaii National Park and Glacier National Park huge hits with outdoor preservationists, the president decided to organize the NPS efficiently, keeping regional requirements in mind. His extensive trip convinced him that a centralized NPS operational program run from Washington didn't make sense. Roosevelt saw to it that the Department of the Interior opened four regional headquarters: in Richmond, Virginia; Omaha, Nebraska; Santa Fe, New Mexico; and San Francisco, California. While the NPS director, Arno Cammerer, would remain in Washington, Ickes gave a great deal of autonomy to the superintendents' four regional offices.[109]

Another beneficiary of Roosevelt's Year of the National Park was West Texas. As the decade reached its midpoint, FDR, under the sway of Ickes, took an intense interest in Big Bend country. Impressed with photographs of the Santiago and Chisos mountains, he sought to establish a vast park along the 118-mile boundary the Rio Grande carved between the United States and Mexico. Since the 1880s a campaign had been under way to protect the sublime beauty of this windswept borderland. Esteemed folklorist J. Frank Dobie had promoted the idea of a Big Bend National Park in *Nature* magazine.

Roosevelt also knew from Miriam "Ma" Ferguson, the first female governor of Texas, that the CCC had been doing tremendous work in Texas Canyons State Park, which the NPS had helped establish in 1933 (it was renamed Big Bend State Park). That summer of 1934, two hundred CCC boys, 80 percent Hispanic, were constructing all-weather access roads and cutting hiking trails in the Chisos Mountains of West Texas. Living in a tent city eighty-five miles from the nearest town, these CCCers, professional transients, facing prejudice in Alpine and Fort Davis, proceeded onward undaunted. This mostly Hispanic corps dug and blasted ten thousand truckloads of earth and rock in the state park. Facing unbelievable summer heat, they also developed a reliable water supply, which made living in the desert furnace bearable.

What Roosevelt envisioned in 1934 for Big Bend was an "international park," comanaged by the United States and Mexico (and modeled on Waterton Lakes–Glacier International Peace Park). Ickes had the NPS write reports about Big Bend that were as rich in "old West" lore as in data about the populations of deer, javelina, and fox. The proposed park FDR supported would consist of 700,000 acres. From hot springs to mile-high peaks, from the arid grandeur of the Chihuahuan Desert to the Rio Grande, Big Bend country was a special place to find peace and solitude, away from the hubbub of modernity. The two main roads entering Big Bend from the north—number 385 from Marathon (thirty-nine miles distant) and number 118 from Alpine (eighty-one miles distant)—converged in the proposed park at Panther Junction, where the future NPS headquarters was built. In November the *New York Times* ran a story about Roosevelt's Big Bend National Park plan, claiming that the semiarid wilderness was "virtually uninhabited by people."[110]

The *Times* was essentially right. Forlorn Big Bend—with no electricity or graded roads—was among the most primitive of America's would-be national parks. It was an ecological island dominated by the spectacularly eroded Chisos Mountains, teeming with desert wildlife, and the Rio Grande River. Turning the remote wilderness into a national park, with help from his Fort Worth millionaire friend Amon Carter, would take a Herculean, Texas-sized effort on Capitol Hill. Roosevelt was undaunted about getting the Big Bend Park job done soon.[111]

CHAPTER ELEVEN

"A DUCK FOR EVERY PUDDLE"

I

Not all of Franklin Roosevelt's conservation work in 1934 was related to the National Park Service. In early January, he appointed three respected conservationists for a blue-ribbon Committee on Wildlife Restoration: chairman Thomas Beck, the influential publisher of *Collier's Weekly*; Jay "Ding" Darling, syndicated cartoonist for the *Des Moines Register*; and Aldo Leopold, the wildlife management professor from the University of Wisconsin who had once worked for the U.S. Forest Service in New Mexico.[1] Roosevelt was eager to have the Biological Survey develop a coherent national wildlife refuge system.[2] By appointing his Connecticut friend Beck as chair, the president could, without fanfare, stay abreast of the direction the committee was taking. In naming Darling (a Hoover Republican) and Leopold (a brilliant midwestern academic) to the committee, Roosevelt hoped to diminish any criticism of liberal partisanship.[3]

What all three conservationists shared was a certainty that the Biological Survey under Hoover had lacked vision, and that a comprehensive $18 million New Deal program to save dwindling North American migratory birds was urgently needed. Of the 120 million acres of marsh and wetlands originally found in the United States, only a rapidly dwindling 30 million acres of waterfowl habitat remained. "I get from a good many sources suggestions that the Biological Survey spends too much time on scientific experimentalism," Roosevelt wrote to Henry Wallace, "and that we ought to have a more practical spirit."[4]

Roosevelt couldn't have chosen three better stewards to make North American wildlife a New Deal priority. Ding Darling was born in Norwood, Michigan, on October 21, 1876. As a child, he moved to Sioux City, Iowa, then still the Wild West to many people back east. Growing up along the Missouri River, traipsing in the vast prairie-scapes of Nebraska and South Dakota, Darling became a self-taught naturalist in the Meriwether Lewis tradition.[5] Irreverent, whip smart, and known for his prankish schoolyard escapades, Darling found an outlet for his imaginative mind as a satirical cartoonist. He took the pen name "Ding"—a contraction of "Darling"—while at the *Sioux City Journal*. By 1917 his satirical cartoons in the *Des Moines Register* were syndicated by the *New York Herald Tribune*, and by the 1920s more than two hundred newspapers carried his drawings. In a career that extended from the presidency of Theodore Roosevelt to that of John F. Kennedy, Darling was beloved for lampooning self-important politicians and promoting environmental conservation.

While Darling's income came from cartooning, his consuming passion was to protect the upper Mississippi River ecosystem from wanton destruction. He credited his Uncle John—who owned a Michigan hay-wheat farm—with teaching him hunting etiquette, proper land management techniques, and the importance of stocking freshwater ponds with fish. Having seen great flocks of golden plovers move across South Dakota during his boyhood, Ding, like his uncle, was alarmed at the prospect of their demise. "It was the disappearance of all that wonderful endowment of wildlife," Darling later recalled, "which stirred the first instincts I can remember of conservation."[6]

After Uncle John died, the Michigan farm deteriorated, eventually looking as if a cyclone had sucked up all the rich black loam soil that had once yielded an easy fifty bushels of wheat per acre. Darling witnessed how quickly rich pastures could become scorched earth, cracked open. "This was my first conscious realization of what could happen to the land," Darling later explained, "what could happen to clear running streams, what could happen to bird life and human life when the common laws of Mother Nature were disregarded."[7]

During Theodore Roosevelt's presidency, Darling lobbied for the creation of the U.S. Forest Service and worked to transform America's slapdash network of federal bird preserves into a scientifically managed, coherent system. When his beloved TR died in 1919, Darling put aside his acid pen and drew his most popular cartoon ever: "The Long Trail Ride," which depicted the Rough Rider on horseback proudly riding off into a Grand Canyon sunset. In the 1920s, Darling became a leader in the Izaak Walton League in Iowa, advocating for wildlife protection.[8] In this capacity he became close friends with Herbert Hoover. Darling, in fact, won the first of two Pulitzer Prizes in 1924 for his caricatures of Hoover—no mean feat, because, as Darling noted, the thirty-first president "had the most average-looking face."[9]

Once Hoover became president, Darling visited the White House twice, holding forth about bass and walleye in Iowa, the upper Mississippi, and the Great Lakes.[10] On one occasion, Hoover and Darling, evading the Secret Service, sneaked off to the Great Smokies for horseback riding, fishing, and hiking. Always a conservation activist, Darling helped organize the Men's Garden Clubs of America and served as its first president. Agitating during the Hoover years for more trained foresters and wildlife specialists in state government, Darling initiated an enterprising partnership with Iowa State College of Agriculture and Mechanic Arts, and established the Cooperative Wildlife Research Unit in Ames, Iowa. The research laboratory soon grew in academic prominence because of its great surf of cutting-edge field research on behalf of midwestern game animals. And thanks to Darling's intense lobbying, the Iowa general assembly created a progressive five-member State Fish and Game Commission; Darling, at the behest of Iowa's governor Daniel Webster Turner, served as the staggeringly high-profile chair of this commission from 1931 to 1934.[11]

Because of his exalted gift for public relations, Darling's reputation soared in North American conservationist circles. Raising the ante for the wildlife protection crusade, Darling was perceived as the "boondocker" who knew more about ducks and geese than most game wardens. Newspaper readers were riveted by his reflections—in the style of Will Rogers—on shooting duck, fishing for black bass, ornithology, gardening, and even eating Roquefort cheese.[12] When, in 1931, the seventy-six-year-old William Temple Hornaday published *Thirty Years War for Wild Life*—lambasting the U.S. government for issuing over six million hunting licenses and thereby causing the "progressive extinction of wildlife"—Darling backed up his reform message. Both Hornaday and Darling knew that the era of commercial hunting had to end. "For most of us—and I speak as a man who has lived in Iowa—the passing of various species of game has been a tragedy," Darling told the American Game Conference. "It has hurt us to the heart. I am not a great shooter, a very frequent shooter, but just the same the passing of the birds has made a deep impression on me. It has filled my soul with the thought that something could be done about it."[13]

Darling served as a delegate to the 1932 Republican National Convention, wanting to see Hoover reelected to a second term. The GOP conservatives in Chicago urgently lobbied Darling—Iowa's most revered celebrity—to run for an open U.S. Senate seat. Darling, though flattered, rejected these overtures with a gruff "Get lost." Such directness was Darling's most elemental asset as an antipolitician who nevertheless engaged in GOP politics. When FDR defeated Hoover in 1932, a disappointed Darling joked that the New Deal was actually going to be more of a Raw Deal.

It is therefore a testament to Roosevelt's cunning that he selected Darling for the Committee on Wild Life Restoration. Darling had ridiculed Roosevelt from 1928 to 1932, portraying him in cartoons as a Hudson Valley patrician detached from everyday life. Though dismissive of the Roosevelt administration's belief in "big government" as the solution to the Great Plains drought of the early 1930s, Darling did hope the comatose Biological Survey could be resuscitated by Roosevelt. Roosevelt cultivated other Republicans, who agreed with him on the need for conservation measures, even or especially in the midst of a human crisis, such as the Great Depression. Another was Peter Norbeck, a senator from South Dakota. A potent supporter of the sculpture then in process at Mount Rushmore, Norbeck fought for parks and preservation measures. "Senator Norbeck is tremendously interested in game preservation," FDR wrote to Darling. "He can be counted on to help in every way."[14] The ardent Republican cartoonist made a deal with the Democratic devil—Roosevelt—for the sake of helping migratory birds thrive in a federal system of prairie-pothole ponds, grasslands, and marshes by means of an innovative natural wildlife refuge system. The ecological damage to the land had to be reversed, and Roosevelt had the mettle to be innovative and to experiment.

Aldo Leopold examined the skin of a Hungarian gray partridge chick in 1943. Leopold, a professor at the University of Wisconsin at Madison, was a committed conservationist, who wrote in the early 1940s that "there is a basic antagonism between the philosophy of the industrial age and the philosophy of the conservationist." FDR's philosophy, however, was that the two had to coexist, and his job was to prove that they could.

The third member of the FDR's committee, Aldo Leopold, was also an Iowan, born on January 11, 1887, in Burlington. Leopold recalled that his businessman father, a partner in what became the Leopold Desk Company, was a utilitarian conservationist like Pinchot. The family lived in a well-built home on a scenic bluff in Burlington that overlooked the broad Mississippi River. Summers were spent at Marquette Island, Michigan, at the northern end of Lake Huron. When Aldo turned eleven, he wrote in a school notebook, "I like to study birds," inventorying thirty-nine Mississippi River Valley species. But it was family trips to Estes Rock, Colorado, and Yellowstone that led Leopold into game management as a profession.

In 1906, Leopold entered the Yale Forestry School, cofounded by Pinchot; there, books like Charles Darwin's *The Formation of Vegetable Mould Through the Action of Worms* and Theodore Roosevelt's *The Deer Family* enthralled him. On earning his master of science in forestry, he joined District 3 of the U.S. Forest Service in July 1909. Insatiably curious about the natural world, he moved to Springerville, Arizona, to become forest assistant in Apache National Forest. That August, Leopold led a reconnaissance crew in mapping land and surveying timber. The following month he shot into a pack of Mexican gray wolves (*Canis lupus baileyi*) and killed the mother. The death caused Leopold to slowly reform his ethical promises. "I was young then, and full of trigger-itch," Leopold wrote in his seminal essay "Thinking Like a Mountain," later published in his collection *A Sand County Almanac*.[15] "I thought that because fewer wolves meant more deer, that no wolves would mean hunters' paradise. But after seeing the green fire [in the wolf's eyes] die, I sensed that neither the wolf nor the mountain agreed with such a view."[16] Like Hornaday and Darling, he would come to believe that even North American predators like wolves and mountain lions needed federal protection.[17]

Mediating his commitments between the Forest Service and academia, Leopold developed a devoted following among sportsmen and conservationists alike. The New York State College of Forestry at Syracuse University tried to recruit him to run the Roosevelt Wild Life Forest Experiment Station; he declined. From 1911 to 1915, he printed his own practical newsletter, *The Carson Pine Cone*, in Tres Piedras, New Mexico. Many of Leopold's articles emphasized that wilderness could be saved if only a new ecological consciousness emerged, and his arguments galvanized a new movement to preserve "roadless" public lands in the Southwest and beyond.

Leopold was a scientific forester who thought about landscapes poetically, like Henry David Thoreau.[18] In 1922, he inspected Gila National Forest in New Mexico and recommended that a 755,000-acre portion be designated a "wilderness" area. Two years later, after traveling around the Boundary Waters Canoe Area of Superior

National Forest in northern Minnesota with his brothers and son, Leopold fought to have it protected from encroaching industrialization. "One of the penalties of an ecological education is that one lives alone in a world of wounds," Leopold wrote of his career. "Much of the damage inflicted on the land is quite invisible to laymen . . . in a community that believes itself well and does not want to be told otherwise."[19]

If any individual deserved the designation "father of marsh conservation," it was Aldo Leopold. Besides understanding that America's wetlands supported large duck and geese populations, he knew these ecosystems also acted as filters, cleaning water of impurities before it flowed into rivers and oceans. Wetlands and marshes acted as spongy buffers between land and large bodies of water, absorbing excess water from storms and floods—and often protecting towns and cities.[20]

In 1933 Leopold was appointed professor of game management in the Department of Agricultural Economics at the University of Wisconsin. That May he published his magnum opus, *Game Management*.[21] This book revolutionized wildlife management in New Deal America. While some chapters of *Game Management* may have seemed professorial or overly technical, many passages were literary art. The Boone and Crockett Club, whose members were aristocratic hunters, endorsed *Game Management* enthusiastically; the Biological Survey used the text to help the new Roosevelt administration scientifically address wildlife management on public lands. According to Leopold, "that land is a community is the basic concept of ecology, but that land is to be loved and respected is a basic extension of ethics"—a creed echoing FDR's own.[22]

Not only was Leopold a gifted teacher, but his prodigious mind was virtually a bank of facts and figures. "In a field where myth, generalization, and enthusiasm were often liable to outweigh results," biographer Curt Meine wrote, "he was constantly pushing himself and others to examine and reexamine their aims, means, and results."[23] His informal, inquiry-driven style of teaching, which he often mixed with hands-on outdoors experience, allowed his theories of "game management" and a "land ethic" to be pragmatically tested in field laboratories.[24] These field assignments, including a stint with the CCC in the Southwest in 1933, epitomized what decades later would be termed "experiential learning." Unlike many academics, Leopold was not tendentious or conceited. His generosity of spirit was legendary in Wisconsin. Not only did he believe hunting was a noble pastime, but he also vigorously argued that the "fair chase" of game connected a man to the outdoors in a healthy way. Buttoned-down, precise, and a stickler for scientific exactitude, Leopold was the sage the U.S. conservation movement needed as species came under pressure or even went extinct.

It was important to Roosevelt, Leopold, Darling, and Beck that the scope and character of wildlife management broaden—from the meager measures of the Coolidge and Hoover administrations to Hornaday's more comprehensive approach. Hunting in flyways and rookeries had to be curtailed. Milliners had to turn to bows and ribbons instead of feathers. Meatpackers had to focus almost exclusively on farm-produced poultry (like chicken or turkey), not on wildfowl. Taking a cue from *Game Management*, the Roosevelt administration, under the banner of a "New Deal for wildlife," wasn't going to tolerate the destruction or fundamental alteration of any more wetlands. Not all of the damage to habitat was immediately apparent. In the 1920s, someone introduced the common carp (*Cyprinus carpio*) to Malheur Lake in southeastern Oregon. It may have been a well-meaning gesture, since the carp is edible, but it transformed the lake, which stretched for ten miles or more as a clear, plant-filled paradise for birds in need of a stop on the Pacific Flyway. Over the course of decades, the carp proliferated in Malheur Lake, ultimately eradicating the plants and turning the water cloudy with mud. The carp that were released so innocently into Malheur Lake in the 1920s destroyed it as a feeding ground for fowl. Ignorance was as dangerous as violence when it came to conserving bird species.

To meet the increasing needs of waterfowl populations, Biological Survey refuge management adopted scientific measures. Recreational hunting, though allowed on public lands, had to be strictly regulated by "closed seasons, bag limits, and license requirements."[25] In bayous and swamps, water levels had to be manipulated to ensure that plants would thrive. Invasive species—responsible for changing habitats essential to the survival of ducks—had to be identified and prohibited. The CCC needed to train recruits in digging ponds, killing weeds, and rehabilitating wildlife. The backlog of deferred maintenance projects at federal wildlife refuges was criminally long. Under Roosevelt's leadership, the CCC worked in thirty-six National Wildlife Refuges between 1934 and 1943.[26]

II

Roosevelt's Committee on Wildlife Restoration first convened in Washington on January 6, 1934, and held a press conference, at which Darling joked that the New Deal's conservation objective was a "duck for every puddle." (The allusion was to Herbert Hoover, who, campaigning in 1928, had promised "a chicken in every pot"—a phrase also attributed to Henry of Navarre around 1600.)[27] With waterfowl populations reaching the lowest point in U.S. history, and the drought persisting, the president tasked his three conservationists to write a visionary report offering recommendations for rehabilitation. Worried about the extinction of North American species of ducks and geese, the committee, after consulting with congressional leaders, immediately set a goal of $50 million for the purchase and restoration of submariginal lands for wildlife, with a special emphasis on migratory waterfowl.[28]

To devote their attention to the task full-time, Darling and Beck moved to Washington, D.C., in early 1934. Leopold stayed in Madison to honor his spring semester teaching commitments at the University of Wisconsin. Roosevelt hoped that the committee's final report, due that February, would become an activist blueprint for securing millions of acres of waterfowl habitat—using drought relief funds—and reversing failed drainage projects. It had to be assumed that the waterfowl situation would worsen without emergency government intervention. Praying for rain simply wasn't working. The 1934 *Yearbook of Agriculture* documented the fact that 100 million acres of formerly cultivated land had lost most, if not all, of its topsoil. Only two states—Maine and Vermont—had escaped the ecological disaster.

As Donald Worster wrote in *The Dust Bowl*, the financial repercussions of the drought in 1934 alone equaled 50 percent of all the money Uncle Sam spent in World War I.[29]

At their makeshift Washington office Beck and Darling sifted through two thousand plans submitted by well-intentioned wildlife lovers.[30] Seeking a consensus, they haggled over the report's final recommendations. Darling unearthed scores of problems created by lax state hunting laws, allowing too much "take." The most divisive issue was the morality of artificial insemination and egg incubators as methods to bring back waterfowl populations. Leopold and Darling were vehemently opposed, while Beck thought hatch-and-release wild duck factories were an idea worth considering.[31] "Generally speaking," Darling recalled, "Beck advocated the theory, held by the More Game Birds crowd, that the way to restore ducks was to hatch them in incubators, and turn them loose into the flight lanes, in other words, restocking by artificial methods, and I held to the principle that nature could do the job better than man and advocated restoring the environment necessary to migratory waterfowl."[32]

Searching for the language of consensus and ironing out legal details for the U.S. government was frustrating for Darling, who wrote the committee's final report. Frequently refereeing the "violent squabbles" between Beck and Leopold, Darling stayed focused on federal acquisition of habitats for migratory birds.[33] Beck felt that the Republican Iowans—Leopold and Darling—had ganged up on him because he was editor of *Collier's* and was an aristocratic friend of FDR's. Later Beck complained that he sensed "some little peevishness, some little jealousy, some little selfishness about my coming in on this work" from Leopold and Darling.[34] Beck was mistaken to feel victimized. His notion of a "duck factory"—of raising waterfowl in captivity rather than making an all-out effort to repair marred landscapes—was just wrongheaded. Speaking at the twentieth American Game Conference, which was held in late January, Beck even suggested that the Biological Survey could be abolished, blaming it for undermining habitat acquisition initiatives by tapping resources for endless study and little real estate action. Such shortsighted thinking made Darling and Leopold apoplectic. "I cannot believe that the conservation movement," Leopold wrote, "is naïve enough to stomach such an absurdity."[35]

When Beck, Darling, and Leopold visited the White House for an assessment of the troubles within the committee, Roosevelt's considerable personal charm won the day. "I remember Dad coming back from the trip and telling the family that he was asked to come in and talk with Franklin Roosevelt," Luna Leopold, son of Aldo Leopold, recalled. "He thought that Roosevelt was one of the most impressive men he had ever talked to, even though he didn't agree with FDR."[36] After weeks of compromise, the committee's "threshold document" was outlined for the press. Knowing he had been outgunned by the Leopold-Darling alliance, Beck fell into line, telling reporters that "the time for conservation has passed" and that "the time for restoration has come."[37]

On February 8, 1934, Beck officially submitted the committee's ambitious, twenty-seven-page "National Plan for Wild Life Restoration" to the White House. This was no routine Washington white paper. For the report's frontispiece, Darling drew a cartoon depicting a Noah's ark of North American animals—bear, deer, goose, trout, rabbit, and duck among them—posting, on a dying tree, a banner that read "Help." The drawing set the tone for the lucid and concisely formulated report, which held nothing back as it recommended $50 million in immediate congressional appropriations. The types of land submitted for federal consideration included habitats that had been used and typically abandoned by farmers: natural nesting marshes, broader marshlands rich in food, lands on the shores of lakes or rivers used by breeding birds, low-valued flatlands adapted to nesting, and heavy alkaline lakes that would have to be freshened or drained.[38]

The Beck Report (as it was known), urged the U.S. government to immediately start purchasing land critical to wild animals: four million acres for migratory waterfowl and shorebird nesting grounds; two million acres for restoration of mammals; one million acres of breeding and nesting areas for insectivorous, ornamental, and nongame birds; and, more tentatively, five million acres of lands that were home to "upland game." A new position, commissioner of restoration, was also sought to monitor and optimize all wildlife recovery efforts at the federal level.[39]

Thomas H. Beck, editor of *Collier's* magazine, after delivering a speech in Washington in 1939. Largely on his own initiative, Beck had launched a campaign to reintroduce and manage native animals in Connecticut. To foster the same work on a national scale, FDR appointed him chair of a new Committee on Wildlife Restoration in 1933. The other members were Aldo Leopold and Ding Darling.

Roosevelt, worried about creating interagency havoc, nixed the idea of a commissioner before it could gather any momentum and he slashed the requested $50 million down to a more realistic $8.5 million. But he adopted the report's overarching recommendations to buy marginal stretches of waterfowl habitat and then have the CCC "plant that land with game."[40] Not only would new federal migratory refuges be established in the Atlantic, Mississippi, Central, and Pacific Flyways, but existing refuges would be restored and expanded. Nesting areas where food was once abundant would come back if water was restored, and invasive plants like Japanese knotweed and giant hogweed could be eradicated with enough manpower.[41] If nothing else, the sheer ambition of the Beck Report got veteran conservationists excited about the New Deal's potential to help waterfowl and improve refuges by constructing water-level controls.

Because both Darling and Beck were celebrities involved in the media, the report garnered considerable attention. However, FDR, preoccupied with the New Deal for people, didn't act immediately on the Beck Report.[42] An impatient Darling, perplexed by Roosevelt's stalling, almost threw in the towel. Couldn't the president accord his Committee on Wildlife Restoration more respect?[43] Part of the reason Roosevelt was slow to implement the Beck Report was that Secretary of Agriculture Wallace had told him that the recommendations were "too ambitious to be feasible in the immediate future." He was right. Neither the president nor Wallace had a strategically simple means of funding an extensive program for wildlife habitat. The political climate dictated that human needs were paramount, with the Depression hanging on despite the first wave of New Deal programs. Wallace, however, had an idea that tapping Darling, beloved by the public, to head the Biological Survey would be a windfall for wildlife restoration. But Wallace didn't know if FDR would agree to his unusual request. Wallace later told an interviewer for Columbia University's oral history project that the president's tactics were a mixture of "intuition and indirection." He observed, "I'd say he had a golden heart, but I wouldn't want to be in business with him. . . . He was a truly great man, there's no doubt about that—but very unpredictable."[44] In the case of the new chief for the Biological Survey, Wallace calculated correctly; Roosevelt saw the merit of an unpredicted candidate.

One afternoon in February 1934, Roosevelt unexpectedly telephoned the *Des Moines Register* newsroom looking for Ding Darling. Within seconds, FDR had the legendary cartoonist on the line. The president asked whether Darling would be willing to replace the retiring Paul Redington as chief of the U.S. Biological Survey starting in early March.[45] Even though Darling would lose income by quitting his lucrative syndicated cartoons, he agreed. "A singed cat," Darling later said about accepting the job, "was never more conscious of the dangers of fire than I was of the hazards in trying to get anything done in Washington."[46]

That March, Darling was sworn in as bureau director of the Biological Survey, hoping that he could implement the Beck Report's recommendations. The anti–New Deal cartoonist from eastern Iowa, to the surprise of many reporters, had formed an alliance with the president for the sake of restoring North American wildlife. However, Darling felt duped when he heard a few weeks later that President Roosevelt had absentmindedly lost the Beck Report (he had misplaced it somewhere in his White House bedroom). Because of all the lore about the "Hundred Days" in 1933, Darling didn't comprehend that there was a limit in 1934 to the president's executive power vis-à-vis Congress. A tedious government process had to be followed, with Congress being brought into the discussion, before the report's suggestions could be implemented.[47] Finding a way to do so constituted Darling's job description. And he did, badgering other bureaucrats for money out of various New Deal pots.

III

On March 6, 1934, the president signed the Migratory Bird Hunting and Conservation Stamp Act (commonly referred to as the Duck Stamp Act). It took a combination of manipulation and talent to make this tax feel good to hunters. Promoted heavily by Senator Frederic Walcott of Connecticut and Representative Richard Kleberg of Texas, the new law required all Americans over sixteen years of age to purchase a special $1 stamp before hunting ducks, geese, or swans and to acquire a valid state hunting license. The stamp was to be sold at most U.S. post offices. Bird hunters were required to affix it to their hunting licenses or else risk arrest and a heavy fine for poaching. Approximately 98 cents of every Duck Stamp dollar collected would go toward the purchase, maintenance, and development of "inviolate" wetlands and wildlife habitat for inclusion in what would be known by 1940 as the National Wildlife Refuge System (though the National Wildlife Refuge System Administration Act of 1966 made many improvements). The remainder of the revenue would be spent to hire USDA wardens to stop poachers on public lands from mistreating nature.[48] Furthermore, Roosevelt, to Darling's delight, issued a federal ban on baiting and live decoys.[49] Never were two men more perfectly suited to optimizing an otherwise rather mundane program. The Duck Stamp might easily have been nothing more than a bureaucratic mechanism, about as intriguing as a rubber stamp that inked "Paid" on a license—which is about all that it was. FDR, the passionate philatelist, loved stamps too much to allow each year's duck issue to be anything but irresistible. And Darling, who communicated best through art, was equally determined that the duck stamps not only support the effort to save wild birds but invoke their beauty. At FDR's request, Darling illustrated the inaugural Duck Stamp as two striking mallards in flight descending on a lake. "There was no one else available," Darling recalled modestly, "to make a design."[50]

A Federal Migratory Bird Hunting and Conservation Stamp—better known as a "duck stamp"—issued in 1934. Political cartoonist Jay "Ding" Darling oversaw the first government program to sell such stamps in order to raise funds for habitat conservation. They could be purchased not only to validate hunting licenses but also to support America's migratory birds. Darling drew the stamp (*pictured*) that was used during the first year of the program.

On August 22, 1934, a highly publicized Duck Stamp ceremony was held at the main post office at Twelfth Street and Pennsylvania Avenue in Washington. There is a fine photograph of Franklin D. Roosevelt, eyebrows arched, grinning warmly and pointing at the first sheet of stamps—*Mallards Dropping*—for the benefit of the shutterbugs that August afternoon. He was proud of his own ingenuity, and his eyes danced with a kind of controlled humor. At such public events, Roosevelt was almost luminous, his charisma especially strong. Speaking informally to reporters, showing off his philately, Roosevelt reiterated the importance of wildlife conservation. Over 650,000 of the Duck Stamps—"miniature pieces of art"—were sold within weeks. During the first year, the Duck Stamp added $600,000 to the funds available for the national program. Between 1934 and 2009, more than $500 million went into the fund to purchase approximately five million acres of waterfowl habitat.[51]

Darling looked at a sheet of the first duck stamps at a post office in 1934. Darling was a nationally known cartoonist, but his work on behalf of wildlife gained him the respect of Roosevelt. FDR was not dissuaded by Darling's background as a staunch Republican, and appointed him to the post of chief of the Bureau of the Biological Survey (later the Fish and Wildlife Service).

Around the time the Duck Stamp was being developed, Roosevelt and Darling discussed migratory birds over cigarettes in the Oval Office. As the president told a stretched-out humorous story, Darling wondered if anyone really knew Roosevelt. Darling mustered the nerve to raise a worrisome issue: a promised $1 million appropriation to buy waterfowl habitat still hadn't been deposited into the Biological Survey's account. FDR, with the wave of a cigarette, quickly adopted a by-golly-I'll-fix-it attitude, writing Darling an IOU for $1 million—literally on a scrap of paper. With theatrical aplomb and a Hollywood handshake, he handed the note over. Darling later reflected, "No small boy with a new cowboy hat and Texas boots ever felt more like a big shot than I did walking out of the White House with my first . . . [document] signed with the familiar 'F.D.R.' in his own handwriting!"[52]

But after Darling left the Oval Office, skepticism haunted him. Where could he cash the chit? Had FDR's three-hundred-watt grin conned him? A common refrain among Roosevelt's brain trust of advisers became "Ding is rattling his tin cup again."[53] On learning that

Darling was having a difficult time trying to cash the $1 million IOU, Roosevelt wrote to Wallace, "I hope that in addition to the million dollars already allocated, we can get from land purchase, relief, etc., another five millions. By the way, the Congressman [Kleberg] says the million dollars which I allocated has got lost somewhere. Will you conduct a search party?"[54]

At one tense meeting Darling handed Harry Hopkins the $1 million IOU from FDR. Hopkins, the head of the Works Progress Administration, the Civil Works Administration, and the Federal Emergency Relief Administration, just chuckled at the proffered IOU and then condescendingly said to Darling, "I don't know if we're interested in the relief of birds." Hopkins thought that public zoos—not the Biological Survey—should be the focus of bird recovery programs. The WPA was already building laboratories and exhibit spaces at a half dozen zoos across the country—for instance, at the Toledo Zoo WPA stonemasons and carpenters were building a new "House of Birds"; with the aid of WPA money, the San Antonio Zoo would soon assist sandhill cranes and whooping cranes.

Darling, tired of abrupt shifts and halts, felt his nervous system revving up for a fight. "Harry," he scolded, his resentment rising like flames, "I was a trustee at Grinnell College when you were a student!" He then recounted how Hopkins had urged him to become the head of the Biological Survey and promised the Beck Committee $1 million to buy land. It was too much hectoring for Hopkins to endure. "Where do I sign these papers?" Hopkins asked in retreat. Then he turned to his personal assistant and said with admiration, "See, that's how you get things done in Washington!"[55]

Eventually, Roosevelt was able to allocate $8.5 million in emergency funds to buy migratory bird habitat and construct fences, dikes, and dams. It was divided as follows: $1 million in special funds for the purchase of new waterfowl refuges; $1.5 million allocated from the submarginal retrieval fund; $3.5 million taken from drought-relief funds (for the purchase and development of lands within drought areas); and $2.5 million slated for WPA projects to improve existing refuges. This wasn't the $50 million that Beck, Darling, and Leopold wanted, but in the depths of the Depression it wasn't chump change.[56] And Roosevelt contributed his own idea to help migratory birds recover: the creation of artificial water areas, especially in arid and semi-arid regions of the West.[57]

At the same time that Darling was scrounging funds to the best of his ability, Congress was taking a parallel course, passing the Fish and Wildlife Coordination Act in 1934. It created bird refuges across an array of lands and waterways under the jurisdiction of the Forest Service, the Bureau of Reclamation, and what would eventually be known as the Bureau of Land Management. As an emergency measure, it was effective, but it carried an important caveat: the main economic engine (mining, logging, grazing, etc.) of a given area would not be curbed. The act also sought to do as its name suggested, coordinate the conservation efforts of the Interior and Agriculture departments, though that was a continuing challenge long afterward. The act was a step in the right direction for conservation, but it wasn't as proactive in establishing refuges as Roosevelt's more personal impetus, carried forth by Darling and the Biological Survey.

Darling's hard-won money went toward buying a list of 323 potential waterfowl and upland game refuge sites from the Beck Report's "Summary of Tentative Projects: Migratory Waterfowl." Upper Midwest and northern Great Plains states—North Dakota (110 projects); Minnesota (thirty-one); Montana (twenty-seven); and South Dakota (twenty-one)—became the starting places for FDR's new waterfowl rescue efforts, which worked in harmony with the New Deal's larger assault on drought relief through rural-land management. By late 1935, protection was ensured for White River (Arkansas), Sacramento (California), Mud and Rice lakes (Minnesota), Medicine and Red Rock lakes (Montana), Valentine (Nebraska), Mattemuskeet (North Carolina), and Turnbull (Washington). All fell into place on the national map. The bugling of cranes and nesting of eagles and egrets holding clamorous convention were going to remain, with CCC help, American birthrights.[58]

IV

As the drought in the Great Plains worsened, plows turned up dust as dry as gunpowder, while the nearly constant winds blew across the prairie. In the spring of 1934 Roosevelt sought large-scale federal solutions to the disaster, which had been caused in large part by human ignorance. Over seventeen million acres of U.S. wetlands had been drained; the president wanted at least five million brought back. Trees and other plants also had to be reintroduced. "The [wind] storms are mainly the result of stripping the landscape of its natural vegetation to such an extent that there was no defense against the dry winds," Donald Worster explained, "no sod to hold the sandy or powdery dirt. The sod had been destroyed to make farms grow wheat to get cash."[59]

Giant dust clouds, tar black at the base, rose from the Great Plains fields and traveled with the prevailing winds, rolling and turbulent. According to historian Tim Egan, the "soil was like fine-sifted flour, and the heat made it a danger to go outside many days."[60] Farmers thought the apocalypse was upon them. When dust storms came through a town, they blotted out light from the prairie sky. Wildlife and livestock choked to death. Lung disease was epidemic. In the summer of 1934, the Great Plains baked in a heat wave, with temperatures reaching 118 degrees in Nebraska and 115 in Iowa. In Illinois, temperatures of over a hundred degrees resulted in the death of 370 people. And the East was not spared. That May, a western dust storm actually blew soot all the way to the East Coast, causing great consternation in the Boston-New York-Washington corridor. Nothing could stop these heartland dusters. "Farmers watched their fields disappear before their eyes," Ian Frazier wrote in his 1989 book *Great Plains*. "Tumbleweeds blew up against fences, caught the dust, and were buried."[61]

Roosevelt was faced with the formidable challenge of developing a grasslands strategy for America to help combat erosion, develop a reliable water-supply system, and cultivate more vegetation. The destruction of millions of acres of prairie grasses, especially those crucial to holding soil down on marginal lands, and the widespread erosion gave rise to the Dust Bowl.[62] The Roosevelt administration, using funds from the National Industrial Recovery Act of 1933 and the Emergency Relief Appropriations Act of 1935, purchased 11.3 million acres of fallow or worn-out land for an average price of $4.40 per acre for grassland restoration; the best hope for the revival of the Great Plains and the Midwest would be to bring back the prairie ecosystems.[63]

A dust storm sweeps across Baca County, Colorado, in the mid-1930s. In some parts of the country, such occurrences were natural, but by the 1930s, overcutting made dust a destructive force in places that had formerly been green, carrying good soil away and leaving the land parched.

Convinced that reforestation and reflooding were ways to restore the natural resilience of the land, Roosevelt issued Executive Order 6793 on July 11, 1934, establishing the Prairie States Forestry Project. Better known as the Shelterbelt, the program entailed the planting of trees and shrubs as windbreaks along the borders of croplands and pastures to reduce wind speeds and decrease the evaporation of moisture from the soil. Running along a carefully configured patchwork from the Canadian border to Abilene, Texas, this great American wall of trees would protect crops and livestock and even contain the huge dust clouds.[64] The CCC was directed to establish a "transition zone" between the tallgrass prairies and shortgrass prairies.[65]

Roosevelt touted the Shelterbelt as a way to anchor the soil to the earth.[66] Its boldness and scope were breathtaking. "This will be," Forest Service Chief Ferdinand A. Silcox accurately predicted that July about the Shelterbelt, "the largest project ever undertaken in the country to modify climate and agricultural conditions in an area that is now consistently harassed by winds and drought."[67]

Roosevelt had appointed Silcox—a stalwart of the Boone and Crockett Club and the American Forestry Association—as the fifth chief of the U.S. Forest Service. A native of Georgia, Silcox brilliantly utilized the CCC and WPA in forest improvement projects until his sudden death in 1939. Under his leadership, the Forest Service provided space for more than eight thousand CCC camps, which would employ approximately two million men during the nine-year existence of the CCC. Soon after taking office in 1933, Roosevelt had been persuaded by Silcox to organize two national forests in northern Wisconsin by presidential proclamation.[68] The resulting Chequamegon and Nicolet national forests were commensurate with FDR's idea of a permanently protected upper Midwest tapestry of pine savanna: white, red, and black oaks; aspens; beeches; basswoods; sumacs; and paper, yellow, and river birches. Coniferous trees in the forests included the balsam fir and eastern hemlock.

Like the CCC itself, the Shelterbelt was essentially FDR's "own idea."[69] Under his direct order, the USDA planted six to twelve rows of trees and shrubs around farm structures; grew trees specifically for posts, piles, and wildlife cover; block-planted trees on enormous swaths of public lands; and, most commonly, created farm strips along fields, constituted of ten rows of trees spaced ten feet apart.[70] Scoffing at skeptical soil scientists, state foresters, and state extension services, the president was confident that the federal government could even alter the climate of the sunbaked Great Plains. Not only would the windbreaks reduce evaporation and therefore curtail the loss of moisture from soil, but in winter they would serve as snow traps and shield living quarters and barns from the greatest danger in blizzards.[71]

Roosevelt's Shelterbelt was the most ambitious afforestation program in world history. Unfortunately, it was also certain to offend

Great Plains farmers and ranchers who didn't like the federal government interfering with their land.[72] Roosevelt marketed the Shelterbelt to them on the firm promise that this tree-planting extravaganza, administered by the U.S. Forest Service, would at least provide decent jobs in the Dakotas, Nebraska, Kansas, Oklahoma, and the Texas Panhandle. When critics of the Shelterbelt protested against the federal government's overreach, Roosevelt had a ready answer. "The Nation that destroys its soil," he reminded Governor James V. Allred of Texas, "destroys itself."[73]

Cynics said Roosevelt's Shelterbelt was socialistic and a pseudo-scientific experiment. Conservative farmers and real estate profiteers demanded that the experimental program be abolished before it became an embarrassment to the United States, or a cataclysm. Many of FDR's trees, the rap went, brought with them insects and diseases. Wild cherry encouraged tent caterpillars. Elms were breeding quarters for canker worms. Red cedar brought rust disease. Republican-owned newspapers in the Great Plains ran vicious opinion pieces designed to derail or suppress the innovative New Deal project. "Only God can make a tree," an editorial in the *Amarillo Globe* argued. "If he had wanted a forest on the wind scoured prairies of Nebraska and Kansas, he would have put it there . . . and . . . for FDR to rush in where the Almighty had feared to tread was not only silly, but possibly blasphemous!"[74]

From the air, the Shelterbelt looked like a dense zigzag of forests draped across the flatlands. Working around prairie potholes in such places as North Dakota, sand hills in Nebraska, and the Llano Estacado in Texas, the planting crews placed seedlings of species that grew quickly and didn't require much water—because the soil was so dry. Most government foresters agreed that conifers were the best wind deflectors, but cottonwoods—regardless of their unusual bent and shape—were a more reliable option. "In the long run, overreliance on cottonwoods and under-utilization of conifers meant less effective shelterbelts," Joe Orth explained in *Agricultural History*, "but in the short term their choices helped get more trees in the ground and increased popular support."[75]

After some deliberation, Roosevelt chose cottonwoods over conifers for the Shelterbelt because they grew rapidly, almost like weeds, even in inhospitable conditions.[76] This meant Great Plains farmers who were dubious about the New Deal project would soon see sprouting cottonwoods blocking heavy winds all around their neighbors' homesteads and perhaps participate in the Shelterbelt program. In that regard, the president's choice worked. An increasing number of Great Plains farmers approved the planting of cottonwoods on their farm acreage, making sure they grew beyond the weed and grass inhibitions.[77]

However, Roosevelt didn't want cottonwood catkins to become the Shelterbelt's enduring symbol; they were too gauzy and flimsy. For that honor, he chose the hardy, rugged Austrian pine—a superior tree for windbreak situations. Roosevelt had first been struck by *Pinus nigra* when traveling around Europe as a boy. It was the darkest shade of pine he'd ever seen. Although it easily grew to 150 feet in height, it was of minimal value as a timber tree because the wood was rough, knobby, and gnarled. But Roosevelt knew from his discussions with Gifford Pinchot that these pines could grow on the poorest of American soils. It was therefore the ideal tree for Dust Bowl farms in the New Deal afforestation program.[78]

Tweaking and burnishing USDA directives, the president numbered tree-planting locations like CCC camps. It all started with the "Number One Shelterbelt" in Greer County, Oklahoma. The state forester, George Phillips, had the honor of planting the first tree (an Austrian pine) on the homestead of farmer H. E. Curtis near Willow, Oklahoma. The government paid Curtis for the right to plant rows of pines on his property. At the time, Roosevelt stressed that for the black blizzards to be fully curtailed the Shelterbelt wouldn't be enough. A comprehensive soil conservation program also had to be undertaken by Great Plains farmers.[79]

Critics couldn't dent FDR's confidence in the Shelterbelt. There was a credible body of agricultural science from the 1920s and early 1930s to back up FDR's ambitious antidote to desertification. Indeed, in the long run the Shelterbelt lived up to its bold promise to stabilize soil, reduce dust, ward off winter injury, provide shade for livestock, and restore habitat for wildlife.[80] Embattled Great Plains farmers, struggling with the hand of drought, came to see that. In the meantime, the program injected fresh money into the plains, through jobs for local men.

Disdainful of incrementalism and almost pathologically impatient, Roosevelt prodded Wallace for raw data and monthly updates about the Shelterbelt project. After he read a two-hundred-page technical guide, *The Possibilities of Shelterbelt Planting in the Plains Region*, issued by the Forest Service Lake States Experiment Station (based in Saint Paul, Minnesota), he demanded that the USDA plant whole forests in the plains states, not just clusters of blue spruce, red cedar, ponderosa pine, cottonwood, and Austrian pine.[81]

"I wish the Department would study and give me a report on this whole subject from a large acreage point of view," Roosevelt wrote to Wallace. "In the State of New York alone, for example, over forty million trees a year are produced by the State and planted either on State-owned land or on privately owned land. This means a total of forty thousand acres a year planted to trees. If the State of New York can do this for a very small annual cost, it is time that the Federal Government did it, especially on the vast public domain the title of which is still in the Federal Government."[82]

Roosevelt hired Paul H. Roberts of Nebraska to direct the project.[83] Roberts knew the soil of the Great Plains better than anyone else in government. After earning a BS in forestry from the University of Nebraska in 1915, he was hired by the U.S. Forest Service to combat drought in the West. Well educated and arrow straight, he started off as a forest ranger and eventually rose to become the supervisor of Sitgreaves National Forest in Arizona and New Mexico. In 1934, as chief of the Shelterbelt project, Roberts hired first-rate field directors and capable local staffers in each Dust Bowl state. To Roberts, it was imperative that the Shelterbelt should not seem to be forced on farmers by the Roosevelt administration—even though it was.

Following Roosevelt's directive, Roberts started tree nurseries like those in Saratoga, dedicated to producing the planting stock.

Farmers were encouraged to grab free seeds and saplings. Federal nurseries were quickly founded in Ames, Iowa; Mandan, North Dakota; Stillwater, Oklahoma; Cheyenne, Wyoming; and Pullman, Washington.[84] Men with shovels planted more than 220 million trees and shrubs on thirty thousand farms. Often these tree belts were the only green living things on the exhausted prairie. Roosevelt had found at Hyde Park and Warm Springs that wildlife flourished when the soil and woodlands were healthy. This basic ecological principle was nevertheless ignored by millions of American farmers. The Roosevelt-Roberts team promoted hedgerows, to be planted on western ranches instead of barbed wire; hedgerows were better for wildlife.[85]

Some scientific foresters dismissed FDR's wind-reducing hedgerows plans as opposed to *real* forestry, mere busywork for gullible farmers. To counter detractors the White House cooperated with *Life* magazine and the *Christian Science Monitor* in a Shelterbelt propaganda campaign. Inspirational WPA posters were printed by the Roosevelt administration showing heroic, muscular work-relief farmers joining the Shelterbelt effort. The ad campaign claimed those men weren't *just* planting trees; they were learning about cooperative conservation and national planning. The teachers were, for the most part, Forest Service technicians from land-grant universities in the Midwest and Great Plains. Operating under the aegis of the Forestry Service in 1934–35, the Shelterbelt project won over many farmers, but powerful members of Congress, in league with respected foresters, continued a stubborn opposition. The congressmen dug deep into their bag of tricks to kill the Shelterbelt, but Roosevelt proved that he had a bag of tricks, too—and it was even deeper. With the Shelterbelt on the verge of being struck from the budget, Roosevelt deftly moved it over to the WPA, in which the executive branch held control of disbursements. It lasted there through the early 1940s.

V

At the president's urging, Ding Darling began forming partnerships with state conservation commissioners in the West and Midwest. He continually battled for increased federal appropriations for duck habitat. Scrounging for operational funds was a constant frustration for him. In the spring of 1935, a frustrated Darling would tell the Associated Press that he was "ready to quit" because of Congress's failure to include wildlife restoration in the major work-relief bill and because other agencies in the Roosevelt administration had treated the Biological Survey poorly.[86] Persevering, he called for the Biological Survey to hire land negotiators, surveyors, engineers, draftsmen, and biologists. Darling's most inspired move was his hiring of thirty-two-year-old John Clark Salyer II to identify tracts of land for the government to purchase. He had first encountered Salyer, a redheaded biologist, in 1932 while working with the Iowa Fish and Game Commission. Originally from Higginsville, Missouri, Salyer had an exhaustive knowledge of North American birds. He was pursuing a PhD in ornithology when Darling offered him a crucial post at the Biological Survey.

Once Salyer moved to Washington, D.C., he set about finding 600,000 acres of suitable migratory waterfowl habitat for federal purchase using money from the Duck Stamps. He first made recommendations about which existing federal migratory bird refuges—including Mattamuskeet in North Carolina and Horicon Marsh in Wisconsin—were in urgent need of rehabilitation.[87] There were around 150 species of ducks, geese, and swans, all belonging to the biological order Anseriformes, that would benefit from the land purchases. Ducks were the largest of the three groups and could be divided into at least two subsets: dabbling ducks and diving ducks. It was a dream job for any intrepid wildlife biologist, but there was a hitch: Salyer, though able-bodied, was afraid to fly. So Darling, not wanting to lose talent, offered to lend him a federally owned Oldsmobile sedan to conduct a six-week inspection tour of far-flung waterfowl habitats west of the Mississippi River.

In late 1934 and early 1935, Salyer traveled twenty thousand miles looking for breeding grounds in Montana and the upper Midwest. Speed was very essential: if the Biological Survey didn't buy land by March 31, 1935, then the money would revert back to the WPA. Salyer worked like a fiend, living on pocket money, hopping out of his car at regular intervals to identify bird species and scribble notes.

"Salyer wore out government cars at a great rate," George Laycock recalled in *The Sign of the Flying Goose*. "He drove with his mind on wildlife, and the list of government workers who refused to ride with him grew rapidly. In those months he only saw his wife and infant son on rare occasions."[88]

John Clark Salyer II, photographed in the field during the 1930s. Salyer was appointed by Darling to develop the Division of Wildlife Refuges along the unbending rules of Leopold. Salyer was well matched to the position, spending much of his time inspecting refuges in person. When he arrived in his job in December 1934, the nation counted 1.5 million acres in refuges; by the time he retired twenty-seven years later, there were more than twenty-eight million acres under protection.

A week passed and then another and another. As Salyer's deadline approached, he began scrambling. A few choice lands still had to be appraised in the Dakotas' prairie pothole regions. It was a do-or-lose situation. He had long passed the point where he felt that sleeping in his car was undignified. After a mad race back to Washington, D.C., Salyer holed up in his Biological Survey office with his secretary and rushed to get the paperwork finished by the deadline. But an unexpected obstacle nearly derailed Salyer's hard work. Secretary of Agriculture Wallace, who had to give final approval for Salyer's proposed habitat purchases in the Dakotas, was out of town. After a few hours of weekend fretting and pacing about the corridors of the USDA, Salyer grabbed a fountain pen and forged Wallace's authorizations. "I could have gone to prison," Salyer later recalled. "That was the longest weekend I ever spent."

Filled with fear and guilt, Salyer paced outside Henry Wallace's door on Monday morning, steeling himself to confess his forgery. When Wallace finally arrived Salyer came clean. As Wallace listened to Salyer's audacious story, there wasn't the barest hint of a smile on his face—but no reprimand was forthcoming either. Instead, Wallace instructed his underling to get back to his desk and keep working.[89]

Salyer's forgery had paid off. At Wallace's behest, Salyer ironed out USDA purchase agreements, making sure all of the boundaries were legally exact. Wallace had only one order: no lawsuits. Darling was also fully supportive, as the dream of the Beck Report finally began to be realized. Quite taken with the Souris River Basin in northern North Dakota, Salyer selected three potential refuges in the region; the area was a critical breeding ground along the Central Flyway. The Souris River crossed the border from Saskatchewan, following a pronounced horseshoe curve through North Dakota for three hundred miles, before running back into Canada. Two of Salyer's sites would be wildlife refuges, while the third would be a wildlife management district. Between August 22 and September 4, 1935, President Roosevelt signed executive orders establishing all three.[90] At the smaller of the two refuges, Upper Souris, CCC crews built dams primarily to create a healthy environment for birds . . . and fish, frogs, and other animals. The lakes that those dams created, including one named after Ding Darling, were also engineered to relieve droughts downriver, if necessary. An unplanned benefit was the ability of the improved river basin to reduce flooding downstream as well. The complex of preserves saved breeding grounds just as migratory bird populations were in crisis.

Determined to save all North American birdlife from extinction, Salyer became indispensable at the Biological Survey during the Roosevelt years. Whenever a refuge manager learned that Salyer was coming for an inspection, everyone on the staff would jump to get things right (employees even baked pies for their boss to enjoy). One former Survey employee called him the General George C. Patton of wildlife protection. Salyer always knew the needs of each refuge and easily bonded with fellow biologists. Denny Holland, who worked at Santee National Wildlife Refuge in South Carolina, remembered that, in a fitting tribute to their boss, the employees there named the biggest red oak tree on the grounds the Salyer Tree. "Whenever he came for a visit," Holland recalled, "he'd say 'Let's go see my tree.' "[91]

Besides acquiring land for refuges, a major priority for Darling and Salyer in the mid-1930s was the enforcement of antihunting laws in federal refuges. Many of the Biological Survey's field wardens had grown lax about busting poachers, to avoid the rigmarole associated with long-drawn-out court hearings. When Darling became chief, the Biological Survey had fewer than thirty wardens for the entire nation. The first place Darling sought to reverse the backslide was along Maryland's waterfowl-rich eastern shore. Poachers there had been slaughtering canvasback and redhead ducks for ornamental display in nightclubs, restaurants, and bars. It was standard practice. Arrests were ordered. Within a year, the Biological Survey busted forty-nine illegal duck shooters in the mid-Atlantic states. "It became suddenly apparent that the Biological Survey was under new management," David L. Lendt noted in *Ding*, "and that the new boss meant business."[92]

FDR knew from living in Hyde Park, where he often participated in Audubon's Christmas bird count in Dutchess County, that establishing federal migratory waterfowl refuges was only a small piece of a much larger conservation effort. During his formative years, FDR had been taught by Dr. Frank Chapman about the "citizen bird" movement, which attempted to get millions of Americans involved in putting out birdseed and birdbaths at their homes. The New Deal picked up on this backyard approach to bird stewardship, directing the Departments of the Interior and Agriculture to issue "bulletins" about how to build birdhouses, dig duck ponds, establish bird-feeding stations, and eradicate feral cats. These bulletins also suggested ways to transform picnic areas, cemeteries, sylvan school lots, and the acreage surrounding reservoirs into prime habitat for birds.[93]

The Fish and Wildlife Coordination Act also passed Congress in 1934, allowing certain lands and waterways under the jurisdiction of the Forest Service, the Bureau of Reclamation, and what would eventually be known as the Bureau of Land Management to become migratory bird refuges. There was, however, an important caveat:

the main economic engine (mining, fishing, logging, grazing, etc.) of a given area couldn't be interfered with or curbed.[94]

Because of the Beck Report, the Duck Stamp, and the president's policies of habitat acquisition and refuge management, fueled by congressional appropriations and the USDA-Interior public awareness campaign, migratory waterfowl would increase in numbers from thirty million in 1933 to more than a hundred million by the onset of World War II.

After three years in office, Roosevelt had done more for wildlife conservation than all of his White House predecessors, including Theodore Roosevelt, establishing forty-five new wildlife refuges.[95] By the end of fiscal year 1935, the Biological Survey had acquired 1,513,477 acres—surpassing all prior Biological Survey refuge land acquisitions—especially in the upper Midwest. "Our restoration efforts," Salyer said, "have been felt in almost every State of the Union."[96]

CHAPTER TWELVE

"SOONER OR LATER, YOU ARE LIKELY TO MEET THE SIGN OF THE FLYING GOOSE"

I

After designing the first Duck Stamp, Ding Darling was still bursting with creativity. In another indelible image for use by the U.S. government he drew a flying blue goose (*Chen caerulescens*), with a feathered back as flat as an ironing board. Beginning in 1935, this logo, with its stylish elegance, welcomed visitors to federal wildlife sanctuaries throughout America.[1] Of all the New Deal emblems, the "Blue Goose," was the most aesthetically well conceived. "If you travel much in the wilder sections of our country, sooner or later, you are likely to meet the sign of the flying goose—the emblem of the National Wildlife Refuges," biologist Rachel Carson later wrote. "You may meet it by the side of a road crossing miles of flat prairie in the West, or in the hot deserts of the Southwest. You may meet it by some mountain lake, or as you push your boat through the winding salty creeks of a coastal marsh."[2]

There were several reasons Darling chose the mysterious blue goose, which became the official symbol of national wildlife refuges in 1940. It was a variant species, a dark color form (or "morph"), of the snow goose (*Chen caerulescens*). Since 1916, North American hunters were banned from killing snow geese, which wintered primarily in Louisiana and Texas, because of dwindling populations. The blue goose was even scarcer, a truly endangered creature. E. A. McIlhenny, the Tabasco heir, wrote movingly about the blue goose in *The Auk*, warning that the species was in danger of extinction.[3] Another reason Darling chose the blue goose was that it migrated along four

North American flyways in small armadas, stopping in lakes, farm fields, brackish marshes, and sandbars.

Workers installed signage designating the border of a national refuge, circa 1938. Darling designed the "Blue Goose" insignia still found on all National Wildlife Refuge Service signs.

The first official Blue Goose sign—a dark blue bird against a white background—was ordered by John Clark Salyer for the Upper Souris Migratory Waterfowl Refuge in North Dakota in 1935. The upper Souris River Valley became a living laboratory for FDR's waterfowl revivification strategy. Reporters were driven to the Souris Valley to see the Biological Survey and CCC hard at work on behalf of migratory birds. In all, eighty thousand acres were reclaimed as waterfowl habitat. "It is a magnificent thing," Dr. Ira Gabrielson, head of the Biological Survey's Division of Wildlife Research, said to the press, about the Souris River Basin refuges, "and will be a better nesting area than it ever was in its native condition."[4]

Most of FDR's so-called Blue Goose refuges of 1935 were in North Dakota. Salyer had wisely bought prime bird breeding grounds for the Biological Survey on the cheap. On August 22, Roosevelt signed Executive Order 7154-A establishing Des Lacs NWR (from the French *Rivière-des-Lacs*, "River of the Lakes"), located on the Canadian border near the town of Kenmore, North Dakota. The picturesque 19,500 acres were designed for migratory birds, but buffalo, antelope, and elk also found safe pasture there. Building on the North Dakota model, Salyer did the same along the Atlantic Flyway, posting Blue Goose signs in FDR's latest refuges, such as Pea Island (North Carolina), and Carolina Sandhills (South Carolina). Phenomenal aerial cascades of ducks, along Coheknig's Central Valley and up the Mississippi River into the Chesapeake Bay, would return under this initiative. "Whenever you meet this sign, respect it," Rachel Carson soon wrote. "It means that the land behind the sign has been dedicated by the American people to preserving, for themselves and their children, as much of our native wildlife as can be retained along with modern civilization."[5]

Symbols like the Blue Goose were personally important to Franklin Roosevelt. As a collector of naval prints and postage stamps, he responded to illustrated posters and strong graphic designs. Cognizant of this, Darling drew a cartoon of FDR as "Doc Duck," a mallard wearing a white top hat and flashing a sign that read "Duck Restoration." Metal signs featuring Doc Duck were prominently posted by the CCC on trees and fence posts all over the Midwest and Great Plains.[6] Children seemed to particularly enjoy the cartoon. It was a fun-loving way to say "Keep Out!" And the CCC often employed local women's garden clubs to post the Doc Duck sign in federal bird sanctuaries. In the spirit of using birds to promote the New Deal, Roosevelt personally designed a red-and-blue stamp featuring an American bald eagle (*Haliaeetus leucocephalus*) used to distinguish between airmail and regular mail.[7]

Salyer, the primary architect for Roosevelt's waterfowl restoration program, pushed for the U.S. Biological Survey to break ranks with the CCC and allow women to work at federal wildlife refuges. Eleanor Roosevelt agreed. But Henry Wallace never moved on this matter. It wasn't until 1946, with Wallace out of the picture, that Salyer was able to employ (under the reorganized U.S. Fish and Wildlife Service) his first female research biologist: Elizabeth Beard Losey, of Marblehead, Massachusetts, who had recently graduated from the University of

Michigan and was an expert in waterfowl and marsh management.[8] Losey later recalled, "Mr. Salyer . . . stuck his neck out" by hiring her, a woman, for the position.[9] Her first assignment for the survey was to document the importance of beavers in waterfowl management at Seney Migratory Waterfowl Refuge, a huge, 95,238-acre tract established by FDR in 1935 in Michigan's Upper Peninsula.[10]

Seney National Wildlife Refuge was also an important spot for the recovery of the Canada goose (*Branta canadensis*), then a threatened species. Instead of simply saving habitat at Seney, biologists had built a fenced-in pond area to contain a crop of goslings and then trained them to establish a migratory pattern. If one were to place a historical marker on a single location where the Canada goose made its remarkable comeback from a threatened to a ubiquitous species, Seney was the spot.[11]

Each NWR established in 1935 saved entire ecosystems from corruption by humans. Most had a particular species of waterfowl as a focus. The reason that Rice Lake NWR was situated in a bog area of northern Minnesota was to help ring-necked ducks (*Aythya collaris*) hatch. Then there was Medicine Lake NWR in Montana, where the white pelican (*Pelecanus erythrorhynchos*) nested. In Arkansas, mallard ducks (*Anas platyrhynchos*) wintering in the Mississippi Flyway had a 160,756-acre home at the White River NWR. All of these landscapes would have been doomed to exploitation if the U.S. Department of Agriculture hadn't acted decisively in 1935.

II

Even though Roosevelt was in the middle of crafting his Emergency Relief Appropriation Act of 1935, which created the Works Progress Administration (WPA), he wrote to Henry Wallace and Harold Ickes early that year about "a deficiency" in feed hay and cottonseed cake for the elk in Jackson Hole and the grizzlies in Alaska.[12] Roosevelt also instructed Robert Fechner to make sure the CCC boys were planting trees with the specific goal of helping wildlife prosper in the Midwest. "What would you think of a conference, starting under my auspices, between Biological Survey, Forestry Bureau, Reclamation Service, to try to work out a definite plan between these and possible other Bureaus?" FDR wrote to Wallace that February. "If you think it wise let me know who should be at such a conference."[13]

At the beginning of 1935, Roosevelt sent a message to Congress on the topic of New Deal conservationism. It was as clear an expression as he ever made of his environmentalist ideals. "During the three or four centuries of white man on the American Continent, we find a continuous striving of civilization against Nature," Roosevelt said. "It is only in recent years that we have learned how greatly by these processes we have harmed Nature and Nature in turn has harmed us." Scolding fellow Americans for violating "nature's immutable laws," Roosevelt praised conservationists who fought to keep the North American forests alive with wildlife and the shorelines clear of industrial detritus. "In recent years little groups of earnest men and women have told us of this havoc of the cutting of our last stands of virgin timber; of the increasing floods, of the washing away of millions of acres of our top soils, of the lowering of our water-tables, of the dangers of one-crop farming, of the depletion of our minerals," Roosevelt said, "in short the evils that we have brought upon ourselves today and the even greater evils that will attend our children unless we act."[14]

Five days later, the Society of American Foresters (SAF) presented President Roosevelt with the Sir William Schlich Memorial Medal for his efforts with the CCC. He was the first recipient of the award. In accepting the honor, Roosevelt took the opportunity to warn that forests had to be "lifted above mere dollars and cents" and the "selfish gain" of industrialists. Americans, Roosevelt charged, have "often heedlessly tipped the scales so that nature's balance has been destroyed."[15]

The president didn't mention the CCC or tell any stories about the Shelterbelt, as the SAF members might have expected. Instead he spoke of trees, which brought beauty to the American landscape and controlled the flow of water in scenic rivers. "The forests are the 'lungs' of our land, purifying the air and giving fresh strength to our people," Roosevelt said. "Truly, they make the country more livable. There is a new awakening to the importance of the forests to the country, and if you foresters remain true to your ideals, the country may confidently trust its most precious heritage to your safekeeping."[16]

In early February 1935, Ding Darling urged the president to regulate hunters so that wildlife could recover in the public domain. In the American West, families hunted on public lands; by contrast, the majority of hunting in the eastern United States took place on private property. In a confidential memorandum to the president, Darling made an impassioned plea for fish and game reform: "Game has been the orphan child without asylum in the conservation world. No provision has been made for its permanent home," he wrote. "There has been no Government agency entrusted with its custodianship. It has subsisted on the crumbs dropped from the table of forestry, parks, reclamation and advancing civilization. That it has escaped total extinction is through no foresight or comprehensive plan."[17]

Determined to institute real reform, Darling urged the president to reduce the hunting season for waterfowl to thirty days, prohibit baiting, limit automatic and repeater shotguns to three shells (at one loading), ban all sink box and sneak boat hunting, and abolish (essentially) the use of live decoys. Roosevelt agreed, and new restrictions were announced on August 1. Certain avians were given outright protection—it became illegal to shoot bufflehead (*Bucephala albeola*), ruddy (*Oxyura jamaicensis*), brant geese (*Branta bernicla*), or wood ducks (*Aix sponsa*). The greater and lesser scaup (*Aythya marila*, *Aythya affinis*, respectively) were also given increased federal protection—only eight of each per hunter were allowed—with a ban on shooting the species at all to become law in 1937.[18]

Both Roosevelt and Darling thought that states shouldn't determine "shooting seasons." Instead the U.S. government, amending the Migratory Bird Treaty Act of 1929, now dictated the precise dates for specific birds and regions. Most sportsmen were expecting strict rules and understood that the need to save species was urgent. A few, of course, bitterly complained. Joseph Pulitzer, publisher of the *St. Louis Post-Dispatch*, objected to the administration's proposed hunting regulations—particularly the one prohibiting the shooting of ducks from baited water—in an editorial that caught the president's attention. "I am forced to the conclusion that some sacrifice by the hunters will have to be made which will reduce the number of ducks killed this season," Roosevelt wrote to Pulitzer after reading the editorial, "and that such restrictions must continue until such time as natural conditions with the aid of our restoration program may allow a liberalization of the regulations."[19]

Using Iowa State's Cooperative Wildlife Research Unit as a model, the Roosevelt administration recruited nine land-grant colleges to collaborate with Uncle Sam to help restore waterfowl populations in the Mississippi Flyway. Darling, putting his celebrity to good use, was masterly at inspiring state conservation divisions to work closely with the USDA. "The correctives for these [land abuse] evils include measures to fasten the loose soil in place by planting trees, shrubs, grasses, and other suitable vegetation to nullify wind-action, conserve moisture, and reduce the runoff of water from rainfall and melting snow," Darling wrote. "These are wildlife conservation measures, too!"[20]

Throughout 1935, Darling sought to develop public-private partnerships to rescue North American wildlife by habitat restoration. To this end, he served as impresario for a gala at the Waldorf-Astoria that April, attended by leading industrialists and munitions makers. Darling hoped companies that benefited from the hunting culture would join in common cause with the Biological Survey. Reminding corporate leaders of their happy boyhoods spent hunting in the great American outdoors, Darling made his pitch: he wanted the CEOs of such corporations as DuPont, the Hercules Power Company, and the Remington Arms Company (the venerable gun manufacturer based in Ilion, New York) to hold a nationwide conference on wildlife (cosponsored with sportsmen's clubs). Indeed, C. K. Davis, president of Remington, decided to contribute to the Biological Survey's wetlands restoration efforts after attending the dinner at the Waldorf-Astoria. Davis, an avid duck hunter, didn't like the New Deal's "unreasonable" seasonal restrictions on hunting, but he sympathized, in general terms, with the waterfowl conservation cause. The event at the Waldorf-Astoria also led to some new nonprofits, organized for the long fight ahead.[21] "Out of this single meeting," Darling said, "there emerged, either directly or indirectly, the Cooperative Wildlife Research Unit Program, the American Wildlife

Institute, the National Wildlife Federation (NWF), and the North American Wildlife Conference—a rather productive three hours of dinner conversation."[22]

During his first two years in office, Roosevelt had walked point on purchases of waterfowl habitat. By 1935, however, his efforts had expanded to include the protection of other wildlife: bison, antelope, grizzly bear, Kodiak bear, elk, moose, caribou, sage grouse, and wild turkey. These species were all fighting for the freedom to roam unmolested by hunters, cars, canals, railroads, and factories. Of particular interest to FDR was the desert bighorn sheep (*Ovis canadensis nelsoni*) making a heroic comeback. "Thirty years ago, Theodore Roosevelt established a permanent government policy for forests," Darling wrote to FDR. "A like permanent policy on wildlife is long overdue. It looks to me like an opportunity."[23]

FDR appreciated the analogy with TR. His New Deal conservation program aimed especially to save big mammals living on public lands in the American West. On March 20, 1935, with the president's full support, the New Deal's first grazing district opened in Rawlins, Wyoming. Those with livestock could use federal land within the districts and in addition, reserves for wild animals would be established.

Anti–New Dealers initially derided FDR's grazing district as the "closing of the West," but the president knew that the reseeding would rehabilitate abused grassland and burned-out rangeland. The U.S. government now forbade free grazing access to the public range. At a Salt Lake City conference, Roosevelt defended the grazing districts as a "new conservation movement" for the American West that "promises historic significance."[24]

Secretary of the Interior Ickes appointed a Colorado rancher, Farrington Carpenter, to head the newly established Grazing Service and to persuade farm communities that the regulation of grazing of federal lands was a good thing. A full-time cattleman with degrees from Princeton and Harvard Law School, Carpenter tried assiduously to bridge the gulf between freedom-loving westerners and the New Deal Washingtonians who were trying to regulate the range. He was a sensitive conservationist who recognized that the land had to be shared with other species—and that westerners had overgrazed the open land to the extent that livestock ranching was barely viable. Through an unending schedule of local meetings, he brought the majority of cattlemen into compliance with the new grazing law, but Ickes didn't like his ameliorative style. In fact Ickes didn't like Carpenter, and vice versa. On the day after the midterm election in 1938, Ickes fired him. (This sent a broad message throughout the Roosevelt administration not to mess with Ickes.) While Carpenter returned to ranching, the value of his work can be seen in the fact that the grazing districts he designated in cooperation with locals in the 1930s are still recognized today. Carpenter's replacement was Richard H. Rutledge of Utah. Rutledge (who would serve until 1944) placed an even greater emphasis on wildlife conservation than did Carpenter.

The implementation of the Taylor Grazing Act of 1935—the closing of the public domain from further entry via an FDR executive order—helped restore fragile grassland and prevent soil deterioration. Within a couple of years, FDR's fifty-nine grazing districts encompassed 168 million acres of federal land and an additional 97 million acres of private or state-owned land. Like the New Deal system of migratory bird refuges, protected grazing districts in the West had become a permanent part of American life.

The Taylor Grazing Act notwithstanding, no federal program had been specifically devised for what Carl D. Shoemaker, an investigator with the Senate Special Committee on the Conservation of Wildlife Resources, called the "perpetual preservation of [America's] national wildlife assets." In a letter of March 1935 to Roosevelt, Shoemaker pointed out that while the public domain was in the process of being apportioned to the grazing districts some acreage should also be set aside by executive orders for the desert bighorns. "The important item for consideration," Shoemaker wrote, "is that the game areas be withdrawn before and not after the last of the Federal domain is given away."[25] Roosevelt agreed.

Darling and Shoemaker weren't the only advocates of huge wildlife reserves whom the president consulted in 1935. When FDR took a cruise to Florida that May—his springtime ritual—he noticed that a palmetto thicket normally twenty feet from shore had been replaced by concrete. Jungle had been uprooted, wetlands had been filled, and the fishing was poor. So concerned was Roosevelt about the coastal environment of Florida that he asked John Baker of the National Audubon Society to investigate the seemingly diminished brown pelican (*Pelecanus occidentalis*) population from Jacksonville to Key West. FDR was sentimental about these comical birds. After investigating the situation, Baker told him that a large pelican colony was thriving in Brevard County on the eastern coast of Florida. "Many thanks for that interesting letter about the Brown Pelicans," Roosevelt responded. "I am happy to know that they are apparently not decreasing in recent years. I became somewhat worried because on our recent trip through the Bahamas we saw practically no Pelicans at all."[26]

The theme of "management efficiency" with regard to natural resources dominated Roosevelt's thinking about the public lands of Florida. Encouraged by Baker, the president determined that relevant government agencies (the Biological Survey, National Park Service, Forest Service, and so on) needed to work in partnership with wildlife conservation nonprofits such as National Audubon and the Izaak Walton League to launch effective wildlife protection efforts. It was Darling's dream that Roosevelt might host a huge "Wildlife in Peril" conference in Washington. "Do you think you could get the President to call a Natl. Congress for Conservation next January," Darling prodded Wallace, "if I'd do all the work and see it properly financed and managed?"[27] An agreeable Roosevelt seized on the idea. "Tell Mr. Darling," Roosevelt wrote back to Wallace, "that I think a National Conference for Conservation would be excellent and that I heartily approve."[28]

III

On May 9, 1935, Roosevelt established the WPA by Executive Order 7034. The guiding spirit of the WPA was Harry Hopkins, supervisor of the Federal Emergency Relief Administration (FERA) and the Civil Works Administration (CWA). Hopkins would hold the purse strings of an initial government appropriation of $4.88 billion, or nearly 7 percent of the nation's GDP in 1935. With the exception of six "federal" projects run in Washington, D.C., the WPA's unemployment relief initiatives began at a grassroots level all across America. Along with Ickes, the president had no closer adviser than Hopkins. In fact, Roosevelt was fond of pitting the two against each other as a way to milk the best out of both. In fact, Ickes was convinced that Hopkins had chosen the acronym WPA to create public confusion with Ickes's own PWA, a view he held until the end of his life.[29]

Although the WPA had a more urban focus than the CCC—New York City claimed one seventh of all WPA expenditures—Hopkins also sent workers to assist in public lands projects.[30] While Harold Ickes and Robert Fechner were building state parks with CCC boys, Hopkins's WPA directed relief workers to construct and improve roads leading to woodsy campsites and swimming lakes. Hopkins also served on Roosevelt's first-term Great Plains Drought Area Committee in 1936. Between 1935 and 1943, over $10 billion came through WPA coffers to stimulate the economy and provide jobs to more than thirteen million Americans. At national parks, the WPA was tasked with installing sewage systems and water tanks, as well as reinforcing riverbeds. Working with the Biological Survey, the WPA began correcting the harm caused by the high number of marshland drainage projects in the 1920s.[31] And the WPA constructed seventeen new zoos—most with outdoor pavilions instead of cages—while brilliantly redeveloping the existing zoos in Central Park, San Diego, Dallas, and Buffalo.

Knowing that the CCC, PWA, and WPA were all ready to help improve federal wildlife refuges—as they were doing in state parks—Darling sought more federal funds. Vast parcels of mainly tax-delinquent land were being acquired by the U.S. government at an astonishing rate and Darling hungered to get in on the action. By late 1935, the U.S. government under FDR had acquired more than twice as much acreage in forestlands as had been bought in the prior history of the national forests.[32] Why not do the same for the federal wildlife refuges? Therefore, Darling wrote to the president in July 1935 to ask for another $4 million to purchase waterfowl habitat, arguing that the Biological Survey did more than the Forest Service or National Park Service. "Others just grow grass and trees on it," Darling told Roosevelt. "We grow grass, trees, marshes, lakes, ducks, geese, furbearers, impounded water and recreation."[33]

In this letter, Darling explained to the president that most of the Biological Survey's earlier appropriation had gone toward buying lands in the Okefenokee Swamp of Georgia and acquiring ranchlands near Jackson Hole, Wyoming. Another million or two had been spent on habitat for pronghorn antelope and elk around Oregon's Hart Mountain. "I need $4,000,000 for duck lands this year and the same bill which gave us the $6,000,000 specifically stated that at your discretion you could allocate from the $4,800,000 money for migratory waterfowl restoration," he pleaded. "We did a good job last year. Why cut us off now?"[34]

Roosevelt was amused by these pleas. He admired Darling's audacity in trying to shake the money tree. But he also thought Darling, always rattling the tin cup for more, was too zealous. "As I was saying to the Acting Director of the Budget the other day—'this fellow Darling is the only man in history who got an appropriation through Congress, past the Budget, and signed by the President without anybody realizing that the Treasury had been raided,' " Roosevelt wrote to Darling. "You hold an all-time record. In addition to the six million ($6,000,000) you got, the Federal Courts say that the United States Government has a perfect constitutional right to condemn millions of acres for the welfare, health, and happiness of ducks, geese, sandpipers, owls, and wrens, but has no constitutional right to condemn a few old tenements in the slums for the health and happiness of the little boys and girls who will be our citizens of the next generation! Nevertheless, more power to your arm! Go ahead with the six million dollars ($6,000,000) and talk with me about a month hence in regard to additional lands, *if* I have more money left."[35]

While Darling didn't get an extra $4 million, Roosevelt did instruct Ickes to transfer $2.5 million from the Public Works Administration to the Biological Survey.[36] A pleased Darling purchased another 526,800 acres of migratory waterfowl breeding areas in the Dakotas, Minnesota, Wyoming, Montana, Oregon, and Washington. Roosevelt justified the money under the authority of the 1934 Coordination Act, which provided, "Whenever the Federal Government, through the Bureau of Reclamation or otherwise, impounds water for any use, opportunity shall be given the Bureau of Fisheries and/or the Bureau of Biological Survey to make such uses of the impounded waters for fish-culture stations and migratory bird resting and nesting areas as are not inconsistent with the primary use of the waters and/or the constitutional rights of the States."[37]

Throughout 1935, Darling paid keen attention to Aldo Leopold's writings in periodicals such as *American Game*, *American Forests*, *Living Wilderness*, and *Wilson Bulletin*. Both Darling and Leopold understood that no longer could American conservation be only about "monumentalism," that is, saving the largest mountains, sequoias, or rock formations. A new caretaking ethic for *all* land—like Roosevelt's ethic at Hyde Park—had to be the ecological basis of the future.

To Roosevelt, the reclamation of rural lands was the heart and soul of the environmental New Deal. If one were to choose Roosevelt's three overarching priorities for land policy in 1935, they were stopping the American farmers' landslide, keeping cattle out of federal grasslands, and helping waterfowl and big mammals rebound. However, when Ickes argued that the panhandles of Oklahoma and Texas, devastated by drought, shouldn't be resuscitated for agriculture—that instead, the land should be allowed to "re-wild"—Roosevelt shrugged him off. Where Ickes wanted the return of tall grass and buffalo, Roosevelt imagined a brave new era of agriculture in Texas by way of water power and massive irrigation projects. As Rexford Tugwell observed, FDR "always did, and always would, think people better off in the country and would regard the cities as rather hopeless."[38] It sickened Roosevelt that Big Agriculture capitalists had slashed and burned their way across the continent with no regard for the ecological balance between "men and nature."[39]

Whenever possible, President Roosevelt returned to Springwood to inspect his tree farm and to start broadcasting hemlock seed in Tamarack Swamp. He had a hunch that hemlock would grow tall in the moist soil. "I have long been interested in comparing the results from acorns gathered from trees of different ages," Roosevelt wrote to a forester in Syracuse. "Would it be worthwhile to gather acorns from three age types—one lot, say, from a tree or trees about eighty years old, another lot from a tree or trees about one hundred and fifty years old? We have all of these trees, both red oak and white oak, on our place and the experiments could be tried with species of oak."[40]

Whenever Roosevelt felt trapped in the White House, he dreamed of the Hudson River Valley. The yesteryears that had passed hadn't dimmed his fond memories of childhood, when his legs were strong and the scooting summer clouds were dreamy and the autumnal colors fired in the woods. Roosevelt's assistant, Bernard Asbell, recalled how excited his boss was about all things related to Dutchess County. Whenever Roosevelt was at Springwood, he would interrupt Asbell with comments like, "You see that knoll over there? That's where I did *this-or-that*."[41]

Always trying to save Dutchess County's historic sites, Roosevelt sought to persuade the Norrie family to donate their tree-rich grounds along the Hudson River in Staatsburg to the New York State park system.[42] Roosevelt worried that the bacteria-laden Hudson, thick with algae and pathogens, was being polluted beyond redemption. In the West, rivers were often part of the public domain, but the Hudson was the responsibility of the New Yorkers living along its course. "When I was living in Albany I spent many hours trying to persuade municipalities to put in sewage disposal plants," Roosevelt wrote to Scott Lord Smith of Poughkeepsie. "As a matter of state government policy we undertook . . . to eliminate all sewage running from State institutions into the Hudson River. . . . The problem is, of course, wholly one for the municipalities and not for the Federal Government."[43]

This yearning for the Hudson Valley was most apparent in letters written to his distant cousin and confidante, Daisy Suckley, a sounding board whom he routinely took on drives to his favorite scenic spots. "Why is it," Roosevelt once asked her, "that our River and our countryside seem to be part of us?"[44] The intimate Roosevelt-Suckley letters are filled with pastoral references to soft summer grasses and the dancing snowflakes of winter. It was almost as if Suckley believed her job was to report all of nature's happenings along the Hudson to the overworked president, who was starved for details about his home. "*Would* you like some news," Suckley wrote in August 1935, in a teasing update, "It's *raining*! And rumbling in the distance—that bowling alley in the Catskills seems to be open 24 hours of the day—and practically *every day*."[45]

In the mid-1930s, Roosevelt and Suckley took on several silviculture projects in Hyde Park and Rhinebeck. The quirkiest was a scheme to grow California redwoods along the Hudson River, in an experimental forest plot of one hundred trees. Having seen this tree, the king of flora, on the West Coast—some individual redwoods had been alive when Christ walked in Galilee—FDR hoped to establish a grove in his own backyard. "The Redwoods have come!" Suckley wrote to Roosevelt at the White House. "But in the form of an envelope full of seeds! With instructions as to their care! They should appear above the ground within 17–20 days! I'll start them right off."[46]

Through her new syndicated "My Day" column, that appeared daily from 1935 to 1962, Eleanor Roosevelt also wrote about idyllic Dutchess County with noticeable regularity. Pastoralism was the most consistently pronounced sentiment in her writing, even more so than feminism and Democratic liberalism. "We entered our cottage at Hyde Park on Friday night," she mused in typical fascination, "sat on our porch, looked at the reflection of the sunset on the water and basked in a feeling of complete peace and quiet."[47] At Val-Kill, the First Lady hosted picnics, bird-watched, swam, and hiked. "The peace," she wrote Franklin, "is divine."[48]

IV

With soil-expert Hugh Bennett as evangelist, most residents in the American West supported the president's New Deal measures aimed at helping struggling farmers improve their soil conservation. But

Roosevelt visited the Everglades in the 1920s and was awed by the experience. As president, he signed the enabling legislation to establish the swamp as a national park. This photograph of the Everglades was taken in 1937 by George A. Grant, the first chief photographer for the National Park Service.

RIGHT: Roosevelt declared 1934 the "Year of the National Park." On his visit to Yellowstone in September 1937 he enthused about seeing the Old Faithful geyser. Photographer Ansel Adams took this landscape shot while working for the Interior Department during the Great Depression.

LEFT: The grandest of Roosevelt's 1943 national wildlife refuges was Chincoteague NWR, on the Virginia side of Assateague Island. The president's principal rationale for saving Chincoteague was that the greater snow goose (*Chen caerulescens*) needed a sanctuary in the mid-Atlantic. This refuge was the setting for Marguerite Henry's 1947 children's book *Misty of Chincoteague*, which made the island's wild ponies its most famous residents.

With Eleanor Roosevelt at his side, FDR traveled the famed Going-to-the-Sun Road in Glacier National Park. He soaked up thick forests and jagged mountain peaks that rose thousands of feet above the valley floor, declaring, "There is nothing so American as our national parks."

When FDR visited the Grand Canyon, he wasn't awestruck. "I like my green trees at Hyde Park better," he told his wife when standing on the South Rim. "They are alive and growing." Nevertheless, as president, Roosevelt had the CCC turn the great chasm into a world-class tourist destination.

A longtime lover of rustic architecture, Roosevelt had the WPA build the wonderful Timberline Lodge in Mount Hood National Forest using native rock, hewn timber, and rough-sawn siding, with heavy roof shakes. At the lodge's dedication ceremony on September 28, 1937, Roosevelt predicted tourists would soon flock to Mount Hood to ski.

TOP LEFT: On August 2, 1937, Roosevelt signed Presidential Proclamation 2246, which established Capitol Reef National Monument (upgraded to national park status in December 1971), thereby saving a Utah wonderland of slickrock canyons, buttes, and ridges.

TOP RIGHT: In the spring of 1937, Roosevelt went fishing in the Gulf of Mexico with an eye to find acreage to save as waterfowl habitat. After anchoring near Port Aransas, Texas, and catching tarpon, Roosevelt decided that 47,261 acres of the Gulf Texas habitat, where whooping cranes wintered, should become a federal preserve. On December 31, 1937, he signed Executive Order 7784 to establish Aransas National Wildlife Refuge. At present, the 115,670-acre NWR contains five management units: Blackjack Peninsula, Lamar, Matagorda Island, Myrtle-Foester Whitmire, and Tatton.

BELOW: After serving as president for only six months, Roosevelt, encouraged by Secretary of War George Dern, signed Presidential Proclamation 2054 on August 22, 1933, establishing Cedar Breaks National Monument in Utah.

TOP RIGHT: Spurred on by environmental activists Rosalie Edge and Irving Brant, President Roosevelt toured Washington's Olympic Peninsula in 1937 to decide how best to preserve the forest-clad mountains. While on an automobile tour of the region, he spotted a blighted swath of timbered land and blurted out, "I hope the son-of-a-bitch who logged that is roasting in hell."

CENTER: In 1937, Roosevelt opened Patuxent National Wildlife Research Center in Maryland to study how the feeding habits of mammals, the migratory patterns of birds, and pollution impacted nature. A sophisticated bird-banding program ensued from Roosevelt's cutting-edge outdoors laboratory. Here an indigo bunting (*Passerina cyanea*) is being appropriately tagged.

TOP RIGHT: An often-overlooked aspect of Roosevelt's leadership was his saving of Alaska's wilderness. On December 16, 1941, he signed Executive Order 8979, establishing 1.1 million pristine acres as the Kenai National Moose Range. Here is a cow moose on the range with her calves.

CENTER RIGHT: Even though Congress refused to fund any more waterfowl refuges during World War II, President Roosevelt found ways to circumvent their desires. Determined to protect coastal areas along the Atlantic coast, the president saved this Massachusetts barrier beach and dune habitat by establishing Monomoy National Wildlife Refuge on June 1, 1944.

BOTTOM RIGHT: Around Christmas 1938, Roosevelt, with the help of photographer Ansel Adams and conservation activist David Brower of the Sierra Club, accelerated the Kings Canyon National Park legislation in Congress. Undaunted and determined in spite of entrenched opposition by the Forest Service, the effort by FDR and the Sierra Club gained momentum on Capitol Hill. "Reverting to the subject of Kings Canyon," Roosevelt wrote to Secretary of Agriculture Henry Wallace.

FACING PAGE, BELOW: In order to save trumpeter swan (*Cygnus buccinator*) populations, Roosevelt established Red Rock Lakes NWR in 1935. This Montana refuge was one of dozens that Roosevelt created in the American West. Just before Pearl Harbor, in December 1941, Roosevelt kicked the 10th Mountain Division out of Henry's Lake, Idaho, because the artillery at ski maneuvers were disturbing the small congregation of trumpeters. "The verdict is for the Trumpeter swan and against the Army," Roosevelt wrote to Secretary of War Henry L. Stimson. "The Army must find a different nesting place!"

Roosevelt saved this spectacular wilderness in Nevada on May 20, 1936, when he signed Executive Order 7373, establishing the 1.5 million–acre Desert National Wildlife Refuge. By doing so, Roosevelt helped desert bighorn sheep (*Ovis canadensis nelsoni*) rebound from the brink of extinction.

After a fierce fight, Roosevelt established Jackson Hole National Monument in Wyoming in 1943 with Presidential Proclamation 2578. Republicans accused Roosevelt of behaving like Adolf Hitler by using the authority granted him under the Antiquities Act of 1906.

BELOW: During his second White House term, Roosevelt grew determined to enact stringent federal regulations to protect the American bald eagle (*Haliaeetus leucocephalus*). When Roosevelt read the Emergency Conservation Committee (ECC) pamphlet *Save the Eagle: Shall We Allow Our National Emblem to Become Extinct?*, he endorsed its "convincing and persuasive" recommendations. At various national wildlife refuges that Roosevelt established—such as this photo from Seney NWR in Michigan—bald eagle protection was mandated. Seney was also an important spot for the recovery of the Canada goose (*Branta canadensis*), then a threatened species. Instead of simply saving habitat and impounding water at Seney, biologists built a fenced-in pond area to get a crop of goslings and then train them to establish a migratory pattern that would conduce to survival.

Roosevelt grew personally interested in the botany of America's deserts. Thousands of saguaros (*Carnegiea gigantea*) in southwestern Arizona were brought back from the brink of disease, thanks to the CCC.

RIGHT: On April 26 1938, Roosevelt signed Presidential Proclamation 2281, establishing the Channel Islands National Monument (it became a national park in 1980). Roosevelt knew from friends at the Smithsonian Institution that the Channel Islands, known as the "Galápagos of North America" for their diversity of rare plants and wildlife, would become a premier destination for whale watchers, birders, beachcombers, and fisherfolk. The archipelago was the only place along the Pacific coast where warm and cold ocean currents commingled.

Roosevelt promoted "greater Cape Hatteras" as an American historic zone due to the adventures of Sir Walter Raleigh in the sixteenth and seventeenth centuries; the aviation experiments of the Wright brothers at Kitty Hawk; the legends of lighthouses and shipwrecks; and the famous exploits of the U.S. Life-Saving Service (the precursor of the U.S. Coast Guard). The *New York Times* lauded FDR's Cape Hatteras National Seashore as "one of the most important conservation measures of all time." It was, in fact, the opening salvo of Roosevelt's second-term battle to save extensive ocean frontage. By protecting about a hundred square miles of North Carolina's Outer Banks, Roosevelt had set into motion a new protocol: federal marine conservation.

In August 1937, Roosevelt signed Proclamation 2221, establishing Zion National Monument, 49,150 pristine acres in the Kolob region of Washington County, Utah. (It was incorporated into Zion National Park on July 11, 1956.) Utah was getting so much work-relief help from Roosevelt that a field representative from the Federal Emergency Relief Administration dubbed it "the prize 'gimme' State of the Union." An astounding 116 CCC camps were run by the Forest Service in Utah. Two visionary Roosevelt administration experiments in Utah—the Widstoe Project (located in the Sevier River drainage near the Escalante Mountains) and the Central Utah Project (later to become the Intermountain Station's Benmore Experimental Range)—aimed to irrigate arid valleys to help communities thrive.

Roosevelt deserved credit for trying to save treasured landscapes in the Southwest with his four national monuments of 1936–1939: Joshua Tree (California), Capitol Reef (Utah), Organ Pipe Cactus (Arizona), and Tuzigoot (Arizona). During the New Deal years, protecting Joshua trees (*Yucca brevifolia*), pictured above, became an Interior Department priority.

the heroic efforts of the federal government—planting trees, rehabilitating grasslands, closing the open range—paled in the face of Mother Nature. That spring of 1935, galloping dust clouds swept through Oklahoma and Texas. "All day it's been quite hot and a queer haze is in the sky," CCCer Belden W. Lewis wrote in the diary he kept while working at Zion National Park in Utah. "I suppose this haze is dust from that dust storm reported last night to have been in Kansas City and Texas."[49]

On April 14, which was Palm Sunday but would henceforth be known as Black Sunday, a huge "black blizzard" blew south from the Dakotas, picking up speed and wiping out topsoil. Fear so engulfed the Texas Panhandle that residents received last rites and prayed for salvation. No amount of Shelterbelt plantings or national grassland designations could have prevented this environmental catastrophe, which destroyed over fifty million acres across the Panhandle of Texas, Oklahoma, and Nebraska; western Kansas; southeastern Colorado; and northeastern New Mexico—the area known in 1935 as the Dust Bowl.

A Denver-based Associated Press reporter, Robert Geiger, journeyed through the areas of Oklahoma and Texas most affected by the Black Sunday dust storm. In a dispatch of April 15 to the *Washington Evening Star*, Geiger inadvertently gave a name to the prolonged drought that caused the worst agricultural crisis in U.S. history. "Three little words," the article began, "achingly familiar on a Western farmer's tongue, rule life in the dust bowl of the continent—'if it rains.' " Almost overnight "Dust Bowl" entered the American vernacular. Probably, the label caught on because it was a play on college football: California's Rose Bowl, Florida's Orange Bowl, and the Southern Plains' Dust Bowl. Once Roosevelt's Soil Conservation Service set up a special office in the "Dust Bowl" and had the label printed on maps, it became part of the grim history of the Great Depression.[50]

The fog-like dusters went from bad to worse during the summer of 1935. Children suffered from "dust pneumonia," a respiratory illness that plagued the western Plains and the Southwest. The winds had sucked up millions of bushels of topsoil, and the prolonged drought devastated agriculture in the lowlands. Despite the good soils of the Texas Panhandle—dark brown to reddish-brown sandy loams and clay loams, good for farming—crops there were mortally vulnerable to the whipping winds and dust clouds. Folk singer and activist Woody Guthrie was holed up in the Texas panhandle. The deforestation he saw inspired him to write a pamphlet called *$30 Wood Help*, in which he complained about lumber barons stripping the land bare.

Representing marginalized western farmers, Guthrie hammered away at the extraction industries for causing the Dust Bowl. He blamed greedy agricultural and corporate oil interests for the chronic land abuse of the Plains. He also wrote a song about it:

If I was President Roosevelt
I'd make groceries free—
I'd give away new Stetson hats
And let the whiskey be.
I'd pass out new suits of clothing
At least three times a week—
And shoot the first big oil man
That killed the fishing creek.[51]

The Roosevelt administration grew desperate to reverse the damage caused by the drought and erosion in the Great Plains and Southwest. Once the grasses were destroyed or removed, submarginal soil would blow away just as troubadour Woody Guthrie described in his folk songs. Throughout 1935 Roosevelt had promoted federal land reclamation. Planting grass and trees to stabilize ecosystems like the Dakota prairie, Cimarron in southwestern Kansas, and the Comanche lands of Colorado became a major New Deal priority. The Agricultural Adjustment Act of 1933, which created the Agricultural Adjustment Administration (AAA), sought to bring stability and a reasonable measure of prosperity to a segment of the economy that had been troubled even before the Depression. In addition to widespread measures to repair farm-commodity pricing and to educate farmers about soil maintenance, the AAA also addressed severely distressed land, including the Dust Bowl, permitting the federal government to buy and restore damaged acreage and relocate hard-luck families.[52]

Roosevelt wisely asked Chester C. Davis, the Iowa-born Grinnell College graduate who was administrator of the AAA, to create a Program Planning Division that would give government experts the chance to execute a national land-use program. After much deliberation, Roosevelt decided that in some situations, notably the Dust Bowl, education wasn't enough. The U.S. government needed to acquire submarginal lands, consolidate farms, relocate inhabitants, restore land, and return the reclaimed land for "commercial use" under "the watchful eye of Uncle Sam." This tall order was intended to ecologically restore cropland to grass and thus halt the dusters and revitalize agriculture. Here was an unexpected foray by Roosevelt into government planning and soil conservation on a very large scale.

By 1935, the AAA had three projects running in the Southwest—one in wind-eroded New Mexico, and two in dust-laden Colorado, respectively known as the Mills and the Southern Otero and Southeastern Colorado Land Utilization Projects.

When Woody Guthrie sang about the West, it wasn't just the fields and creeks of the Dust Bowl that he wanted saved. He was enraptured by the rugged mountain grandeur and Chihuahuan Desert flora around the Big Bend of the Rio Grande in southern Texas. To Guthrie the purple Chisos Mountains, where the summits and drifting clouds seemed otherworldly, were a land of deep mystery. "Sandy cactus of every kind, slim and long, fat and thick, wide and low, high and skinny, curly, twisty, knotty, sticker, thorny, daggery-knifed, razor sharp; hot darts, burning needles, fuzzy needles, cutting edges, stinging leaves, and blistering stems," Guthrie wrote in his memoir *Seeds of Man*, about his youthful days wandering around Big Bend country. "I saw a whole world of sword and dagger weed and limb—a new world to my eyes. The feel and the breath of the air was all different, new, high, clear, clean and light."[53]

Another writer FDR admired who pushed hard for a Big Bend

National Park in the mid-1930s was folklorist J. Frank Dobie from Live Oak County, Texas. In 1910, Dobie had gotten a job teaching high school English at Alpine, a gateway town to Big Bend. Inspired by the "big sky" landscape, he began writing a string of marvelous books—starting with *A Vaquero of the Brush Country*—that documented Texas life. Once the CCC built roads in the Chisos Basin, Dobie lived in the New Deal camp while he collected cowboy stories for his classic book *The Longhorns*. By day, Dobie compiled information about the Big Bend region, and by night he sat with the CCCers to eat chili con carne, frijoles, and fried potatoes and then smoke hand-rolled cigarettes, holding forth until midnight on perpetual drought, rattlesnakes, scorpions, javelinas, and that Texas icon, the longhorn. The Big Bend National Park movement had in Dobie an able promoter almost comparable to John Muir, with access to big-time editors in New York.[54]

On June 20, 1935, to the delight of Dobie, Congress authorized the creation of Big Bend National Park. There was, however, a catch: the Big Bend Act (49 Stat. 393) stipulated that Texas had to give the federal government title to the acreage before the park could be established officially. This provision was inconvenient because the Texas state legislature, which met biannually, had adjourned before the law passed Congress.[55] In fact, it would take years to secure the park, once and for all. Roosevelt spent considerable time trying to persuade Texans to allow the federal government to save the rugged wilderness. "I have heard so much of the wilderness and the beauty of this still inaccessible corner of the United States and also, of its important archeological remains that I very much hope that some day I shall be able to travel through it myself," Roosevelt wrote to a Texas congressman. "I feel sure that [Big Bend] will do much to strengthen the friendship and good neighborliness of the people of Mexico and the people of the United States."[56]

Many West Texas ranchers loathed the Big Bend National Park idea. Such a federal wilderness refuge, they complained, would be a "predator incubator," providing a sanctuary to which mountain lions and coyotes could sneak off after preying on sheep and cattle. Big

Bend cowboys fretted that the national park was a seizure by the federal government of the Chisos range. The arduous task for the state of Texas of procuring title to over 200,000 acres portended county courthouse disputes. The fact that Roosevelt wanted Big Bend to be administered in some fashion by both the United States and Mexico raised xenophobic fears in West Texas border towns.[57]

Soon after Big Bend was (tentatively) established, Ickes started lobbying to change its name to Jane Addams International Park, after the social worker and women's suffrage leader from Chicago, who had been the driving force behind the secular Hull-House settlement and had won the 1930 Nobel Peace Prize (the first American woman to do so). That May, the seventy-four-year-old Addams had died. Ickes had greatly admired Addams and felt that an "International Park" in West Texas, with corresponding land across the Rio Grande, would be the perfect way to honor her. Roosevelt, when approached by Ickes, was "enthusiastically in favor of the proposition."[58] But Senator Tom Connally of Texas nixed the idea. There was no way his good old boy West Texas constituents were going to tolerate Big Bend country being renamed after a female Chicago social worker.

On September 6, 1935, Roosevelt established Hart Mountain National Antelope Refuge in southeastern Oregon, which saved 422 square miles as a permanent range for herds of pronghorns. Able to run at sixty miles an hour, pronghorns are considered the fastest mammals in North America; as they ran away their large rump patches of long white hairs could be seen for miles. All over the West this unique animal was being hunted toward extinction. Roosevelt, in touch with William Finley, the great Auduboner and wildlife conservationist of Oregon, boldly deeded the entire Hart Mountain ecosystem to wildlife—notably, pronghorns.

Finley had been trying to improve the plight of the pronghorns since 1924. The only complaint Finley had with Roosevelt's September executive order was that it specified that four thousand antelope be maintained by the Biological Survey at Hart Mountain; he wanted eight thousand. "I know you do not have time to go into the details of the matter of this kind," he wrote to FDR, "but I hope you will call in some federal experts who have a heart for your wildlife interests."[59]

Unfortunately, the chances for a herd count of eight thousand pronghorns were nonexistent at the time. Although Hart Mountain was designated as a refuge, two-thirds of its total 278,000 acres were still leased for grazing by sheep and cattle. It would be fifty years before the livestock was banished and the Oregon refuge was finally restricted to wildlife. All the while, pronghorn herds averaged a few thousand, at best. Roosevelt had the correct impulse, but Hart Mountain demonstrated that he could not fully resuscitate threatened species with a sweep of his pen on an executive order. The commitment had to be widespread and it had to last generations. Nevertheless, he was rightly proud of the start he made on behalf of antelope.[60] In numerous letters he promoted Hart Mountain National Antelope Refuge as the New Deal model for protecting endangered mammals—give them a huge habitat reserve in which to prosper. "This refuge will leave our grandsons and granddaughters an inheritance of the wilderness that no dollars could recreate," William O. Douglas wrote after visiting the expansive Hart Mountain sanctuary. "Here they will find life teeming throughout all the life zones that lead from the desert to alpine meadows."[61]

V

On September 14, 1935, Franklin Roosevelt arrived in Lake Placid, New York—a resort town that had hosted the 1932 Winter Olympic Games—to speak about conservation. He stressed that the CCC was a triumph, even though, during the 1932 presidential campaign, Herbert Hoover had scorned the "tree army" as a "crazy dream" and merely a "political gesture." In 1935, *Business Week* backed the president's assertion, reporting that "hundreds of communities have discovered since the CCC was organized two years ago that the neighboring camp is the bright spot on their business map."[62] Surrounded by about two hundred uniformed CCC enrollees, the president spoke about the Adirondacks and Catskills, his enduring friendship with Gifford Pinchot, and the art of planting trees. Referring to himself as a "very old conservationist," he told his listeners that "the spreading of the gospel of conservation was God-ordained work."[63]

Later that day, FDR was driven to the summit of Whiteface Mountain to dedicate the Memorial Highway he had supported since the late 1920s. Sensitive to charges that his highway had marred the Adirondacks, he ably defended himself by explaining that the scenic road would allow the elderly, infirm, poor, and busy to see the beauty of the place. "You and I know that it is only a comparatively small proportion of our population that can indulge in the luxury of camping and hiking," Roosevelt explained to the crowd. "Even those [citizens] who engage in it are going to get to the age of life, some day, when they will no longer be able to climb on their own two feet to the tops of mountains."[64]

It is fair to interpret Roosevelt's Whiteface Mountain address as a defense of all of the New Deal parkways he had authorized. Top wilderness conservationists criticized Roosevelt for floating proposals to build scenic drives atop the summits of the Presidential and Green mountain ranges in New Hampshire and Vermont; along three hundred miles of the Blue Ridge Mountains running from Virginia's Shenandoah National Park to the Great Smoky Mountains; and in the High Sierras of California. The hardest-hitting criticism came from the 1934 article "Flankline vs. Skyline" by Benton MacKaye—father of the Appalachian Trail—who was decidedly aghast at the very ridgetop roadways the president loved.[65] Even within the Roosevelt administration, Bob Marshall became a one-man Paul Revere, warning whenever he could that the *roads are coming.* "The Bulldozers are already rumbling up into the mountains," Marshall wrote U.S. Forest Service chief Ferdinand Silcox weeks after FDR delivered his Whiteface Mountain address. "Unless you act very soon on the seven primitive area projects I presented to you a month ago, eager CCC boys will have demolished the greatest wildernesses which remain in the United States."[66]

A few weeks after the dedication at Whiteface, Roosevelt, preparing to doff his jacket for a Hawaiian shirt, asked Ickes to join him on a Pacific Coast fishing trip. While casting line after line, the two men discussed the creation of Big Bend National Park in Texas, and what executive orders could be drafted to save Utah's slickrock canyons along the Colorado River. Neither Wallace nor Darling had been invited. Ickes, who was amazed at Roosevelt's "wide information" on the natural world, had clearly become the president's indispensable man on everything related to public lands and conservation of natural resources.[67] Ickes pressed his position, cajoling Roosevelt to strip the Forest Service and Biological Survey from the USDA, give them to Interior, and then rename it the Department of Conservation overseeing all of America's public lands. Roosevelt liked the idea, but had other fish to fry for the time being.

On November 15, 1935, Darling resigned from the Biological Survey after eighteen months in office, blaming his deteriorating health. His worried doctors at the Mayo Clinic had urged him to slow down or risk dying. "My engine got overheated and my valves began to leak," Darling lamented to Salyer. "Anyway it's been a great war. I've done the best I know how and had it not been for you I could not have done it at all."[68]

Conservationists lamented Darling's departure from Washington in late 1935 but were grateful for his efforts in elevating the importance of such treasured landscapes as Red Rock Lakes in Montana and White River in Arkansas. "I don't like it at all," Ickes complained to Darling about his retirement. "I feel a distinct loss in your going. You had been doing a fine job, and now that you are no longer head of the Survey, I haven't nearly the interest that I did have in attaching it to the Interior Department."[69]

Although much has been written about the friction between Roosevelt and Darling, in truth, they got along just fine. With Roosevelt injecting $14 million into the Biological Survey's habitat purchasing program, Darling was able to start building a meaningful wildlife refuge program. Roosevelt regularly laughed at Darling's pro-conservation cartoons, of which he was sometimes the butt. Once when a Darling cartoon did not amuse, however, Roosevelt sent it back with a note saying, "Some day, Ding, you'll go too damn far."[70]

Dr. Ira Gabrielson was photographed releasing a duck at a sanctuary near Washington on March 8, 1940. Gabrielson joined the U.S. Biological Survey in 1915. Succeeding Darling as its head in 1935, he remained at the renamed Fish and Wildlife Service until 1946. Engaging and well informed, Gabrielson was one of Washington's preeminent champions for conservation.

As a final gesture Darling fought for Ira N. Gabrielson to be his replacement. Gabrielson's boyhood and youth were spent in the "duck country" of northern Iowa. Transfixed by the V's of geese in autumn, he had developed a boyhood penchant for wildlife photography.[71] His childhood hero was Frank Chapman of the American Museum of Natural History—the same ornithologist who had mentored young FDR. The editor of the *Wilson Bulletin*, an ornithological journal, published Gabrielson's drawings of wild turkeys and ruddy ducks. This encouraged Gabrielson—stocky, with long arms and lumberjack shoulders—to become a wildlife biologist.[72]

On his graduation from Morningside College in Sioux City, Iowa, in 1912, Gabrielson joined the Bull Moose Party. Deeply proud of his

Scandinavian heritage, he was an expert in both Great Plains wildlife and pioneer life in Iowa. Hired by the Biological Survey, Gabrielson moved to Washington, D.C., to oversee the bird food habits laboratory. He earned national recognition for his research on rodent control. An extremely modest, humble, and self-sacrificing devotee of Aldo Leopold, Gabrielson spoke in a deliberate, focused way, as he canvassed America from 1915 to 1931, teaching farmers how to properly poison rats and attract birds to irrigation ponds.[73]

In February 1936, Roosevelt summoned influential leaders of state wildlife agencies, conservation organizations, and sportsmen's groups to a symposium in Washington, D.C. It was the meeting Darling had suggested the previous June, but as it came to fruition, Gabrielson helped FDR in the planning. The North American Wildlife Conference aimed to educate the public about the relationship between forestry management and wildlife rehabilitation. Ghosts of extinct passenger pigeons (*Ectopistes migratorius*) and great auks (*Pinguinus impennis*) in mind, Roosevelt especially wanted the attendees to protect imperiled species—the trumpeter swan (*Cygnus buccinator*), whooping crane (*Grus americana*), pronghorn antelope (*Antilocapra americana*), and white-tailed deer (*Odocoileus virginianus*)—with coordinated policy initiatives. The White House was frustrated at the piecemeal way in which the thousands of existing conservation clubs and federations approached urgent problems. "Our President's call was, then, a national one and international one," the chief of the Forest Service, F. A. Silcox, told the attendees as he opened the conference as its official chairman. "It recognized the broad wildlife plight, and the urgency of it. Through the medium of this conference and the open covenants which he hopes it may bring forth, his call provides an opportunity to remedy that plight."[74]

Because 1936 was an election year, Roosevelt also wanted the disparate groups to organize into a voting bloc in support of the New Deal. "It has long been my feeling that there has been a lack of full and complete realization on the part of the public of our wildlife plight, or the urgency of it, and of the many social and economic values that wildlife has for our people," President Roosevelt told conference attendees in a written message on February 3. "This, and the firm belief in the ability of the American people to face facts to analyze problems and to work out a program which might remedy the situation, impelled me to call the North American Wildlife Conference."[75]

To Roosevelt's surprise, and pleasure, the event ballooned to include all sorts of ardent ecologists, fishermen, able-bodied hikers, gardeners, nature photographers, Boy Scouts and Girl Scouts, 4-H Clubs, Audubon societies, and wildlife enthusiasts of all stripes—around one thousand attendees in all.[76] Exhibition space was provided for vendors at the Mayflower Hotel, with gun manufacturers, aquarium dealers, and nature photographers showing off their wares. Delegations from Canada and Mexico justified the title of the conference. A roll call of kindred New Deal spirits who spoke from the rostrums included Arno B. Cammerer of the National Park Service, Kermit Roosevelt of New York, William L. Finley of Oregon, Harold Ickes of the Department of the Interior, and Henry A. Wallace of the USDA.

For five days, wildlife—and its symbiotic relationship with public and private lands—was analyzed with an unprecedented degree of sophistication. All participants acknowledged that the primary culprit in wildlife extinction in North America was predatory man. The Labrador duck (*Camptorhynchus labradorius*), heath hen (*Tympanuchus cupido cupido*), Eskimo curlew (*Numenius borealis*), and Carolina parakeet (*Conuropsis carolinensis*) had already vanished forever thanks to human recklessness. Canvasback and redhead ducks would be doomed to extinction if the Roosevelt administration didn't create even more bird refuges, and stop the CCC from using poison in their predator and rodent control program; it too often killed birds.

At the conference, Henry Wallace, with misplaced archness, publicly belittled Darling: "I mean there are some folks who think that human beings are just as important as ducks," Wallace said, looking mockingly at Darling. "I have trouble with that, because I have a boy of some 17 years of age who thinks that wildlife is absolutely the most important thing in the world, and I have spent two years trying to convince him that the habits of human beings were just as interesting as the habits of wildlife with absolutely no success. . . . [Ding] comes round with all kinds of absurd ideas as to how you can raise a thousand dollars' worth of muskrat fur off an acre, and things like that. I take all these figures of [his] with a grain of salt. I think he is an awfully good cartoonist. I don't think much of him as an economist."[77]

The *New York Times* reported that Wallace's insults "created a tense atmosphere" at the event. Darling, defending himself admirably, argued for even more federal and state wetlands acreage; new measures, undertaken with Mexico and Canada, to protect migratory birds; and increasing law enforcement efforts in existing refuges.[78] He complained that the U.S. government, while owning 400 million acres of public land, had *no* authorized jurisdiction over wildlife resources and *no* legalized custody of the species roaming within its borders. He was making an important point. If a buffalo or trumpeter swan went outside the boundaries of Yellowstone National Park, for example, hunters could shoot it with impunity. "And you folks worry," Darling sarcastically scolded, "about your little fish pond and your little river!"[79]

Harold Ickes stole the show at the summit. Clearly influenced by the Wilderness Society, he declared that conservation was "the most vital" function of government and dramatically vowed to resist all further road projects in national parks. Ickes proclaimed that FDR was the most environmentally conscious president in American history. "We have in the White House today a President who is practical as well as a theoretical conservationist," Ickes said. "His stand is far in advance of any of his predecessors. What other President in our history has done so much to reclaim our forests, to reclaim submarginal lands, to harness our floods and purify our streams, to call a halt to the sinful waste of our oil resources? In his conservation program he ought to have the enthusiastic assistance of every true conservationist."[80]

Another star at the conference was John Clark Salyer, who warned about a "perilous gap" between endangered birds and tougher hunting regulations. Having bought land for forty-five migratory bird refuges, an energetic Salyer boasted that the Biological Survey, once primarily interested in coastal marshlands, had saved the interior wetlands of America—namely the middle portion of the Mississippi River Valley; the Illinois Valley; the Ohio River Valley; and the valleys of the Platte, Arkansas, Pecos, Rio Grande, Colorado, and Sacramento rivers.[81] And even with that impressive list, Salyer's government career was only beginning. He would go on to serve as chief of the Branch of Wildlife Refuges for twenty-seven years. When he died in 1966, newspapers described him—not Darling or Gabrielson—as the "father of the wildlife refuges."[82] In truth, the refuges were a New Deal group effort. "With almost 200 refuges now established in the United States along every flyway and at every concentration point for waterfowl, the Bureau of Biological Survey will soon be in a position to have the most comprehensive waterfowl-counting or inventory system obtainable," Salyer reassured the audience, "and we will soon eliminate the guesswork of waterfowl numbers and utilization at least within the border of the United States."[83]

One message the White House pushed at the wildlife conference was that conservation education should begin in primary school. A favorite quip of FDR's was that there was no such thing as an "ex"-conservationist. Once an appreciation of nature was instilled in children, and woodcraft was learned, no bulldozer, snake-oil salesman, or commercial developer could shake their faith. The wildlife that the federal government was rehabilitating as part of the New Deal belonged to *all* Americans for all time. Fish released in public streams from federal facilities like Ennis National Fish Hatchery (Montana) and Dexter National Fish Hatchery (New Mexico) were public property whether they were found in streams running through private lands or public lands.

The Washington wildlife conference was a victory for New Deal conservationism. Out of its minutes the National Wildlife Federation was born. An argument can even be made that the modern-day concept of "conservation reliance"—coined by U.S. government biologist J. Michael Scott in 2005—was really born at the North American Wildlife Conference.[84] The doctrine of conservation reliance stipulated that for species to come back, they would have to be bred by biologists, protected by wardens, and constantly "rejiggered" to establish an "asymmetrical balance" that would prevent extinction. From approximately the time of the conference onward, the federal government was in the business of "gardening the wilderness," and the line between conservation and domestication was "blurred."[85]

Back in the Theodore Roosevelt era, bison were saved as a species by breeding a herd at the Bronx Zoo and then transferring it to the newly established Wichita Mountains Wildlife Refuge near Lawton, Oklahoma. They bred successfully, and before long southwestern Oklahoma was exporting bison far and wide. But in 1936 FDR's Biological Survey applied numerous methods to keep the bison free of brucellosis, a disease that affected milk production and fertility. Other large mammals—including the desert bighorn sheep and the pronghorns at Hart Mountain—were also facing extinction and in desperate need of federal intervention.

VI

The Southwest had never been the president's favorite part of the United States. America's five great deserts didn't afford much in the way of recreation for a wheelchair-bound fisherman who loved forests. Primitive conditions, scorching heat, and lizards of all kinds scurrying across the baking wastes made him thirst for the Hudson River Valley. In his mind, the mesas and slickrock canyons of the Colorado Plateau could never hope to compare to the forest-studded beauty of the Adirondacks or Great Smokies. Eleanor Roosevelt comically recalled traveling with her husband to see the Grand Canyon for the first time. Holding hands on the South Rim, they peered across the ten-mile-wide abyss. "I thought it the most beautiful, and majestic sight I had ever seen," she wrote. "But my husband said: 'No, it looks dead. I like my green trees at Hyde Park better. They are alive and growing.' "[86]

Even though FDR, the Dutchess County provincial, wasn't overwhelmed by the majesty of the Grand Canyon, he launched a New Deal environmental revolution in the way the U.S. government protected ecosystems in the arid Southwest. It had started with Utah's Cedar Breaks National Monument back in 1933 and moved forward with the authorization of Big Bend National Park on June 20, 1935. Roosevelt in 1936 turned to the crisis of the desert bighorn sheep, a species struggling to survive in the deserts stretching from West Texas and Mexico to California, Nevada, and Utah.

Admired by Native Americans and lovers of nature alike throughout the West, desert bighorns ranged in color from chocolate brown to tan to beige. Only the males, however, boasted the magnificently curled horns. As "charismatic" North American species went, these sheep were in the same league as grizzlies, manatees, and mountain lions. The cliff-dwelling sheep had long managed to survive in the remote landscapes of the Southwest after arriving via the Bering Land Bridge about 300,000 years ago. When Thomas Jefferson negotiated the purchase of the Louisiana Territory in 1803, there were around two million bighorns in North America. But a ruinous combination of factors—diseases introduced by domestic livestock, habitat loss to agricultural concerns, the fouling of water sources by humans, and excessive hunting—had caused their numbers to decrease to a mere seven hundred animals.[87] Of the four recognized varieties of North American sheep, the desert bighorn was in the greatest peril. Two of the Roosevelt family's favorite wildlife conservationists—William Temple Hornaday and Charles Sheldon—had each written convincingly about the need to save the desert bighorn sheep.[88]

Harold Ickes informed the president in 1936 that these wild sheep had vanished from Washington, Oregon, Texas, North Dakota, South Dakota, and parts of Mexico; their only remaining strongholds were in Nevada, Utah, and Arizona. The desert bighorns' dwindling numbers had also led to social disruption and aberrant behavior within those herds that managed to survive.[89] Queried about what to do, Ickes warned that if the Southwest's desert bighorns weren't given *huge* reserves, they would perish as a North American species.

Taking a cue from Theodore Roosevelt's conservationism, FDR wanted to establish a large-acreage desert bighorn preserve in the Southwest. Senator Key Pittman of Nevada—a leader at the recent Washington wildlife conference—believed that an isolated swath of desert wilderness, just twenty miles north of Las Vegas, would make a mighty preserve for these keen-eyed, surefooted big-game animals. The rugged mountains of the Desert Game Range, as it would eventually be designated, were characterized by red rocks, steep cliffs, canyons, mesas, natural arches, and bottomlands. In autumn, rams often engaged in dramatic confrontations, facing each other, charging at full speed, and then slamming together.

Most Americans at the time viewed southern Nevada, where the Desert Game Range was located, as an arid ecosystem largely uninhabitable for humans. A small number of homesteaders, cowpokes, rogue miners, gold prospectors, and horse wranglers were able to withstand the blistering summer heat in hopes of striking it rich. It was common in mid-1930s Nevada to see road signs announcing city populations of 7 or 18 or 39. But Ralph and Florence Welles, a married couple employed by the Park Service, were environmentalists completely at home in the wild tangle of desert ridges and canyons. Together they wandered the loose-rock gulches, rust-tinged mesas, and canyonlands of Nevada to study desert bighorn sheep. The Welleses gathered a wealth of information about the sheep, publishing articles about springtime lambing seasons, horn-locking duels, and foraging habits.[90] They helped convince Harold Ickes and Ira Gabrielson that the ledge-loving animal embodied the "wild spirit" of the American West and needed federal protection from overhunting and reckless ranching activities.[91]

Because of Nevada's powerful Senate delegation—Key Pittman and Pat McCarran—federal appropriations poured into the state. More than thirty-one thousand young men, many from Arkansas and Missouri, were assigned to CCC camps in Nevada. While most of the serious work was aimed at devising flood control strategies or establishing military outposts, Roosevelt wanted Nevada's state parks and wildlife refuges to flourish as well. Working in conjunction with the Division of Grazing, the CCC developed mountain springs with underground storage tanks and troughs. While they were sold to Nevadans as a way to help ranchers tend cattle, these water facilities also doubled as key measures in the saving of desert bighorn sheep.[92]

On May 20, 1936, the president established a Desert Game Range of more than 1.5 million acres by Executive Order 7373.[93] This act protected not only desert bighorns but also vast swaths of the Mojave and Great Basin ecosystems.[94] A few years later, Roosevelt added a

320-acre parcel at Corn Creek to serve as the range's administrative headquarters.[95] (Eighty years after its creation, the Desert National Wildlife Refuge would remain the largest refuge in the "lower forty-eight.") The range was to be jointly run by Interior and Agriculture. Pack mules were used by biologists to spread feed for the bighorns during drought. When FDR acted with foresight in 1936, there were only three hundred bighorn sheep left in Nevada; thanks to this executive action, they numbered 1,700 by 1939.[96]

Six years after Desert National Range was established a passenger plane carrying Hollywood star Carole Lombard crashed into mountains at the refuge. The Biological Survey's best, most surefooted mule, "Madam Sweeney," was pressed into service to find the wreckage. Lombard was found dead. "Madam Sweeney" brought the body down to an overwrought Clark Gable, the late actress's husband. Visitors to the Desert National Range immediately made the mule a celebrity and a tourist attraction equal to the rams.[97]

CHAPTER THIRTEEN

"WE ARE GOING TO CONSERVE SOIL, CONSERVE WATER, AND CONSERVE LIFE"

I

A common way for Franklin Roosevelt to organize his mind was to grab a scratch pad and do a little math. On January 30, 1936, thinking about his chances for reelection, he calculated that the Democrats could secure 325 electoral votes, with the Republicans getting 206. It wasn't going to be easy. "We are facing a very formidable opposition on the part of a very powerful group among the extremely wealthy and the centralized industries," he wrote to an acquaintance. "Ours must be a truth-telling and falsehood exposing campaign that will get into every home."[1]

President Roosevelt, pruning a tree in Warm Springs during one of his 1936 getaways to Georgia. Up for election that year, Roosevelt traveled the nation, preaching the gospel of forestry. On the preservation front in 1936 he established the Okefenokee NWR (Georgia) and Joshua Tree National Monument (California).

FDR was lowering expectations. Although Alf Landon of Kansas was intellectually impressive, wealthy, and a twice-elected governor, he didn't have enough populist appeal to thwart Roosevelt's bid for reelection. While Landon, a GOP moderate, approved of certain New Deal programs—those that provided a safety net for the elderly and poor—he was opposed to excess wildlife refuges, national monuments, shelterbelts, and grazing districts. Believing FDR was "park drunk," and reckless in his insistence on establishing one federal migratory waterfowl refuge per month, Landon championed rights-of-way for utility transmission lines and permit rights for operation of service facilities on public lands. A large complaint of Landon's was that U.S. government spending had risen from $697 million in 1916 to $9 billion by 1936; he considered this an outrage. "National economic planning—the term used by this Administration to describe its policy—violates the basic ideals of the American system," Landon charged. "The price of economic planning is the loss of economic freedom."[2]

Governor Landon's attack on New Deal planning and conservation of public lands was surprising, coming from a Kansan. Between 1930 and 1932, eighty-seven banks had failed in that state alone. Compounding the economic problem, the state's farmers had grossly overharvested the land. As a result of all this, and the historic drought, Kansas was in bad shape. In 1930, a bushel of wheat sold for 63 cents; a year later, a struggling farmer would be lucky to receive 33 cents per bushel.[3] For Landon to run against federal planning efforts was patently hypocritical; in 1935 he had complained to Robert Fechner, director of the CCC, that Kansas wasn't getting its fair "quota of CCC camps." This surprised Fechner, for Kansas had twenty-seven CCC camps at that time, while neighboring Nebraska had only nineteen.[4] New Deal work-relief was the best thing to happen to Kansas in the thirties.

Disingenuousness aside, Landon wasn't wrong to advocate for more CCC camps in Kansas. As the *Topeka Daily Capital* noted, the CCC was orchestrating superb reforestation efforts at Fort Leavenworth and Fort Riley. And there were also successful Soil Conservation Service (SCS) camps in Kansas, many on private land. The

CCC's Company 788 in Kansas was nicknamed the Fire Devils because of its success in extinguishing prairie wildfires. Likewise, Kansas's CCC–Indian Division helped the Iowa, Sac (Sauk), Fox, and Kickapoo tribes combat typhoid fever and tuberculosis, as well as optical and respiratory damage from dust storms.[5] Landon's criticism of big government under FDR fell flat in 1936 with most Kansans, who knew that the president had come to the aid of struggling farmers—more so than Hoover Republicans ever had. Further flummoxing Landon in July were newspaper headlines claiming that the drought was worse than expected. Around one thousand American counties—one third of the United States—had been agriculturally devastated by dry weather that year.[6]

The same year, 1936, a group of New Dealers released a report, "The Future of the Great Plains," which championed Roosevelt's Shelterbelt effort.

The headquarters of the two-year-old Shelterbelt project were in Lincoln, Nebraska. Not only were farmers in towns from Devil's Lake, North Dakota, to Childress, Texas, pleased with the program, but state colleges embraced it because it provided jobs for graduates. While some ecologists worried that planting trees where they didn't naturally grow led to unwelcome competition for water with native species, they had to admit that the trees chosen for the project grew quickly.[7] If Roosevelt got the Shelterbelt right, then he missed the need to promote intentional burning of the Kansas prairie and pasture, which would have improved conditions on the grasslands. The report, which found favor with Roosevelt, mistakenly assumed that controlled burns damaged grass and killed mulch that protected the soil.[8]

For Landonites the real cause of the Dust Bowl wasn't deforestation or prairie fires, but the black-tailed jackrabbit (*Lepus californicus*). A publicity film issued by Kansas Emergency Relief noted, "It is estimated that a single jackrabbit will do $10 worth of damage in a normal year" (nearly $175 in 2015 currency).[9] To kill the pest, huge drives were undertaken at county or state expense. V-shaped fences were erected to corner rabbits, herded by club-wielding citizens.

Western Kansas towns had competitions for the most kills. Around Garden City in 1935, one rabbit drive had massacred six thousand animals. In Dighton, Kansas, ten thousand citizens slaughtered forty thousand rabbits. "Old and young men, women, and children had lots of fun on these drives," a Kansas reporter wrote. "They carried sacks and were armed with clubs of various sorts and sizes, which were swung madly and at times hurled at the rabbits. The war whoops and yelling would increase when a rabbit tried to break the line."[10] In total, two million jackrabbits were killed in thirteen counties in western Kansas during these drives.[11] For all of the Soil Conservation and Forest services grasslands strategies, the Great Plains had regional customs of their own, ones that the Ivy Leaguers back east couldn't get their heads around.

II

In late March 1936 Roosevelt headed to Florida for rest and relaxation. With him on the excursion were his twenty-eight-year-old son James, Ross T. McIntire (his personal physician), Frederick Delano, Pa Watson, and Captain Wilson Brown (his military aide). After receiving an honorary doctorate from Rollins College in Winter Park, the president journeyed to Islamorada, where, over Labor Day weekend the previous year, 259 World War I veterans had been killed by a fierce hurricane while rebuilding a road for the WPA that connected Florida's upper keys to Key West.[12] Some people wondered why more hadn't been done to protect those in the government camps from the storm—the strongest ever recorded on American territory.

Visiting the memorial designed by the Florida division of the Federal Art Project, Roosevelt was solemn. During this trip he spoke informally with Audubon Society leaders about establishing a network of birding trails, and he fished around Boca Grande and Useppa Island. Unfortunately, because of time constraints Roosevelt could not visit Fort Jefferson National Monument, established in the Dry Tortugas island chain by a 1935 executive order, but he vowed to explore the combination old fort–marine sanctuary soon.[13]

After President Roosevelt started collecting exotic fish as trophies, a special room next to the Oval Office was designated to hold them. Filled with mounted fish, mementos, books, tackle boxes, and other paraphernalia, the space was dubbed the Fish Room (it was later renamed, by Richard Nixon, the Roosevelt Room).[14] Roosevelt also had a large aquarium installed to enhance the aura he was trying to achieve: that of a Florida rod-and-reel club. Starting in 1936, FDR began studying the life histories of cartilaginous fishes like sharks and rays (each of those two categories consisted of approximately four hundred species). In the frontispiece of a ship's log, Roosevelt scrawled an old sportsman's saying: "Allah does not deduct from the allotted time of man those hours spent fishing."[15]

The president wasn't alone in his enthusiasm: in 1937 Marineland, an oceanarium in Saint Augustine, Florida, opened to great fanfare. Giant tanks and pools filled with sharks, rays, and dolphins attracted a huge influx of tourists. The success of Marineland made FDR hope that every American city would develop its own aquarium, in order to educate the public about marine ecosystems and increase support for ocean conservation efforts.[16] Ding Darling, who wintered on Florida's Gulf Coast, started a campaign to save marine ecosystems around Sanibel Island as a *living* oceanic parkland. They were waters FDR knew well. "My husband," Eleanor Roosevelt wrote in her column "My Day," "likes the ocean from the deck of a ship, even when the vessel rolls and pitches so much that most people retire to bed. My own appreciation of the ocean is always enhanced by being on dry land. I have a thrill when I drive up the coast of Maine and the road runs close to the beach, or high above it, with a view of the bays and tree covered lands, or limitless stretches of waters."[17]

Once back at the White House, FDR dispatched Ickes to Florida, to further propel the Everglades toward becoming a national park. However, Ickes found that a new obstacle had cropped up: several greedy landowners in the Everglades had raised the price of their land to $5 per acre once the idea of a national park took hold. Ickes made a public statement in response, saying that the value was $1 per acre and accusing these owners of holding the land for ransom. Ickes worked with the *New York Times* to run a profile—accompanied by stunning photos—about Florida's great watery wilderness.[18]

Determined to keep up the momentum for Everglades National Park, Roosevelt had the Biological Survey consult with Audubon official Dr. Cushman Murphy and John Baker, as well as Theodore Roosevelt Jr. and Percy Morris of Yale University, about making a documentary film of Everglades wildlife. He also asked the Soil Conservation Service and the Florida Fire District to sponsor a joint study on the water quality there. When legislators in Tallahassee blocked the study, Roosevelt chastised Democratic governor Fred P. Cone for managing natural resources recklessly in the state. Meanwhile, the Audubon Society's John Baker was successful in cajoling a few additional congressmen to support the acquisition the land deeds required for Everglades National Park.[19] "I agree with you as to the urgency of action in this matter," Roosevelt wrote to Baker, "and I will cooperate so far as possible in expediting this joint study."[20]

As Roosevelt campaigned around America for reelection in 1936, it became clear that the New Deal was especially popular in the counties hardest hit by dust storms. As a follower of Hugh Bennett, Roosevelt learned that the blowing topsoil from Oklahoma was red; from western Kansas, brown; and from Texas, a strange yellow haze. At a conference in Des Moines, FDR doubled down on the Shelterbelt as a revolutionary "break" for high winds and a protective measure against the crippling dust storms. Continuing to the Missouri River Valley, he wore his water conservation accomplishments proudly. When he spoke of the New Deal having built "literally thousands of ponds or small reservoirs" in the upper Midwest along with community lakes and new wells, he wasn't exaggerating. Public lands, Roosevelt knew, were the source of most of the water in the eleven coterminous western states, providing over 60 percent of the natural runoff in the region. His policies were helping to contain the menace of floods in the central Mississippi states, including the Ozark region of Arkansas and Missouri. Rivers—such as the Ohio, Arkansas, Red, White, Ouachita, and Mississippi—were being bolstered by Roosevelt's national forests and the CCC's levees. "We are going to conserve soil," FDR promised Americans in a fireside chat that September, "conserve water, and conserve life."[21]

Because Japan occupied Korea, Manchuria, and parts of China, it had access to the natural resources needed to build an industrial base adequate to gain control of the entire Far East. Foreign policy in the Pacific sphere was a major concern for Roosevelt and Landon throughout 1936. Nevertheless, at campaign rallies that fall, the president elevated conservation as a defining election-year issue. Using the slogan "Green Pastures!" Roosevelt pleaded guilty to Landon's charges that he had become a broken record with regard to wise-use conservation of natural resources. In Charlotte, North Carolina, he dubbed himself the heavyweight champion of protecting America's natural heritage.[22]

On April 16 the new Interior Department Building in Washington officially opened. It had been financed largely by the PWA, and most of its innovative special features were the handiwork of Ickes. To propagandize New Deal conservation, Ickes even convinced FDR to green-light a radio studio.[23] "I think every American who loves this country ought to take heart in the earnest and sensible pleas of the Secretary of Interior for a vigorous, continuing national policy of conservation," Roosevelt said at the dedication. "As for myself, I am dedicated to the Cause. And the Department of the Interior, as now constituted, was fully alive to the imperative necessity of protecting and preserving all of our natural resources. Without a national policy of conservation, a Nation less bountifully endowed than ours would have ceased to exist long ago. The remarkable thing was that the people of the United States were so complacent for so long in the face of exploitation, waste and mismanagement, yes, and even larceny of the natural wealth that belongs to all the people."[24]

That mid-April afternoon, with the cherry blossoms in bloom and the Potomac River looking glorious—even though the temperature had been unseasonably warm (eighty-six degrees) the day before—FDR connected the dots of his conservationist thinking. Praising activists in the Thoreauvian tradition whose cries from the wilderness warned about the "ravaging of our forests, the waste of our topsoil and our water supplies," the president credited his distant cousin with starting the modern-day conservation movement. "Theodore Roosevelt, for one, when I was a very young man, rose up and battled against this squandering of our patrimony," FDR said. "He, for the first time, made the people as a whole conscious that the vast national domain and the natural resources of the country were the property of the Nation itself and not the property of any class, regardless of its privileged status." Declaring national parks "birthrights," the president promised to forge forward with the New Deal to conserve America's "God-given wealth" of nature.[25]

That same April, the president attended the semiannual Gridiron Dinner at the Willard Hotel in Washington.[26] Because the humongous new Interior Department Building—a seven-story complex covering a two-block area of Foggy Bottom of Washington near the State Department—had been dedicated two days earlier, poking fun at FDR's mania for conservation was part of the roast. At one point the press corps sang Irving Berlin's "We Saw the Sea" with altered lyrics:

We joined the New Deal
To see the world
But what did we see—the CCC.[27]

That spring, Roosevelt signed into law the Soil Conservation and Domestic Allotment Act, which, in accordance with the AAA, authorized the federal government to pay farmers to shift production from soil-depleting crops such as cotton, wheat, and tobacco to erosion-arresting crops like grasses and legumes. As part of this New Deal "sound farming" strategy, grantees would have to adopt proper "soil conservation" stewardship protocols. "The recurring dust storms and rivers yellow with silt are a warning that Nature's resources will not indefinitely withstand exploitation or negligence," Roosevelt said. "The only permanent protection which can be given consumers must come from conservation practiced by farmers."[28] Farmers willing to shift away from the production of "unneeded" surpluses of soil-depleting crops to the production of "needed" soil-building ones would be given federal subsidies. "Every farmer takes pride in the productivity of his soil," Roosevelt said in promoting the act. "Every farmer wants to hand on his farm to his children in better shape than he found it. The conservation payments offered by the Government in accordance with the Act will help him to do this."[29]

Roosevelt's travels in 1936 confirmed his impression that the CCC must become a "permanent institution."[30] Eleanor Roosevelt, doing her part for the campaign, toured nearly every state in the nation, speaking about conservation at CCC camps in Grayville (Illinois) and Cheyenne (Wyoming).[31] Seeing the farmlands of the Midwest stricken by drought made her heart break. Blaming "Big Agriculture" for the crisis, she offered a green vision for the Dust Bowl region. "Now we are faced with lands that must be returned to buffalo grass or trees in order to prevent the suffering which many of our people are enduring today," she wrote. "All of us will gladly help in the emergency but I hope we will do more than that, and set ourselves to studying things which have caused these conditions and never rest until our government remedies them."[32] And the first lady laid down a challenge to the CCC: "There is a plan afoot for every girl to plant two trees and if she not only plants but cares for them," she said, "our state governments may find in the Camp Fire Girls, rivals of the CCC in conservation work."[33]

At the dedication of a memorial to the war hero and pioneer George Rogers Clark in Vincennes, Indiana, the president lamented the loss of primeval forests and untilled prairielands throughout the Midwest. Nature had blessed the region with "bounteous gifts," he said, but the waves of pioneer settlers had thoughtlessly stripped the land bare. He considered the "tragic extermination" of buffalo and deer a historical abomination. "Yes, my friends," Roosevelt said, "because man did not have our knowledge in those older days, he wounded Nature and Nature has taken offense. It is the task of us, the living, to restore to Nature many of the riches we have taken from her in order that she may smile once more upon those who follow after us."[34]

Roosevelt delivered a speech dedicating the Shenandoah National Park on July 3, 1936. Before a crowd of fifty thousand people, he said, "In bygone years we have seen the terrible tragedy of the age—the tragedy of waste. Waste of our people, waste of our land. . . . This park, together with its many sisters which are coming to completion in every part of our land, is in the largest sense a work of conservation." He then described the effect of national parks in reclaiming the land, the ethic of the local workers, and the spirit of people all over the nation.

In early July, after wrapping up his tour of the Great Plains and the Midwest, the president returned east, stopping at Shenandoah National Park for its official dedication. Three years earlier, he had visited the park to mingle with CCC boys; now he wanted to inspect their craftsmanship. What Roosevelt admired was the durability of their best work. The stone bridges, chiseled and fitted with perfection, would still be around decades after the New Deal faded, when the entire Great Depression was finally long gone. It was touching to

Roosevelt. With promise in their hearts, unemployed youngsters left a mark of betterment on their land with the only things they had to offer: hard work and pride in a job well done. "The drive is a beautiful drive, most of the way it was familiar to me, but the actual Skyline Drive is new," Eleanor Roosevelt wrote about her trip to Shenandoah National Park. "The CCC boys have done a wonderful piece of work; at the top of the hill there is a delightful picnic grounds where we all ate our luncheon. The view on both sides is perfectly gorgeous over miles and miles of forest and farm country in the valley."[35]

When Roosevelt had spoken at Glacier National Park in 1934, he hoped to boost tourism. Now, in the Blue Ridge Mountains, he stressed, with dramatic flair, the patriotic connection between the CCC and environmental protection. "Our country is going to need many other young men as they come to manhood," he said, "need them for work like this—for other Shenandoahs."[36] Hardworking Americans, Roosevelt said, could find spiritual renewal and solitude in America's national and state parks. The CCC boys weren't merely relief workers; they were an army of land healers; a national poll gave the corps an 80 percent public approval rating.[37] He implored all Americans to explore America's national parks. "Those people will put up at roadside camps or pitch their tents under the stars, with an open fire to cook by, with the smell of the woods, and the wind in the trees," Roosevelt said. "They will forget the rush and the strain of all the other long weeks of the year, and for a short time at least, the days will be good for their bodies and good for their souls. Once more they will lay hold of the perspective that comes to men and women who every morning and every night can lift up their eyes to Mother Nature."[38]

The scuttlebutt in Washington during the campaign was that FDR was far more dependent on his inner circle of advisers, especially Harry Hopkins, than on his cabinet. The loyal Ickes was the exception to this rule. Photographs of Roosevelt and Ickes on fishing trips showed them as conservationist compadres, bonded by a guiding ethic, smoking the same brand of cigarettes, wearing similar Tilly hats, and each playing the bon vivant for a few rum-filled days in the sun. The intellectual shorthand and the comfort level between them were natural. Ickes was able to persuade the president to beautify Chicago by erecting lily ponds and rock gardens in Lincoln Park and to plant trees along Lake Shore Drive. And at long last their joint determination to establish Joshua Tree National Monument came to fruition.

Since October 1933, when President Roosevelt made his land withdrawal in southern California, the Office of National Parks, Buildings, and Reservations had struggled to iron out the details of the monument's final boundary and obtain land from myriad private owners. Warding off the land claims of the Southern Pacific Railroad and established miners was "a tortured process," according to NPS historian Lary M. Dilsaver.[39] It had been a long waiting period for grassroots activist Minerva Hamilton Hoyt of the International Deserts Conservation League. But once the Desert Game Range was added to the Biological Survey's portfolio in May 1936, she prodded Ickes to finish tying up the loose ends for the designation of her long-hoped-for Joshua tree sanctuary. On August 10, 1936, Roosevelt signed Presidential Proclamation 2193, establishing a 825,340-acre national monument.[40] Though this was less acreage than Hoyt had originally wanted, Joshua Tree National Monument was proof that Roosevelt had made good on his promise to her. His executive action was yet another demonstration of his unshakable faith in his (and Ickes's) ability to accomplish the near-impossible; at the time, there was more privately owned land in Joshua Tree National Monument than in the rest of the NPS holdings combined (excluding Boulder Dam).

Together Ickes and Roosevelt evoked a new consciousness in the United States that desert ecosystems mattered—the Sonoran, Great Basin, Mojave, Chihuahuan, and Colorado, among others. The record was not quite as pure in the case of New Deal dam programs that proved to be destructive to western rivers. However, Roosevelt was able to stop the unregulated mining and grazing in the Southwest that the chief forester of the U.S. Forest Service, Ferdinand Silcox, called a "cancer-like growth."[41]

While Ickes argued against turning national parks into overly accommodating tourist mills, his high-mindedness had no relevance in the Black Hills of South Dakota. There, the sculptor Gutzon Borglum had been chiseling away to form colossal sixty-foot-high carvings of U.S. presidents George Washington, Thomas Jefferson, Theodore

Roosevelt, and Abraham Lincoln. Under FDR's orders the National Park Service took Mount Rushmore into its family of heritage sites in late 1933. The memorial, representing 130 years of American greatness, was exactly the kind of project that captured Roosevelt's optimistic belief in the can-do spirit of America. All four presidents' faces were completed while FDR was in the White House. When visiting Mount Rushmore in August 1936, Roosevelt first spoke about both Borglum's patriotic work and the importance of protecting the surrounding Black Hills ecosystem. "I am very glad to have come here today informally," Roosevelt remarked. "It is right and proper that I should have come informally, because we do not want formalities where Nature is concerned. What we have done so far exemplifies what I have been talking about in the last few days—cooperation with Nature and not fighting against Nature."[42]

While FDR and Ickes were on a roll, Roosevelt struggled with placating Wallace. Gutsy, methodical, and shrewd, Wallace continued to find FDR's circuitous, intuitive, and unpredictable style irritating. His frustration with FDR's "kill them with kindness" attitude sometimes caused him to condescend to Roosevelt. "Now in this letter I would like to write you as frankly," he wrote to FDR, "as though I were speaking to you face to face."[43] Wallace never seemed to understand that a cabinet secretary should *always* speak frankly to the president. By contrast, a chief executive seldom needed to speak frankly to his underlings. Wallace also grew frustrated that the president was easily swayed by ethereal photographs of America's treasured landscapes; if the president got a handsome print of a pristine forest or an alpine lake, then he would inevitably want it to be preserved like Adirondack Park. That ardent environmental activists like William Finley and Irving Brant had direct access to Roosevelt, thanks to Harold Ickes, infuriated Wallace. And leaving pockets of America as wilderness zones left him cold. "Why oh why," an exasperated FDR once chided Wallace, "can't we let original nature remain original nature?!"[44]

In 1936, Roosevelt went to Dallas to urge Texans to get behind Big Bend National Park. Because Theodore Roosevelt remained widely popular in the Lone Star State, having adopted its cowboy customs as his own and famously registering volunteers from San Antonio for the Rough Riders, to fight in the Spanish-American War, FDR told the audience in Dallas a story about TR. It could be construed as a threat of what might happen if the land deeds to Big Bend weren't soon acquired by the state of Texas to donate to Uncle Sam. "A young lady that I was engaged to, also a member of the family, and I were stopping in the White House, and the then President Theodore Roosevelt—this was after supper—was visibly perturbed and was stamping up and down in front of the fireplace in the Oval Room upstairs," FDR told the luncheon. "The various members of the family did not know what was the matter with T.R., and finally somebody said, 'What is the trouble tonight? . . . Oh,' he said, 'you know that bill for the creation of a large number of national parks? I am not going to be able to get it through this session because there are a lot of people up there that cannot think beyond the borders of their own States.' And then he clenched his fist and said, 'Sometimes I wish I could be President and Congress too.' "

When asked to explain what he would do with *that* much power, TR, according to FDR, said, "I would pass a law or a Constitutional Amendment . . . something making it obligatory for every member of the House, candidate for the House, candidate for the Senate . . . to file a certificate before they can be elected, certifying that they had visited in every State of the Union."[45]

The Texas trip, and every trip in 1936, was part of the reelection campaign for the intrepid FDR. In Pennsylvania, Pinchot predicted a landslide for FDR that fall. While Pinchot confided to his diary that the USDA wasn't keeping "a growing forest on the land always," he esteemed FDR's conservation-mindedness. On November 3, Roosevelt did crush Landon, 523 electoral votes to 8. Even Kansas voters rejected Landon in favor of FDR. The Democrats also increased their majorities in both houses of Congress. "FDR wins by ten million," Pinchot gloated in his diary. "I take it as a tremendous defeat for concentrated wealth and for States Rights big fellows. It means more and better national security legislation, conservation, labor, and corporation control."[46]

And it meant more National Wildlife Refuges comparable to the Hart Mountain Antelope Reserve. On December 11, 1936, Roosevelt issued Executive Order 7509, establishing the Fort Peck Game Range (renamed the Charles M. Russell NWR by President John F. Kennedy in 1963) in eastern Montana. At nearly one million acres, it fulfilled Roosevelt's promise to protect wildlife around Fort Peck Dam. FDR was fascinated that the Lewis and Clark expedition had named many of the features in the refuge. Ding Darling, who had gone back to cartooning, was thrilled beyond words that Roosevelt was protecting a vast Missouri River ecosystem (although he was disappointed that livestock would be allowed to graze in some sections). The real reason that Roosevelt established the refuge was to protect the largest population of Rocky Mountain bighorn sheep (*Ovis canadensis*) outside the Rockies. And William L. Finley had written to the president about the few thousand prairie elks ("Roosevelt elk," *Cervus canadensis roosevelti*), the largest herd in America, which desperately needed protection from overhunting around Fort Peck.

The establishment of wildlife refuges and game ranges in the West—such as Fort Peck Game Range and the Sacramento Migratory Waterfowl Refuge (established February 27, 1937, via EO 7562)—caused wildlife conservationists to start comparing FDR to TR.[47] On a number of occasions he dismissed the comparison, saying he was just a concerned citizen.[48] Early in 1937, upon accepting an award from the New York Rod and Gun Editors Association for the greatest individual contribution to the U.S. conservation movement, Roosevelt explained his New Deal environmental philosophy. "Long ago, I pledged myself to a policy of conservation which would guard against the ravaging of our forests, the waste of our good earth and water supplies, the squandering of irreplaceable oil and mineral deposits, the preservation of our wildlife and the protection of our streams," he said. "We must all dedicate ourselves for our own self-protection to the cause of true conservation."[49]

There was a side of FDR in early 1937 that surprised even his friends. When Hendrik Willem van Loon, a Dutch historian and journalist, suggested on WFAF-Radio in New York City in early 1937 that the United States could feed three billion hungry people, the president disagreed; America could feed only 300 million citizens. Sounding like an agronomist, Roosevelt said he believed that van Loon hadn't factored in the deleterious effects of large-scale wildfires in the West, deforestation in the South, climatic violence in the Rocky Mountains, and soil blown bare in the sun-beaten Great Plains. "For century after century your Netherlandish ancestors and mine, whether they lived near the mouth of the Rhine or further up the river or still further east, could count on an almost complete lack of erosion for the very simple reason that the European rains drop constantly but gently from Heaven and the wash of top soil from the cultivated land which replaced the forest has been on the whole negligible," Roosevelt replied to van Loon. "In almost every part of the United States, on the other hand, an equal amount of rain comes from the heavens in vast torrents and has taken away in three hundred years on this East Coast, about half of the original topsoil. I forget my figures, but I think that I am approximately correct in saying that it takes one hundred years to restore one inch of topsoil through reforestation. Perhaps that is over optimistic."[50]

III

Throngs of spectators descended on Washington, D.C., on January 20, 1937, for Roosevelt's second inauguration. Standing at the East Portico of the Capitol, Roosevelt promised to "solve for the individual the ever-rising problems of a complex civilization." The president didn't speak that day about saving wilderness or establishing national parks, but did lament the grim fact that millions of his fellow citizens were being denied education and opportunities for recreation. Ickes thought that FDR that day "delivered his message as well as I ever heard him speak."[51] The crowd cheered Roosevelt's castigation of "blindly selfish men," even though he admitted that the "happy valley" hadn't quite been found in his first term. In words borrowed from the British poet and herpetologist Arthur O'Shaughnessy, Roosevelt told his assembled supporters, " 'Each age is a dream that is dying, or one that is coming to birth.' "[52]

During his first term, the president, out of respect, had made sure that William Temple Hornaday, the first director of the Bronx Zoo, was regularly briefed on New Deal conservation efforts. At eighty-three, Hornaday had just published *Migratory Waterfowl Abandoned to Their Fate*. Bedridden with neuritis that made his legs nearly useless and very painful, he wrote a letter in January to the president. He was in so much discomfort that he had to dictate it to a secretary sitting by the bed. Hornaday asked the president to effect a three-year ban on all hunting of waterfowl and described the urgent situation facing other species. What he wanted, too, was the chance, despite his infirmities, to travel to Washington to speak directly to Roosevelt. Even he knew that was impossible, though.

After reading Hornaday's ten-page letter of despair, Roosevelt felt an upsurge of affection for the grand old man of the wildlife protection movement. No one at the previous winter's North American Wildlife conference had offered any ideas that Hornaday hadn't already advocated for decades. In fact, it annoyed the president to realize how Hornaday's heroic defense of northern fur seals and bison in the 1910s had been minimized in recent years. "My dear Dr. Hornaday," the president replied, "it is with feelings of great regret that I read the note accompanying your letter of January fourth and learn of your suffering. I hope it may afford you some consolation to know that I have the greatest admiration for your courage and for your continued devotion in the presence of physical pain and weariness to that cause to which you have devoted your years."[53]

Hornaday was elated by the president's kindheartedness, telling his grandson it was "one of the most charming and sympathetic letters" that a president could send to "an old broken campaigner" who wished to "score once more in a public cause" of wildlife protection "before closing his account."[54] Hornaday died that March. Roosevelt, unable to attend the funeral in Connecticut, nevertheless wanted to properly memorialize Hornaday.[55] After long reflection, he prodded the Boy Scouts of America to rename their annual Wildlife Protection Medal the William T. Hornaday Award. In another deeply thoughtful gesture, Roosevelt also had a beautiful peak in Yellowstone National Park's northeastern section, overlooking the Lamar River Valley, renamed Mount Hornaday.[56]

Most conservationists were ecstatic that Roosevelt had won reelection in a landslide. This boded well for scenic landscapes such as Big Bend, Isle Royale, and Jackson Hole. But then, on February 5, 1937, just two weeks after his second inauguration, Roosevelt shocked the nation by sending a bill for the "reorganization of the judiciary" to Congress. Boldly unwary of GOP blowback, the president called for the legal ability to appoint up to six additional justices to the United States Supreme Court (one for each justice then on the bench who was over seventy years and six months of age). The four septuagenarian justices he most wanted to replace on the bench—George Sutherland, Pierce Butler, Willis Van Devanter, and James McReynolds—were antiregulation, anti–Social Security and lackluster on conservation. Always game for a bureaucratic reorganization scheme and tired of gridlock, Roosevelt insisted—somewhat speciously—that having the option of adding up to six justices to the Court would improve the efficiency of the judicial branch and reduce the (alleged) backlog of cases from the "nine old men."[57] An uproar ensued. Critics, including many Democrats, lashed out at FDR, calling his brazen attempt "court packing," dictatorial in intent. An unrepentant Roosevelt locked horns with Republicans trying to derail his second-term New Deal agenda. The Supreme Court, he argued, had willfully derailed his New Deal policies.

The battle of 1937 was under way. No matter how brazenly Roosevelt was acting, no matter how excessive his use of executive authority was, his plan to enlarge the Court wasn't unconstitutional; the number of Supreme Court justices wasn't specified in the Constitution and had fluctuated several times in early U.S. history. Congress had determined the current number of justices—nine—by default.[58] But when Chief Justice Charles Evans Hughes delivered a rare public statement that the Court wasn't behind in its docket, the Roosevelt administration (if not the president himself) was forced to admit its "mistake" and abandon its fabricated ageist argument. No matter how decisive Roosevelt's victory in 1936 had been, Americans—including many Democrats—weren't eager to give the White House uncontested authority to meddle with the Supreme Court, and clamorous opposition to his "court packing" plan intensified. It was a bridge too far for moderate southerners and fiscal conservatives exhausted by the hectic pace of the New Deal in the previous four years.

Grassroots conservationists, by and large, stuck by the president during the judicial brouhaha. The Emergency Conservation Committee (ECC), for example, operating out of New York City, under the leadership of Willard Van Name, Rosalie Edge, and Irving Brant, saw Roosevelt as a gutsy hero for trying to safeguard the New Deal from a Court that had consistently sided with "big industry" on environmental protection issues. Therefore, if public lands were to be protected from private interests, environmentalists argued, fresh blood had to be added to the Supreme Court—preferably in the form of liberals from the American West who had a heightened sensibility regarding public lands.

Enter William O. Douglas, who loyally backed Roosevelt's "court packing" plan from his perch at the Securities and Exchange Commission in Washington, D.C. Douglas, a committed environmentalist, once boasted that he had seen firsthand the Civilian Conservation Corps "work miracles with men" in his home state, Washington. Douglas, like many westerners in the 1930s, thought one of the greatest accomplishments of the CCC was connecting pauperized urbanites to the public lands of the American West, teaching self-sufficiency like Daniel Boone's in the motorized age of Henry Ford. In his memoir *Of Men and Mountains*, Douglas credited the CCC with instilling a strong work ethic in its recruits. He recounted befriending a CCC corpsman from Brooklyn, New York, in the heavily forested backcountry of the Walla-Wallas. According to Douglas, this tough-talking young Brooklynite had been reborn on joining the CCC. Although at first the greenhorn recruit struggled with the bugle-call discipline—the playing of reveille at sunrise, the calisthenics before breakfast, and so on—he was now a proud public servant and outdoorsman devoted heart and soul to improving the land of America.

"It was two years in the woods that changed his character," Douglas wrote. "He poured out his story. . . . Things were different in the woods: 'No use getting sore at a tree.' He had found how great and good his country was. He was going to try and repair it for what it had done for him. The CCC had paid great dividends in citizenship of that character. He was not an isolated case. I heard the same story repeated again and again by supervisors of CCC camps."[59]

Douglas was first introduced to the president in 1935 by the head of the Securities and Exchange Commission, Joseph Kennedy. Scrappy, industrious, and with burning, messianic blue eyes, Douglas had overcome a hardscrabble youth and a bout with polio early in life, managing to work his way through Columbia Law School with flying colors. With Theodore Roosevelt and Gifford Pinchot as his principal heroes, Douglas saw life as an adventure in slaying dragons. A prodigy of judicial prudence, he taught at Yale Law School until 1934, when Kennedy hired him to regulate financial markets at the SEC.[60] Douglas, an ardent New Dealer and a champion of the underdog, believed that many large corporations were "a menace to the ideals of democracy."[61]

As a talent scout, Roosevelt admired the legal wizard from Yakima whose skin was thick as rawhide. Like many of the President's intimate circle, the craggy Douglas was deeply interested in the outdoor life. Working with Finley, he had played an important auxiliary role in getting Hart Mountain National Antelope Range established. If he seemed too ambitious or too calculating, that was all right by Roosevelt. He was a young chip off the old block. Whenever FDR raised a concern about a powerful financier or GOP obstructionist, Douglas would say, "Piss on 'em."[62]

On evenings when Roosevelt needed a diversion from work, he would sometimes slip out of the White House to the home of Morgenthau to play poker and mix rum cocktails. Douglas, a master card bluffer bursting with gossip about Wall Street and the Supreme Court, became a popular fixture at these games.[63] The swing and sway of lively conversation on these evenings impressed Douglas. The president learned that the sandy-haired Douglas felt as attached to his hometown—Yakima, Washington—as he himself was to Hyde Park. Even when dealing cards Douglas, an encyclopedia of ecological information, regaled the president with stories about hiking in the Cascades and Bitterroots, or the need for an Olympic National Park, or horseback adventures in counties around Coulee Dam. The acuteness of his geographic knowledge of the Pacific Northwest, Idaho, and Wyoming was undeniably impressive. If FDR mentioned a particular species of fish or bird, Douglas could respond in kind. In addition, because Douglas returned to Yakima every summer, he became a valuable source of information regarding the CCC's performance in southern Washington.[64]

William O. Douglas was recruited to Washington in 1934 to work at the Securities and Exchange Commission. When Roosevelt named him to the Supreme Court in 1939, Douglas was only forty. In his work and personal life, Douglas was a fanatic for wild America, writing extensively about his experiences.

As noted, Douglas stood by FDR in the spring of 1937 during the "court packing" debacle. That loyalty was rewarded two years later, in 1939, when Justice Louis Brandeis announced his retirement from the U.S. Supreme Court. Ickes, determined to get an environmentalist on the bench, led a silent campaign to have Douglas fill the vacancy. FDR initially favored Lewis Schwellenbach, an easterner, but eventually warmed toward the idea of Douglas. Not only had he run the SEC effectively since being appointed chairman in 1937, but because he was from the Pacific Northwest, he would give the Supreme Court some geographic diversity. Roosevelt summoned Douglas to the White House for a private chat. Douglas was dreading the prospect that FDR would offer him the opportunity to head the Federal Communications Commission. "I have a new job for you," Roosevelt told Douglas. "It's a mean job, a dirty job, a thankless job." Douglas, heart sinking, waited for the president to give him the FCC post. After the tease went on for a while, the president at last said, "Tomorrow I am sending your name to the Senate as Louis Brandeis's successor."[65]

Douglas was floored with joy. As a justice, he remained loyal to FDR's New Deal conservation initiatives, especially the establishment of Olympic National Park, but opposed the damming of western rivers. In the coming years, Douglas would also become a strong voice in the conservation movement, playing a crucial role in the establishment of Alaska's Arctic National Wildlife Refuge in 1960, the passage of the Wilderness Act of 1964 (a hallmark of the roadless movement), and the rapid development of environmental law. From the bench, Douglas fought for roadless wilderness protection—like Aldo Leopold and Robert Marshall—until his retirement in 1975.

Recognizing that the president was under assault, Ickes and Hopkins put aside their differences and joined forces as the progressive defenders of the New Deal. As White House counsel Tommy "the Cork" Corcoran put it, Ickes and Hopkins found they had to "stand in a circle with tails touching and horns facing the common enemy."[66] Together they backed Roosevelt's attempt to reorganize the judiciary branch. Nevertheless, Roosevelt, under pressure from Congress, ordered sweeping cuts to Hopkins's WPA and virtually halted all PWA construction activities. The president's actions were, as historian J. Joseph Huthmacher notes, "the first steps in a concerted drive to make good his often-repeated promise to bring the budget into balance."[67] The year 1937 had been a difficult one for Ickes—he had a heart attack in May and was still grieving for his wife, who had died in a car crash in New Mexico in the summer of 1935 (he remarried in 1938). Nevertheless he worked with Frederic Delano excitedly drawing nearer to his long-held dream of transferring the Forest Service to Interior. FDR made it clear that he wanted a Department of Conservation to avoid "ridiculous interlocking and overlapping jurisdiction" among New Deal conservation projects.[68] When Ickes tried to organize a Governors' Conference on Conservation, Wallace went nuts. "He insisted that conservation had passed the propaganda stage," Ickes wrote, "and if there was to be a conference on anything, it ought to be on the constitutional situation or the international situation."[69]

Douglas spoke for many New Dealers when he observed that Wallace was never a hail-fellow, nor much of a mixer. "By and large, he stayed aloof and remote," Douglas wrote. "While he had a popular following in the country, he had few political friends in Washington."[70] Notwithstanding that unhappy trait, Wallace, to be fair, was more concerned about how the "three bandit nations"—Germany, Italy, and Japan—were turning into belligerent global menaces than whether Isle Royale, Big Bend, and Mammoth Cave should be national parks. Ickes, of course, also had foreign policy concerns. But he didn't think New Deal conservation should retreat because of the rise of global fascism.

After the success of the CCC, the Soil Conservation Service, and the Duck Stamps, a host of New Deal environmental nonprofits had formed. The Wildlife Society, an organization of professional wildlife biologists, was founded in late 1936. Other nonprofits, like the American Fisheries Society, the Society for Range Management, and the American Society of Mammalogists, started by publishing scientific journals.[71] The influential Ecological Society of America continued its mission of studying "organisms" in relation to their "environmental conditions."[72] None of these groups, however, had as much political muscle with the Roosevelt administration as the small and relentless Emergency Conservation Committee (ECC) did.

Dr. Francis Harper and his wife, Jean, also had a direct influence on FDR. Their years of lobbying paid off when, on March 30, 1937, the president created the Okefenokee National Wildlife Refuge by Executive Order 7593. It was by far the largest wildlife refuge in the eastern United States. Although Roosevelt personally preferred the national monument designation, the Biological Survey came up with the cash first, paying $1.50 per acre, so Wallace's USDA took charge of the Georgia swamp.[73] After decades of ecological damage, the draining, plowing, and subdividing of the wild wetlands would stop. All told, according to the essayist Edward Hoagland, there were forty kinds of mammals and fish, thirty-five species of snake, fourteen types of turtles, eleven varieties of lizards, and twenty-two kinds of frogs and toads protected by Roosevelt's action.[74]

Although Ickes was perturbed—he believed that only the National Park Service could oversee the Okefenokee's "inviolability"—the good news was that, with the swamp firmly under the jurisdiction of the

Biological Survey, federal wardens were able to ban most hunting, trespassing, and timbering there.[75] Two "swampers," Sam Mizell and Jesse Gay, were asked to stay as wildlife wardens. Under their watchful eye, the plants and animals in the swamp were allowed to thrive.[76] Roosevelt would remain vigilant about the Okefenokee, insisting during the remainder of his presidency that it should be left in "pristine condition."[77]

The Okefenokee wasn't the only Georgia landscape that had captured Roosevelt's attention. A week later he signed another Executive Order, this one creating the Piedmont National Wildlife Refuge. The Piedmont, an upland forest in central Georgia, had once been dense with loblolly pine and hardwoods but was devastated after decades of bad timber-cutting practices by farmers, tenants, and laborers. It may have seemed an unlikely candidate for NWR designation, but Roosevelt, whose Little White House was only an hour's drive away, saw the abused region as rich with possibility. After all, at Warm Springs he had purchased a thousand acres of second-growth-timber mountain land—valued at $3 an acre—so as to dabble in forestry and cattle ranching.[78]

The Piedmont Forest project was actually selected for its abused condition.[79] It was an experiment: if the USDA and the Department of the Interior could repair exhausted and eroded soil and promote thoughtful conservation practices in the blighted Piedmont landscape, then wildlife, the president predicted, would return. Indeed, the red-cockaded woodpecker (*Leuconotopicus borealis*) rebounded after the forest was fully rehabilitated.[80]

Throughout 1937 Roosevelt remained frustrated that the Great Smoky Mountains National Park, officially organized in 1934, was still not officially opened to the public. FDR insisted that its boundaries be settled so that tourists could enjoy the parkland the CCC had been readying. In a conversation with William P. Wharton, a respected naturalist who had surveyed the Everglades in 1932 with Frederick Law Olmsted Jr., Roosevelt intimated that he didn't want his plans for the Great Smoky Mountains to devolve—like those for the Everglades—into a prolonged standoff, trying everybody's patience. "In regard to the Great Smoky Park, I understand that North Carolina has made all of its necessary land purchases and that we are held up by Tennessee's failure to complete its acreage," Roosevelt wrote to Ickes. "Would it be possible, under the law, to open the North Carolina side of the park? Perhaps this would encourage Tennessee to complete their purchases as agreed on."[81]

IV

The story of how FDR rescued Arizona's organ-pipe cactus (*Stenocereus thurberi*) in the spring of 1937 remains a marvelous example of New Deal desert protection during the second White House term.

It was E. D. McKee, a naturalist at Grand Canyon National Park, who had first come up with the idea of an Organ Pipe Cactus National Monument in southwestern Arizona, on its border with the Mexican state of Sonora. McKee was awestruck by the leafless organ-pipe cactus, with stems ranging from five to twenty feet high, which was found in abundance there. He wrote a series of reports advocating the creation of a monument that would focus on the beauty of organ-pipe cacti. Every spring, they bloomed in a dazzling burst of brownish-green or pinkish-lavender flowers, and around midsummer, spherical fruits appeared on them. They were surrounded by a breathtaking array of purple escobita, blue lupines, and Mexican gold poppies.

McKee's plan gained steam throughout the 1930s. The Tucson Natural History Association led a local movement to push the preservation plan onto Roosevelt's desk. Emphasizing the plant's range around the Sonoyta Valley, the association's board pressured the Department of the Interior into moving quickly to save the rare cactus. Meanwhile, Roosevelt himself had heard about the organ-pipe cactus from Ickes, who was quite bullish on a vast Arizona desert park. As projected, the Organ Pipe Cactus National Monument would encompass more than five hundred square miles of poor-growth mesquite and prickly pear west of the Tohono O'odham (Pagano) Indian Reservation. Although Tucson residents hoped that the national monument would draw tourists from California, the Roosevelt administration envisioned a wilderness zone—a vast borderland that resembled a desolate moonscape more so than a tourist attraction. There one could find exotic plant species—the Mexican jumping bean (*Sebastiania bilocularis*), the pudgy elephant tree (*Bursera microphylla*), the Mexican nettlespurge (*Jatropha cinerea*), and the Senita "old man" cactus (*Lophocereus schottii*)—not on view anywhere else in the United States. The organ-pipe cactus, for

A car driving through the Organ Pipe Cactus National Monument in Arizona in 1944. The original newswire caption read: "Penetrating deep into the desert, this car appears lost in a huge growth of Sinitas. The plant at the extreme left is an Organ Pipe [cactus]." FDR used his prerogative through the Antiquities Act to create the monument in 1937, balancing the dramatic scenery of many other national parks with the more subtle ecology of the Sonoran Desert.

which the proposed monument was named, provided the necessary impetus for the administration to begin the arduous paperwork involved in protecting the vast acreage.[82] Ickes dispatched National Park Service officials, based in Ajo, Arizona, to carry out a survey of the entire Sonoran Desert. These biologists were astonished at the scenic wilderness, rugged mountains, cactus-studded slopes, and desert flats. There were hedgehog cacti (*Echinocereus triglochidiatus*) and brittlebushes (*Encelia farinosa*) in bloom everywhere. A pair of natural arches resembled formations in Utah canyonland. A true desert paradise had been found.

After ironing out a few legal issues, FDR issued Presidential Proclamation 2332 on April 13, 1937, establishing Organ Pipe Cactus National Monument. It was created out of 330,000 acres owned by the federal government, according to the law of public domain. The president's action saved this large tract of exquisite Sonoran Desert landscape.[83] The Department of the Interior tried to make this new national monument sound mystical, going so far to bill it as a lusty land somewhere between "the habitable and uninhabitable world." The very strangeness of the organ-pipe cactus was marketed by the state of Arizona as the monument's primary attraction.[84] Although these cacti were protected species, Ickes made sure to reserve the right of the Indians of the Papago Reservation to pick the fruits of the organ-pipe and other cacti.[85] And organ-pipes weren't the only plants to benefit from the new federal designation. The sagebrush (*Artemisia tridentata*), part of the sunflower family, grew in abundance within the boundaries, as did rabbitbrush (*Chrysothamnus paniculatus*) and four-wing saltbrush (*Atriplex canescens*).[86] A 1.5-acre pond in the national monument alone attracted twenty-two different species of birds, including five kinds of sandpipers.

Gateway towns to the new monument—such as Why (at the monument's north entrance) and Lukeville (a town on the Arizona-Mexico border)—became hubs for desert scientists, wanderlusters, and stargazers from all over the world. The largely inaccessible land reminded visitors of Palm Springs country (near Los Angeles) without the movie stars. "I spent three winters as a ranger in Organ Pipe Cactus National Monument in southwestern Arizona," environmental activist and writer Edward Abbey would recall decades later in *Abbey's Road*, "a lovely place swarming with rattlesnakes, Gila monsters, scorpions, wild pigs, and illegal Mexicans. The only useful work I did there was rescuing rattlesnakes discovered in the campground, catching them alive with my wooden Kleenex-picker before some tourist could cause them harm, dumping them in a garbage can and relocating them by stuffing them down a gopher hole six miles out of the desert."[87]

In 1938 Organ Pipe Cactus National Monument received helpful national attention when a *Los Angeles Times* reporter, Lynn Rogers, wrote a rapturous piece about the desert wilderness and the odd organ-pipe cactus that "never-failingly delighted the eye."[88]

The organ-pipe wasn't the only species of cactus the National Park Service rescued during Roosevelt's second term. The saguaros (*Carnegiea gigantea*) in Saguaro National Monument east of Tucson were in similar peril, after a particularly vicious cold spell hit the region, with temperatures falling below twenty-five degrees. Founded in 1933, Saguaro (later designated a national park) was beloved by Arizona residents as an iconic landscape of the frontier West. But the soft rot that invaded the cactus forest following a winter cold snap reached epidemic proportions in the mid-1930s, despite an attempt by Interior to arrest the spread of the blight by removing and burying the afflicted portions of the plants. Scientists determined that some of the large columnar cacti, which generally lived for more than two hundred years, had bacterial infections. Luckily, Interior biologists knew how to differentiate between plants with bacterial rot and those damaged by climate conditions. This distinction helped federal botanists figure out which plants could be saved. Thousands of saguaros were brought back from the brink, while CCC boys removed those dying of disease.[89] (In 2009, Saguaro was declared one of America's most "imperiled" national parks, due to climate change.)

For Roosevelt, to save treasured Southwest landscapes like Cedar Breaks, Big Bend, Joshua Tree, and Organ Pipe Cactus was to protect some of God's best work—a religious obligation. To thoughtlessly destroy the Mojave of California or the Great Basin of Nevada and to reject proper land stewardship was both un-American and heretical. "To him," Frances Perkins wrote in *The Roosevelt I Knew*, "man's relationship to God seemed based on nature."[90]

PART THREE

CONSERVATION EXPANSION, 1937–1939

CHAPTER FOURTEEN

"WHILE YOU'RE GITTIN', GIT A-PLENTY"

I

The Louisiana delegates at the North American Wildlife Conference of 1936 boasted that because of their state's geographic position on the Gulf of Mexico its wetlands were the winter home of giant flocks of migratory birds. The very name Louisiana was evocative of mighty bayous and cypress hung with Spanish moss. Many wealthy New Yorkers—including John D. Rockefeller Jr. and Mrs. Russell Sage—recognized this and donated world-class bird refuges to the state of Louisiana. Furthermore, Louisiana had long produced more fur pelts (muskrat, otter, mink, raccoon) than Canada and Alaska combined. But its wildlife populations were being depleted. While the state of Louisiana had a decent conservation department, the wicked combination of drought, floods, disease, pollution, soil erosion, and deforestation was turning parishes into ecological disaster zones. Just as Roosevelt tried to help Florida save the Everglades and Georgia preserve the Okefenokee, he hoped to have the Biological Survey establish migratory waterfowl refuges in southern Louisiana; these refuges would also help furbearers and fish stocks repopulate and would make the Pelican State a "sportsman's paradise."

To get things rolling in 1935, Roosevelt had established Delta National Wildlife Refuge, just east of Venice, Louisiana, along the Gulf of Mexico. These forty-nine thousand acres were thick with tens of thousands of wintering waterfowl attracted to the diversity of fish and shellfish species. In the waters around the Delta NWR were speckled trout, redfish, flounder, blue crabs, and shrimp in large quantities. Just as winter sports enthusiasts would travel to Vermont and Idaho to ski, Roosevelt thought that Louisiana's "ragged boot heel" could become a recreational hub for anglers. At a time when the banks of the Mississippi River from Baton Rouge to New Orleans were lined with oil and gas fields, refineries, warehouses, and petrochemical plants, the notion of saving abused marshland as inviolate wildlife sanctuaries was visionary.

As Roosevelt prepared for his combination fishing and inspection trip to the Gulf South in the spring of 1937, he learned that the National Audubon Society had established twenty bird refuges in Louisiana and Texas. The impressed president, smelling an opportunity, called for the Biological Survey and the private sector to collaborate to enhance the Mississippi Flyway. A joint initiative was undertaken, birds were caught without injury in cage traps, and numbered metal bracelets were "banded" to their legs. The sex, established age, and weight were all estimated. The Paul J. Rainey Wild Life Sanctuary—a twenty-six-thousand-acre oasis along a seven-mile strip located along Louisiana's Gulf Coast in the marshes of Vermilion Parish—served as a demonstration plot for the modern management of waterfowl. Building on the Rainey model, Roosevelt now also hoped to scout locations for new federal migratory waterfowl refuges along Gulf Louisiana and Texas—stopover safety zones where great flocks could feed, breed, and nest unmolested. The difference was that they would be staffed with wardens supplied by Audubon, making the new refuges a public-private undertaking.[1]

On April 29, 1937, Roosevelt's eleven-day fishing trip in the Gulf of Mexico started at Biloxi, Mississippi. There, with Governor Hugh L. White by his side, he stopped by Beauvoir, the plantation home where Jefferson Davis wrote his memoir *The Rise and Fall of the Confederate Government*. The 608-acre estate, replete with oaks and cedars, operated as a home for Confederate veterans of the Civil War. Five of the ten Confederate veterans, all octogenarians, wore gray uniforms and shakily stood to salute the commander in chief when his car drove up. Eighty-nine-year-old J. C. Cain, who served at the Battle of Shiloh, came forward to present the president with a magnolia blossom.[2]

From Beauvoir, FDR headed to New Orleans for food, fishing, and an inspection of public works projects. At Antoine's restaurant, established in 1840, the president ate oysters and pompano, washed down with beer. The WPA had built a "Roosevelt Mall" in New Orleans's

City Park—complete with elegant fountains and miles of lagoons—and the president toured the grounds, impressed by southern live oaks (*Quercus virginiana*) colorfully named "Suicide Oak," "Dueling Oak," and "McDonogh Oak."[3]

The principal reason that Roosevelt had come to New Orleans, above and beyond fishing, was to commemorate the tenth anniversary of the great Mississippi flood of 1927 and to discuss New Deal flood-control policies. That March, shortly after his second inaugural, the entire 981-mile Ohio River, swollen by torrential winter rains, had climbed to record heights. Over one million people were forced to flee their homes. The WPA erected makeshift levees and used sandbags to prevent inundation. The so-called thousand-year flood took more than four hundred lives and caused more than $500 million in damage. This superflood caused Roosevelt to think even more seriously about ways to prevent future calamities.

To Roosevelt, the finest new flood-prevention structure on the Mississippi was the Bonnet Carré Spillway, a project of the Army Corps of Engineers twelve miles south of New Orleans. This engineering marvel was intended to spare New Orleans the ravages of superfloods like the recent Ohio–Mississippi River deluge by allowing floodwaters to flow into Lake Pontchartrain and then drain into the Gulf of Mexico.[4] By explaining this connection between the spillway and the recent disaster in Pennsylvania, Kentucky, Ohio, and West Virginia, Roosevelt reassured the nation that the federal government wasn't asleep at the wheel. The Bonnet Carré Spillway had been opened by federal engineers in March, redirecting some of the excess water from upriver. The citizens of New Orleans were proud that it had worked and that FDR now did them the honor of visiting it. The president implied in his speech that more such spillways would be constructed in coming years along American riverways, but that proper land management would obviate many of the problems of flooding and droughts. Some people might blame God for the Ohio–Mississippi flood of 1937, but Roosevelt blamed deforestation.[5]

Roosevelt marveled at Louisiana's vast wetlands, but he was apoplectic about the state's reckless clear-cutting of trees. During the nineteenth century, over 80 percent of Louisiana had been forested, with vast swaths of virgin oak (*Quercus robur*), loblolly (*Pinus taeda*), longleaf (*Pinus palustris*), slash (*Pinus elliottii*), and shortleaf (*Pinus echinata*) pine. By the time of the Ohio-Mississippi flood of 1937, almost all of these forests were gone. The Louisiana lowlands, once blanketed with oak, gum, and cypress, had met a sad fate: machine saws that hissed and coughed and cut up trees for "forest products" like lumber, basket veneer, shingles, handles, hewn railroad ties, fuel, fence posts, telephone poles, and piling. Parishes that had once been considered botanical crossroads—filled with swamp tupelo and delicate water lilies—had been drained until only muck holes remained. Even Louisiana's biologically blessed bayous were an ecological disaster in the making.

In 1936 Louisiana had the smallest state park system in America; its holdings totaled only twenty-four thousand acres. This bothered Roosevelt. While the CCC had established a foothold in the state, planting huge numbers of trees in several parishes, the "boys" hadn't been welcomed as in Pennsylvania, California, or Georgia.[6] At least at the Kisatchie National Forest, the CCC boys were tending Louisiana's longleaf pine forests for the sake of future generations.[7] But correcting the balance between human needs and stewardship of land and water was an uphill struggle in Louisiana. The destruction of the state's woodlands had been perpetrated by powerful companies—such as Long-Bell Lumber—that routinely stripped woodlands bare. Many politicians in Baton Rouge were beholden to the timber, oil, and gas industries. Trying to make ends meet during the Great Depression, people in Louisiana, and in the adjacent "big thicket" of Texas, seldom cared about wise-use conservation or replant schemes.[8]

II

A throng gathered at the New Orleans wharf to bid "skipper Roosevelt" bon voyage as the destroyer USS *Moffett* left port down the Mississippi to Gulf Texas, where the USS *Potomac* awaited the president. Wearing his good-luck fishing hat, Roosevelt watched for pelicans and raptors as he headed downriver to the new Delta National Wildlife Refuge. Venice, where the Great River Road ended and began, was the last town on the Mississippi River that Roosevelt studied as he traveled downstream from New Orleans. Now, thanks to the New

Deal, the colorful fishing village was the gateway community for the Delta NWR. There was a "last chance" mentality in Venice, reminiscent of Key West and Provincetown, which Roosevelt liked.[9]

With the president on the ten-day fishing jaunt were his personal secretary Marguerite LeHand; his twenty-six-year-old son Elliott, who was then living in Fort Worth; Captain Paul H. Bastedo of the U.S. Navy; and Dr. Ross McIntire, the president's personal physician. McIntire's job was to make sure that FDR was rested and hydrated, and that he found relief from his chronic sinusitis. According to Dr. McIntire, Roosevelt's rehabilitation from polio had been so successful that he could walk nearly fifty yards with only the aid of an attendant's arm—quite an accomplishment. Occasionally during the Gulf trip, McIntire would dilate the president's eyes and check his pulse (from the artery in the back of the neck). Exercise was ordered daily. But, for the most part, the doctor's prescription for the boss was to have "no humorless nights" and to inhale sea air. "As far as his upper body was concerned," McIntire wrote, "he developed it to a degree that would have shamed a heavyweight champion."[10]

The great conversationalist of the voyage was Edwin M. "Pa" Watson, FDR's favorite personal aide. Watson had been born and raised in Virginia's tobacco region, and the proudest day of his life was entering the U.S. Military Academy; in the same class (1906) were World War II heroes George S. Patton and Jonathan Wainwright. Deep-voiced, blessed with a marvelous Virginia drawl, Watson wasn't the smartest or most ambitious of the West Point cadets, but he was extremely likable (it was at West Point that he was given the nickname Pa). Watson later served nobly in the Philippines and Mexico, and shortly before World War I, President Woodrow Wilson made him a junior military aide. Wilson only reluctantly let Watson leave the White House to serve with the American Expeditionary Force in France, where he earned a Silver Star for gallantry in battle.

Beginning in 1933 Major General Watson served as Roosevelt's senior military aide. Almost exact contemporaries, they got along famously, talking a mile a minute, heartily ribbing each other as if brothers. It was Watson who often helped Roosevelt strap on his steel braces many mornings. In coming years, when Roosevelt wanted to get out of Washington, he often paid a visit to Watson's well-manicured Kenwood estate in Charlottesville, Virginia. The manor had been built by FDR's cousin, architect William Delano (known for designing the balcony on the South Portico of the White House).[11]

On this Gulf South trip Roosevelt and Watson served as self-appointed federal scouts looking for the most essential migratory bird habitats to withdraw for protection by the Biological Survey. Taking advantage of the momentum gained by the Delta NWR, Roosevelt now sought to establish a sizable federal wintering region for birds in the Gulf South that would extend from the mouth of Florida's Saint Marks River through the Mobile Bay region of Alabama, the delta and Cameron regions of Louisiana, and all the way down the east coast of Texas through Galveston, Port Aransas, and Laguna Larga, along the Sabine River in Gulf Texas and eastward to Calcasieu Lake, Louisiana. What fascinated Roosevelt about Texas was that three major migratory flyways converged within its borders; an astounding 625 species of birds had been cataloged.

In 1935, Roosevelt had established Muleshoe Federal Migratory Bird Reservation in Bailey County, Texas, near the New Mexico line. This was the first refuge for sandhill crane (*Grus canadensis*) in the United States. Smaller than whooping cranes (*Grus americana*), but similar in proportion, these largely gray cranes were gregarious, noisy, and conspicuous, and flew in large flocks. The Muleshoe grasslands offered ideal habitat for them. Unlike the whooping cranes that migrated on the edge of the Gulf, the adaptable sandhill had moved inland and upland to survive in sanctuaries like Muleshoe.[12] In central Wisconsin in 1937, Aldo Leopold wrote his elegant tribute, "Marshland Elegy," to the sandhill cranes. "When we hear his call we hear no mere bird," Leopold wrote. "We hear the trumpet in the orchestra of evolution. He is the symbol of our untamable past, of that incredible sweep of millennia which underlies and conditions the daily affairs of birds and men."[13]

On the heels of the Muleshoe sanctuary came Roosevelt's second 1935 refuge in Texas, this one largely for the Canada goose (*Branta canadensis*) and the greater white-fronted goose (*Anser albifrons*). Located on the Big Mineral Arm of Lake Texoma on the Red River between Oklahoma and Texas, Hagerman National Wildlife Refuge saved an important transitional zone between two different ecosystems known as blackland prairies and Eastern Cross Timbers. While both Muleshoe and Hagerman were in Texas, neither was along the Gulf of Mexico. So on his voyage, Roosevelt hoped to find other spots in Texas to establish federal wildlife sanctuaries near the sea.

The *Moffett* cruised past Galveston toward Rockport, the "birding capital of North America." A basic schedule was soon established for the Gulf holiday that the press could follow. Every night Roosevelt went to bed at 10 p.m. He then woke up at dawn. Over breakfast, he'd read newspapers, catch up on correspondence, and scan government reports that were shuttled to him by plane and boat at his various anchorages. (For instance, on May 1, his first day on the *Potomac*, Roosevelt received thirty-eight letters and documents that required his signature "for the first time in Central Standard Time Zone," after an airplane brought them from Galveston.)[14] Then, with Watson by his side, he'd talk to an array of guests about coastal conditions while he relaxed in the sun. One person FDR met was a rookie Democratic congressman, Lyndon Baines Johnson, a self-proclaimed "land conservationist." Deeply impressed by his political cunning, FDR, on returning to Washington, told adviser Tom Corcoran, "I've just met the most remarkable young man. Now I like this boy, and you're going to help him with anything you can."[15]

Many easterners mistakenly assumed that south Texas was a flatland of overgrazed cattle pastures, oil fields, caprock formations, and punishing desert heat. But Roosevelt, the master angler, knew better. No fewer than fifteen major rivers flowed through Texas before emptying into the Gulf of Mexico. This made its tidal flats nurseries for all sorts of creatures. Much of the coast was sheltered by barrier islands, such as Galveston, Matagorda, Mustang, and Padre, which were renowned as fishing spots. On this spring trip, everything from redfish (red drum or channel bass) to speckled trout (spotted weakfish) to flounder was at risk when Roosevelt's keel came near. But the trophy FDR most hoped to haul from the Atlantic was a tarpon (*Megalops atlanticus*) for the White House's Fish Room.

What attracted Roosevelt to Gulf Texas in the first place were the jetties and surf of Port Aransas, renowned in sportsmen circles for the Tarpon Rodeo, the most popular fishing tournament in the

Gulf of Mexico. Also, Elliott Roosevelt had fished in Port Aransas the previous year and convinced his father that the waters were thick with fish.[16] Once the *Moffett* arrived at PA (as the locals called it), the presidential party was transferred in a dinghy to the steel-hulled USS *Potomac*. This switch allowed Roosevelt more freedom of movement; the modernized *Potomac* had an elevator that allowed FDR quick and easy access to both decks.[17]

On board the *Potomac*, Roosevelt selected safe harbors and anchorages, performed low-grade chores, and directed the crew's activities. Amid banter and storytelling on the deck chairs with Pa Watson, he trolled for marine trophies. When he was bored there was always bottom fishing. One rule on the yacht was that FDR, usually wearing dark sunglasses to protect his eyes, played the role of commodore. After the first flash of luck with mackerel, the seas turned still and the fish stopped biting for the president. Refusing to give in to failure, he improvised supper for his crew. A nearby shrimp boat was waved over to visit with the floating White House gang. Flabbergasted to meet the president in the flesh, the fishermen donated two

Roosevelt (*seated*) reeled in a seventy-seven-pound, four-foot-eight tarpon off Port Aransas, Texas, on May 10, 1937. His guide, Don Farley (*far right*), held the fish with help from FDR's son Elliott.

buckets of jumbo Gulf shrimp to the *Potomac* potluck. A U.S. Navy bulletin reported: "Easterly wind not good for big fish but President and Captain Bastedo got 350 shrimp."[18]

Strapped in a skipper's chair, sitting with a straight back, elbows propped on his knees, with a saltwater rod in his hand, a light breeze gathering as the knots increased, Roosevelt was in his element on the *Potomac*. He was determined to experience the furious runs and jumps of a tarpon—a fish that can weigh hundreds of pounds. Built for speed, but with beautiful raiment, tarpon preferred brackish inlets and bays to open water. In short order Roosevelt caught four handsome kingfish (twenty-three pounds of the party's fifty-one-pound catch), but not a single tarpon. At such moments, Roosevelt's physical disability seemed irrelevant. Relaxing over card games with his son, he recalled seeing tarpon at the American Museum of Natural History that were more than seven feet long and weighed three hundred pounds.[19]

While fishing in gulf waters, the president received news that the *Hindenburg*, a German zeppelin, had caught fire and was destroyed within seconds when attempting to dock at Lakehurst, New Jersey. After writing Chancellor Adolf Hitler a letter of condolence, Roosevelt went ashore to study the birdlife on Saint Joseph's Island, an unpopulated barrier island twenty-one miles long. Numerous species of thrushes, warblers, vireos, tanagers, and grosbeaks popped up in the underbrush with surprising regularity. Ospreys hovered above the president's head, scanning for prey. Onshore his host, Texan oilman Sid Richardson, took FDR hunting for black-tailed jackrabbit in the sweet bay brush—there was no word about whether the president shot any.[20]

Resting from time to time at fishing shanties, soaking in the ambience of the vast coastal marsh, Roosevelt was as happy as a kid playing hooky. There were about seven hundred species of grasses in Texas—half of the number of varieties throughout the entire United States—and a swarm of butterflies to enjoy.[21] From the piney woods and bayous near Galveston to the prairielands and unspoiled sand beaches near Port Aransas to the mesquite brush of the Hill Country, near Austin, the Texas outdoors exceeded expectations. A couple of farmers and fishermen were stunned to see the president in such a remote part of America, scanning the Gulf coastal prairie with binoculars around his neck. After being exposed to so many species of birds there, Roosevelt, with his "map mind," determined that much of Mustang Island—where Port Aransas was located—should become a federal waterfowl sanctuary. Besides its pintails, royal terns, coots, green-winged teal, and redhead ducks, the island was appealingly secluded and rich in shrimp.

Roosevelt had allowed a reporter and a photographer from *Life* magazine to document his Gulf holiday and, he hoped, catch him reeling in a tarpon. To guarantee a catch for the cameras, fishing guide Don Farley, a boatbuilder from Port Aransas, told the president they should fish from the small launch—*Hangover II*—and use live mullet as bait. On May 10, after casting overboard with a sinker, they drifted with the tide. Then the president, from his "fighting chair," hooked a seventy-seven-pound tarpon; any fish weighing more than fifty pounds was classified as a "giant."[22] Because this silver king wasn't fully mature, FDR was able to handle his rod without assistance.[23] After an hour-long fight, Roosevelt, sweating profusely, reeled in the fish. *Life* ran a ten-photograph sequence of FDR in the May 24 issue, documenting his triumph. In one of the photos, Farley can be seen wielding a gaff in an attempt to haul the tarpon into the launch. The last photo shows a tanned and invigorated Roosevelt smugly smoking a cigarette, clearly proud of his conquest.[24] "I knew him as Mr. Roosevelt, the fisherman in khaki clothes, relaxed, happy, eager, enthusiastic, knowing he was among people who loved him," Farley recalled. "He put himself on the same level as others, desirous of learning things he didn't know, willing to share his wonderful personality with everyone. He was an attentive listener when others were talking. He was a considerate, thoughtful, and compassionate man."[25]

One afternoon, Roosevelt fished from the banks of the Brazos River. He was impressed by the open oak savannas and low sandy prairie openings.[26] As his fishing holiday wound down, the president spoke in Galveston, Houston, and College Station (before fifteen thousand at Texas A & M). The mayor of Galveston presented FDR with an expensive rod and reel. "The fish have been as good to me as the people of Texas," he said as he accepted them. To the roar of laughter he told how Pa Watson caught a twenty-four-inch amberjack, which was "now weighing thirty-five pounds."[27]

From the Gulf South, the president traveled by train to Fort Worth for a barbecue with Elliott at the ranch of Amon Carter, a lifelong Democrat who was the publisher of the *Fort Worth Star-Telegram*.[28] Roosevelt admired the fifty-seven-year-old Carter for championing the Big Bend National Park project. For Texas's centennial in 1936, Carter also helped to arrange for the CCC to construct replica buildings at three state parks: Fort Parker, Fort Griffin, and Mission Nuestra Señora del Espíritu Santo de Zúñiga at Goliad State Park. Bubbling over with fellowship, he and FDR got along famously, chatting about politics, the recent death of Carter's good friend, humorist Will Rogers (who was killed in a plane crash), and how the CCC was faring in the twenty-nine state parks it was building in Texas.[29] Carter disliked FDR's pro–labor union policies, but felt that the New Deal's land rehabilitation in Texas was brave and righteous.[30] And the PWA granted Carter $888,000 for Fort Worth's Southwestern Exposition and Livestock Show buildings, which Jim Farley dubbed "Amon's cowshed."[31]

Guided by the professional architects of the National Park Service, the Roosevelt administration dispatched more than fifty thousand CCC recruits to Texas to build trails, campgrounds, cabins, dance pavilions, and even an adobe hotel and motor court. When Herbert Hoover was president, Texas had only eight hundred acres of state parks, but by 1942, thanks to the efforts of FDR, that number had soared to sixty thousand acres. Outdoor recreation in Texas—from camping in Palo Duro Canyon and swimming in the spring-fed pool at Balmorhea to fishing among the cypress knobs on Caddo Lake—was economically viable. The CCC was responsible for helping bolster Texas tourist revenue and received high marks by politicos in Austin for its use of native stone, rocks, logs, and timber in the construction efforts. Because Texas was like a sovereign nation, with many different regions, local vernacular was adhered to. In Bastrop State Park, buildings were made of stone and timber; in Corpus Christi they were made of sand and oyster shell.[32] At Buescher State Park in Smithville, a red sandstone pumphouse was built.

Texas, more than any other state, saw the CCC as advance agents tied to local economic revitalization. Company 817 in Stephenville, Texas was embraced fully by the Chamber of Commerce, which disseminated copies of the CCC camp's *Blue Eagle News* throughout the city. Denison, Texas Company 857 had local merchants purchase advertisements in their camp newspaper *857 Log.*[33] These camp newspapers in Texas stood out from other states' by the raw humor presented—as in Bastrop's *The Pine Box* (Companies 1805 and 1811), which ran the risqué joke, "A pal of ours landed a soft job. He's in a bloomer factory now pulling down a couple thousand a year."[34] Only Roosevelt's attempt to teach "progressive" conservation classes at CCC camps was met with derision in Texas. "I have constantly fought the attempts of long-haired men and short-haired women to get in our camps," Colonel Duncan Major said. "We're going to be hounded to death by all sorts of educators."[35]

The long reach of the New Deal in Texas could also be found in projects other than the CCC's. In 1936 and 1937, the Roosevelt administration oversaw three experimental farm projects in Texas—Ropesville (near Lubbock), Sabine Farms (near Marshall), and Dalworthington Gardens (in east central Tarrant County)—that provided small agricultural plots to people who were employed but still had trouble feeding their families. These projects were hugely successful. Although Republicans called them a "socialistic collective," FDR thought they represented community-level Americanism at its best. Much of the Ropesville Farms acreage—around fourteen thousand acres total—remained in cultivation well into the twenty-first century.[36]

Before returning to Washington, D.C., by train, Roosevelt explained to reporters in Fort Worth how his fishing vacations helped him remove clutter from his mind. Many of his best ideas were hatched at sea. Because Roosevelt regularly went on outdoor vacations, he understood the importance of bays, marshlands, cays, lagoons, and swamps. Getting out of official Washington allowed him to consider wildlife conservation and forestry in more tangible ways. From this trip he learned that if the wetlands were healthy, two million ducks and 750,000 geese would spend the winter in the midcoast Texas region.[37] "The objective of these trips, you know, is not fishing," he told the press. "You probably discovered that by this time. I don't give a continental damn whether I catch fish or not. The chief objective is to get a perspective on the scene which I cannot get in Washington anymore than you boys can. . . . You have to go a long ways off so as to see things in their true perspective."[38]

What the president left behind in Texas—besides a signed fish scale for the Tarpon Inn at Port Aransas—was a rumor that proved true. All the old-timers who gathered at Cap Daniel's store in Austwell, Texas, gossiped about what FDR was *really* doing in their wildlife-rich backyard. "I hear the government is buying up 'the Blackjacks' for a pile of money just to protect a couple of them squeaking cranes," *Audubon* magazine reported the old men saying. "They tell me they ain't bad eating but there's no open season on them." To which another man said, "If you can't shoot them, what the blankety-blank good are they?"[39]

The squeaking birds were whooping cranes (*Grus americana*), the tallest birds in North America, which frequented the east shore flats of Aransas and Refugio from November to May. As a result of unregulated hunting, wetlands drainage, habitat loss, and bobcat attacks, these statuesque waders—pure white with black-tipped wings and a seven-foot wingspan—were at the brink of extinction. Rice farmers hated them. The only remaining wild populations bred in Wood Buffalo National Park of Canada (on the border of the Northwest Territories and Alberta) and wintered along the Gulf coast of Texas. With only around fifteen or twenty of these cranes remaining in the wild (and two in captivity), quick action was needed to save them.[40] If possible, Roosevelt wasn't going to let the whooping crane be "doomed to extinction" while he occupied the White House.[41] The laborious process of establishing what is known today as Aransas National Wildlife Refuge was begun by Roosevelt that spring. His order to Wallace was that the principal wintering grounds of the whooping cranes in Gulf Texas should be run by the Biological Survey. He wanted an executive order ready for him to sign by the year's end.

III

Sitting behind a large flat-topped desk in mid-June 1937, still tan from his trip to Texas, President Roosevelt eagerly studied naval logs from the War of 1812 at the National Archives in Washington, D.C. He rested his elbows on the desktop and read historical documents for a couple of hours, head tilted eagerly forward. An electric fan was plugged in nearby to keep him cool and blow away the heavy cigarette smoke he exhaled. Amid the turmoil and tumult of the Great Depression, the president had overseen the construction of the National Archives building in 1934 as America's temple of history, referring to it as "my baby." Roosevelt regarded microfilming as the best way to save storage space. And he took a special interest in the methods by which film footage was being preserved. His eyes, calm and self-possessed, settled most frequently on the logbooks of famous U.S. naval vessels. When amiably chatting with archivists, Roosevelt said with a certain earnestness, "When my term is up, I'm coming here to work."[42]

Helping the National Archives prosper was part and parcel of President Roosevelt's almost touching belief that the federal government had a sacred obligation to save treasures for future generations of Americans: crucial documents, whooping cranes, rare artifacts, bighorn sheep. In 1926, Congress had authorized the construction of a national archive, but funds for the project dried up during the Depression. A groundbreaking did occur in September 1931, with Herbert Hoover laying the cornerstone, but it wasn't until FDR became president that the project received the funding it required.

As Roosevelt envisioned it, the National Archives and Records Administration (NARA) was necessary for the centralization of public and government documents. Working with Congress, he got NARA established in 1934. Its District of Columbia holdings were classified into "record groups." Roosevelt appointed Robert Digges Wimberly Connor as the first archivist of the United States. And it was Roosevelt who told Connor that the three "charters of freedom"—the Declaration of Independence, the U.S. Constitution, and the Bill of Rights—would permanently reside, with free public viewing, in the National Archives rotunda. The president appointed WPA workers to build NARA's regal headquarters north of the National Mall on Constitution Avenue. In later years, NARA would set up regional offices in Atlanta, Boston, Chicago, Denver, Fort Worth, Kansas City, Riverside, and Seattle. But the grand "temple of history" was in the nation's capital.

True to form, the president couldn't refrain from discussing landscape planning with the National Archives employees he met that June afternoon. At one point, he surprised the archivists by saying that, after his death, he wanted a simple marker erected on their grounds: a rectangular block of marble resting in a grassy triangle and engraved only, "In Memory of Franklin Delano Roosevelt." The inscription would face Pennsylvania Avenue. He didn't want anything as grand as the Washington Monument, the Lincoln Memorial, or the Jefferson Memorial (the construction of which he was then sponsoring). "I should like it to consist of a block about the size of this (putting his hand on his desk)," he later told Associate Justice Felix Frankfurter. "I don't care what it is made of . . . whether limestone or granite or whatnot."[43] (On April 12, 1965, the twentieth anniversary of FDR's death, Frankfurter and other veteran New Dealers would unveil a seven-and-a-half-ton Vermont marble monument at the National Archives building.)[44]

That summer of 1937, energized by his trip to the Gulf South, Roosevelt was eager to establish a series of national seashores (a designation partially of his invention), still "searching out" as he put it "bits of original wilderness."[45] One afternoon FDR invited a conservation-obsessed friend, the reporter Irving Brant, who had recently published *Storm over the Constitution*, to the White House to discuss possible marine sanctuary sites in Back Bay (Virginia), Cape Meares (Oregon), and Morro Beach (California). Brant, then moonlighting as the publicist for the Emergency Conservation Committee in New York, was lobbying for national seashores in the mid-Atlantic and in Gulf Texas. He believed that increased federal protection for coastal areas would go a long way toward permanently helping threatened shorebirds. If the White House didn't act fast, commercial and private residential development would erase marine ecosystems such as Cape Hatteras (North Carolina) and Matagorda Bay (Texas). In his 1988 memoir *Adventures in Conser-*

vation with Franklin D. Roosevelt, Brant claimed that he had plotted with the president to have "the entire front of the Gulf of Mexico, from Avery Island 350 miles westerly . . . owned by the government."[46] FDR's commitment to protecting America's coastlines stunned Brant. "I don't know what the New Deal will do for the nation generally," he wrote to a friend, "but in conservation it is certainly time to apply the motto of the Indiana lady 'While you're gittin', git a-plenty.' "[47]

The idea of protecting American seascapes entered Roosevelt's consciousness with a report on "potential shoreline parks" commissioned by NPS Director Cammerer and authorized by Harold Ickes in 1934. The survey, released in 1935, recommended twelve undeveloped coastal sites to become National Seashores.

Roosevelt's spokesman for selling an American national seashore system to the general public was the indefatigable Ickes, who went on a barnstorming tour throughout 1937. Echoing the president, he declared that beaches and shores belonged to *all* Americans to enjoy. "When we look up and down the ocean fronts of America, we find that everywhere they are passing behind the fences of private ownership," Ickes said, on behalf of Roosevelt. "The people no longer get to the ocean. When we have reached the point that a nation of 125 million people cannot set foot upon the thousands of miles of beaches that border the Atlantic and Pacific Oceans, except by permission from those who monopolize the ocean front, then I say it is the prerogative and the duty of the Federal and State governments to step in and acquire, not a swimming beach here and there, but solid blocks of ocean front hundreds of miles in length. Call this ocean front a national park, or a national seashore, or a state park or anything else you please—I say the people have a right to a fair share of it."[48]

Despite Ickes's best intentions and the strongest possible commitment from FDR, the initial seashore effort yielded depressing results. Of the original twelve recommended sites, ten were subsequently developed commercially. Only one was saved by FDR as a National Seashore: Cape Hatteras.

Roosevelt had come to know North Carolina's spit and barrier islands well on his Atlantic cruises. The salt marsh ecosystems were rich with eelgrass (which the Canada goose ate) and widgeongrass (preferred by black ducks). Many man-made features of Cape Hatteras that FDR had once seen while sailing—like the Bodie Island Coast Guard Station—had been destroyed by hurricanes. One of the great symbols of the Atlantic seaboard in Roosevelt's estimation had been the black-and-white-striped Cape Hatteras lighthouse, 208 feet tall, half a mile from the Atlantic coast. After the horrific hurricane season of 1933, the lighthouse was so close to the ocean (thanks to weather-driven erosion) that waves were crashing on the tower's foundation. The United States Lighthouse Service was forced to close the beloved marine beacon after the beach erosion became crippling.[49]

The plight of Cape Hatteras lighthouse moved Frank Stick—a popular commercial artist whose illustrations appeared in *Sports Afield*, *Field and Stream*, and the *Saturday Evening Post*—to write an article in the *Elizabeth City Independence* promoting his "dream" of a national seashore park along the Outer Banks of North Carolina. Besides being an artist, Stick was a wealthy real estate salesman, who prized the seventy-mile stretch from Bodie Island to Ocracoke Island for having the most "beautiful beaches in the world." He owned several of them. A bona fide New Dealer, with Cape Hatteras his sanctified place, Stick went on the warpath to protect the coast from erosion of its sand dunes, hurricane damage, and misguided private development.

As a first step, instigated by the efforts of Stick, the CCC was dispatched to help tackle the problem of shoreline erosion. Instead of constructing beach groins and jetties, which naturalists considered eyesores, these CCCers erected sand barrier dunes on the beaches. WPA workers joined the cause, planting six hundred miles of brush fences to arrest beach erosion.[50] From his sailing experience, Roosevelt knew that the sand dunes of the cape—known in the seafaring world as the "graveyard of America" for its vicious currents and shoals—offered the best frontline protection against hurricanes. The presence of the CCC and WPA workers led New Deal administrators to visit Cape Hatteras, in order to survey the ongoing work. While they were there, they couldn't help but be awestruck by the beautiful and delicate shoreline. One of those New Dealers was Roger Wolcott Toll, the NPS administrator who had filed influential reports leading to the creation of Joshua Tree National Monument. In November 1934 he was visiting potential parkland around the country and gave a positive assessment of Cape Hatteras, noting that "the area is primitive in character. There are no summer homes nor public resort development in the area."[51] Sadly, in February 1936 Toll died in an automobile wreck in Arizona while looking over the Ajo Mountains. Another New Dealer who advocated for a Cape Hatteras park was Conrad Wirth, assistant director of land planning for the National Park Service. He became a convert after traveling to Cape Hatteras to see the work of the CCC boys.

Wirth was born in Hartford, Connecticut, in 1899. He became a top-tier horticulturist, park planner, and accomplished government administrator; during the Hoover years he fell into line with Frederic Delano on the National Capital Park Planning Commission. Once Roosevelt became president, Horace Albright tasked Wirth with having the National Park Service help develop federal, state, and local parks.[52] A devotee of Alexander Jackson Downing and Frederick Law Olmsted, Wirth often bucked bureaucratic protocol while trying to put across his new conservation ideas. In North Carolina, Wirth allied himself with Stick, despite the fact that the two did not get along, but they were intent on saving Cape Hatteras as a seashore park, including a hundred miles of the state's barrier islands. Roosevelt believed the area's historical importance was another justification for national seashore status. After all, the Wright Brothers National Memorial at Kitty Hawk was already contributing to a vibrant tourism trade.

Roosevelt's thoughts were starting to cohere. With his uncanny sense of timing, FDR knew that if a few more hurricane seasons passed without heavy storm damage, then the "trauma factor" would dissipate. Developers would start building stilt houses along Cape Hatteras, and the chance of creating a national seashore in this area would be considerably diminished. The intrepid work of the CCC and WPA workers only increased the probability of development, since the shore was more fully protected than ever from Atlantic storms. Reasonable residents in the Outer Banks concluded that increased tourism was indeed the best economic solution to the problems the Great Depression had brought them.

That presented a problem that delayed NPS designation for Cape Hatteras. A national seashore, as it was conceptualized, welcomed the fact that a seashore could also be a beach—a recreational area open to humans just for pleasure and complete with beach balls, suntan lotion, picnic food, restrooms, and roads. To many conservation purists, that was unacceptable. They had always been uncomfortable with the line between total preservation and the accommodation of humans, but eventually, they made their peace with the fact the respectful visitors were all right (just barely). Hedonistic tourists were anathema. The plan for the national seashore, however, invited the tourists: people interested in nothing more than soaking up the sun and having a splash in the water. In addition to the philosophic debates that swirled around the status of the cape, FDR had to consider that bitter squabbling would occur between Agriculture and Interior about which department would oversee Cape Hatteras. In the meantime, Republican opposition to Roosevelt's New Deal spending hardened in his second term, in part because Supreme Court rulings against programs such as NIRA (and its more visible arm, the NRA) inspired his opponents to question his every White House initiative. Funding for new land acquisition was problematic, but fortunately Stick had been able to convince many landowners within the proposed national seashore to donate their land. In terms of the operating expenses, Republicans gearing up to stop the NPS from controlling the Outer Banks were too late; FDR had already turned over all administrative duties in the area to the NPS. The *New York Times* reported that FDR had even given the NPS its first airplane, specifically in order to check daily on the erosion control work being done along the North Carolina shore.[53]

In spring 1937 Wirth circulated a bill creating the Cape Hatteras National Seashore. Just reaching that point was a tribute to FDR's vision and the stubborn efforts of Stick, Wirth, and several North Carolina politicians who found an answer to every objection. At the president's insistence, the lighthouse was part of the proposed national seashore. Even though the economy slumped alarmingly in the spring of 1937—an event that led to a thirteen-month recession—Roosevelt kept pushing for more land to be set aside for recreation. When Congress rejected his bid to make the CCC a permanent agency that May, the president responded by having the existing CCC camps work even harder to protect his favorite spots, including Cape Hatteras, from private developers.[54]

Meanwhile, as the last-minute legalities regarding Cape Hatteras were being dealt with in the spring of 1937, Roosevelt quietly established Bombay Hook Migratory Waterfowl Refuge in Delaware. Until this action, Delaware had no large federal land holdings. The conservation movement there had been spearheaded by Coleman DuPont, a businessman and politician who was a disciple of Gifford Pinchot. DuPont almost single-handedly forced the creation of a State Forestry Commission, and then jump-started it with donations of good land. Unfortunately he died in 1930, and with no one of similar passion to take his place, Delaware's reclamation and preservation efforts slowed to a snail's pace.[55]

So it was doubly significant that FDR instructed the federal government to purchase the 14,850 acres of Bombay Hook in order to designate them as a sanctuary for greater snow geese (*Chen caerulescens atlantica*), American black ducks (*Anas rubripes*), blue-winged teal (*Anas discors*), and various shorebirds. Having passed "the Hook" numerous times as a seafarer, FDR knew that protecting this salt marsh was essential for the continued prosperity of the Atlantic Flyway. "The most substantial dividends of the refuge investment come after a period of five to ten years has passed," biologist Rachel Carson wrote. "The Bombay Hook Sanctuary in Delaware is a good example. Here the number of waterfowl using the area increased more than 400 percent since the refuge was established in 1937. Official reports show that it was used by 30,000 wild fowl in the fall of 1937, by 60,000 in 1942, and by 137,000 in 1945."[56]

On July 1, 1937, with the Senate nearing a vote on the "court packing" scheme, Roosevelt signed an executive order establishing Moosehorn Migratory Bird Refuge in Maine. Eventually, the National Audubon Society would record 218 species of birds in the refuge. Rachel Carson (who would later become famous as the author of *Silent Spring*) considered Moosehorn one of the most intact marine habitats on the entire Atlantic seaboard. These 22,565 acres of wetlands at the eastern tip of Maine were as isolated as Campobello Island, but without anchorage for large vessels. More than forty square miles of forest were interspersed with lakes, streams, fields, and marshes. There were also a few miles of rugged seacoast. Birders were overjoyed that the refuge would provide protection for the famous courting performance of the male woodcock, which was known to take place there.

At the end of July, the Senate voted against FDR's plan to revolutionize the Supreme Court, a tactical defeat for Roosevelt. He'd seemed invincible until then. Nonetheless, Roosevelt refused to have his conservation effort derailed. On August 17, 1937, Roosevelt established Cape Hatteras National Seashore. There were congressional stipulations on the new entity that deviated from the Organic Act of 1916: national seashores weren't as well protected as national parks. But they were, in spirit, biologically intact recreation areas.[57]

Because Roosevelt was scheduled to visit North Carolina on August 18, the timing of his signature on the seventeenth was either very lucky or very cunning. It may have been the former. The president made the announcement quietly, even though he was touring coastal sites, including Roanoke Island. "The issue was of less interest to reporters than the President's visit," noted NPS historian Cameron Binkley, "which was probably the biggest event to take place on the small island since the Civil War."

Roosevelt arrived at Roanoke Island with panache. Dressed in a white linen suit and sporting a colorful paisley handkerchief in his pocket, he was a larger-than-life figure. That beautiful day, even his hat was bright white. At Fort Raleigh, FDR stood on a podium with Senator Josiah Bailey, a Democratic senator from North Carolina, and mentioned the new designation before reverting to his prepared remarks. Roosevelt was no doubt proud, though, that as of that day, American coastal areas would begin to enjoy protection from private developers. "Realizing that the priceless asset of the ocean shore on both the Atlantic and Pacific was rapidly being appropriated for private use and for public exploitation, the National Park Service recently completed an exhaustive survey to determine what remaining unspoiled area was most suitable for preservation as a national seashore," the *New York Times* reported. "It was found that the remote reefs and islands off the North Carolina coast presented the best opportunity, with their immense stretches of the fine ocean beach, untouched by modern development."[58]

Seemingly unconcerned with the lingering controversy over his "court packing" plan, Roosevelt watched *The Lost Colony*, a new play by Pulitzer Prize–winner Paul Green about the vanished Roanoke Colony of the sixteenth century. His excitement began with the rise of the curtain. The Wayside Theater, where the performance took place, had been built with WPA money from the Federal Theater Project, and the actors were on the WPA payroll. Meanwhile, the CCC was encamped all around the island, fighting beach erosion.[59]

Two weeks later, the *New York Times* ballyhooed Cape Hatteras as "one of the most important conservation measures of all time." It was, in fact, the opening salvo of Roosevelt's second-term battle to save extensive ocean frontage.[60] By protecting the Outer Banks, Roosevelt had set into motion a new protocol: federal marine conservation. Building on his success with Cape Hatteras, Roosevelt also established the Salem Maritime National Historic Site in Massachusetts, which consisted of a dozen historic waterfront sites—including Derby Wharf Light Station and the Custom House.[61]

Ickes, to personally celebrate 1937 as the Year of American Seashores, spent late August and early September at Homans House in Acadia National Park. The coast of Maine, known for its rocky beaches, was one of Ickes's favorite natural places. The house was park property, and Ickes worked there on his political autobiography. The Atlantic made its presence felt, by sight, sound, and smell, every second of every day. "The Homans House is beautifully located, well above the water and at the foot of a high, rugged and partially wooded ridge," Ickes wrote in his diary. "It has a wonderful outlook. Directly across the bay is Schoodic Mountain and the bay itself is dotted with rockbounded islands."[62]

IV

Another of Roosevelt's historic conservation initiatives became public law over the summer of 1937. The Bankhead-Jones Farm Tenant Act directed the secretary of agriculture to launch a program of "land conservation and land utilization" on submarginal acreage unsuited for cultivation.[63] The northern plains, Southwest, and Deep South were the principal beneficiaries of the act. Here was legal justification for Roosevelt to replant cut-over areas from Virginia to Arkansas to Louisiana; establish wildlife refuges in the Dakotas and Montana; and rehabilitate grasslands by means of the Mills project in New Mexico; the Morton County project in Kansas; the Cimarron project in Oklahoma; the Dallam County project in Texas; and the southeastern and southern Otero projects in Colorado (in 1960, these became the Kiowa, Cimarron, Rita Blanca, and Comanche National Grasslands).[64] Title III of Bankhead-Jones established national grasslands in a manner almost identical to the establishment of national forests. In these protected federal preserves, hunting, grazing, mineral extraction, and recreation were all allowed, but under *strict* government environmental oversight.

Following the Bankhead-Jones Act, which helped wildlife prosper, Roosevelt scored big for the conservation movement. After eighteen months of arduous negotiations, the president successfully pushed a hunting tax through Congress. His chief political allies and principal sponsors were Senator Key Pittman of Nevada and Representative A. Willis Robertson of Virginia. Both Democrats chaired their chamber's committees on wildlife conservation. Pittman, in particular, was invaluable to Roosevelt in the fight to get the act passed. A senator since 1913, Pittman was generally opposed to letting the Department of the Interior treat western public lands as one gigantic national park; however, in spite of his disagreements with Interior, he wanted to preserve the wilderness character of his state. This meant securing protection for Nevada's desert bighorn sheep, mountain goats, and pronghorn antelopes.[65]

At the president's insistence, the revenue raised by what was popularly known as the Pittman-Robertson Act would be allotted by the Treasury Department to aid states in funding wildlife restoration programs. To rectify the disparity among the states in land areas and population density, a protocol would be established. Each state would make a calculation of how much money it should receive, taking into account the size of the state and the number of licensed hunters registered there. States were eligible to receive up to 75 percent of total project costs from the Pittman-Robertson fund, with the expectation that they would provide the remaining 25 percent themselves.[66]

On September 2, 1937, President Roosevelt signed the Federal Aid in Wildlife Restoration Act—the actual name of the Pittman-Robertson Act. This major legislation had grown out of the North American Wildlife Conference. The act levied a 10 percent excise tax on guns and ammunition specifically used for hunting. All revenue would provide federal financing of state-owned refuges and public shooting grounds and state-run wildlife-restoration projects subject to approval by the Biological Survey.[67] As conservationists had noted at the wildlife conference, sportsmen would eagerly pay a gun and ammo tax if it meant that robust numbers of game animals were available to hunt. Pittman-Robertson would bring back white-tailed deer to New England, Pennsylvania, and New York; and elk to the Rocky Mountains; and beavers and wild turkeys south of the Canadian border.

The fountain pen Roosevelt used to sign Pittman-Robertson became a museum attraction at the U.S. Fish and Wildlife National Conservation Training Center in Shepherdstown, West Virginia. The pen was displayed, alongside the typewriter Rachel Carson used while working for the Fish and Wildlife Service, as an artifact of importance in environmental history.

President Roosevelt realized, with considerable pride, that he had achieved a legislative miracle with Pittman-Robertson. It was one of the New Deal's finest moments. And the tax yielded results: thanks to Pittman-Robertson, the American deer population swelled from fewer than one million animals to almost thirty million by the twenty-first century. People cursing deer for eating their backyard flowerbeds or running in front of their automobiles had Roosevelt to blame. And between 1938 and 1948 twenty-eight states, spurred on by Pittman-Robertson, acquired nearly 900,000 acres of refuges and wildlife management areas where dams, dikes, and water diversion projects were built that helped the larger hoofed species by offering reliable water sources.[68]

Just how did Roosevelt, under fire from Republicans, pull off a hunting tax in the fall of 1937? Frederic C. Walcott, the former senator from Connecticut who became president of the American Wild Life Institute, thought that the 1936 wildlife conference was the turning point. Hours after Pittman-Robertson was signed, Walcott wrote Roosevelt a note full of appreciation for the "outstanding boosts" he had given wildlife.[69] "Naturally I was all for the objectives of the Wild Life Bill," Roosevelt replied. "There were strenuous objections to it from the Treasury and Budget, however, because it set up what amounts to a continuing appropriation and this is contrary to the laws of the Medes and Persians among the experts! However, the pros outweighed the cons."[70]

Pittman-Robertson became law on July 1, 1938. Within two years, white-tailed deer, wild turkeys, and wood ducks all started to make startling comebacks. That year, almost $3 million in Pittman-Robertson revenue was allocated by the federal government to the states—real money during the Great Depression. A catastrophic situation had been reversed by Roosevelt's environmental activism and, in fact, the law proved so successful that, in the 1950s, similar legislation was enacted for fish populations.[71]

Congressman Absalom Willis Robertson of Virginia, a Democrat, soon complained to Roosevelt about new USDA restrictions on the hunting of sora (*Porzana carolina*), the small, secretive rail bird of freshwater marshes. Robertson, chairman of the House Select Committee on Conservation of Wildlife Resources, considered these small waterbirds as delicious as Cornish hens or wild turkey. Why rob

Americans of a fine country meal? Most important, Robertson's constituents in the Seventh District wanted the hunting regulations for all game birds reduced.

Roosevelt was fond of sora, but not as a meal. Their call—a slow whining *ker-whee*—was a soothing sound of summer in Campobello. And the game birds *were* damn good eating. But Roosevelt ultimately chose the survival of the sora over Robertson's palate. "I have reason to know that such sportsmen, well known to you, as former Senator F. C. Walcott, Dr. George Bird Grinnell, and Mr. Thomas H. Beck share the Department's apprehension for the future sport of sora shooting," Roosevelt replied. "The Federal bag limits are as liberal on all species of migratory game birds as their present status will admit and I feel, in view of the widespread sentiment throughout the country that shooting of all migratory game birds should be prohibited, at least for a year or so, that the real sportsmen should be willing to accept with good grace reasonable limitations on their hunting."[72]

V

Suffering from an abscess in his mouth, his temperature sometimes running up to 103 degrees, FDR was bedridden for some of the fall of 1937. Once he recovered, he asked Assistant Attorney General Robert H. Jackson to join him, Ickes, and Pa Watson on an excursion to the Florida Keys and the Dry Tortugas island chain. Jackson, a fellow upstate New Yorker, though a much younger man, had caught FDR's attention and would be appointed to the Supreme Court in 1941. On November 29, the Roosevelt party left Miami in high spirits for a 165-mile run on the *Potomac* through the waters off southern Florida.

One of the highlights was inspecting Fort Jefferson National Monument on the island of Garden Key. Roosevelt had designated the crumbling hexagonal ex-military prison a national monument on January 4, 1935, as a way to protect portions of the Dry Tortugas ecosystem of precious coral reefs. (The monument was expanded in 1983 and redesignated as Dry Tortugas National Park in 1992.) Somewhere along the line Roosevelt had learned the history of the fort. When the Union Army ran it as a prison for Confederate soldiers during the Civil War, the moat around it was stocked with sharks to prevent inmates from escaping. Dr. Samuel Mudd, who treated John Wilkes Booth after the assassination of President Lincoln, was sent to Fort Jefferson in 1867; there, he treated fellow prisoners who had contracted yellow fever.[73] Just as he had when visiting Beauvoir, Roosevelt enjoyed holding forth about this kind of Civil War lore.

Ickes was proud that Fort Jefferson was in his Interior portfolio, wishing the entire Dry Tortugas chain was designated either a national park or a national seashore. "This fine old historic fort," Ickes had written in his diary, "ought to be maintained for all time."[74] FDR had known about the Dry Tortugas since childhood because of the account of the islands given by the character Billy Bones in Robert Louis Stevenson's *Treasure Island*. With a twinkle in his eye, the president told Watson that he would have to report to the army brass at Fort Jefferson on arrival. Watson, not realizing Roosevelt was playing a joke on him, rowed to shore. There wasn't a single person there to greet him. "Ickes, thereupon, charged Watson with trespassing upon *his* grounds and we held a mock trial at which the President, of course, presided and I was appointed to defend Watson," Jackson wrote in his diary. "The President, of course, found him guilty."[75]

What had caused Roosevelt to save the Dry Tortugas islands, besides the birdlife, were the six hundred varieties of fish in the blue-green waters. Between Garden Key and the Everglades, the president had saved one of the most beautiful marine gardens on the planet. Just looking into the clear waters from the *Potomac*, Roosevelt could see dolphins, sea turtles, devil rays, and colorful coral reefs. On a few of the Tortugas islands, noddy and sooty terns nested each year. By establishing Fort Jefferson National Monument, the president had once again increased America's ocean heritage.

Calling themselves the "four horsemen," Roosevelt, Ickes, Watson, and Jackson played poker, bet on fishing catches, cooked grouper, looked for shipwrecks, mixed martinis, discussed the Civil War, and did a little work related to the Federal Trade Commission. "Life had been most informal in dress and in conduct," Jackson wrote in his diary. "[Roosevelt] was of course treated respectfully by all but with perfect informality. In fishing contests, playing cards, and conversation he was and wanted to be an equal. He asked no favors and granted none. He played the game on its merits. He was able to avoid all pose. He was away from curious eyes. We were completely isolated."[76]

Roosevelt ended 1937—his year of oceanic adventure—in style. Once back from Florida in early December, he signed Executive Order 7780, establishing two migratory waterfowl refuges—Lacassine (31,858 acres) and Sabine (142,850 acres)—to protect southern Louisiana's waterfowl.[77] Then, on December 31, 1937, as a gift to himself, the president signed Executive Order 7784, establishing the 47,200-acre Aransas Migratory Waterfowl Refuge in Texas. (It would become known as Aransas National Wildlife Refuge in 1940 and would eventually grow to encompass 115,000 preserved acres.)[78] Roosevelt was starting the job of protecting Louisiana's bayous and the Texas coastline. Not only were the geese and ducks protected on these southern marshes but so too were alligators, muskrats, raccoons, and deer.

Since the president's spring visit to the Gulf of Mexico, the Biological Survey had snapped up tracts of land in parts of Aransas, Refugio, and Calhoun counties, using $463,500 in revenue from the sale of Duck Stamps. The creation of the Texas refuge was a blessing for the nation, as it ensured the survival of the whooping crane. No hunting was allowed in Aransas, and freshwater ponds were built by the CCC and stocked with food suitable for ducks, geese, and cranes. The new refuge manager, Charles A. Keefer, soon documented more than 250 species of birds in Roosevelt's refuge, including the vermilion flycatcher and American bald eagle. No more reckless drainage, clearing fires, cattle grazing, plowing, or plume hunters would touch Aransas.

According to the *Washington Post*, FDR's commitment to save coastal treasures like Port Aransas, Cape Hatteras, Bombay Hook, Moosehorn, and the Dry Tortugas was an unrivaled "object lesson" in what can be done for American conservation "under proper management."[79] FDR had done more to protect America's coastlines, marine sanctuaries, and barrier islands than all of his White House predecessors combined.

CHAPTER FIFTEEN

"I HOPE THE SON-OF-A-BITCH WHO LOGGED THAT IS ROASTING IN HELL"

I

Not all of Franklin Roosevelt's American adventures in 1937 were on the open sea or in coastal areas. That summer, the president decided to build a new house at the easternmost end of his Springwood estate, on what Daisy Suckley had called "the nicest hill" in Dutchess County.[1] He designed a Dutch colonial home, built of stone and surrounded by brush and briar. He dubbed it "Top Cottage." From this wheelchair-friendly house, the president could take in views of three ranges: to the northwest, the Catskills; to the west, the Shawangunks; and to the southwest, the Hudson Highlands. Closer at hand, meadows and woodlands, which fanned out below the front veranda, would have pleased Andrew Jackson Downing and all the Delanos.

When FDR drove the three miles from the main house in Hyde Park to Top Cottage—*his* Val-Kill—he could "escape the mob." Wooded, serene, and unmarred by the prefabricated clutter of the industrial mid-twentieth century, it was a place where the president could relax without disruption. In fact, Roosevelt hoped someday to write his political memoirs at Top Cottage. All of his personal papers would be safeguarded at Springwood, but transporting selections of them to Top Cottage would have been an easy enough task.

Because Daisy Suckley influenced the retreat's location, some historians have speculated that Top Cottage was the place designed for trysts. However, the house primarily served as a safe-haven for FDR, away from the public gaze, somewhere he could gossip about Livingstons and Vanderbilts, make ham-and-cheese sandwiches for lunch, and mix martinis at sunset. At Top Cottage he never worried about facial stubble or a wrinkled shirt. He was secluded, and surrounded by the sanctified land that invariably gave him a sense of spiritual renewal. "I was driving through the middle part of [Dutchess County] the last time I was here in early August," Roosevelt told his neighbors at a picnic, "and I was struck by the number of lovely streams we have in the county, not only the larger creeks, like the Wappinger, but also the Krum Elbow and a lot of the smaller creeks, and it occurred to me what a wonderful escape we had."[2]

Throughout 1937, Roosevelt continued to plant trees at Springwood under the supervision of Syracuse-based forester Nelson Brown, who reported that he was introducing Asian chestnuts (*Castanea crenata*) on an experimental plot of Hyde Park.[3] Silviculturists hoped this species of chestnut was blight-resistant. Although FDR had objected to the Asian variety being planted in the Okefenokee, he approved its addition to his personal estate; still, he instructed Brown to give the American chestnuts (*Castanea dentata*) priority. "When you next make a check on the woods I wish you could have someone from the college go through the woods, especially the woods around the oldest white pine grove back of the farm to see if there are many young chestnuts growing," Roosevelt wrote to Brown. "I have seen a number of them—possibly forty or fifty—ranging from five feet to twenty feet in height. Back on the top of the hill at the extreme east end of the place I think there are some others."[4]

Tracts slated for commercial harvest included thousands of Norway spruce, European larch, and red and white pine. Brown gave regular reports on the trees to FDR, who listened with delight. "With the recent heavy rains I am very much encouraged and believe that results of planting should be very successful," Brown reported. "Even the trees stricken by the drought last year have shown excellent survival due to the emergency watering which we did during July."[5] Roosevelt, in turn, passed these reports along to the nation with the happy relish of a country farmer. At a spring 1937 press conference, the president boasted of his arborist ambitions:

The President: I was just showing the Dean [longtime reporter John Russell Young] my bill which has just come in for 26,000 trees which are going to be planted next month.

Q: Spruce trees, aren't they?

The President: In others words, I am practicing what I preach.

Q: Are these the Christmas trees?

The President: Yes. 23,000 Norway spruce, 2,000 balsam firs and 1,000 Douglas firs. That is experimental. That is an awful lot of trees.

Q: Those are the ordinary Santa Claus trees?

The President: Yes. They are on another ten acres of waste land. That is stopping erosion.

Q: You did not tell us the amount of the bill.

The President: $130.00.

Q: Did you say, sir, whether they are to be planted in Georgia or in New York.

The President: At Hyde Park.

Q: $130.00 for all of them?

The President: Yes, $5.00 a thousand.

Q: Are they seedlings?

The President: They call them three-year old transplants. In other words, they have been transplanted once already from the original bed.

Q: Do you get them from the State Conservation Department?

The President: Yes.

Q: You had the others at Valkill?

The President: In the back of the cottage, yes. Outside of that I don't think there is any news at all.[6]

By bringing his own Hyde Park tree planting into a press conference, Roosevelt kept the spotlight on the New Deal's reforestation programs. Roosevelt's Forest Service, in fact, had established the four greatest tree nurseries in the world, headquartered in Michigan, Wisconsin, Mississippi, and Louisiana. From those and other facilities came the saplings the CCC "boys" planted across America.[7] Because FDR had turned Hyde Park and Warm Springs into successful demonstration farms, he hoped rural folks might be more inclined to follow suit. Moreover, after his landslide reelection in 1936, an emboldened FDR planned on protecting a new set of landscapes. The more that extraction industries tried to block new national parks in the West, or new migratory bird refuges in the South, the more determined Roosevelt was to win. By 1937 state governments and private enterprises in the Far West were no longer hostile to the Forest Service overseeing huge tracts of public lands. Regulating commercial access to the national forests made sense even to timber barons. But the national park, as an idea, was loathed by Pacific Northwest businessmen, who derided it as aimed only to "please the women."[8]

To the chagrin of Eleanor Roosevelt, Congress halted funding to all female-oriented, federally funded, work-relief programs in 1937. Adding insult to injury, no summary report of how the 8,500 women performed was written. "The CCC camps with their millions of dollars for wages, educational work, travel, and supervision constantly reminded me of what we might do for women," Hilda Worthington Smith, pioneer social worker, complained to the first lady. "As is so often the case, the boys get the breaks, the girls are neglected."[9]

At the heart of Roosevelt's primary conservation confrontation in 1937 was the American Forestry Association (AFA), an industry group representing, in large part, lumber companies. Its board of directors was disenchanted with Roosevelt, a member of AFA, because he was adopting the NPS view of absolute preservation of forests instead of the "wise-use" Pinchotism of the Forest Service. The fight boiled down to the same old thing: turf. AFA president G. K. McClure, for example, complained directly to the White House about Ickes's determination to protect the primeval wilderness of the Hoh Rain Forest in the Olympics under the auspices of the National Park Service. McClure wondered how FDR, the proponent of Natchez Trace Parkway and Grand Coulee Dam, could suddenly oppose timbering in the national forests and parks. The AFA saw Roosevelt's transformation as a "matter of deep concern." [10]

Roosevelt nobly defended himself in a letter to McClure. Explaining his longtime disdain for clear-cut-lumbering, he admitted that his senses were assaulted every time he saw the remains of treetops or the massacre of an evergreen stand juxtaposed with the emerald green of an adjoining forest. The president matter-of-factly told McClure that great forests of "rare and exotic trees" were "desirable to preserve," and that their natural splendor should be forever "classified among the wonders of nature." The president announced that against the encroachments of the bulldozer, the federal government had an obligation to protect western forests. "I refer, for example, to the Giant Sequoias in Southern California, to the Sugar Pines in the same general area, to the Redwood Forests of the California Coast, and to the Douglas Fir of the Columbia River region," Roosevelt wrote to McClure. "In the case of all these varieties, with the exception of the Douglas Fir, it is too much to expect that we can cut and renew these forests on a yield basis." According to Roosevelt, "the only way" to preserve "these marvelous trees" was to offer federal or state preservation.[11]

FDR's letter infuriated McClure. If farmers were allowed to harvest the biggest ear of corn or the largest pumpkin on their land, then why shouldn't lumbermen be allowed to chop down the biggest trees—the ones worth the most money—with impunity? After all, the president himself had dabbled in commercial forestry, harvesting Norway spruce from his Springwood estate in the 1930s and 1940s and earning thousands of dollars.[12]

The American Forestry Association learned the hard way that Roosevelt could turn on a dime, transforming himself from a tree farmer to an incurable Hudson River romantic infatuated with the feel-good pastoralism of John Burroughs and the environmental idealism of Bob Marshall. Most troubling to the AFA was the way that Ickes, the most powerful secretary of the interior in American history, continued to grow in prominence—while the power and influence of the Forest Service had correspondingly declined—during Roosevelt's first term. Between 1933 and 1947 the NPS doubled in size.[13] This anti–Forest Service critique was overwrought. FDR remained a staunch champion of the Forest Service and its director, Frederick Silcox. In the western national forests, the president had the CCC boys reseed thousands of acres of grazing land and built truck roads to help loggers prosper. The New Deal had already seized twenty-two million acres of forested land for both outdoors recreation and harvestable groves.

By Roosevelt's second term, he had to face head-on the escalating feud over public lands management between the Forest Service and the National Park Service; their long-standing rivalry had intensified into hatred. Each agency considered the other unrealistic. The Forest Service worked closely with the AFA and its member businesses. The National Park Service was the darling of New Deal preservationists. To the Forest Service, the smartest defense against Ickes's takeovers was to argue that in national parks wilderness would be "desecrated" by overdevelopment and commercialization—due to a never-ending quest to attract hordes of tourists. Advocates of the NPS countered that the Forest Service's primitive areas had no genuine security and could be abolished at the stroke of a pen by an unelected administrative officer. By contrast, national parks had federal protection (though not necessarily as wilderness).[14] The burning question of 1937 was how far FDR would lean toward one or the other when pressured.

That summer, Nelson Brown traveled to Europe, where he would inspect a variety of managed forests. The president was envious: if his appointment calendar weren't so full, he would have joined his personal forester. Roosevelt urged Brown to specifically visit the community-owned forests in France, Germany, and Austria. The president still fervently wished that American farmers could learn how European towns made admirable profits from selling lumber, so he suggested that Brown lend a hand. "If you could get up a little book in popular vein with photographs and a catchy title and cover, it would sell like hotcakes," FDR said. "The important thing is to confine the story to small communities which have not got much capital to invest. I hope you have a wonderful time."[15]

II

On September 22, 1937, President Roosevelt began a fifteen-day trip through the western states, ultimately to inspect the work being done at three dams in the northwest: Fort Peck (on the Missouri River in Montana); Bonneville (on the Columbia between Oregon and Washington); and Grand Coulee (on the Columbia in northeastern Washington).[16] There was also a fierce turf struggle between the U.S. Forest Service and the NPS over forestland in the Olympics that he hoped to resolve. Leaving Hyde Park and traveling by train, with Eleanor coming along for the adventure, the president arrived in central Iowa and delivered a radio address from the village of Marshalltown about the stability of crops. A few days later, the Roosevelts arrived in Wyoming with a retinue of government clerks, stenographers, press handlers, and secretaries.

FDR gave speeches in Cheyenne and Casper about the nation's public lands. Genuinely curious, he asked locals for updates on drought conditions and forest fires and was gratified to see that his first-term New Deal policies had helped the state cope with soil erosion. "The grass is better this year than it was last year," Eleanor Roosevelt observed; "there has been a little more rain. In consequence the cattle and the sheep look better and the people themselves look more cheerful."[17]

Moving north, the Roosevelts detrained at the Gardner entrance to Yellowstone National Park, for a stay of two days and nights. Immediately inquiring about a series of recent forest fires, FDR was impressed that two thirds of Yellowstone still had healthy lodgepole pines (*Pinus contorta*). The historic Mammoth Hot Springs Hotel at Yellowstone was in the midst of remodeling, and Roosevelt chatted with workers about the architectural plans for a new lobby, recreational hall, barbershop, hair salon, and dining room.[18] He was particularly delighted by a large wooden map of the United States he was shown; it had been carved from fifteen different types of local wood in preparation for his arrival.[19]

Rustic cabins were also under construction by CCC workers, and the president met with the CCC "boys" who were assigned to Yellowstone.[20] The first family, of course, watched the famous geysers—Daisy and Old Faithful—erupt like clockwork. "I was not disappointed," an enthralled Eleanor Roosevelt reported. "The water shooting up in the air, or in the case of the Daisy out sideways, was

Roosevelt stops at Artists' Point in Yellowstone National Park in the back of a touring car in September 1937. With temperatures in the park at or near freezing, the president wrapped himself up in a robe, sitting in the open air. Roosevelt's western tour was intended in part to inspire other Americans to visit their national parks, so he did as they might, using a car to travel from one destination to another.

most graceful and the rainbows added to the beauty. I think the colors, looking down into some of the hot pools seemed more beautiful than almost anything else. Nature combines so many colors and has so much to teach us where this is concerned. If only we realized it, it is the shades that matter, almost any colors go well together."[21]

Superintendent Edmund B. Rogers escorted the Roosevelts, who continually turned their heads to see wilderness scenes from their open-air automobile—for instance, a herd of buffalo grazing on grass, sedge, and forb. No speeches were given; they were just sightseeing. A picnic lunch was held at Fishing Bridge, where supposedly more fish were caught each year than in any other landlocked spot in America.[22] Later in the tour, they enjoyed the spectacle of a great elk with huge antlers tending his harem; only a few months earlier, Ickes had inaugurated an intensive program in Yellowstone aimed at permanently protecting the elk population. "A number of deer and antelopes came quite near us and three bears were fed on the trip into the park by the President," Eleanor Roosevelt recalled. "One of them became a little too friendly and put his paws up on the side of the car right next to my husband and immediately the superintendent of the park, Mr. Rogers, ordered the car to move on. The thought of a nice tear from his claw on the President's coat was too great an anxiety to allow us to loiter any longer, but the bear held up all the other cars by standing in the middle of the road."[23]

But Eleanor Roosevelt wasn't pleased with the overcommercialization in Yellowstone. The hotel's curio shop, to her chagrin, sold wooden bear statuettes made by Swiss wood-carvers. "I can't help feeling we might encourage some of our North Carolina mountain carvers to do a number of the park animals," Eleanor wrote, "as they would be more interesting as souvenirs if done by American talent."[24] And the first lady was appalled at the way tourists had vandalized Yellowstone's natural features, tossing garbage into the thermal pools, chiseling off chunks of rock, and carving initials in lodgepole pines. "I suppose it's all a matter of education and self-control," she concluded, "and older people are just as guilty as young people, so we will just have to wait until the nation grows up and in the meantime we will have to guard our beauty spots the best we can."[25]

Leaving Yellowstone, the Roosevelts headed southwest by train into Idaho, passing Henrys Lake in Caribou-Targhee National Forest, where the trumpeter swans wintered. Turning westward, they rode through Idaho Falls and Pocatello, before crossing the Owyhee desert to the capital city, Boise. Traditionally a Republican state, Idaho had voted for FDR in 1932 and 1936, largely owing to his personal appeal and to federal irrigation programs that drew a large bloc of farmers into the New Deal.

By the time Roosevelt visited Boise via Pocatello on September 27, the Department of Agriculture oversaw fifteen national forests entirely or partly in Idaho, totaling more than twenty million acres—the largest area in any state. Within the national forests, vast swaths of land that had formerly been logged had been reclaimed and were covered with second growth. As admirable as the progress in forestry was, Idahoans still distrusted governmental intervention in anything related to business. A number of anti-New Deal Republicans were angry that FDR's protection of forests along the Elk River had caused the Potlatch Lumber Company to downsize, damaging local communities. Now, on his tour, Roosevelt promoted tourism—hunting, fishing, skiing, and white-water rafting—as the best available economic measure for Idaho. To that end, its great pine and fir forests should be protected as wilderness zones. That year Bob Marshall, always bursting with ideas, pushed for two new forest regulations (U1 and U2) that were aimed at prohibiting roads, logging, and mining in designated "wilderness areas" in the West. In accordance with Marshall's vision, the Forest Service soon established the Selway-Bitterroot Primitive Area, a 1.3 million-acre wilderness.[26]

The Roosevelt administration also oversaw, in Idaho, five Native American reservations run by the Bureau of Indian Affairs: Kootenai, Duck Valley (Shoshone-Paiute), Coeur D'Alene, Nez Percé, and Fort Hall (Shoshone-Bannock). The five tribes were strongly pro-Roosevelt because of the 1934 Indian Reorganization Act, which helped to modernize reservations and return some disputed land to Native Americans. Thanks to the Indian Division of the CCC, which employed seventy-seven thousand Native Americans during its first six years of existence, reservations soon had additional homes, schoolhouses, sewage treatment facilities, telephone lines, reservoirs, firebreaks, and truck trails. Life on these reservations had improved dramatically thanks, in part, to the Indian New Deal.[27]

In Boise, the president spoke extemporaneously about the glories of Idaho after motoring around the North and East Ends and downtown. After visiting WPA sites and irrigation projects sponsored by the Bureau of Reclamation—Idaho ranked fifth in New Deal expenditures per capita from 1933 to 1939—Roosevelt basked in optimism.[28] "When I look back on today's visit to Boise I shall think chiefly of two things: first your beautiful, tree-lined streets and, secondly, your children," he said before a rally of fifteen thousand. "And I take it, being a Roosevelt, that you are following the Rooseveltian creed, and that the population is not going to die out. There is something about children and trees that makes me think of permanence and the future. It is not by any means the sole task of the Presidency to think about the present. One of the chief obligations of the Presidency is to think about the future. We have been, in one-hundred-and-fifty years of constitutional existence, a wasteful nation, a nation that has wasted its natural resources and, very often, wasted its human resources."[29]

A slogan for Idaho during the Great Depression could have been "Great Heart of the CCC." There were 163 CCC camps in the state—more than in any other state except California.[30] Over thirty thousand unemployed Idahoans were receiving CCC paychecks for improving the state's mountain scenery. Another sixty thousand CCCers had come from such places as New York and the Midwest to help transform Idaho.[31]

Because many Idaho boys had grown up in the outdoors, they proved to be ideal CCCers. As children, they had learned how to ski cross-country, fish, and hunt for wild game. Now many of them had the opportunity to help bring recreational opportunities to the Idaho wilderness. In western Idaho, the CCC built roads in the primary drainage areas like the middle and south forks of the Boise River, the south fork of the Payette, and the south fork of the Salmon.[32] In Boise National Forest and Heyburn State Park, recreation rooms were built and equipped with pool tables and freestanding weights. Mess halls in Idaho CCC camps served up fresh trout and venison for lunch and dinner. Little cabin libraries were established in even remote counties. Personal disagreements between CCC recruits in Idaho were often settled in the boxing ring. "When the fight ended," Kenneth Hart of Company 195 recalled, "that was the end of it. Everything was settled."[33]

The CCC's accomplishments in Idaho were impressive: 236 forest lookout houses or towers built; ninety-one diversion dams constructed; 3,034 miles of telephone lines strung; twenty-eight million trees planted; and many historical sites restored. White pine blister rust and gypsy moths were finally brought under control, and lakes and numerous streams were stocked with fish. In Owyhee County, as Roosevelt proudly pointed out, the CCC had made the largest contiguous irrigated area in America even bigger. And Idaho was the very first volume in the WPA American Guide Series, complied by novelist Vardis Fisher, state director of the Idaho Federal Writers' Project and author of the well-received *Dark Bridwell*.[34]

Sometimes the CCC and the WPA took away from the wilderness more beauty than they brought to it. Elers Koch, who had been a forester since the early days of the national forest program in 1903, worked at Lolo National Forest in Idaho and was a progressive, enlightened conservationist. At first, he welcomed the CCC, but he soon regarded it as the "hammer" of bad forest policy. In the February 1935 issue of *Journal of Forestry*, Koch wrote a scathing "memorial" for his beloved region. "The Lolo Trail is no more," Koch began. "The bulldozer blade has ripped out the hoof tracks of Chief Joseph's ponies. The trail was worn deep by centuries of Nez Percé and Blackfeet Indians, by Lewis and Clark, by companies of Northwest Company fur traders, by General Howard's cavalry horses, by Captain Mullan, the engineer, and by the early day forest ranger. It is gone, and in its place there is only the print of the automobile tires in the dust."[35]

But Roosevelt was in Idaho to take credit for the CCC's and WPA's successful work-relief programs. Warmly greeting Boise residents with his usual "Hello, neighbor"—just as he would greet his friends in Dutchess County—and declaring that he felt like Antaeus (a figure in Greek myth whose strength doubled whenever his foot touched the ground), Roosevelt reminded the enthusiastic crowd that "the saving of our timber" was one of the most significant public works of the New Deal.[36]

The president sampled many local foods on his western trip in 1937. The White House chef, Henrietta Nesbitt, had learned before that the president was interested in the provenance of his food. Such culinary details mattered mightily to Roosevelt, who had developed pet theories about the best soil for various food crops. His favorite dish was salmon (another attraction of the Pacific Northwest), and Nesbitt recorded how he liked it prepared: cleaned expertly, wrapped in a cloth tied with string, and boiled for forty minutes in a kettle of water to which half a cup of vinegar and two tablespoons of salt had been added.[37]

The question for Roosevelt and millions of other people who enjoyed salmon was whether the species could survive overfishing along the Pacific coast. Large-scale fishing traps, normally set up in the shallows of the sea, utilized an extensive system of nets arranged in a V-pattern to collect live salmon. The fishermen took advantage of the salmon's instinct to head inwards toward the rivers to spawn. So huge and effective were these traps that a single one could snare hundreds of thousands of salmon, or sometimes a million salmon, in a four-month season. Theodore Roosevelt had highlighted the problem with the traps and their hyperefficiency in his 1908 State of the Union address. Oregon and Washington had subsequently banned the traps, but serious problems with the salmon population still remained in those states and especially in the U.S. territory of Alaska, where the traps were still legal. The outlook for the species—which now numbered only a tiny fraction of the salmon that were present when Lewis and Clark visited the Pacific Northwest in 1805—was a serious issue when the Roosevelts arrived in Idaho.

Alaskan tribes such as the Snoqualmie and Skokomish couldn't compete with huge conglomerates from Seattle that used traps. Native Americans depended on the traditional method of using seine nets, which stretched from one side of the bank to the other. A delegation of two tribal fishermen from Alaska had come to visit Eleanor Roosevelt at the White House the year before, hoping she might persuade her husband to use his executive authority to save their food source. In her newspaper column, she sided with the Native Americans in the dispute: "In a few years, history will say that there were once salmon along these coasts and in these rivers but fish traps destroyed them all."[38]

During this western trip, the president got personally involved in the dire salmon situation. Once he was back at the White House, he coaxed the Biological Survey into increasing the number of fish ladders at Grand Coulee, but that was only a short-term remedy. Within months, Roosevelt would allocate money to protect Alaskan coastal areas from overfishing. He didn't stop there. "It occurs to me," Roosevelt wrote to R. Walton Moore, a State Department official who was a staunch defender of fisheries, "that a Presidential Proclamation closing the sea along the Alaskan coast to all fishing—Japanese, Canadian, and American—a kind of marine refuge essential to end the depletion. I do not know what Japan could well say in the event of such a proclamation and I am reasonably certain that the Canadian Government would approve and do the same thing along their British Columbia coastline."[39]

The following day, Roosevelt drafted a memorandum for Secretary of State Cordell Hull aimed at saving the great salmon runs and preserving the pristine coastline of southeastern Alaska as a federal marine preserve. While Harold Ickes thought it a grand idea, the State Department worried that it would unnecessarily antagonize the Japanese. Recognizing that the world's main supply of sockeye salmon (*Oncorhynchus nerka*) came from the Bristol Bay area, FDR pestered Hull for more information about how Alaska's fish population could best be protected from irreparable damage. He asked for "a map showing the depth contours of the Alaskan Coast" and "an estimate from the experts as to which contour depth could be chosen as affording complete protection." He agreed with Hull's assertion that "far more than the Bristol Bay area is involved" and argued in favor of safeguarding "the entire shore line of the whole of Alaska." Roosevelt suggested to the State Department that any efforts to save salmon fisheries for the future could be sold to the public by stressing "not only the investment in this American industry but also its relationship as a large factor in the American food supply."[40]

From Idaho, the Roosevelts traveled to eastern Oregon to the region of the Owyhee Dam, dedicated by President Hoover on July 16, 1932. This concrete arch-gravity dam, once the tallest in the world, had served as the prototype for Boulder Dam and others. Its reservoir held water for an elaborate system that provided irrigation to parched farmland throughout the region. Eleanor Roosevelt wrote in her column "My Day," "If someone had said to me that I would see a desert one minute with sagebrush the only visible vegetation, and the next minute some of the best farming land that I have seen anywhere, I would have thought they were telling me a tale! As I looked more carefully I saw the irrigation ditches with their companion drainage ditches."[41] She scoffed at the New Deal's detractors who claimed that the chief beneficiaries of dams like Owyhee were the "sage brush and jackrabbits."[42]

After the Owyhee Dam region, the Roosevelts made their way to Portland. On September 28, they drove to Mount Hood to see Timberline Lodge, built the year before by the WPA at an elevation of 5,960 feet within Mount Hood National Forest. The Adirondacks-style lodge was made of local timber and stone and had intricately carved decorative elements courtesy of the Federal Art Project.[43] It received rave reviews from architects across the nation.

When Roosevelt dedicated the lodge, he explained his vision of a "Magic Mile" ski lift that would surpass anything similar in the Swiss Alps and a ski season that would be longer than any in the Colorado Rockies. "This Timberline Lodge marks a venture that was made possible by WPA emergency relief work, in order that we may test the workability of recreational facilities installed by the Government itself and operated under its complete control," Roosevelt told the enthusiastic audience. "Here, to Mount Hood, will come thousands and thousands of visitors in the coming years. Looking east toward eastern Oregon with its great livestock raising areas, these visitors are going to visualize the relationship between the cattle ranches and the summer ranges in the forests. . . . Those who will follow us to Timberline Lodge on their holidays and vacations will represent the enjoyment of new opportunities for play in every season of the year."[44]

The president gazed at the varied topographical features of Mount Hood, including vast evergreen forests, the Columbia River Gorge, and the hot springs (which would have made fine retreats for polio patients) of Multnomah Falls. Fascinated by Pony Express history, Roosevelt commissioned a stamp to honor the legendary mail-delivery speedsters and directed the CCC to reconstruct a section of the old Barlow Road in Oregon as the Pioneer Bridle Trail.[45] There was something uplifting and American and musical about the names of CCC camps assigned to the Mount Hood area—Wyeth, Cascade, Locks, Parkdale, Friend, Bear Springs, Zigzag, Summit Meadows,

Oak Grove, and High Rock, to name a few. They would have delighted novelist Thomas Wolfe, who would embark on a grand tour of the West's national parks in 1938. There was even a guard station called the "Little White House," named after the president's retreat in Warm Springs, Georgia.[46]

III

While Roosevelt's dams and the CCC were popular in the Pacific Northwest, the National Park Service, which had expanded rapidly over the first five years of the New Deal, wasn't—at least not with the lumber companies.[47] The timber lobby was preparing for a battle royal over the Olympic Peninsula of Washington to determine the fate of this beautiful ecosystem. In 1897, President Grover Cleveland designated 2.2 million woodlands acres on the peninsula as a forest reserve, closed to logging, mining, and ranching. In the ensuing years, that designation evolved and the extent of the preserve shrank. In 1908, the area around 7,969-foot Mount Olympus became a national monument, bordering Olympic National Forest, which was administered by the USDA Forest Service. In the 1930s a movement grew to consolidate the government's Olympic holdings, add more acreage, and emerge with one of the most ecologically rich national parks in America.

The impetus for establishing the Olympics region as a national park resulted from a hike made through the peninsula by Dr. Willard Van Name, a curator at the American Museum of Natural History. He had heard that the Forest Service was allowing extensive commercial use of the national monument, including logging. In 1934 Van Name wrote a call to arms in the form of a booklet, *The Proposed Olympic National Park: The Last Chance for a Magnificent and Unique National Park*. Because Van Name's employment contract with the museum discouraged his activist writings on conservation, the booklet carried the byline of his colleague on the Emergency Conservation Committee (ECC), a society of New York City natural scientists and park advocates organized in 1930 to reform the Audubon Society. The booklet was distributed widely in Seattle and Washington, D.C., making a mark in both places. Before long it inspired a campaign to transfer the lands from the Forest Service to the National Park Service. An enthusiastic set of Puget Sound residents supported the ECC's plan for Olympic National Park. Representative Monrad Wallgren, of Everett, introduced a bill in Congress calling for a 728,860-acre park. The lumber industry, outraged at Wallgren's bill, forced a second congressional bill that significantly reduced Wallgren's proposal.

The fight over the Olympics triggered yet another turf battle between the USDA's Forest Service and Interior's National Park Service.[48] Ickes went on the warpath against the Forest Service and its relationship to the pulpwood industry, charging that it was protecting far too few acres as wilderness areas across America. Ickes determined that national environmental needs trumped the local economic argument on the peninsula. To bolster his case, he directed his staff to draft a bill that would give FDR the authority to designate wilderness areas within both national parks and national forests. The proposed legislation stipulated that once such an area had been designated, it could be modified only by an act of Congress. After the bill was introduced and Ickes's team mustered congressional support for it, the House scheduled a hearing.

Among those testifying was Bob Marshall. Ever since he worked for the Bureau of Indian Affairs as the chief forester, Marshall had been friendly with Ickes, but now he was running the Forest Service Division of Recreation and Lands. Marshall was also the founder, leader, and chief benefactor of the nonprofit Wilderness Society, and he had directed it to send to relevant personnel at both the Forest Service and the National Park Service a questionnaire about how they would protect the areas surrounding Mount Olympus. On the basis of the responses, Marshall testified on Capitol Hill that he was in favor of the National Park Service taking control of the Olympics. The Forest Service, he concluded, was protecting far too little acreage as "pristine" wilderness, and the primeval groves could be logged with the "stroke of a pen" by a future U.S. president. By contrast, logging wasn't permitted by Congress in any National Park. This took courage, as Marshall's boss at the Forest Service was now Secretary of Agriculture Henry Wallace. Marshall's testimony caught the president's attention. FDR knew that since Ickes had been able to recruit

Marshall to his side, then national park designation for the Olympics really was the best route. Sensing victory, Ickes and the House committee chair promised to drop the bill if Congress would establish a large national park on the Olympic Peninsula. "Secretary Ickes sought to deflect the 'parks-will-be-overdeveloped' Forest Service defense by announcing his policy that the wilderness of the new parks would not be developed," historian Doug Scott explained, "while he lambasted the weakness of the Forest Service's wilderness efforts."[49]

The Emergency Conservation Committee, consisting of Rosalie Edge, Irving Brant, and Van Name, was a key ally for Ickes in securing Olympic National Park.[50] At the urging of Ickes, the ECC lobbied for the trees below the timberline on the Olympic Peninsula—stands of Douglas fir, western hemlock, red cedar, and Sitka spruce—to be saved in perpetuity. As the timber companies vociferously objected, the ECC argued that the point of establishing Olympic National Park was not only to protect nearly treeless mountain peaks but also to save the rain forests at their bases.[51] "The development of our national parks," Brant had written to Roosevelt in 1932, "at the expense of the national forests is demanded by the more fundamental needs of our country—permanent preservation of magnificent primeval forests which cannot be replaced for centuries if once cut down as the Forest Service intends they should be, and preservation of wildlife through large, permanent sanctuaries."[52]

While Van Name and Edge were New Yorkers, Brant was a seasoned journalist from the Midwest. Born on January 17, 1885, in Walker, Iowa, Brant worked as a reporter or editor for various regional newspapers including the *Iowa City Republican*, *Clinton Herald*, *Des Moines Register and Tribune*, and *St. Louis Star-Times*.[53] His heroes were the Supreme Court justice Louis Brandeis and writer Henry David Thoreau.[54] In 1918, Brant, as an editorial writer and military analyst for the *St. Louis Star*, hired his wife, Hazeldean Toof, to write book reviews. Whenever they had a free moment, the Brants spent time with "books and birds, birds and books"; they were especially enthralled by Enos Mills's 1920 classic, *The Adventures of a Nature Guide*.[55] Mills, a disciple of John Muir, was credited with convincing the Wilson administration to establish Rocky Mountain National

Park in Colorado in 1915. After reading Toof's glowing review of his book, Mills invited the Brants to spend a week at Long's Peak Inn, the hotel he had built near the national park. This trip spurred Irving Brant into becoming a full-time activist for the National Park Service. In particular, he hoped to establish an Ozark River National Park, even writing a "folk drama" about the region to create public interest. Meanwhile, Brant was also researching a historical novel about the maritime fur trade; over the seven years it took him to complete his novel, he visited the forests of the Olympics a half dozen times.[56]

Irving Brant, assistant to Harold Ickes, was traveling with Roosevelt on his western tour when he snapped a photograph of fishermen in the Queets River in northwest Washington State in 1937. Roosevelt's tour of the Olympic Mountains changed the future of the region, giving him a firsthand understanding of the abuse wrought on the rain forest there and a resolve to stop it.

During a winter in the Olympics, Brant snowshoed along Hurricane Ridge, hiked in the moss-covered forests, and watched whales at scenic Pacific beaches. With great freneticism, he drove around the American West, reporting on the public lands for newspapers. The shabby conditions of the Petrified Forest National Monument in Arizona in particular enraged him. "Day by day, month by month," Brant charged in a *Saturday Evening Post* article, "the Petrified Forest of America is being looted and smashed to pieces by the motoring public of America."[57]

By 1927, the Brants had settled in Berkeley, California, where

Brant wrote conservation articles drawn from his experiences on America's public lands. In *Scientific American*, he covered the plight of some forest rangers who had been struck by lightning; in the *New York Times* he advocated for national parks to include more swaths of wilderness. He also wrote about how tourists were feeding Yellowstone's bears and elks, effectively taming them (FDR was later guilty of that temptation); it sickened him that the "welfare of the concessionaire" had become the "prime consideration of the National Park Service." But it was Brant's fiery article "What Ails the Audubon Society" that seized Ickes's attention. Brant ably exposed the extent to which Audubon had become entangled with ammunition manufacturers. Brant's piece also outraged Rosalie Edge, a venerated World War I–era suffragist. Her nonprofit organization, ECC, went on the warpath against the Audubon Society.[58]

Along with FDR's son-in-law John Boettiger and uncle Frederic Delano, Brant became the president's eyes and ears on public land and wildlife protection issues. Starting in the mid-1930s, often writing in the *New Republic*, Brant sounded the New Deal alarm about lumber and power interests in the Pacific Northwest region that were ignoring federal laws. Brant soon became the preeminent champion of expanding Mount Olympus National Monument to become Olympic National Park by incorporating 400,000 acres of adjacent national forest—plus the addition of a sizable eastern unit along the coast of Washington state.

In 1934 Van Name's pamphlet *The Proposed Olympics National Park* caught Roosevelt's close attention. Van Name believed that the Olympics were on par with Yellowstone or Yosemite for grandeur. Spurred on by Van Name, Roosevelt wrote to his friend Albert Z. Gray in 1935 that he was mulling over an Olympics National Park. "There has been no official proposal by the Secretary of the Interior to establish this area as a national park," Roosevelt wrote to Gray. "However, if such a recommendation is made, I shall be glad to give the matter every consideration."[59] Less than a year later, the president had asked Attorney General Homer Cummings whether he could create this national park by presidential proclamation. Could he cite the Antiquities Act as his authority to seize lands from Olympic National

Forest, join them to Olympic National Monument (where logging was prohibited), and make the giant national park Ickes envisioned?

The entire Olympic Peninsula was a tree lover's paradise. These thick woods often received 160 inches of rainfall per year. Some fifty glaciers existed in these mountains, including the Ice River, Blue, Humes, and University glaciers. These glaciers formed the headwaters of long streams—including the Hoh, Quinault, Queets, and Bogachiel rivers—that flowed westward toward the Pacific.[60] Their water was as pure and clean as the rain and snow that fed them. Salmon and steelhead ran these pristine rivers in the fall and winter, cutthroat trout in the summer. The peninsula was also home to an unmanaged herd of Roosevelt elk (*Cervus canadensis roosevelti*), which too often were being slaughtered by rogue hunters. Nevertheless, after careful investigation, Cummings determined that Roosevelt's plan to combine Olympic National Forest and Olympic National Monument was an overreach of executive power.[61]

In February 1936, Brant, then editor of the *St. Louis Star-Times*, talked with the president at the White House about how to stop the conflict between the National Park Service and the Forest Service. Brant reported to Ickes, "The matters about which I spoke to the President were the preservation of the virgin forests around the present Mount Olympus National Monument and the sugar pine grove near Carl Inn, on the edge of Yosemite Park." On the subject of logging that continually threatened the Olympic region, Brant argued that sacrificing irreplaceable trees, which might be a millennium old, was shortsighted. At best, it would keep sawmills running for another few years.

Five days later, in a memorandum with "Mt. Olympus National Monument" printed in bold lettering, Roosevelt wrote to Wallace (and copied Ickes): "I understand that there is a forest area immediately adjacent to the national monument. Why should two Departments run this acreage? If the forest portion is not to be used for eventual commercial forestation and cutting, why not include the forest area in the national monument?"[62] Later that same afternoon FDR, still under the influence of Brant, pestered Wallace about federal protection for Yosemite's sugar pine groves. "Can this land be saved," the president asked Wallace, "by a trade of lands?"[63] Wallace was greatly annoyed by Ickes's and Brant's attempts to bring both the Olympics and Yosemite under the Department of the Interior, as timbering there would bring revenue to the USDA. He wrote disingenuously to the president that the sugar pines were not worthy of preservation.[64]

The fate of federal lands on the Olympic Peninsula remained unclear. Wallace had jumped into the fray in July 1937, claiming that a "primitive area" of 238,930 acres of Olympic National Forest would be established to help the Roosevelt elk prosper. Ickes was still hoping that FDR would transfer much of the national forest into a new Olympic National Park. Meanwhile, Brant was dazzling the president with photos of the last virgin forests in the Northwest. Wallace angrily realized that the Forest Service was being outfoxed in the Pacific Northwest by Ickes and the ECC activists Brant, Edge, and Van Name, who had direct access to the president. It seemed to Delano and Ickes, along with the ECC, that the sensible solution would be to combine both federal sites and enlarge the acreage under the aegis of the National Park Service. But the Olympic Peninsula was a prime lumbering region, and timber businesses and ranchers on the peninsula had formed an immovable alliance against national park designation. That was the cacophonous situation that stayed in FDR's mind, as he approached the lands over which so many were willing to fight.

On September 30, 1937, Roosevelt left Seattle on the destroyer USS *Phelps*, bound for Victoria, British Columbia, to speak with Canadian officials about forestry.[65] Heading south again that afternoon, battling stormy weather, with gulls circling and gliding above the deck, *Phelps* crossed the Strait of Juan de Fuca, rounding Ediz Hook Spit, and was greeted by a twenty-one-gun salute, fired by the U.S. Coast Guard cutter *Samuel P. Ingham*, on approaching Port Angeles, Washington. A huge delegation of local dignitaries gathered at the pier to greet the president, who was accompanied by his daughter Anna Roosevelt Boettiger of Seattle and her family.

The week before, when Roosevelt was in Yellowstone, he had claimed he was "not sure" about the idea of an Olympic National

Park. Now, in Port Angeles, he seemed to have fully endorsed it. As the Roosevelt caravan crept slowly up Lincoln Street, pausing periodically for the president to wave at the starstruck crowd, the cheers and chants grew. "Babies and toddlers were held aloft for the president to see," Mary Lou Hanify—who, as a teenager, had been among the spectators—later recalled. "Most people felt sure that they had been individually singled out for the famous Roosevelt smile."[66] As the convoy crawled slowly to the courthouse, where thousands of schoolchildren had congregated on the lawn, the anticipation grew. A huge banner was draped across the face of the courthouse, reading, "Mr. President, we children need your help. Give us Olympic National Park."

As the clock struck 6 p.m. Mayor Ralph Davis of Port Angeles introduced Roosevelt to the cheering throng, and the Theodore Roosevelt High School band played a rousing Sousa-like version of "The Star-Spangled Banner." A hush fell over the gathering as Roosevelt began to speak. Here, in person, was the uplifting voice that had transfixed them through many dozens of fireside chats. "That sign is the appealingest appeal I have ever seen in my travels," Roosevelt declared. "I am inclined to think that it counts more to have the children want that park than all the rest of us put together. So, you boys and girls, I think you can count on my help in getting that national park, not only because we need it for old people and you young people, but for a whole lot of young people who are going to come along in the next hundred years of America."[67]

From Port Angeles, Roosevelt drove to Lake Crescent Tavern (now Lake Crescent Lodge), where he hashed out the bedeviling issues of jurisdiction, size, and loopholes for a new national park with representatives of both the Forest Service and the NPS. "The Olympic Peninsula will in the future be as popular as Yellowstone is now and we must provide for future generations to come," Roosevelt told the group. "The western hemlock is a beautiful tree and eastern people want to see it." Reflecting on the school banner he saw earlier that day, he added, "Why not call it 'The Olympic National Park?' This would tie in with the Olympic Peninsula and mean something. Mount Olympus is too hard to say."[68]

Usually, FDR's approach to conservation valued getting negotiations started rather than holding an all-or-nothing line. But the Olympics were different. The fight had gone on far too long. Building on the momentum of his 1936 reelection, Roosevelt made clear at Lake Crescent Tavern that the time was right for the proposed Olympic National Park. If Roosevelt could sell Congress on it, he would be responsible for having created the third-largest national park in the country. Tomlinson wrote to his superior, National Park Service director Arno B. Cammerer, about FDR's commitment to a major Yellowstone-style national park on the Olympic Peninsula: "During the entire discussion, the President left no doubt in the mind of anyone present that he favored a large national park and that he especially desires the preservation of typical stands of timber," Tomlinson reported to Cammerer. "He emphasized the need to save the western hemlock."[69]

Overnight the word traveled far and wide that Roosevelt was going to fight for Olympic National Park. In the excitement of the moment, a Port Angeles schoolboy, Willis Welsh, had caught a trout in Barnes Creek especially for the president to eat in celebration. Once again delighting in a regional cuisine, Roosevelt dined on Olympic blackberry jam, Olympia oyster cocktails, Dungeness crab, young Puget Sound turkey, and prime rib of Washington beef, and capped off the dinner with Grays Harbor cranberry sherbet and wild blackberry pie.

At 9:15 a.m. on October 1, Roosevelt drove the 105-mile scenic road that skirted Lake Crescent and Lake Quinault to inspect the Olympic Peninsula's Sitka spruce (*Picea sitchensis*), western hemlock (*Tsuga heterophylla*), and western red cedar (*Thuja plicata*); some of these giants reached 250 feet in height. The awesome beauty of the multilayered canopies, standing snags, and fallen trunks astonished him. In every direction was a satisfying picture. The president's visit around the disputed area was overseen by the Forest Service and its allies in the lumber and pulp mill industries, all of whom were intent on convincing him that a national park, especially an extensive one, would devastate the already struggling economy. The Forest Service excluded all National Park Service officials from the invitation list.

Timber lobbyists kept telling Roosevelt about the need for industrial development in the region. They even moved a sign marking the national forest boundaries, giving the impression that a heavily logged area—several square miles of burned stumps, a picture of voracious commercial logging—wasn't on federal land. An acute sadness fell over the president as his motorcade passed miles of overlogged hillsides. He blurted out, "I hope the son-of-a-bitch who logged that is roasting in hell."[70]

All afternoon, Roosevelt met with people from towns west of the Olympic Mountains, under gray sky drizzling moisture. Special demonstrations, such as tree plantings and ax throwing, were held in his honor. At Snider Ranger Station, Roosevelt watched a simulated forest fire and a demonstration of emergency firefighting. The Indian CCC erected a pair of huge totem poles on the northern boundary of the Quinault Reservation to honor FDR. Schoolchildren from the Taholah Indian Agency held canoe maneuvers at great speed, to the president's delight. The boats were hewn by hand out of cedar logs. Salmon were toted from the river for lunch. Governor Clarence D. Martin of Washington talked with Roosevelt about how best to establish the national park. Those two days proved to be the turning point.

As Irving Brant wrote to the National Park Service's associate director A. E. Demaray four days later, "With President Roosevelt publicly advocating the park, there no longer seems to be any danger of being confronted with a 'this or nothing' choice dictated by the Forest Service and lumber interests."[71] Negotiations over the details of Olympic National Park—including the fate of the largest old-growth forest in the Pacific Northwest—continued for the next several months.

IV

From Lake Quinault, the Roosevelts headed to Tacoma and then Grand Coulee Dam. On this return visit Roosevelt studied the Columbia River on topographical maps. He felt confident about Grand Coulee—a town ninety miles west of Spokane—where the New Deal worksites were a beehive of activity. Roosevelt kept promising that expansive farmlands in the area would soon be irrigated by the dam.

He spoke publicly about all the recreational opportunities Grand Coulee was creating. "When the dam is completed and the pool is filled, we shall have a lake 155 miles long running all the way to Canada," Roosevelt remarked at the dam on October 2. "You young people especially are going to live to see the day, when thousands and thousands of people are going to use this great lake both for transportation purposes and for pleasure purposes. There will be sailboats and motor boats and steamship lines running from here to the northern border of the United States into Canada."[72]

One leading New Dealer opposed to Pacific Northwest dams was the future Supreme Court justice William O. Douglas from Yakima, Washington. On the Columbia River, he complained, severe damage was being done to the populations of salmon and steelhead that migrated upriver to spawn. The fish ladders were frying the fish as they went through the turbines. "I was so concerned that I visited the dams to see how the fish ladders were working, and on seeing FDR after one of these visits, I asked him for the job of counting the fish at Bonneville, the only job I ever asked him to give me," Douglas recalled. "He took it as a joke and roared with laughter saying 'You've got yourself a new job.' "[73]

Lake Roosevelt, the reservoir of the dam created in 1941 by the impoundment of the Columbia River, immediately became the largest lake in Washington state. But it also had many unintentionally deleterious effects as Douglas had warned. Most notably, while building the Grand Coulee may have brought electricity to the rural poor of Grant and Okanogan counties in Washington state, it also resulted in the destruction of eleven towns along the Columbia River. Several were Native American villages representing a way of life dating back more than ten thousand years. When the dam came, the river as they knew it vanished. Overall, three thousand people had to abandon their homes and businesses.[74] And in coming decades man-made Lake Roosevelt would be mired in lawsuits related to pollution.

Professor Aldo Leopold of the University of Wisconsin likewise roundly criticized the New Deal for allowing technical solutions to overwhelm cultural, economic, and ethical issues. The push for hydropower, Leopold argued, ignored the damage building colossal dams had caused to salmon runs and the Columbia River Valley ecosystem as a whole. In a 1938 essay, "Engineering and Conservation," Leopold charged that Roosevelt's penchant for building dams failed to resolve the "standard paradox of the twentieth century: our tools are better than we are, and grow better faster than we do. They suffice to crack the atom, to command the tides. But they do not suffice for the oldest task in human history: to live on a piece of land without spoiling it."[75]

Concerned about complaints from conservationists like Leopold, who were usually allies of the administration, Ickes decided that Grand Coulee Dam needed some positive publicity. He at first relied on *Look* and *Life* to promote the recreational benefits of the proposed dam-reservoir, but ultimately hired Woody Guthrie to write songs celebrating the Bonneville Power Administration's construction of Grand Coulee Dam on the Columbia River. While it's not known whether FDR admired his music, Guthrie wrote twenty-six enduring songs for the Department of the Interior, including "Roll on Columbia," "Pastures of Plenty," and "Grand Coulee Dam." This most unlikely New Deal propagandist fell in love with the Columbia River Valley. "I can't believe it," Guthrie wrote of the surrounding wilderness. "I'm in paradise."[76]

On October 3, heading eastward to Washington, D.C., Roosevelt arrived in eastern Montana to inspect Fort Peck Dam, just as he had promised to do in 1934. A large crowd gathered to hear him speak about the virtues of hydroelectric power. While it was wonderful that the construction project employed 10,500 people and spawned new communities, there was still something dreadful going on at the site. The dam had been clumsily constructed, and cracks appeared, resulting in an accident that killed eight workers and injured many more. The following year, an engineer would pinpoint the problem as an incorrectly elevated pipeline. Though the dam was not quite finished, one got the feeling that perhaps Douglas and Leopold had been right, that in the end man couldn't—or shouldn't—mess with nature on such a gargantuan scale.

After extolling Fort Peck Dam, the president gave a gallant conservationist speech in Grand Forks, North Dakota, about bettering the abused land. He had seen firsthand how the Biological Survey had saved the regional type of wetlands known as prairie potholes. On October 4, Roosevelt brought his New Deal road show to Saint Paul, Minnesota. Lashing out at the Supreme Court for gutting his minimum wage and his eight-hour workday, Roosevelt sounded like a socialist. Claiming that Minnesotans weren't "wild-eyed radicals" as Wall Street believed, Roosevelt championed the little guy, the working folks whom the Supreme Court was giving a raw deal.[77]

As recently as August, Roosevelt had signed a bill to establish Pipestone National Monument in southwestern Minnesota. The significance of Pipestone (as noted in Chapter 10) had to do with the northern Great Plains Indians, who mined the soft red clay from the Minnesota quarry to make long, flutelike ceremonial peace pipes. To members of the Yankton and Lakota people of the Great Sioux Nation in South Dakota and Minnesota, the pipestone quarry was a holy place. Its soft clay so impressed western artist George Catlin in the 1830s that he painted a panoramic picture based on the religious ceremonies held there.

Roosevelt was particularly elated about the work the Indian CCC had done in Minnesota. By 1937, the Consolidated Chippewa Indian Agency—headquartered at Cass Lake in Minnesota—used the CCC as a way to completely revamp six reservations: Fond du Lac, Grand Portage, Leech Lake, Mille Lacs, Nett Lake–Vermilion Lake, and White Earth. The Chippewa grew wild rice, dredged canals, and built roads. For the first time, hospitals and recreation centers were constructed on Minnesota reservations. A major feature of these reservation camps was night school. Courses were taught in conservation principles and biology. No other president had ever helped Native Americans prosper with the heartfelt conviction of FDR. The New Deal encouraged Indian self-rule, the restoration of tribal government, and the resuscitation of native culture and religion. But this wasn't government paternalism or welfare. Native American CCCers were pulling themselves up by the bootstraps even while many of the sixty Indian languages spoken in the West, from Apache to Zuni, were going dormant or extinct due to homogenization. And for the most part, the CCC funded initiatives through which Native Americans would improve 52 million acres of Indian country agricultural lands. In the opinion of the historian Donald Parman, "probably at no time before or since the founding of the United States have Indian forests and lands been in better condition than in 1942."[78]

On October 5 in Chicago, the president spoke at the dedication of the Outer Drive link bridge and delivered his "quarantine speech," outlining U.S. foreign policy measures in a troubled world. He decried the "epidemic of lawlessness" around the world yet never mentioned Nazi Germany, imperial Japan, or fascist Italy by name. This foreign policy speech implied that potential external threats to America were brewing everywhere.[79] Some reporters, however, were shocked that Roosevelt had spoken with more passion on the Olympic Peninsula about conservation than in Chicago about foreign affairs. Willing to publicly express his hope that despoilers of land were "roasting in hell," he chose to stay largely mum about Hitler and Mussolini.[80]

Roosevelt's Western trip convinced him that more than ever the United States needed a National Highway system. Automobiles in the West should be able to drive safely and quickly across the Continental Divide in the Pacific Northwest or the Mojave Desert from Los Angeles to Phoenix. Just weeks after Roosevelt returned to Washington, he began consulting with highway engineers. When interstate road visionary Thomas N. MacDonald visited the White House, the president pulled out a map of America and proceeded to draw a grid of lines in blue pencil. Three went from coast to coast. Others followed the migratory bird routes from Canada to the Gulf of Mexico. Here was the birth of what in the 1950s became the Eisenhower system of Interstate and Defense Highways. At forty-seven thousand miles long and four lanes wide, Roosevelt's public works roads project, which led to interstates such as 10, 40, and 75, forever changed America. That such super-roads were bad for wildlife wasn't contemplated. "Franklin Roosevelt," historian Earl Swift concluded in *The Big Roads*, "had a greater hand in its creation than Eisenhower did."[81]

V

Back in Hyde Park in mid-October, Roosevelt attended the dedication of a new post office the WPA had designed using the Dutch-style architecture he so loved. Afterward, he held a press conference in which he expressed his delight at the various forests he had enjoyed in Wyoming, Montana, Idaho, Washington, and Oregon. With great glee, he revealed how he had procured sugar pine, western cedar, sequoia, and lodgepole pine seeds to plant at Springwood. "We are having a planting ceremony tomorrow and we are also taking the cones apart and trying them in the greenhouse," he said. "You remember in Yellowstone, those hillsides where there were those perfectly straight trees? They are lodgepole pines. They may not grow in this climate, for they came from a level where they have snow and ice, but we hope they will grow here."[82]

While Roosevelt turned his attention to pressing world affairs in late 1937, the indefatigable Brant lobbied Congress to pass legislation creating Olympic National Park. It was a tough sell. Sixty percent of the Olympic Peninsula's population was directly connected to the logging and milling industries.[83] But Brant was so dazzling on Capitol Hill that Roosevelt hired him away from the *St. Louis Star-Times* to become a White House speechwriter and consultant. FDR was frustrated that Congress wouldn't allow the Hoh and Bogachiel valleys, which remained privately owned, to enter into the debate over the national parks. "I am disturbed," Roosevelt said, "by three matters which we cannot accomplish under the recent Olympic National Park Bill. The first of these relates to the preservation of the Pacific Shoreline." Another irritation for the president was that the forests of the Quinault Indian Reservation weren't part of his new park. "I think," he wrote, "we should have legislation for the preservation of all the remaining timber in the Quinault River Valley."[84] From an ecological perspective, Brant knew that the president was indeed correct. Emergency funds were needed for the government to make land purchases. In short order, $1.7 million was raised. The new boundaries made the proposed park one of the largest in the system and the crown jewel of the Pacific Northwest.

"Big timber," sensing that the 1938 midterms would be problematic for Roosevelt, continued to try to derail the national park. The complexity of land deeds in the Pacific Northwest was daunting and Congress moved glacially on the Olympic issue. But the bill managed to survive. Finally, in June, a compromise was reached. At 500,000 acres, the park's boundaries would be smaller than environmentalists—and President Roosevelt—might have wished (300,000 acres drawn from the existing national monument and the rest from Olympic National Forest). According to the bill creating the park, though, the president had the authority to expand it by proclamation at a later date. The legislation was on FDR's desk for signature on June 29, 1938. "It is the intention to keep this park, so far as possible, in a wilderness area," Harold Ickes later wrote in his diary. "It is truly a wonderland of nature and it is more than I can understand how people who pretend to be interested could be opposed to its creation as a national park."[85]

Roosevelt was overjoyed on signing the authorizing legislation. "In the future the new Olympic National Park may be extended in area by adding lands acquired by gift or purchase of additional lands from the Olympic National Forest," he said. "The establishment of this new national park will be of interest to everybody in the country. Its scenery and the remarkable tree growth are well worth seeing, and it is a worthy addition to the splendid national parks which have already been created in many parts of the country."[86]

CHAPTER SIXTEEN

"PERPETUATED FOR POSTERITY"

I

On January 3, 1938, Roosevelt delivered his State of the Union address, linking conservation with economic recovery from the Great Depression. About two thirds of the economic gains achieved since March 1933 had vanished. The stock market had lost half its value. The unemployment rate hovered at 17.2 percent. Harry Hopkins and Henry Morgenthau—Keynesians both—had warned Roosevelt that work-relief projects like the WPA and CCC weren't enough, that the pump still needed priming. Aware that his New Deal recovery was losing steam, Roosevelt blamed the shrinking economy on America's irresponsible corporations, which were slow to invest in new production and growth; and on the extremely wealthy, who were neglecting to participate in commerce in favor of stockpiling their fortunes. The president excoriated "economic royalists" and "selfish interests" and championed minimum wages and maximum work hours. A core premise of the State of the Union address was that the industrial order had been careless with America's natural resources, and now the New Deal was rectifying the damage with creative conservation. "We went forward feverishly and thoughtlessly until nature rebelled," Roosevelt said, "and we saw deserts encroach, floods destroy, trees disappear, and soil exhausted."[1]

Putting his words into action, the president spent 1938 signing executive orders and issuing presidential proclamations galore, establishing new National Wildlife Refuges (NWRs) from coast to coast. It was almost as if he were collecting wildlife sanctuaries as a hobby. In northern Montana there were Black Coulee NWR and Hewitt Lake NWR along the Canadian border where pronghorns (*Antilocapra americana*) ranged and waterfowl bred. Alabama got Wheeler

NWR, the first federal refuge overlaid on a multipurpose reservoir (in this case formed by the TVA). No fewer than ten endangered species lived at Wheeler.

The intensity of Roosevelt's interest in wildlife protection became apparent on February 14, when he proclaimed that March 20 to 27 would be National Wild Life Week. Declaring all animals our "inarticulate friends"—even salamanders, tortoises, and gila monsters—Roosevelt argued that by saving species in peril, Americans were choosing life over death, beauty over destruction, God over the devil. "To this end I call upon all citizens in every community to give thought during this period to the needs of the denizens of field, forest, and water and intelligent consideration of the best means for translating good intentions into practical action in behalf of these invaluable but inarticulate friends," Roosevelt said. "Only through the full cooperation of all can wild life be restored for the present generation and perpetuated for posterity."[2]

Buoyed by his successful NWRs in Montana and Alabama, Roosevelt moved decisively to protect numerous offshore islands in the lower Florida Keys, a marine ecosystem he considered worthy of being a national park. Looking beyond Fort Jefferson National Monument, Roosevelt selected other islands as federally protected sanctuaries for the great white heron (or great egret, *Andea alba*), which was, he said, an "almost extinct species that occurs only in southern Florida," hunted by bird poachers for its feathers.[3] After consulting with Ernest Coe, Florida's most energetic environmentalist, the president signed Executive Order 7993 on October 27, establishing Great White Heron National Wildlife Refuge, protecting 117,683 acres of tidal flats, white-sand beaches, and turquoise water especially for this charismatic species.[4] (In 1975, Great White Heron NWR, Key West NWR, and National Key Deer Refuge were combined under the National Wilderness Preservation System.)[5]

Another marine ecosystem Roosevelt protected that spring of 1938 was the Channel Islands of California, known as the "Galápagos of North America" for its rare plants and wildlife—145 species found nowhere else. Roosevelt knew from friends at the Smithsonian Institution that these five unique islands just off the coast of Ventura County were a premier destination for whale watchers, birders, beachcombers, and fisherfolk like himself. The archipelago was the only place along the Pacific coasts of the Americas where warm and cold ocean currents commingled.[6] Stephen Albright, the nephew of the former NPS director, had sent the president photographs of whales—humpbacks, grays, and blues—taken around the archipelago. Roosevelt thought that a Channel Islands marine sanctuary would be a wonderful national park, a place where West Coast residents could snorkel and scuba dive in offshore kelp forests and canoe along the gorgeous coastlines. Furthermore, Roosevelt's favorite recent movie, *Mutiny on the Bounty*, starring Charles Laughton and Clark Gable, had been filmed, in part, in the Channel Islands, giving the location cachet with the Hollywood crowd.

But a consortium of special interests rejected the preservation effort. California speculators saw the Channel Islands area as a promising offshore oil field and led a campaign to quash the idea of a national park. The movement for a park was further damaged when H. C. Bryant, an assistant director at the NPS who had seen the islands only through binoculars from Ventura, on the mainland, decided that they didn't meet NPS's high aesthetic standards. Others dismissed the islands as desolate and too difficult for tourists to reach. Such anticonservation sentiments made Ickes's blood boil. Owing to "selfish oil interests," as he put it, the southern California coast hadn't been able "to keep conservation laws" on the statute books in Sacramento.[7]

Coming to the rescue of the Channel Islands, however, was noted biologist, botanist, etymologist, and paleontologist T. D. A. (Theodore Dru Alison) Cockerell, who, since his retirement from the University of Colorado in 1934, had been spending winters in California studying the flora and fauna. In 1937 he published an article, "The Botany of the California Islands," a scholarly natural history essay, that Ickes brought to the president's attention.[8] Other grassroots environmentalists wrote to the Department of the Interior about the huge harbor seal (*Phoca vitulina richardii*) and sea lion (*Zalophus californianus*) populations of the Channel Islands.

About two thousand pairs of brown pelicans—favorites of Roosevelt—nested on the islands.[9] Reports that San Miguel Island had the finest assemblages of seals and sea lions in California was enough to cause FDR to act decisively. Cleverly bypassing Congress, Ickes drew up a presidential proclamation for Roosevelt to sign, establishing Channel Islands National Monument with archaeological research as a prime justification. On April 26, Roosevelt, in signing it, proclaimed that the islands contained certain "fossils of Pleistocene elephants and ancient trees, and furnish noteworthy examples of ancient volcanism, deposition, and active sea erosion, and have situated thereon various other objects of geological and scientific interest."[10] This was more in line with the traditional use of the Antiquities Act for archaeology. To circumvent potential congressional inquiries over funding the new national monument, Roosevelt had the superintendent of Sequoia National Park also manage the productive and diverse Channel Islands marine sanctuary.

Once the national monument was a fait accompli, Roosevelt had the Department of the Interior inventory the rare species of the ecosystem for the Smithsonian Institution. Roosevelt's instinctive notion that the Channel Islands were indeed North America's Galápagos was vindicated a few years later when a field naturalist with the Department of the Interior wrote that he never had seen so much teeming wildlife in a single day. "Boy," he wrote in an official report, "we've got something out there in the Channel Islands."[11] (Two future presidents—Harry Truman and Jimmy Carter—enlarged the monument by adding islands.[12] In 1980 Congress voted to upgrade Roosevelt's "Galápagos" to national park status.)[13]

After the hassles over Key West, Channel Islands, and Cape Hatteras, Roosevelt decided it was easier to save America's coastal areas by executive orders establishing national wildlife refuges.[14] Because of his exhaustive naval and nautical expertise, Roosevelt didn't need advice from Ickes or Brant on the choice of coastal locations. With Congress ignorant of his intentions, Roosevelt went on a spree. Among the coastal marine ecosystems Roosevelt saved in 1938 was the gorgeous Cape Meares along the Oregon coast, which formed a high, steep bluff on the south end of Tillamook Bay. Ornithologist William Finley had sent photographs of the vertical coastal cliffs and rock outcroppings to FDR, including images of a remnant of coastal old-growth forest. The Oregon Coast Trail passed through the rolling Cape Meares headlands and the Audubon Society had documented thousands of brants, pelagic cormorants, common murres, tufted puffins, western gulls, and black oystercatchers populating the cape's dramatic landscape. If all that wasn't enough to make Roosevelt pick up his pen, the nineteenth-century Cape Meares lighthouse was a famous reference point for Pacific Ocean sailors. Executive Order 7957 on August 19, 1938, brought the president as much joy as establishing Great White Heron in Florida and the Channel Islands in California.

Another natural feature that Camp Meares offered, one that surely appealed to Roosevelt, was the "Octopus Tree," a giant old Sitka spruce (*Picea sitchensis*), which hadn't developed a massive single trunk like most of its kind along the Oregon Coast. Because of its candelabra branching and large six-foot base, this spruce was revered by Native Americans as the "Council Tree." It was possibly two thousand years old, and Roosevelt, with his executive order, saved it from becoming lumber.[15]

On the Atlantic coast the president created Tybee National Wildlife Refuge at the mouth of the Savannah River in South Carolina, signing Executive Order 7882 on May 9, 1938. While this island was only a dollop of land, just a hundred acres, it was the home of piping plovers (*Charadrius melodus*) and wood storks (*Mycteria americana*). Manatee (*Trichechus*) swam up the coast to Tybee in the summer months. When Johnny Mercer wrote the lyrics for Henry Mancini's "Moon River" in 1961, this was essentially the romantic Atlantic landscape he was celebrating.[16]

II

River conservation was very much on Roosevelt's mind in 1938. Utility companies and coal suppliers sought congressional investigations into the legality of the TVA, while environmentalists from the Pacific Northwest complained that Grand Coulee Dam, due to open in 1941, was devastating riverine and riparian wildlife. Even with fish ladders, the "tamed" Columbia aborted the spawning runs of salmon and steelhead. A geographer would be hard-pressed to find a major western river that Roosevelt didn't want to dam. And FDR wanted not the small earthen plugs that the CCC built across streams to stock water or raise bass but colossal dams to steal the power of the Columbia, the Tennessee, the Sacramento, the Snake, the Red, and the Colorado.[17]

In his defense, Roosevelt saw that remaking rivers by means of dams was a middle course between reckless exploitation and extreme environmentalism. As a corollary of this principle, he also wanted to keep America's rivers and reservoirs from biological wreckage. Waste treatment plants, he said, were needed all across America. Tough laws to protect salmon, trout, bass, and pike were encouraged. "As you know," Roosevelt wrote to Kenneth Reid of the Izaak Walton League, "I am vitally interested in the protection of our streams." At the president's prodding, the PWA started purifying the Potomac River. All towns and cities along the river would get sewage treatment plants. Ickes called the PWA program the "freeing of the Potomac."[18] Roosevelt especially wanted Rock Creek and the Anacostia River cleaned up in a grand effort to beautify Greater Washington.[19]

Indeed, Roosevelt fancied himself America's caring riverkeeper. To Roosevelt, "water conservation" in the East meant protecting rivers like the Hudson and Delaware from overdevelopment. For the West it meant constructing dams with reservoirs, thereby guaranteeing year-round water resources. One such project was the Grand River Dam in Oklahoma City, authorized in 1935. At a large rally there, Roosevelt said, "The Grand River Project is a good illustration of the national aspect of water control, because it is a vital link in the still larger problem of the whole valley of the Arkansas—a planning task that starts far west in the Rocky Mountains, west of the Royal Gorge, and runs on down through Colorado and Kansas and Oklahoma and Arkansas to the Mississippi River itself and thence to the sea. The day will come, I hope, when every drop of water that flows into that great watershed, through all those states, will be controlled for the benefit of mankind, controlled for the growing of forests, for the prevention of soil erosion, for the irrigation of land, for the development of water power, for the ending of floods and for the improvement of navigation."[20]

What infuriated Roosevelt most was the stream pollution from mines; he considered this the primary source of noxious pollution in America. Discharge from metal, clay, and coal mines were devastating fish in streams, large and small. In the Susquehanna and Delaware river watersheds, the mining of anthracite coal was poisoning and blackening streams. Pulp and paper mills were almost as bad, coughing out acidic washes. Because rivers and streams were by nature part of an interstate system, they needed federal regulation. The problem, as Roosevelt saw it, was that federal jurisdiction over water pollution was almost solely investigatory. The White House had the power to order abatement of discharge of solids into navigable waters where such refuse interacted with navigation. There was also some power to control discharge of oil in waters where the tide ebbed and flowed. But, by and large, the federal government was impotent in matters related to the pollution crisis. Fourteen states had no pollution control laws whatsoever. Twenty-six offered only partial or ineffective control. Recreation, public health, the development of fish life, and shellfish culture were all suffering because of untreated waste. Even on the Potomac River watershed, two factories were discharging industrial waste greater than the sewage flow of the District of Columbia. Working closely with Ickes, FDR started a long push for a Clean Rivers Act, which, at last, came to fruition in 1965, when President Lyndon B. Johnson got Congress to pass the Water Quality Act.

Yet, for all Roosevelt's good intentions, the New Deal often ecologically damaged America's rivers. The Army Corps of Engineers, for example, turned the Mississippi River into one necklace of dams and levees. Likewise, for nearly one thousand miles, the Ohio River became a series of impoundments. Almost all of California's rivers were plugged in hundreds of spots to fuel the giant boom of agricultural and urban sprawl. In the Tennessee Valley, the New Deal was building dams—Norris (1934), Pickwick Landing (1935), Chickamauga (1936), Kentucky (1938)—at an astonishing rate. For all the public electricity generated, the once-bucolic landscape was marred beyond recognition.

That spring of 1938, Franklin Roosevelt invited Carl Carmer, a thirty-five-year-old writer from Cortland, New York, to discuss Hudson River history with him in Washington. Refreshingly,

Carmer, whose best-selling 1934 book *Stars Fell over Alabama* was seen by Roosevelt as a gift to America in spirit with WPA guides, wasn't interested in New Deal politics or the rise of Hitler; he merely wanted to interview FDR about New York history for a book he was writing, *The Hudson*, in the new "Rivers of America" series.[21] Taking off his glasses, his blue eyes peering at Carmer, Roosevelt told colorful anecdotes about Dutch immigrants in Dutchess County; his grandfather, who sang him the nursery rhyme "Trip-a-trop-a-tronjes"; his ancestors who served in the Revolutionary War; and the life-affirming beauty of the mystical Catskills.[22] Carmer divined that, gifted as Roosevelt was at politics, if the burden of the Great Depression wasn't on his shoulders the president would have retired to the Hudson River Valley to spend what the Hindu sages call the "forest years," reflecting on New York history.

Roosevelt and Carmer were likewise kindred spirits when it came to promoting Dutch architecture in upstate New York. Both men sought to restore buildings from the eighteenth century and early nineteenth century in the Hudson River Valley style. And by insisting that citizens deserved "a river of clean water,"[23] both men were really progenitors of what became the "Scenic Hudson" movement of the 1960s.

Although Roosevelt had far more English blood in his veins than Dutch, he preferred being seen as a Dutch American. Proud of his family's genealogical ties to the Netherlands, he oversaw the construction of five Dutch-style post offices in the mid-Hudson, taking a direct role in determining the architecture, layout, materials, and artwork. Sketching building designs on White House stationery, he communicated his preferences for the craft- and stonework and the layout of the interior, complete with local history murals painted by Olin Dows and others. Not even the minutest detail of his WPA post offices in Rhinebeck, Wappingers Falls, Beacon, Hyde Park, and Poughkeepsie escaped his attention.[24] Not only did Roosevelt choose an artist for the post office's interior murals, but he also shaped the patriotic language chiseled on its walls and engraved on bells. "Instead of poetry on the Poughkeepsie Post Office bell," the President wrote to the project's supervisor."I suggest the following: 'Ring the Perpetuation of American Freedom.' "[25]

By 1938, the Works Progress Administration (reorganized as the Work Projects Administration the following year) had done much to educate Americans about their local history and geographic points of interest. The agency installed new parks, bridges, roads, and schools, all built by unemployed men and women—around three million in 1938 alone—who were paid for this work-relief. Eleanor Roosevelt encouraged her husband to extend the benefits of the WPA to unemployed artists, actors, writers, and musicians in America. Federal Project Number One was the result of Eleanor's lobbying. Its five branches—the Federal Art Project, the Federal Music Project, the Federal Theatre Project (the first lady's favorite), the Federal Writers' Project, and the Historical Records Project—employed more than forty thousand people from 1935 to 1938.[26]

Eleanor Roosevelt toured a Works Progress Administration worksite in Des Moines, Iowa, on June 8, 1936. She was in the city to deliver the commencement address at Drake University, but made time, as usual, to see what local people were doing and how they were faring. Mrs. Roosevelt visited three WPA sites that day, including a sewing room for unemployed women and the Negro Community Center. At the third, shown in the picture, WPA crews were employed turning a former city dump into a park along the Des Moines River.

In 1938, in the aftermath of the "court packing" scandal, the "Roosevelt recession," and the "Clean Potomac" campaign, there was increased blowback against New Deal conservationism from the extraction industries. Quietly circumventing Congress, however, Roosevelt increased the boundaries of several existing national forests, such as Ottawa (Michigan); Green Mountain (Vermont); Tongass (Alaska); Columbia and Snoqualmie (Washington); and Ouachita (Arkansas). Roosevelt stressed that national forests should have names evoking Americana or Native Americans, such as Apalachicola in Florida, Cumberland (later Daniel Boone) in Kentucky, and Chattahoochee in Georgia.[27] While the first lady rallied behind Federal Project 1, the president kept pushing forward on forests, rivers, shorelines, and wilderness.

With the bill for the proposed Olympic National Park in Congress that spring, Roosevelt maneuvered to get Mammoth Cave (Kentucky) and Isle Royale (Michigan) into the National Park Service. Large national parks, he believed, were desperately needed in the eastern states as a cushion against overurbanization. "There is no national park in New England except Mount Desert [Maine], which should be called a 'monument' rather than a park," Roosevelt wrote to Daniel Bell, the acting director of the Bureau of the Budget, "because it is used by very few people each year and benefits principally the rich summer residents."[28] Using surrogates, the president thwarted the passage of two pieces of legislation before Congress that would have violated the ecological sanctity of Yellowstone National Park. "I learned something of the plans of the Snake River Valley water-users which are now reflected in this proposed legislation," Roosevelt wrote to a friend in New York City. "I am opposed to any measure affecting the national park and monument system which would modify this vital and inviolable principle."[29]

That summer of 1938, Ickes spoke to the Northwest Conservation League at an event in Seattle. He said that once the new Olympic National Park was "rounded out by proclamation under the power given to the President to add additional territory," it would "take its place with the greatest parks." Ickes promised to keep much of the Olympics in wilderness condition. "When a national park is established, the insistent demand is to build roads everywhere, to build broad trails, to build air fields, to make it possible for everybody to . . . go everywhere—*without effort*," Ickes lamented. "These last two words are what cause the trouble. It is characteristic of the American people that they want everything to be attainable *without effort*. Too many of us want a predigested breakfast food for our stomachs and a previewed national park for our eyes. Nine people out of ten, visiting our national parks, stay within a half a mile of the motor roads and the hotels. [Many] feel that they are roughing it if they twist their necks in a sightseeing bus, or expose their adenoids to the crispy air while gazing through field glasses at some distant scene."[30]

Never before—or since—did a secretary of the interior speak out so forcefully for primeval zones in national parks. As if Bob Marshall were whispering in his ear, Ickes's voice was almost completely in unison with the Wilderness Society. Nevertheless, possibly with FDR in mind, Ickes talked about the need for "physically handicapped" people to have access to visitor centers and scenic lookouts in parks. They, too, had a birthright to escape noisy civilization to get lost or inhale solitude in America's remaining wildlands. "Limit the roads," Ickes said. "Make the trails safe but not too easy, and you will preserve the beauty of the parks for untold generations. Yield to the thoughtless demand for easy travel and in time the few wilderness areas that are left to us will be nothing but the back yards of filling stations."[31]

Buoyed by the creation of Olympic National Park, Ickes pushed for a new California "wilderness" (roadless) park south of Yosemite and high in the Sierra Nevada. The battle to establish Kings Canyon National Park was nearly sixty years old when Ickes took up the crusade.[32] The Forest Service oversaw the controversial Kings Canyon lands, insisting it was a more reliable caretaker than the National Park Service. All Roosevelt cared about was enlarging General Grant National Park to prevent the logging of sequoia trees in the Kings Canyon area. Long fascinated with *Sequoiadendron giganteum*—which could survive wind and fire, had heartwood and bark infused with tannic acids to protect against fungal rot, and warded off wood-boring beetles—Roosevelt deemed it a moral crime to cut ancient sequoias down for the sake of grape stakes, shingles, and fence posts. From Irving Brant he learned that the U-shaped canyon carved by the two principal forks of Kings River was clearly in need of federal protection. The wild, free-flowing Kings River rushed and sparkled as it coursed between the high walls of this magnificent canyon. In Kings Canyon the president had a perfect showcase for the New Deal's forest preservation and riverkeeping, and for the noble growth of NPS.

In the 1880s, miners, loggers, and ranchers had prevented a national park at Kings Canyon. During Calvin Coolidge's presidency there was serious interest in a national park, but the movement stalled when advocates of "pure wilderness" clashed with irrigation and hydropower interests, which wanted to build a dam like Hetch Hetchy. Nine years later, in 1935, Ickes had involved himself in the fate of Kings Canyon. He decided to be a maverick and blindsided both supporters and adversaries of the park with regard to his plans. Consequently, the bill he drafted was met with strong resistance in California, particularly from residents of the San Joaquin Valley. Even the Sierra Club, the Wilderness Society, and the National Parks Association couldn't muster much enthusiasm for Ickes's proposal; these nonprofits preferred that Kings Canyon be a wilderness area within the Forest Service instead of in Ickes's NPS portfolio.

In need of a high-profile ally, Ickes traveled to San Francisco to secure the Sierra Club's endorsement. This was quite an amazing moment in U.S. environmental history; a secretary of the interior was pleading with the esteemed Sierra Club to back a national park in the California high country—not the other way around. At a Bohemian Club luncheon in San Francisco on October 21, Ickes, in an hour-long speech, explained to Sierra Club leaders that the Roosevelt administration wanted to designate Kings Canyon National Park a roadless primitive area. This was important to Ickes because neither the Wilderness Society nor the National Parks Association wanted Kings Canyon National Park.[33] The Sierra Club, impressed by Ickes's sales pitch for wilderness, agreed to change course, abandoning the Forest Service and supporting the national park if—and only if—

it would be "devoted to roadless wilderness." The club's president, Joel H. Hildebrand, also procured a guarantee from Ickes barring all "roads or hotels like those in Yosemite."[34] Kings Canyon had to be free of commercialization. Hildebrand would soon write to Ickes that the Bohemian Club appeal was "a grand triumph."[35]

In 1936, Congress held acrimonious hearings over Kings Canyon. The Sierra Club, sensibly, had asked photographer Ansel Adams to serve as its key lobbyist promoting national park status. Wearing a Stetson hat, armed with his gorgeous photos of Kings Canyon's mountains, lakes, and waterfalls, Adams persuaded many in Congress to join the preservation cause.[36] Two years later, Ansel Adams published his second book of photography, *Sierra Nevada: The John Muir Trail*, limited to only five hundred copies. His ethereal images of Kings River Canyon area were masterpieces of landscape photography; clearly Adams's deeply held Muirian conservationism was the wellspring of his timeless art.

At a time when the Forest Service and National Park Service were fiercely squabbling over the Kings Canyon region, the power and irrigation interests in the San Joaquin Valley united in an attempt to end any chance of national park designation. In this situation, Adams—having no real money for lobbying—sent a copy of his book as a gift to Harold Ickes, whom he had once met at a conference on the future of state and national parks. Ickes, impressed with its artistry, showed the book to FDR on January 8, 1939, at a private lunch.[37] Mesmerized, Roosevelt seized the handsome book and kept it for himself. "Dad had to then send Ickes another copy," Michael Adams, the son of the photographer, recalled. "Getting the Kings Canyon fight front and center of Roosevelt was an important step."[38]

Once Adams's *Sierra Nevada* had ensnared FDR's imagination, the Sierra Club redoubled its efforts for Kings Canyon. "While Ansel would have been quick to protest that he was only one of many advocates on Kings Canyon's behalf," biographer Mary Street Alinder wrote, "it would not likely have become a national park when it did were it not for his visual testament and his tenacity."[39]

With the John Muir Trail—a hiking path along the crest of the

Sierra Nevada through General Grant National Park—completed, the Sierra Club led burro and knapsack outings in the high country to generate publicity. Once again Irving Brant acted as the unofficial troubleshooter for Roosevelt and Ickes. In this campaign, Brant exploited the divisions within the Forest Service to help move Kings Canyon National Park forward. Ferdinand Silcox, director of the Forest Service, despised Brant as an environmental radical and was reluctant to cede control of Kings Canyon's ancient sequoia groves to Interior. Dan Bell, acting director of the Bureau of the Budget, concurred with Brant that 415,000 acres of the "most scenic and inspirational portion" of the Kings River Canyon ecosystem deserved to be a national park.

Entering the fight was Bob Marshall, the rangy and athletic Forest Service preservation purist. Breaking with his own agency over Kings Canyon, Marshall, collaborating with Ickes, thought the Forest Service lacked a solid preservation plan for Kings Canyon. A step ahead of Pinchot and Silcox in terms of vision, Marshall emerged in the mid-1930s as the boldest forester America ever produced. He was an old-style explorer like Jim Bridger—with rugby shoulders, muscular legs, electrifying brown eyes, and sometimes an unkempt beard—and his belief that Interior, under Ickes, would best preserve the Kings Canyon ecosystem as a primeval wilderness carried some weight. "If I was outside the government, and was shown this plan of the Forest Service, I would swallow all my prejudices against the Park Service and root for Kings River National Park merely to keep out these commercial desecrations and the roads which will go with them from as glorious a wilderness as remains in the United States," Marshall wrote in an official comment. "If the Forest Service was to keep the Kings River country as a National Forest, I think it should burn the Kings River report and write a radically new one."[40]

III

Having worked tirelessly that June of 1939 to establish the Federal Works Agency (which merged the PWA, WPA, and other New Deal agencies) and promote New Deal environmentalism, the president was ready to go to Hyde Park for a reprieve from the Washington shuffle. On July 1, Roosevelt announced plans for a Presidential Library at Hyde Park, which would serve as the repository for all his papers and memorabilia. This caused many in the White House press corps to mistakenly speculate that Roosevelt wouldn't be running for reelection in 1940.[41] With almost childlike excitement, he directed his friends in the press corps to "Dutchess Hill," five miles from Springwood, so they could see Top Cottage, claiming he was its chief architect. (In reality, Henry Toombs of Georgia, who had also built Val-Kill, acted as Roosevelt's "ghostdrafter.")

With construction on Top Cottage well under way, Roosevelt was planning to embark on an excursion to the Galápagos Islands to study marine biology, with the Smithsonian Institution as sponsor. The voyage would cover almost five thousand miles in about twenty-four days. Between July 16, when he'd leave for the island chain from San Diego, and August 9, when he'd arrive back in Florida, FDR would be at sea for most of a month.[42] (He would also check the suitability of Cocos Island, off Costa Rica, as a site for a new U.S. naval base.) Intensely proud of his seafaring ancestor, Captain Amasa Delano of Massachusetts—who had written a captivating account of the Galápagos Islands before Charles Darwin published *On the Origin of Species*—Roosevelt planned to approximate Delano's voyage.

As Roosevelt prepared for his seafaring trip, Ickes raised the idea of a park jointly administered by the United States and Ecuador with his boss. Secretary of State Cordell Hull opposed the idea, but Gifford Pinchot, who had headed a 1929 expedition to the Galápagos for the Smithsonian, discovering a giant "sea bat" of the genus *Manta*, continued to promote the bilateral policy. He believed that the Galápagos should be permanently protected as a World Heritage site (that designation, however, did not come until 1978). A collaborative wildlife protection project between the United States and Ecuador would be conservationist diplomacy at its finest. Kermit Roosevelt, TR's second son, had also led a scientific delegation to the Galápagos (for the American Museum of Natural History in 1930) and wanted the chain biologically preserved. It was rare to find Ickes, Pinchot,

Dr. Waldo Schmitt, a zoologist attached to the Smithsonian, holds an iguana while visiting the Galápagos Islands. As a curator of the Department of Marine Invertebrates, he made frequent trips to the waters of South America, and was chosen to accompany Roosevelt on a 1938 fishing trip to the eastern Pacific islands of Clipperton, Cocos, and Galápagos.

and Kermit Roosevelt all in step on a conservation issue. Hull nevertheless continued to be a spoiler: "This whole matter [is] so delicate," he confided to Ickes, "that I should hesitate very much to take it up even informally with the Ecuadorian authorities."[43]

Excited about the expedition, Roosevelt tapped Professor Waldo L. Schmitt of the Smithsonian to serve as the primary marine biologist. The president unexpectedly called Schmitt at the Smithsonian one afternoon, telling him, as if it were an order, "I know we'll have a good time."[44] Schmitt was obsessed with marine life. Born in 1887 in Washington, D.C., he received his PhD from George Washington University in 1922, then worked for the federal government as a fisheries expert. His studies on the spiny lobster and king crab were groundbreaking. While working at the Carnegie Institution's marine laboratory in the Florida Keys, Schmitt received acclaim for identifying an unknown subspecies of crustacean in the stomach of a fish. He had already visited the Galápagos three times; on one of these trips he drew widespread media attention after stumbling on a group of utopian colonists living on Floreana Island.[45]

The Roosevelt-Schmitt expedition departed from San Diego on July 16 on USS *Houston*, the heavy cruiser that had taken the president to Hawaii in 1934. Captain G. N. Barker commanded a crew of six hundred. Schmitt had brought along seines, bottom samplers, traps, and hand nets to obtain all sorts of exotic marine animals. FDR reassured the press that White House business would be conducted from the *Houston* even on the open sea. Roosevelt's seafaring party included press secretary Stephen Early, marine photographer Frederick Allen, and "Pa" Watson—all of whom happen to have been good card players. A few of his shipmates recalled the president saying that he was finally living his childhood fantasy of visiting the Galápagos. The chief objective of the trip was marine biology. "Throughout the cruise," Schmitt recalled, "the President took an active part and a live interest in all our collecting."[46]

The *Houston* dropped anchor at Clipperton Island, nearly a thousand miles south of the tip of the Baja peninsula. Much to Schmitt's surprise, the President demanded to be briefed about the island's unusual tree species. At Clipperton the Smithsonian procured numerous flowers, ripe seeds, and seedling plants of great rarity for to use in propagation experiments back in Washington.[47] A rare palm tree—"not only a new species, but also a new genus"—was also found.[48] Throughout the expedition, the Roosevelt party had ideal weather for collecting.[49]

On such oceangoing voyages Roosevelt, a celestial navigator, was aglow with what psychologist Wallace Nichols has termed "blue mind": the power of seas, oceans, lakes, and rivers to rid the mind and spirit of anxiety. Dismissive of landlubbers, often studying maps and charts, blessed with an intuitive feel for favorable currents and perfect fishing grounds, FDR was in top form out in the blue-green waters of the Pacific. Relishing the fact that nobody in the press knew his exact location, he deemed his time on the ocean helpful in making momentous decisions about the fate of America.[50] There's a marvelous photograph showing Roosevelt in his element, sitting comfortably on a deck stool with a tiger shark, hooked in Sullivan Bay, hanging from a rope in front of him. Flashing a cocky smile, floppy white hat in place, his cabana shirt half open, Roosevelt looks in ruddy good health.[51]

Once the *Houston* anchored in the Galápagos FDR, in private conversation, cracked jokes about Theodore Roosevelt hunting for grizzlies in the Bighorns of Wyoming while he himself was seeking out tiny crustaceans. When he wasn't dealing with government matters or helping Schmitt collect specimens, Roosevelt delved into William K. Vanderbilt's travel book *To Galápagos on the Ara, 1926*. Vanderbilt and his party had discovered two new species of shark on their voyage.[52] Early in the trip, Schmitt and Roosevelt lightheartedly discussed what they were hoping to accomplish for the Smithsonian.

"Is there any particular thing or animal that you would like to find?" Roosevelt asked.

"Oh yes," Schmitt replied, "I have been on it two years, and the one thing I am searching for in these waters of Mexico and the islands of the Pacific [is] a burrowing shrimp."

"Well Dr. Schmitt, why leave Washington?" Roosevelt chuckled, "Washington is overrun with them. I know that after five years!"[53]

In the bright Pacific wind, Roosevelt and Schmitt's team made fourteen stops to collect specimens in and around the Galápagos, in hopes of finding unusual crustaceans to bring back to Washington. To FDR's delight, they did locate a rare burrowing shrimp along Isla Socorro, a hundred miles or so off the coast of Mexico. The president immediately renamed *Neotrypaea californiensis* the "Schmitty Shrimp" and joked that he would issue an executive order changing the name of the Smithsonian Institution to the "Schmittsonian."[54]

By the time the *Houston* dropped anchor along San Salvador Island, Pa Watson was worried that Roosevelt was getting sunburned; he saw to it that canvas was spread above the *Houston*'s deck to protect the president's skin. Key areas of the Galápagos—Santa Cruz, San Cristóbal, Isabela, and Floreana—were explored by Roosevelt's group. Schmitt filled his packing boxes not only with exotic fish, but also with specimens of plants, birds, snakes, and even a few fossils. The president, who normally bragged about big catches, was now keenly invested in small ones. To be searching for eight-inch white salema instead of eight-foot tarpon was, to put it mildly, a change of pace for Roosevelt. Schmitt named one of the two new species of tiny gobies caught after the president: *Pycnomma roosevelti.*[55]

Roosevelt joined the crew most evenings to feast on fresh seafood. He also performed the time-honored ceremony when the *Houston* crossed the equator. With theatrical aplomb, the president inducted new staffers and crewmen who had never before crossed the equator ("pollywogs") into the Ancient and Honorable Order of Shell Backs. He found time to write a series of letters to Eleanor detailing his marine antics and bird-watching. In return, she reported to him her own tale of witnessing "a bird sweep down this morning and apparently grab a fish under the surface of the water . . . but I am afraid it was only a common swallow catching a gnat."[56]

The temperature stayed steady, between sixty-five and seventy-five degrees, but when the skies were overcast, everyone wrapped up like a mummy to ward off the cutting winds.[57] FDR was normally a dry correspondent, but his letters to Eleanor from the Galápagos were full of joy. "I have had good weather, excellent fishing on the Mexican Coast & Socorro Island. And at Clipperton Island we were fortunate enough to have a smooth sea and Dr. Schmitt the Smithsonian scientist whom I brought along was able to land with a picked crew and collect many fine specimens of marine and plant life and birds—and to shoot a wild pig which we duly ate!" He even found time to mention the topography of the Galápagos Islands. "The two we saw today are very barren but full of color, for they are all volcanic and there are lava flows and all kinds of weird twisted rocks."[58]

While Schmitt hunted for smaller species of fish, the president cast off the ship for whoppers, reeling in thirty- to forty-five-pound tuna that were cooked for supper. One afternoon, he hauled in five sharks, some of them quite large. Other species he caught included California yellowfin, black sea bass, broomtail grouper, bluestripe snapper, Pacific sierra, and barred sand bass. "It is always cool though directly on the Equator, for the Humboldt Current from the Antarctic passes through the Islands," FDR wrote to his wife. "The water is cold too, so the fish are excellent eating!"[59]

Roosevelt aboard the USS *Houston* with a sixty-pound shark that he had caught in Sullivan Bay, Galápagos, in July 1938. Roosevelt's twenty-four-day ocean voyage, from San Diego to the Galápagos, and then through the Panama Canal to Florida, gave him a chance to relax with the sailors of the *Houston*, some of whom may be seen in the background.

Days passed with welcome serenity. The hot topic of conversation aboard the *Houston* was Charles Darwin's voyage on the *Beagle* in 1835. Unlike Darwin, however, the president was unable to ride one of the two-hundred-pound Galápagos tortoises, because of his paralysis. Taking advantage of the scientists on board, he inquired about all the driftwood, bamboo, and other South American debris on the islands. He was astonished that so much had washed up when they were six hundred miles away from the mainland. The power of the tides never ceased to amaze him. "A very successful week in the Galápagos from every aspect—good fishing, many specimens of all kinds for the Nat. Museum, and my only complaint is that the weather here on the Equator has been too cool," FDR wrote to Eleanor on July 31. "Also one has no feeling of the tropics—no lush vegetation—it might be Nantucket Island—only not so green. Still it is all interesting and colorful—especially remembering that the tortoises, iguanas, etc. are the oldest living form of the animals of 15,000,000 years ago!"[60]

Unself-consciously, Roosevelt would point to seabirds, a pair of binoculars hanging around his neck. He used them to study a huge (and famous) albatross colony from the safety of the *Houston* during a period of rough seas. After a large number of strange crustaceans were netted in Magdalena Bay, no one wanted the trip to end, but Roosevelt needed to return to the White House. The president's final treat was getting to inspect the Panama Canal. In 1912 FDR, then a state senator, had visited the canal before the water had been let into Gatun Lake (the largest man-made reservoir in the world). Back then, he had been able to climb to a high vantage point and observe steam shovels and railcars in the middle of the great "cut." Now, in 1938, the Panama Canal was fully operational, and the enormous effort involved its construction was fading into history.

Touring the Panama Canal in 1938 wasn't merely a lark for the president. His visit had implications for national security. He knew that the U.S. Navy would have to defend the canal at all costs from enemy sabotage if another world war erupted. Nothing would disrupt American commerce more, Roosevelt believed, than letting the Panama Canal be blown up or decommissioned.[61]

During his trip there were no press reports about President Roosevelt's scouting of Cocos Island, off the Pacific Coast of Costa Rica, as a site for potential American naval bases—only puff pieces about the fish FDR had caught. However, there is ample historical documentation to prove that the U.S. Navy was stockpiling war materials in the Canal Zone to eventually use to build bases in the Galápagos and Cocos chains. Roosevelt saw numerous signs that war might be imminent. Italy and Germany had already made aggressive moves in Europe. The Third Reich had annexed Austria and was making claims on territory belonging to Czechoslovakia. The situation was likewise looking grim in Asia: Japan had invaded China the previous year. While FDR was aboard the *Houston*, Chiang Kai-shek, in the face of daunting military pressure from the Japanese, moved his government farther inland. When the *Houston* finally dropped anchor in Pensacola, Florida, ending the expedition, Roosevelt headed to Dutchess County, full of worry for the world.

No sooner did he arrive at Springwood than he had to repack his bags for a trip to Canada. On August 18, 1938, the president traveled to Kingston, Ontario, to receive an honorary degree from Queen's College and dedicate the Thousand Islands Bridge, connecting New York and Canada. This was a dream come true for Roosevelt, who had supported such a bridge since 1911 and long been enamored of the Saint Lawrence River. In honor of FDR's visit, Canadian and American young people planted white birch trees on Hill Island and Wellesley Island, respectively, and a flotilla of yachts planned to pass under the American side of the bridge.[62]

The president delivered his dedication address in front of a six-acre knoll that formed a natural amphitheater. Praising Canada as a great democracy, Roosevelt, to the surprise of many, spoke more about the grandeur of the Saint Lawrence River than the engineering marvel that had been built above it.[63] "The St. Lawrence River is more than a cartographic line between our two countries," he said. "God so formed North America that the waters of an inland empire drain into the Great Lakes Basin. The rain that falls in this vast area finds outlet through this single natural funnel, close to which we now stand."[64]

That fall, the president threw a lighthearted dinner at the White House for his *Houston* crewmates. At the reunion, crewmembers recounted their discoveries on shores, on reefs, and in tide pools. Schmitt also lunched privately with FDR to discuss the future of the world's saltwater fisheries. At an impromptu press event, Roosevelt said that pretending to be a marine biologist was a "source of great satisfaction" for him. Suggesting that there would be future collaborations between the U.S. Navy and the Smithsonian, Roosevelt bragged about his discoveries of the burrowing shrimp, gobies, and rare palms.[65] "We cannot know too much about this natural world of ours," Roosevelt wrote. "We should not be satisfied merely with what we do know."[66]

Schmitt also compiled an inventory of what Roosevelt's expedition had accomplished in marine biology. Having surpassed their wildest expectations, he told the *Washington Post* that the trip "couldn't be beaten" and that "scientist" Roosevelt had helped the museum tremendously.[67] "I got an awful lot of specimens out of that cruise," Schmitt told reporters. "I didn't go fishing with the President but would talk about the fishing spots with him. Then I was free to go on a launch with five men and do a lot of collecting with a small boat dredge." The *Washington Post* praised FDR for being a "benefactor of science."[68] The statistics were impressive: eighty-three different species of fish were hooked, gaffed, iced, photographed, and shipped back to the Smithsonian. "The fish were unpacked there still frozen hard," Schmitt recalled. "When thawed out in tanks of water, they returned to practically the identical fresh condition in which they had been placed in Cold Storage. Many of the fish still retained much of their original coloration, having apparently undergone little or no change from the time they were brought aboard ship."[69]

To Roosevelt's great delight, the Smithsonian planned to mount a golden grouper (*Mycteroperca rosacea*) he had caught in the Pacific. Reports about the echinoderms, sponges, annelids, and plants the team collected during the expedition reached the president's desk. Routinely he showed off photographs to White House visitors of him playing Darwin. "I'm about halfway through with the collection of crustaceans," Schmitt wrote to Roosevelt. "Among the starfishes and their relatives five new species have been discovered. The manuscript describing the new genus and species of palm will soon be finished, together with a number of interesting illustrations. In the collection of shells fifteen new species or varieties have been noted."[70]

FDR hoped to write a magazine article about the Galápagos, as Theodore Roosevelt would have done, but the troubled world required too much of his attention that fall. There was a midterm election on the horizon; there was a crisis in Czechoslovakia; and there were ongoing atrocities against Jews in the Third Reich, including *Kristallnacht*.

The next time FDR had an extended stay at Hyde Park, he regaled his Home Club with wild fishing tales from the far-flung Galápagos. In a rare unguarded moment, however, Roosevelt told his friends and neighbors that the real motivation for visiting the Galápagos was indeed to inspect America's naval defense in the Pacific. "I was very happy to note that the American defenses of the Canal had improved since I was there three years before," Roosevelt said. "We are getting airplanes, and submarines, and anti-aircraft guns, and various other things, to try to make reasonably certain that in case of war—which we are trying to avoid in every possible way—we shall still be able to maintain the link of the Panama Canal between the Atlantic and Pacific."[71]

IV

A lot of wildlife habitat Roosevelt and his administration had saved between 1933 and 1938 was still involved in controversy. After the Okefenokee swamp received federal protection, the Biological Survey started haphazardly planting Asian chestnut trees there. A little detective work by Francis and Jean Harper led to a troubling revelation: the Biological Survey was disregarding its pledge to preserve the "pristine quality" of the swamp for pond lilies, bladderworts, and hardhead grass. Jean Harper, testing friendship, bitterly complained to Eleanor Roosevelt about this gross betrayal of the USDA's commitment to preservation. "We have just received word of another and far greater project of insane vandalism—to cut down all the big pine trees on Chesser Island," she wrote to the first lady. "This is one of the very few islands in the swamp where the timber has remained largely untouched, and we and other naturalists have made very special efforts to keep it so."[72]

Jean Harper was what the 1960s generation would have called an "eco-warrior." Enclosed in her letter to Eleanor were damning reports of the Biological Survey's malfeasance in south Georgia. She complained that a federal road was being planned for Chesser Island—a road which would ruin the wilderness value of the Okefenokee. The worst accusation in her commentary was that with regard to deforestation the president's own CCC boys were the chief culprits. "The Biological Survey, as presently organized, is a thoroughly unfit guardian of our wilderness areas and wildlife," Harper wrote. "And no CCC camp should be tolerated in a place like the Okefenokee. May I implore you to do whatever you can to preserve natural conditions in the Okefenokee from this new and senseless attack?"[73]

Harper's allegations about Chesser Island, while true, were somewhat misleading: the CCC was doing valuable work elsewhere along the Atlantic seaboard during FDR's second term. CCC crews had eradicated undesirable plants like cattails and diligently spread seed to attract migratory waterfowl at refuges such as Blackwater (Maryland), Swanquarter (North Carolina), and Saint Marks (Florida).[74] Small islands resembling muskrat lodges were built in shallow ponds to create nesting and preening areas for waterfowl.[75] In Georgia the CCC cut firebreaks throughout the Okefenokee wilderness, preventing a repeat of the disastrous 1932 conflagration.[76]

But the CCC did make one undeniable mistake in Georgia, apparent to anyone who drove around the back roads. The CCC boys were responsible for planting kudzu (*Pueraria lobata*), a Japanese climbing vine first introduced to America at the 1876 Centennial Exposition in Philadelphia. Among those boys was Lee Brown of Bainbridge, Georgia, who joined the CCC as a sixteen-year-old to help his grandmother survive the Depression. At an all-black camp in Sumter County, he worked as a surveyor's helper, rock man, and meteorological data keeper (measuring rainfall). "Our goal," Brown recalled, "was to plant as much kudzu as possible. We were like the Johnny Appleseeds of kudzu."[77] But in planting kudzu the CCC inadvertently wreaked havoc on the Georgia landscape. Although it was good feed for livestock, kudzu caused what botanists call "interference competition": it suffocated native plants by blocking their access to sunshine. Kudzu was therefore a plant-killing menace—and the CCC was largely responsible for its proliferation.

Whenever Roosevelt visited Warm Springs in the late 1930s he made an effort to hang out with CCC Company 4462 in Chipley and Company 1429 in Pine Mountain. One afternoon the president gave an impromptu exclusive interview to Donald Burns, the editor of the newspaper at the latter camp, *Pine Mountain Progress*. "I used to be an editor myself—of my college paper," Roosevelt told Burns. When asked about the future of the CCC as a nonemergency relief agency, Roosevelt answered, "As long as there are men and boys whose people need relief, they will be given first consideration." As the president and Burns good-naturedly discussed the merits of Margaret Mitchell's *Gone with the Wind*, press secretary Marvin McIntyre grew nervous. Roosevelt let his guard drop when he got comfortable with a CCC boy. "The President shouldn't have done this," McIntyre told Burns. "He doesn't normally give exclusive interviews. . . . Every camp and college paper in the country would be waiting for one."[78]

In the decades before federal acquisition, the Okefenokee had been depleted of its abundant fish, but by late 1938 the jackfish and gars were back. One Georgia farmer who took advantage of the Okefenokee's bounteous wildlife was a future president of the United States: Jimmy Carter. His father, James Earl Carter—a conservative Democrat, land conservationist, and peanut farmer—resented bossy New Dealers in Washington, D.C., who told him what to do on his Sumter County acreage. But he loved the fact that the Roosevelt administration had protected the Okefenokee and he took his ten-year-old son there for the fishing expedition of a lifetime. Using long cane poles, the Carters caught large bream, redbreast, warmouth, and bluegills (which Carter called "copperheads" because of the hue generated by tannins in the water). The future president not only saw alligators, but also caught a baby gator and kept it as a pet.[79] "The Okefenokee stands out as perhaps the best time I had with my father," Carter later recalled. "We bonded in the swamp like never before."[80]

Roosevelt had adopted Georgia as his second home. His sense of the state's geography was astounding. Only nine days after Jean Harper's appeal about Chesser Island was mailed to Eleanor Roosevelt, the president threatened to put Henry Wallace's head on the chopping block. He was angered that the USDA had reneged on its agreement to be a wilderness custodian of the swamp. Roosevelt had wanted the

Okefenokee to remain pristine, i.e., roadless. Travel would be done via existing canals and runs, and push boats were Roosevelt's vessels of choice. And logging was to be prohibited on the island; in fact, the president had the CCC replant stands of pine and add gum trees. "I am told that the Biological Survey is 'at it again' in the Okefenokee Swamp—and that this time they are about to cut down the big pine trees on Chesser Island," Roosevelt wrote to Wallace. "I hope the report is not true for, as you know, you and I have agreed that the Swamp is to be kept definitely in its untouched pristine condition. Will you let me have a report? I do not want any trees cut down *anywhere* in the Swamp."[81]

It was a disquieting experience for Henry Wallace to be reprimanded. But he soon found out that the president's sources were correct. The USDA, it seemed, was leasing woodlands on Chesser Island and getting rid of rotting trees throughout Okefenokee Swamp. Wallace lamely explained to Roosevelt that the Chesser Island pines had been cut to build USDA administrative headquarters around Camp Cornelia. While Wallace acknowledged that "interested conservationists" were indeed up in arms, he reassured the president that the USDA had the situation under control. To further ease Roosevelt's mind, Wallace dispatched Dr. Ira Gabrielson of the Biological Survey to conduct a personal inspection of the Okefenokee and then draft a plan for managing the 438,000-acre swamp. Roosevelt's reply was that he wanted to be kept informed of all "proposed plans" related to the refuge going forward.[82]

Just two days later, Roosevelt again chided Wallace, insinuating that USDA game wardens had turned a blind eye when the War Department disregarded U.S. waterfowl protection laws. Roosevelt railed against the army and navy brass for ignoring his conservationist rules on military bases and gun ranges. Soldiers didn't get an exemption. The shooting of migratory birds on federal bases was a punishable offense, and Roosevelt wanted Wallace to investigate abuses. If the sanctuaries for migratory waterfowl had been violated by the War Department—as Roosevelt's conservationist spies intimated—then there would be consequences. "The fact that the canvasback duck and indeed other ducks have so diminished in numbers that we have had to prohibit all hunting on them," Roosevelt wrote to Wallace, "and the fact that the Government is engaged in bringing back the numbers of migratory birds to a point well above the danger of extinction lead me to believe that . . . all government owned or controlled areas should be closed to hunting where such closing would benefit the general policy."[83]

At one point in 1938 the president demanded information from Wallace about the situation of canvasback ducks in Maryland's Susquehanna flats. An exasperated Wallace assured FDR that the flats were being adequately protected by the Biological Survey, but Roosevelt wanted more details. He had heard that the crucial Garrett Island stopover was vulnerable to industrial development. (In 1942, FDR established the Susquehanna Flats National Wildlife Refuge by Executive Order 9185.)[84] Wallace was alternately amused, perplexed, and annoyed by—and curious about—the president's insatiable interest in lesser-known reaches of the country.

With midterm elections looming in November 1938, the president knew he would have to give stump speeches for Democratic candidates. He managed to knit in a conservation message. Earlier in the year he had visited the Texas Panhandle to sell soil conservation to farmers. Now, campaigning around the East, he spoke against the evils of water pollution. At a press conference at Hyde Park on October 7, FDR recounted how PWA loans had built some five hundred sewage disposal plants, at a total construction cost of more than $1.25 billion. But not in his backyard. He accused residents of New York of dumping raw sewage into the Hudson River and polluting streams to the point where the water was unsafe to drink. When asked point-blank whether he would drink water in his beloved Dutchess County, the president snapped, "I would not drink water in Poughkeepsie."[85]

There were a number of reasons why correspondent Arthur Krock held that the New Deal had ground to a "halt" as the midterm elections approached in 1938.[86] The "court packing" scheme of 1937 had led many to question whether Roosevelt should be reined in. The 1937–1938 economic downturn, which pushed unemployment to perilously high levels, caused voters to reconsider the amount of power they had entrusted to the federal government. A series of strikes by the Congress of Industrial Organizations (CIO) were blamed on the New Deal's empowerment of unions through the National Labor Relations Act of 1935. While Roosevelt was capturing crustaceans that summer with Schmitt, a Gallup poll showed that 66 percent of Americans wanted him to be more conservative.

On Election Day, November 8, the Democrats lost six Senate seats, seventy-one House seats, and a dozen governorships. Political pundits declared that the New Deal had sputtered to a halt. Nevertheless, nonprofits such as the Sierra Club, Garden Clubs of America, and Audubon Society trusted that 1939 would be another banner year for conservation, with the potential for a lot of new NWRs, and national parks.[87]

Another crisis, on a smaller scale, hit FDR right where he lived, as a proud arborist. In fact, nothing pleased Roosevelt more than to hear about the life stories of trees. At the White House he adopted a magnolia planted by Andrew Jackson as his good-luck totem. Given his devotion to trees, it pained the president to be criticized by a group of activists, led by Cissy Patterson—owner of the *Washington Times-Herald*—for the plan to remove eighty-eight cherry trees from the Tidal Basin in order to complete the Jefferson Memorial. Also coming under fire was his uncle Frederic Delano, for having greenlighted the tree removal; that was ironic, because Delano's life had been devoted to planting trees in urban settings. At a press conference that November 8, Roosevelt fielded questions from reporters about the impending removal by the CCC of the cherry trees: "Well, I don't suppose there is anybody in the world who loves trees quite as much as I do, but I recognize that a cherry does not live forever. It is what is called a short-lived tree; and there are forty or fifty cherry trees that die, or fall down, or get flooded out, or have to be replaced [each year]. It is a short-lived tree and we ought to have, in addition to the 1,700 trees we have today, I think another thousand trees. Let us plant 2,700 trees instead of 1,700. . . . That net loss will be made up, not only those eighty-eight, as I hope, but 912 others."[88]

It was a brilliant answer and a solution to the controversy at the Tidal Basin, but it did not assuage Cissy Patterson, who called the

Jefferson Memorial a "meaningless, useless, hideous scramble of cold marble and bronze."[89] With a groundswell of support, she threatened to chain herself to one of the cherry trees marked for removal. Patterson had sixty thousand signatures on a petition to leave the trees alone. A reporter asked FDR what he would do if she executed her plan. "If anybody wants to chain herself to the tree," Roosevelt said, "and the tree is in the way, we will move the tree and the lady and the chains, and transplant them to some other place."[90] The reporters roared with laughter. But Eleanor Roosevelt later confirmed that her husband was quite distraught about the entire incident: part of him sided with Cissy Patterson.

As a young man, FDR would probably have chained himself to Newburgh's Balmville tree, the old cottonwood favored by George Washington, if someone had dared try to remove it. Huddling with Delano and Ickes about the impending act of disobedience, FDR decided to offer free coffee and doughnuts to Patterson and her fellow protesters. And a trap was set. On November 18, the day the trees were to be removed, when the tree rebellion women needed to use a restroom, FDR guaranteed they would be shuttled to use one at a nearby hotel. As a result, no unfortunate photographs were snapped (or arrests made).[91] Roosevelt had outfoxed Patterson. "It is beginning to look very beautiful," Eleanor Roosevelt would later write about the dedication of the memorial; "and some day when the cherry trees around it bloom in great profusion, people will forget that we were afraid of spoiling the landscape around the Basin."[92]

V

On November 25, 1938, Roosevelt took the existing seven-square-mile Arches National Monument in Utah and enlarged it to approximately forty-five square miles via Presidential Proclamation.[93] The sheer size of the Arches enlargement caused the administration to authorize the establishment of a CCC camp near Moab. The two hundred men assigned by the National Park Service to work at Arches lived near the swift-flooding Colorado River.[94] Enduring sunstroke and heat exhaustion, they built monument headquarters, maintenance sheds, a handsome road with switchbacks, and the Moab

Canyon Wash Culvert. Unlike Zion or Bryce Canyon, Arches wasn't concerned about attracting the tourism industry. As a result, Arches National Monument became an exemplary model of the CCC enhancing the wilderness experience without exploiting nature.[95]

Around Christmas 1938, Roosevelt, with the help of Ansel Adams and David Brower of the Sierra Club, accelerated the Kings Canyon National Park legislation in Congress.[96] Undampened and determined in spite of entrenched opposition by the Forest service, the effort by FDR and the Sierra Club gained momentum on Capitol Hill. Henry Wallace remained the most significant holdout in the administration. However, Ickes did some sleuthing and managed to discover that Wallace's father—Henry Cantwell Wallace—had wanted to enlarge Sequoia National Park, just south of Kings Canyon, to include the disputed lands along Kings River while he was serving as Warren Harding's secretary of agriculture.

Ickes gave this information to FDR, who then used it to encourage Wallace to support Kings Canyon National Park. "Reverting to the subject of Kings Canyon," Roosevelt wrote to Wallace early in 1939, "I think you will find that three former Chief Foresters agreed that this area ought to have a national park status. I think you will find that a grand Secretary of Agriculture, by the name of Henry C. Wallace, wrote on January 17, 1924, in reference to H.R. 4095, to add certain lands to Sequoia National Park the following: 'The proposed enlargement of the park and the specific boundaries relating thereto are endorsed by this department.' I think you will find that the area then proposed for addition to the Sequoia Park was even bigger than the area now proposed for the Kings Canyon Park."[97]

Using Wallace's father in this way was perhaps unfair. But FDR's insistence on finalizing Kings Canyon National Park wasn't about favoring NPS, or about his sometimes difficult relationship with his secretary of agriculture. He was dead set on saving the largest remaining natural grove of sequoias in the world. Wallace knew he had been outmaneuvered and, with no visible ill-feeling, joined the Kings Canyon crusade. The crusade was helped by Ansel Adams's books; besides that, the Sierra Club's David Brower, while on high-profile hikes in 1939, had made a long silent film with a sixteen-millimeter Bell and Howell camera. The film, *Sky-Land Trials of the Kings*, lovingly highlighted the gorgeous natural features of Kings Canyon and is widely considered, as Brower's biographer Tom Turner put it, "the first conservation propaganda film ever made."[98]

On January 30, 1939, special celebrations were staged in CCC camps throughout America to honor FDR on his fifty-seventh birthday. Beyond politics, the vast majority of Americans still loved the president, with his broad optimism. They honored him by making his birthday a holiday, dedicated to winter fun and raising money for the National Foundation for Infantile Paralysis (a charity FDR had started to combat polio). In the course of raising millions of dollars each year, there were balls and parties, massive multilayered birthday cakes, and fishing rodeos to pay homage to the president—whom Company 1950 in Upland, California, called the "Champ."[99] To Roosevelt's delight, a number of former CCCers soon became famous in the sports world, including the light heavyweight boxing champion, Archie Moore, and the baseball legend Red Schoendienst of the St. Louis Cardinals. Unfortunately, Congress didn't give FDR the gift he wanted: making the CCC permanent.

Sensing that Congress might try to gut the CCC, Roosevelt took every opportunity to praise the work the "boys" did in helping communities recover from hurricanes and flooding. Nevertheless, in early 1939 Congress stripped the CCC of its independent status and transferred operational control to the Federal Security Agency (as it did with the National Youth Administration, the U.S. Employment Service, the Office of Education, and the Works Progress Administration).

Throughout 1939, President Roosevelt, backed by the Sierra Club, built an alliance of California politicians to help with the effort for Kings Canyon. Key politicians who took the time to visit the area were taken down old logging roads and bridle trails to see the gorgeous vistas. Congressman Bertrand Gearhart, whose district included the proposed park, introduced a bill formally excluding the two reservoir sites that the Sierra Club had wanted inside Kings

Canyon: Tehipite Valley on the middle fork of Kings River and Cedar Grove on the south fork. When the mayor of San Francisco, Angelo J. Rossi turned pro-park, Roosevelt thanked him in writing: "I appreciate your telegram of March 13 expressing your support of the proposed John Muir-Kings Canyon National Park. It is my belief that this area merits national park status and I am happy to know that the project has your endorsement."[100]

The opposition—what Ickes had called "local selfish interests"—kept trying to delay a vote in Congress. The Federal Power Commission had refused to give up control of Kings River since 1920. But Roosevelt, with the 1938 midterms behind him, appealed directly to Clyde L. Seavey, the commission's acting chairman. "The creation of the John Muir-Kings Canyon National Park would protect a unique area in the High Sierra which is magnificently scenic," Roosevelt argued. "It has been shown that this area has no large economic value and that the potential water powers are of doubtful feasibility. Undoubtedly the many beautiful little lakes do afford opportunities for power production, but the inaccessibility of the lakes and the cost of constructing tunnels and power plants in so wild a region would make this area less attractive for power development than many others in California. For these and other reasons, the creation of the Kings Canyon Park is favored by those best informed about this area."[101]

Having secured USDA's compliance, Ickes now found himself fighting against the Federal Power Commission—in spite of FDR's pleas—for much of 1939.[102] The commission agreed to support the bill for Kings Canyon National Park only if it included a rider that "retained power rights for the federal government and state of California."[103] Ickes and Roosevelt were engaged in "one of the fiercest congressional battles on record."[104] "John Muir" was dropped from the name of the park, but the Roosevelt administration remained adamant that Kings Canyon National Park be kept pristine and free of rank commercialism.

The roadless-wilderness provision meant that FDR, wheelchair-bound, would never be able to truly enjoy Kings Canyon National Park. The Sierra Club's William Colby, one of the great advocates for wilderness in American history, had been fighting to preserve Kings

Canyon country since John Muir had died in 1914. His admiration for the work Roosevelt, Ickes, Adams, Brant, and Marshall had done on behalf of the Sierra knew no bounds. Now, with the White House on board, he went into overdrive, working with Congress to create the park. "I feel now that I can die in peace," Colby wrote to Ickes after Congress finally passed the bill, on February 20, 1940. "At least you have added the greatest possible joy to my remaining years."[105]

On March 4, 1940, after years of squabbling, Kings Canyon National Park was finally established; it constituted some 454,000 acres. Two gorgeous canyons—the Tehipite and Kings—were initially left out of the park's boundaries. (But William O. Douglas, with help from the Sierra Club, would persuade President Lyndon Johnson to add these exquisite areas to the park in 1965.)[106] Having helped establish both Olympic and Kings Canyon national parks, Brant, an unsung environmental hero of the late 1930s and early 1940s, resigned as White House speechwriter to begin a six-volume biography of James Madison.[107]

Ansel Adams, in a self-portrait in an antique, convex mirror in 1936. Raised in the San Francisco area, Adams developed an extraordinary proficiency with the technical aspects of black-and-white photography, which he applied to portraits of the natural world, especially in the West.

A pleasant windfall resulted from the creation of Kings Canyon National Park: the Department of the Interior—using "mural project" funds—hired Ansel Adams to photograph America's national parks and monuments. Ickes believed Adams's sublime images would be valuable weapons in the battle to protect pockets of the pristine American West. The elegiac National Park photos that Adams took for Ickes—the Grand Canyon as if a continuous landmass; the jagged Kearsarge Pinnacles of Kings Canyon; the Snake River winding its way through the Grand Tetons; thunderstorms raging in the Rockies; the Big Room of Carlsbad Caverns; and a sweeping panoramic of Glacier from Going-to-the-Sun Road—were astonishing. Ickes was so taken by Adams's close-up shots of leaves found in Glacier National Park that he ordered a mural-sized folding triptych made and placed it just to the left of his office desk at the Department of the Interior.

Adams's only regret about his role in Kings Canyon National Park and his productive stint as a photographer for Interior was that he never got to discuss the natural world with the president. "I never met Franklin D. Roosevelt," Adams wrote in his autobiography. "I photographed him at a distance in Yosemite and was nearly nailed by a Secret Service agent who mistook my Contax telephoto lens for something else. Once in New York City I was given a front row seat when he spoke at the Museum of Natural History. I witnessed how difficult and painful it was for him to move across the stage on the arm of his son and how vigorous he appeared to be at the podium and delivered a strong plea for conservation in his inimitable voice. He was a wily and brilliant politician who made great cabinet appointments."[108]

CHAPTER SEVENTEEN

"TO BENEFIT WILDLIFE"

I

Franklin Roosevelt deserved credit for saving treasured desert landscapes in the West with his four national monuments of the late 1930s—Joshua Tree (California), Capitol Reef (Utah), Organ Pipe Cactus (Arizona), and Tuzigoot (Arizona). What he took from all the archaeological discoveries in these Southwest sites was that civilizations often succumbed to lack of water or contaminated water, or a combination of the two. FDR had the National Park Service and Indian CCC collaborate to stabilize pre-Columbian ruins in the Southwest such as Chaco Canyon, Navajo, Tonto, Wupatki, and Montezuma Castle.[1] He had learned that identifying—and showcasing—an archaeological site, a "charismatic" animal, or a native plant with a Southwest landscape made it easier to galvanize public support for preservation.

After successfully protecting the desert bighorn sheep in Nevada's Desert Range, Roosevelt sought further success in the rock-and-thorn terrain of the Cabeza Prietas and Kofa Mountains in Arizona. The harsh granite mountain range crested with dark lava. The Spaniards had called it Cabeza Prieta ("black-headed") and it was the desert bighorn sheep's last true home in Arizona. A sparse herd had also congregated in the Kofa Mountains—a name coined during early mining days by shortening the well-known claim "King of Arizona"—about ninety-five miles from the Cabeza Prietas on the Mexican border.

Dr. Ira Gabrielson of the Biological Survey had long touted these Arizona desertscapes as the "most promising" sanctuary outside Nevada for the protection of desert bighorns.[2] Driven from their northern latitudes by hunters, the bighorns had taken refuge in sunbaked Arizona's remotest pockets, adapting to the punishing heat, almost nonexistent rainfall, sparse food, and dizzying elevations. However, the bighorns were still threatened by overhunting, diseases spread from domesticated sheep, and dwindling food sources. Whenever the bighorns foraged along the Gila and Colorado rivers to nibble grass, they were easy for sportsmen to kill. At some depots and truck stops in the Southwest, ram horns were stacked behind buildings like cordwood.

Frederick Russell Burnham, an American-born adventurer, celebrated his eightieth birthday with a scout troop at the Carlsbad Caverns National Park. Burnham was instrumental in starting the Boy Scouts with Robert Baden-Powell, whom he first met while seeking his fortune in Rhodesia. A big-game hunter who tried unsuccessfully to import African large animals to the United States to be shot by paying hunters, Burnham embodied self-sufficiency in the outdoors, which was intrinsic to the Boy Scouts.

In protecting desert bighorn, Roosevelt received a boost from Major Frederick Russell Burnham, a legendary figure in the age of exploration. Born to a missionary family on a Sioux reservation in Minnesota, Burnham became an Indian scout in the 1880s, when the U.S. Army was trying to pacify the Apache in the Arizona Territory.

Across sparsely settled western land he had also famously guarded bullion on Wells Fargo stagecoaches traveling in the dusty Southwest and worked as a bounty hunter. During the Boer War in South Africa (1899–1902), he was Lord Robert Baden-Powell's chief scout. In 1933, the magazine *Boys Life* published a series of Burnham's actual adventures, concluding he was in the same survivalist class as Daniel Boone, Davy Crockett, and Jim Bridger.[3] Few hunters of animals (or of people, for that matter) had the cunning of Burnham, one of the founders of the Boy Scouts of America.[4]

Settling again in the American Southwest after his stint as a soldier of fortune in war-torn Africa, Burnham became an expert on the vibrant cultures of the Opata and Yaqui Indians; was a leader in the California Club (where Los Angeles's power brokers met); joined the board of the Southwest Museum; and advocated for "fair chase" hunting. Burnham took to heart President Theodore Roosevelt's stirring oratory at Grand Canyon in 1903—TR had implored the public to leave this natural wonder alone, declaring, "You cannot improve on it. The ages have been at work on it, and man can only mar it."[5] Burnham became a leading conservationist of the Southwest, aligning himself with the American Committee for International Wildlife Protection for the purpose of saving the desert bighorn sheep (as Ralph and Florence Welles were doing in Nevada). Burnham's commitment to teaching frontier survival skills to young Arizonans led him to get involved with the Theodore Roosevelt Boy Scout Council in Phoenix.

Knowing how TR had saved the buffalo from extinction, Burnham was determined to do the same for the bighorns in Arizona's mountains. When Europeans brought domestic sheep to the Arizona Territory, bighorns caught scabies. These mites caused fatal infections that drastically reduced desert bighorn populations. Only sheep in isolated parts of the territory were able to survive the scourge. "Major Burnham put it this way," George F. Miller, who knew Burnham through their work with the Arizona Boy Scouts, recalled. "I want you to help save this majestic animal, not only because it is in danger of extinction, but of more importance, some day it might provide domestic sheep with a strain to save them from disaster at the hands of a yet unknown virus."[6]

Burnham and his Boy Scouts appealed directly to FDR to help protect the wild sheep along the Arizona-Mexico border. Less than 1 percent of Arizonans had ever seen a desert bighorn sheep, but Burnham had made it the state mascot.[7] Because Major Burnham was a living legend, Arizona politicians didn't dare step on his toes. And many other conservation-minded Arizonans heeded the major's call to action. In 1936 Herman Hendrix, the state superintendent of schools, held a "Save the Bighorns" poster design contest with Burnham's help. Winning drawings were made into colorful posters that were displayed in storefronts in Flagstaff, Phoenix, Tucson, and Yuma. Burnham chose a child's sketch of a bighorn sheep head as the official Boy Scouts of Arizona emblem. To help publicize the conservation crusade, Burnham had his troop wear desert bighorn bandanas around their necks. The National Wildlife Federation, the Izaak Walton League, and the National Audubon Society all joined the Boy Scouts in pestering President Roosevelt to designate the Kofa Game Range and Cabeza Prieta Game Range as sanctuaries for the desert bighorns.

FDR received counterpressure from many Arizonans who were opposed to the federal refuges. Likewise, Major Burnham and George Miller received hate mail accusing them of being socialists, Yankees, and claim-jumpers. It was one thing, these detractors argued, to teach Boy Scouts how to hunt bighorns; it was quite another to ask Uncle Sam to grab half a million acres of Arizona—a state that had experienced the second-greatest decline of income in America (only South Dakota was more affected) during the first three years of the Great Depression.[8] Governor Rawghlie Stanford, a Democrat, dismissed the bighorn sheep as "George Miller's billygoats." Quite sensibly, he didn't dare mock Major Burnham. On general principles, the powerful Cattle Growers' Association was opposed to *all* wildlife reserves in Arizona. Since cattle didn't graze in the bighorns' desert—which was five thousand feet high, and rocky—the cattleman's complaints rang hollow to Roosevelt. Predictably, the Arizona Small Mine Operators' Association also opposed the notion of a bighorn sheep preserve.

Ignoring objections from the Tucson Chamber of Commerce—which had taken to referring to Roosevelt as the "Great Seizer" (a pun on Great Caesar)—Ickes asked lawyers with the Department of the Interior to cobble together a compromise deal that would protect Arizona's desert bighorns. Exemptions were made for cowboys who were already tending cattle in the region, and furthermore, the administration would guarantee that when the Kofa and Cabeza Prieta ranges became wildlife sanctuaries, a few longtime miners would be allowed to work their small claims for life. That was, however, the only concession the extraction lobby received from Washington, D.C.

What Roosevelt admired, during the fight in Arizona, was the advocacy of the Boy Scouts of America. He had long credited the BSA with helping to inspire the CCC. His New Deal work-relief programs, he believed, were based on the same public service "fundamentals" as scouting.[9] Speaking once at Ten Mile River Scout Camp in New York, Roosevelt suggested that *all* American men should adopt the ethics of scouting as their own.[10] So when Arizonan Boy Scouts wrote to him movingly about the need to rescue bighorn sheep in the Kofas and Cabeza Prietas, he responded eagerly. What made Roosevelt such an impressive conservationist leader was that he *liked* it when ecology-minded young people engaged in local causes urged him to do more. "The ideals of Scouting are not simply ideals for boys," Roosevelt said. "They are good ideals for man."[11]

In early 1939, R. K. Wickstrum, the president of the Arizona Game Protective Association, sent the president loads of information on Arizona's dwindling bighorn sheep population. He followed up with a telegraphed plea. "I have received your telegram of January 8, urging the establishment of the Kofa and Cabeza Game Ranges," Roosevelt replied. "The administration of these areas have been given careful consideration and I believe the Executive Order, which I have just signed, will solve the problem. This order sets up a limited game range within the grazing district and protects the interests of prospectors as well as those of the stockmen, while at the same time taking care of the matter of the preservation of the Big Horn sheep."[12]

On January 25, 1939, Roosevelt established both the Kofa Game Range (around 660,000 acres) and Cabeza Prieta Game Range (around 860,000 acres) to be jointly managed by the Grazing Service (now the Bureau of Land Management) and the Biological Survey.[13]

The Biological Survey saw Kofa and Cabeza Prieta as field laboratories where it could conduct research about ways to combat bot fly larvae, menaces that caused chronic (and often fatal) sinusitis in Arizona's desert bighorn sheep. This sinusitis had seriously reduced the domestic sheep population in Arizona. Over the years, studies suggested the best way to fight disease was through an extremely healthy habitat, and so the creation of the refuge rescued the species in more than one way. Owing to the administration's foresight, the wild sheep population of Arizona's Sonoran Desert would double in size within three years. And federal protection arrived in the nick of time: *True* magazine had recently published a pro-hunting article extolling the virtues of achieving a "grand slam of rams" (i.e., shooting one of each of the four varities of North American sheep).[14] If not for Roosevelt's and Ickes's actions in Nevada and Arizona, it's unclear whether the desert bighorns would have survived in the American West. Nevertheless, the language in FDR's executive orders about limited ranching and grazing worried Ickes: these were loopholes, approved by Wallace, that Ickes thought should be monitored by Interior.

Setting aside the remote 860,000 acres of the Cabeza Prietas brought other unexpected benefits for wildlife—such as the protection of the endangered Sonoran antelope (*Antilocapra americana sonoriensis*), a subspecies of pronghorn. If the pronghorn is the fastest land mammal indigenous to North America, then the Sonoran is the fastest pronghorn, capable of sprinting at sixty miles per hour.[15] Now, it was in a precarious state, thanks to overhunting, habitat fragmentation, loss of foraging acreage, and periods of extreme drought. If Roosevelt hadn't created the Cabeza Prieta refuge, it's very likely that the Sonoran antelope would have gone extinct. (As it stands, there are fewer than two hundred left in the United States, with slightly more in Mexico.)

At the dedication of the Kofa Game Refuge on April 2, 1939, Major Burnham spoke movingly about the vanishing southwestern frontier he called the *depoblado* (the unpopulated wilderness). Creating the Kofa and Cabeza Prieta game ranges not only saved bighorn sheep, but also furnished protected habitat for peccaries, mule deer, and other desert-dwelling species. And Burnham urged FDR to authorize similar reserves in New Mexico. (In 1941, FDR indeed established yet another desert bighorn preserve in New Mexico—the San Andres.) To further bolster the bighorn's chances of survival in the Southwest, Ickes tasked fifty Native Americans who were working for the Indian CCC on Four Peaks Dam with securing its water supply. They were assisted by some descendants of the Rough Riders: grandsons of men who fought with Colonel Theodore Roosevelt at San Juan Hill in 1898 in Cuba were part of FDR's legendary tree army in Arizona. "They did their work well," *Fish and Wildlife News* declared in the late 1970s, "for the impoundment has not been known to go dry since and the structure is as solid as the day it was finished."[16]

The utilities, railroad interests, cattlemen, and mining companies were not pleased with Roosevelt's passion for "desert wild." Ickes was blamed for the New Deal's sudden focus on protecting desertscapes. An angry Henry Wallace, in fact, tried to maneuver around Interior in order to allow domestic sheep to graze in the Arizona game ranges. When Ickes caught wind of Wallace's plan, he was furious. There was no way that domestic sheep and desert bighorns could properly graze on the same acreage without any incidents of crossbreeding and the spread of disease.[17]

II

In the spring of 1939, Roosevelt dispatched Henry Wallace to dedicate a 2,670-acre tract of southern Maryland woodlands and cultivated fields as the Patuxent Wildlife Research Center. Its mission was to "develop the scientific information needed to provide the biological foundation for conserving and managing the Nation's biological resources most effectively."[18] The *Washington Star* described Patuxent as "the world's first national wildlife experiment station"; there had been others, but none with nearly the scope and the impact of the New Deal facility.[19] Patuxent's biologists would study flora and fauna, migratory bird routes, mass agriculture, predator-prey relationships, nutritional requirements for various species, the effects of fertilizers and pollutants on mammalian reproduction cycles, and how diseases affected animal health. Visitors were asked—in this regard, Patuxent was ahead of its time—to consider global environmental issues and how they affected the wildlife found in their own backyards.

Situated within the watershed of the Patuxent and Little Patuxent rivers, the center, which today sits on a 12,841-acre tract of land and is the largest science and environmental education center in the Department of the Interior, would reinvent the studies of wildlife biology and ecology. Roosevelt wanted bird populations observed to gauge the health of varied ecosystems—and not only for biological reasons. Birds were sensitive even to changes in climate: this meant they were species that could indicate trends. Such genera as plovers, sandpipers, ducks, hawks, and warblers were banded and tracked at Patuxent. The woodsy Maryland compound looked like a handsome Rural Demonstration Area. Located midway between Baltimore and Washington, D.C., Patuxent was a hodgepodge of depleted farmlands under crop cultivation; upland woods that included mixed stands of pitch and Virginia pine; deciduous trees (predominantly oak); a vast array of shrubs; native-plant landscaping; and marsh-swamp lowlands fed by numerous creeks.[20] There were two remarkable trees at Patuxent: an overcup oak with a circumference of fifteen feet five inches, and a river birch with a circumference of eleven feet five inches, believed to be among the largest living specimens of their kind.[21]

Research operations at Patuxent were centered on Snowden Hall, a house built around 1815 and enlarged by the WPA and PWA to provide lodging and a cafeteria. Also erected on the refuge were laboratory buildings, a sawmill, a carpentry garage, a machine shed, large barns, incubator facilities, and employees' apartments.[22] More than two hundred CCC workmen razed forty buildings with bulldozers and then engineered twenty-four, measuring twenty by fifty feet, for one of the center's first extensive studies, focusing on aquatic flora. Early that June, when Wallace arrived for the ribbon-cutting ceremony, a two-month-old red fox (*Vulpes vulpes*) was thrust into his arms. Always the obliging politician, Wallace scratched the kit behind its ears. The assembled photographers happily clicked away.[23]

During the Hoover years, the Biological Survey had worked to exterminate the fox—a predator of chickens—by using poison; now, however, in a crucial turnaround, these canids were being hand-raised in carefully monitored conditions. A new "predator conservation" philosophy was developing at Patuxent. "The projected program for this national wildlife experiment station is intended to benefit wildlife in general," Wallace explained at the dedication, "to find out under what conditions wildlife may be produced or waste-lands retired from agriculture and to determine the inter-relationship of agriculture and forestry practices on wildlife."[24]

Operating on a shoestring, the research center focused on helping to rehabilitate species in danger of going extinct. Phrases such as "as extinct as the dodo" or "gone the way of the passenger pigeon" were common among biologists at the facility, thanks to Hornaday's *Our Vanishing Wildlife*. These government scientists were saddened by the long list of species that had already gone extinct. That list included a dark species of bison (*Bison bison*) that had roamed the American East up to the early nineteenth century; the Maine giant mink (*Neovison macrodon*), twice as large as other mink species, which went extinct in 1860; and the California grizzlies in the Sierras, which had died out in the 1920s.[25]

The Patuxent experiments implemented in Maryland would simultaneously be applied on a grand scale in North Dakota, especially at the Sullys Hill National Game Reserve and Des Lacs Migratory Waterfowl Refuge. No longer was the USDA merely studying the impact wildlife had on human communities; the government would examine how human activity influenced struggling wildlife populations. Roosevelt entrusted Dr. Ira Gabrielson, chief of the Biological Survey, with the administration of the compound. Gabrielson, in turn, chose Dr. L. C. Morley, a respected veterinarian, to serve as the superintendent of the refuge. Although Gabrielson had a friendly personality, no one doubted his professional commitment to ecology.

In the coming years, sitting at a desk in Washington, surrounded by mounds of documents, Gabrielson would write numerous books, including *Wildlife Conservation* (1941), *Wildlife Refuges* (1943), and *Wildlife Management* (1951).[26]

Dr. Gabrielson had personally selected the Patuxent site for the research center because of its proximity to Washington and the Beltsville Agricultural Research Center, with which the campus shared a western boundary. The retired tobaccolands were ideal for experimental farming. Due to the healthy grasses and ponds at Patuxent, one never knew when an interesting bird—a "feathered jewel"—might drop down from the sky.[27] Numerous Biological Survey labs would be relocated from the District of Columbia to Maryland, nearer to the research center (eventually, only the administrative operations of the Biological Survey would remain in Washington). Clarence Cottam, head of the Division of Food Habits, eventually developed lab space in which he could analyze the stomachs of quail and the gizzards of crows.[28] Wild birds from Patuxent such as tree swallows and brown thrashers were fed a variety of seeds in an attempt to find the optimal one. Before long Cottam and his staff discovered that waterfowl died from lead poisoning after ingesting fishing tackle made from lead, proving that lead was toxic to most creatures. (It wouldn't be until 1977 that the U.S. government would ban lead-based paint.) Dr. Cottam and his research were important to Rachel Carson early in her career.

Suspicion about the Patuxent compound abounded in agricultural circles. Unconfirmed reports in *Modern Game Breeding* magazine charged that the research center was a sly cover for duck and geese factories that would undercut poultry farmers already in the breeding business.[29] Many hunters viewed the facility as a way to use science to impose further regulations on their activities.[30] The misgivings of animal rights activists focused on vivisection, including experiments that involved infecting animals with viruses to study the effects. (A reporter found this complaint to have some merit, writing, "Ferret with sniffles put in with normal ferret on 4/25/39 to induct infection by contact.")[31] Many hunters and trappers were convinced that the Biological Survey was getting into the lucrative fur business (because of the photo of that fox cradled in Henry Wallace's arms) and that it would soon be reducing their profits. Manufacturers regarded Patuxent as a backdoor route to factory regulation through public health scares over fertilizers and chemicals. "Pollution does vitally affect birds and animals," Gabrielson said in response to such complaints, "as well as fish and other interests." One of the "other interests" was the human race.[32]

What Roosevelt was popularizing was the belief that wildlife mattered greatly, that animals were priceless natural resources. All animal species native to Maryland—the wild turkey, cottontail, raccoon, squirrel, and woodcock—were studied at Patuxent. A seven-foot-high fence, extending for thirty miles, was erected by the CCC to protect the wild animals from stray dogs, feral cats, and poachers. Rehabilitation of white-tailed deer was an early priority at Patuxent. By the late 1930s, the species was scarce and seemed to be headed for extinction. Entire herds in the Shenandoah Valley of Virginia were captured, placed in tall boxes, and carefully trucked to Patuxent for ear-tagging and biological study. The scientists at Patuxent also tried to find ways to help the Pacific black-tailed deer (*Odocoileus hemionus columbianus*), whose numbers were down to around a thousand; and the tiny Key deer (*Odocoileus virginianus clavium*) of Florida.[33]

The *New York Times* mistakenly called the compound an "animal Eden."[34] Perhaps it was a haven for those animals that could roam freely, but for most of Patuxent's critters, the compound was more like Alcatraz or San Quentin. Caged ferrets, quail, and buzzards rattled their doors, trying to escape. (The animals, of course, had no way of knowing that their individual losses of freedom would help their respective species survive in the long run.) A comical story in the *Washington Evening Star* recounted an episode in which wild turkeys, skunks, raccoons, muskrats, and foxes were released at Patuxent overnight. On waking, one CCC recruit became obsessed with capturing a wily baby fox, but the kit hissed viciously, causing him to flee in terror.

A 475-foot dam was built along Cash Creek at Patuxent to attract waterfowl to the pond grasses. The CCC also constructed rustic bridges across draws and creeks, giving the grounds a distinctly downcountry feel.[35] As the center added staff, its activities expanded in myriad directions. Twin farms were established: one was operated according to wildlife practices recommended by the Soil Conservation Service; the other relied on commonly accepted methods of living among wild animals. The National Herbarium, founded in 1848, collaborated with Patuxent on aquatic-plant research.[36] Thrash seed was shipped from Maryland to Oregon, Washington, and California.[37]

The president's persistent call for rehabilitation of the beaver on a federal level was also prioritized at Patuxent. Thanks to excessive trapping by reckless fur traders, the beaver was scarce in the mid-Atlantic region. FDR, who hoped *Castor canadensis* would make a comeback nationally, believed that beavers often dramatically altered landscapes for the better. Long ago, before contracting polio, he had learned firsthand in upstate New York that a beaver colony could gnaw more than a ton of trees annually and raise the water table. By damming streams, beavers expanded wetlands, offering rich habitat for other species to prosper. Beaver ponds also controlled runoff and reduced erosion, the most dreaded word in FDR's lexicon.[38] At Patuxent, trapped beavers were now given expensive radio tracking collars and released back into the wild.[39]

Four other threatened North American species that attracted Roosevelt's specific attention were the bald eagle (*Haliaeetus leucocephalus*), trumpeter swan (*Cygnus buccinators*), whooping crane (*Grus americana*), and California condor (*Gymnogyps californianus*). Not that FDR was regularly telephoning the center's biologists for updates, but the USDA employees knew the White House cared deeply about these imperiled animals. New techniques such as bird banding, induced birth, public education campaigns, and no-hunt provisions were tested at Patuxent to determine their usefulness in helping these birds prosper once more.

Recognizing that saving endangered species was an interagency undertaking, Roosevelt asked the Patuxent biologists to work with entities on all levels, including the aforementioned Soil Conservation Service, state conservation commissions, 4-H Clubs, state and national parks, chapters of the Future Farmers of America, Audubon societies, sportsmen's groups, and so on. The results of such consultation with Patuxent were varied. Wooden feeding troughs were constructed at the National Bison Range in Montana to help buffalo prosper. In Nevada, pronghorn antelope were herded into traps and shipped by truck, train, and plane to new wildlife ranges. Elk were captured in traps built around haystacks where they foraged for food in the winter. Sometimes government planes dropped blocks of salt into alpine ranges so sheep and elk could thrive. Many new approaches to wildlife survival, untested before 1939, were tried at Patuxent. A hospital for sick ducks was built within the center's disease-control laboratory to end the scourge of botulism (type C), which killed thousands of waterfowl annually; it was determined that the specific strain of botulism was spread by ducks eating decayed food.[40]

Meanwhile, despite Eleanor Roosevelt's influence, the federal government lagged behind academia and even industry in the employment of female biologists. The Bureau of Fisheries had hired Rachel Carson on a part-time basis in 1935. The following year, she wrote a series of highly readable booklets for the bureau, including "Food from the Sea." Carson's primary job was to analyze biological and statistical data about fish populations. Though she never worked at Patuxent, she did work closely with her fellow scientists there. She also wrote research abstracts for publication in *The Progressive Fish-Culturist*, a periodical for ichthyologists and fish hatchery workers.[41] Able to make complex concepts in marine biology comprehensible to the general public, she was asked to write radio scripts about sea creatures. A number of Carson's nature essays were published by the *Baltimore Sun* and the *Atlantic Monthly*. In 1939 Elmer Higgins, head of the Bureau of Fisheries, promoted her to the post of assistant aquatic biologist. When the Bureau of Fisheries and the Biological Survey were combined to form the Fish and Wildlife Service in 1940, she became the public relations voice on all issues pertaining to national wildlife refuges, marshlands, and the sea.[42]

Unusually for a biologist, Carson avoided getting bogged down in arcane scientific data. There was a controlled fluidity in her prose; she conveyed scientific concepts to laymen with grace and ease. In the coming years Carson would write a trilogy of marine conservation classics: *Under the Sea-Wind* (1941); *The Sea Around Us* (1951); and *The Edge of the Sea* (1955).[43] And in 1962, her manifesto *Silent Spring*, a brilliantly reasoned critique of the use of DDT, launched the modern environmental movement.[44]

Carson remained loyal to FDR's conservation legacy. When ecologists criticized the president for building Grand Coulee Dam, nearly destroying the salmon runs, Carson concurred but also added that Patuxent helped establish the Division of River Basin Studies to prevent future damage to fish as a result of federal water projects.

III

On May 10, 1939, to coincide with the opening of Patuxent, President Roosevelt issued Executive Orders 8110 to 8129, which established twenty federal migratory waterfowl refuges in North Dakota.[45] Eighteen months earlier in Grand Forks, Roosevelt had spoken to farmers about proper land stewardship, telling them that uprooting trees and plowing out grass chased away all the migratory birds. With genuine emotion in his voice, he promised that the New Deal would soon help "farm families settle on good land."[46]

Under FDR's plan, North Dakota farmers would provide the lands for these twenty waterfowl sanctuaries as part of the ingenious Limited-Interest Program. Roosevelt told these farmers that if they allowed the Biological Survey to purchase easements on their land to ensure the retention of breeding grounds for migratory birds, then the government would award them cash payments—money that a farm family could use to stave off foreclosure.[47] Additionally, the Biological Survey would offer to any farmers who participated free advice on scientific water management strategies and crop diversification methods from agricultural experts.[48] After procuring these easement rights, the federal government would, in effect, manage certain aspects of the land while the farmers retained ownership of their acreage: this was a socialist-tinged agrarian policy with a distinctly American flavor.

Once North Dakota farmers enrolled in the Limited-Interest Program, they had to allow the WPA, PWA, and CCC to build structures such as "check" dams, reservoirs, and ponds on their lands. As FDR sold the program, it was a win-win: a partnership of conservation and agriculture that continues to the present day. There was also environmental tourism as a side benefit: North Dakota soon became a favorite destination for birders and hunters alike, and this boosted tourism revenue.

The Roosevelt name had a lingering appeal in North Dakota. It was at the Elkhorn Ranch along the Little Missouri River, near the Badlands village of Medora, that Theodore Roosevelt—the self-proclaimed "wilderness hunter"—developed his conservation ethic in the 1880s. To honor TR's outdoors legacy, FDR opened a CCC headquarters in Bismarck and gave instructions to erect another camp in Medora. Shortly thereafter, land for Roosevelt Recreation Demonstration Area in the Badlands was purchased with National Industrial Recovery Act funds. The CCC built dams and dug lakes in the area, which eventually became Theodore Roosevelt National Park.[49] FDR wanted to transform Medora into a popular recreation center—like Jackson Hole, Wyoming, or Gatlinburg, Tennessee—an attractive gateway town where tourists could learn about his famous cousin's Wild West days, visit prairie dog burrows, and watch buffalo graze on rehabilitated grasslands. North Dakota would later be nicknamed the "Rough Rider State" after TR.

On June 12, Roosevelt boldly added fifteen more migratory bird refuges to the ever-expanding system in North Dakota.[50] No other state has ever been so richly transformed in terms of ecology—or so completely micromanaged by Washington officials. Composed of grasslands, ravines, and modest lakes, the refuges formed in June were typical of the mottled prairies of the upper Midwest. Central Flyway species benefiting from FDR's second round of North Dakota proclamations included sandhill cranes, the sharp-tailed grouse, nesting ducks (mallards, gadwalls, pintails, and blue-winged teals), and four types of geese. Biologists at Patuxent immediately stepped up to help these new migratory bird refuges in North Dakota combat outbreaks of botulism and chlamydiosis. USDA wardens distributed supplemental food—such as hardstem bulrush, saltmarsh bulrush, and sago pondweed—to the waterfowl species struggling on the drought-ravaged refuges.[51]

Roosevelt especially wanted the North Dakota waterfowl refuges on the Des Lacs and the upper and lower Souris rivers to become showcases for wildlife protection and water conservation in the Great Plains states.[52] In October, fifty thousand mallards arrived at the North Dakota refuges, clamoring in lakes and bushes, creating hunting opportunities theretofore never seen.[53] It's no exaggeration to say that FDR saved North Dakota from ecological ruin during the Dust Bowl years. "This massive outpouring of federal funds by the Democratic administration in Washington," Elwyn B. Robinson wrote in *History of North Dakota*, "was of the utmost importance to the state, contributing much to its survival and wellbeing."[54] At the time of his second announcement, adding fifteen refuges, more than thirteen thousand North Dakotans were working on WPA projects, and the WPA had spent $23 million in the state.

The drought in North Dakota lasted from 1929 to 1940, although not every area was affected, except in two brutal years, 1934 and 1936. Only rarely did residents see normal, moist conditions. Not only was the state parched, it was abnormally hot in the summer (breaking the state record, with a reading of 121 degrees Fahrenheit in 1936). Dust storms were regular occurrences. And that was not all that nature unleashed on North Dakota: the winters over the same periods were often extreme as well. It is understandable that the state lost population every year for twenty years, starting in 1930. The New Deal held out some hope for North Dakota, if only because it did not forget about a state that was down but not out.

Dozens of North Dakota's picturesque stone post offices and durable city halls were built by New Deal agencies—with custom-painted murals included. Migratory waterfowl refuges were also part and parcel of the New Deal's revitalization of the Great Plains.[55] On the Manitoba border, as a botanical preserve, New Deal work-relief crews built the International Peace Garden (giving North Dakota the nickname "The Peace State"). CCC Company 794 SP-1, "Camp Borderline," laid out stone bridges, constructed a lodge of North Dakota timber and Manitoba wood, and transformed the flatland by planting 150,000 flowers annually.[56] There was even a gigantic floral clock on display. Because this International Peace Garden was so closely linked to the natural wildlife and migratory waterfowl refuges, it became the headquarters for the North American Game Warden Museum.[57]

Under Roosevelt, the National Park Service helped develop sixteen new state park units in North Dakota. FDR thought that North Dakota could attract a booming summer tourist trade by publicizing local legends. Interpretive centers were built along the river to provide history lessons on the state's buffalo hunters, fur traders, and Native American tribes. The best example of the CCC's historic restoration work in the state, however, was Fort Abraham Lincoln State Park, which perpetuated the former home of the Seventh Cavalry, commanded by Lieutenant Colonel George A. Custer (it was from this outpost that he left to fight the Battle of Little Big Horn). Roosevelt thought the entire Missouri River, as it snaked throughout the state, told the story of the nation's expansion westward, and of figures from Lewis and Clark to Sitting Bull to Theodore Roosevelt.

As a result of Patuxent research initiatives, North Dakota's Limited-Interest Program, CCC's can-doism, a network of refuges and new parks, and reasonable hunting laws, the waterfowl population in the United States doubled between 1934 and 1941. North Dakota alone became home to more than a third of America's waterfowl protection areas.[58] Encouraged by this progress in bolstering the future of migratory birds, FDR was determined to enact stringent federal regulations protecting the American bald eagle. Rosalie Edge, one of the founders of the Emergency Conservation Committee (ECC), had become the leading activist for birds of prey. Besides lobbying to establish Olympic National Park, the ECC campaigned on behalf of many species, publishing fact-filled pamphlets like *A Last Plea for Waterfowl* and *The Antelope's S.O.S.: The Extinction of the Pronghorn Antelope Is a Preventable Misfortune That We Are Neglecting to Prevent.* Its hard-hitting campaigns were deservedly effective. Ickes turned to Edge for regular counsel on wildlife protection issues; he was tired of dealing with special interest sportsmen's clubs that prioritized hunting above all else.[59]

Rosalie Edge at the Hawk Mountain Sanctuary, which she created in eastern Pennsylvania. A resident of New York City, Mrs. Edge chafed at what she considered to be the hypocrisy or ineptitude of established conservation groups, uncovering a plot by which the directors of one group sold unrestricted hunting rights on a nature "preserve" in Louisiana. Working mostly on the outside of such conservation groups, she was a force in the movement, buying Hawk Mountain herself, in order to give the raptors of the region a true safe haven.

Edge established a sanctuary for hawks near Kempton, Pennsylvania. No battle over conservation ever fazed her. She didn't hesitate to take on even the Audubon Society after she perceived that it had become corrupted by generous donations from oil and gas interests. In 1939, she fought to destigmatize hawks, which had long been considered dangerous pests by farmers. The following year, "Hawks—Good and Bad," an article in *National Sportsman*, advocated the extermination of "undesirable" raptors. This rankled Edge. The article derided Cooper's hawk (*Accipiter cooperii*), the sharp-shinned hawk (*Accipiter striatus*), and the goshawk (*Accipiter gentilis*) as winged vermin and offered anatomically precise drawings of the supposed menaces so farmers could more easily identify them for wholesale

Franklin D. Roosevelt, with (*left to right*) Robert Fechner, Henry Wallace, and the men of Civilian Conservation Corps Company 350 (*in background*) at CCC Camp Big Meadows in Virginia's Shenandoah National Park, August 12, 1933. The CCC was the greatest peacetime mobilization *ever* of American youth.

Eleanor Roosevelt at a Civilian Conservation Corps camp in Yosemite, California, 1941. In her "My Day" columns she regularly wrote about the wonders of the natural world.

Men from the Civilian Conservation Corps clearing land for soil conservation, 1934. From 1933 to 1942, the CCC had enrolled more than 3.4 million men to work in thousands of camps across America.

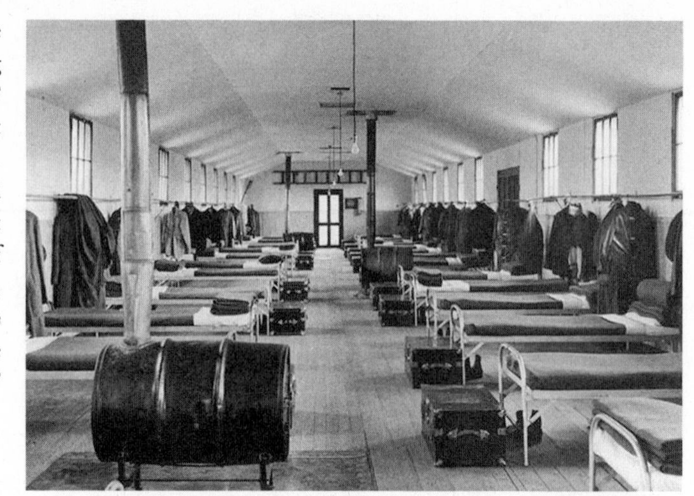

Interior of the bunkhouse at Long Lake CCC Camp F in Wisconsin's Nicolet National Forest. Roosevelt hoped to restore the great North Woods of Wisconsin, Michigan, and Minnesota. "The forests," he said, "are the 'lungs' of our land."

Every morning, at CCC camps, Old Glory was raised in a military-style ceremony. Instilling patriotism and national service in recruits was a central FDR policy mission.

CCC Mobile Unit at New Mexico's Chaco Canyon National Monument, 1938. Roosevelt had the CCC fruitfully collaborate to stabilize pre-Columbian ruins in the Southwest, such as Chaco Canyon, Navajo, Tonto, and Montezuma Castle.

CCC workers clear snow at Rocky Mountain National Park. Roosevelt was a huge booster of winter recreation. The New Deal sought to promote skiing in Colorado, Idaho, and Vermont.

Queen Elizabeth of England looking at the Civilian Conservation Corps exhibit at Fort Hunt, Virginia. *Left to right:* Prime Minister Mackenzie King of Canada, Queen Elizabeth, and Eleanor Roosevelt, June 9, 1939.

The CCC camp located in the Chisos Mountains of Texas, November 26, 1937. Stationed at the camp was Company 1855, under the command of Lieutenant William L. Hagman and Lieutenant Archie L. Murray, second in command. Roosevelt established Big Bend National Park, thereby protecting the Chihuahuan Desert.

The CCC camp from Trail Ridge Road, Rocky Mountain National Park, Colorado. It took six CCC companies to build FDR's "top of the world" road in the Rockies.

A primitive CCC washroom at Camp Rock Creek, California. The CCC was crucial in establishing the first state parks for Virginia, West Virginia, South Carolina, Mississippi, and New Mexico.

Civilian Conservation Corps, States, Indiana, Jackson Camp 558, Brownstown, Indiana. Note the portrait of FDR looming over the CCC worker's shoulder. To the CCCers, the president was an icon. The *Woodpecker*, a CCC newspaper in Magdalena, New Mexico, created an oath: "Roosevelt is my shepherd, I shall not want."

CCC Company 1394, Camp S-68, Weikert, Pennsylvania. After work hours, men were encouraged to read in the camp library. As actor Humphrey Bogart put it, the CCC was "a 14-karat opportunity for young men" to receive an education.

Civilian Conservation Corps, Third Corps Area (Pennsylvania, Maryland, Virginia, and the District of Columbia), plant nursery, Company 5445, Camp A-4, Beltsville, Maryland, 1933.

CCC boys at work on a project at the experimental farm of the Department of Agriculture at Beltsville, Maryland, 1933. Between the dual plagues of soil erosion and foolish farming, one half of the upper Midwest's topsoil was deleted or lost.

Civilian Conservation Corps, Third Corps Area (Pennsylvania, Maryland, Virginia, and the District of Columbia), cabinetmaking shop, Company 1351 C-V, Yorktown, Virginia, 1933. African American corpsmen at this segregated camp made beautiful furniture.

Civilian Conservation Corps, California, March Field District, 1933. White, black, and Hispanic men stand in boxing trunks. Two are wearing boxing gloves. Although the CCC was segregated, athletic competitions between camps were often integrated.

Civilian Conservation Corps, California, March Field District, 1933. Men splitting firewood. The values learned in the Boy Scouts—self-reliance, piety, woodcraft, conservation—were all values of the CCC.

K.P.'s Rock Creek CCC Camp, California, June 21, 1933. The work relief agency's erosion-control programs alone benefited forty million acres of farmland.

CCC "boys" from Camp F-167, Salmon National Forest, Idaho, ready to transplant beaver from a ranch location where they had been damaging crops to a forest watershed where they would help conserve water supply. All these men are from Brooklyn and the New York City area. Wildlife rehabilitation was a major component of the CCC.

Boys from CCC Company 2759V surveying a thinning area of the Black Hills National Forest, South Dakota, 1933. *Time* magazine declared that "more continuously" than any other New Deal project, the CCC had the "respect of the foes as well as the friends of Franklin Roosevelt."

CCC activities included roadside clearing, Boise National Forest, Idaho, 1933.

CCC enrollee planting a tree. Nearly three billion trees were planted on American soil by "the boys" during Roosevelt's presidency.

slaughter. Apparently, the crime of these hawks and the reason to exterminate them was that they raided poultry yards.[60] Edge helped popularize the notion that poultry could be protected from eagles and hawks through means other than wholesale extermination.

When Roosevelt read an ECC pamphlet—*Save the Eagle: Shall We Allow Our National Emblem to Become Extinct?*—he was moved to take executive action. He refused to allow the symbol of the United States to become extinct. FDR, who had designed a postage stamp featuring the bald eagle, endorsed the hard-charging ECC pamphlet: "The case made for the protection of the eagle, if indeed it were necessary to make a case for it, is convincing and persuasive," Roosevelt wrote to Edge, "and I share with you the desire to see this bird adequately protected by law."[61] Working through back channels, Roosevelt was soon able to get the bald eagle the strict federal protection it needed.[62]

As the ECC's influence on the White House demonstrated, New Deal conservation didn't operate in a vacuum in the late 1930s. Researchers from land-grant universities helped keep federal agencies on the right path. For example, Ohio State University's Stone Laboratory had developed a first-rate fish hatchery, located on Gibraltar Island in Lake Erie near the site of Commodore Oliver Hazard Perry's dramatic victory over the British in the War of 1812. In 1936 FDR signed a proclamation establishing the Perry's Victory and International Peace Memorial as a national monument, thus affording further history and environmental protection to a string of Ohio islands, including Gibraltar, which were like stepping-stones to Canada.[63] While Erie was the smallest of the five Great Lakes by water volume, it was the richest in fish biodiversity. Ohioans, deeply proud of their 312 miles of Erie shoreline, accepted federal aid in the 1930s to help refurbish the lake. And Roosevelt, on August 2, 1938, signed Executive Order 7937, establishing West Sister Island Wildlife Refuge as a sanctuary for the largest wading bird nesting colony on the U.S. Great Lakes.[64] It provides a haven for herons, egrets, and cormorants. (Commodore Perry had put West Sister Island as the dateline on his famous message to General William Henry Harrison: "We have met the enemy and they are ours.")

Throughout the 1930s crews on commercial fishing boats in Lake Erie collected yellow perch and lake whitefish eggs, bringing them to Stone Lab to hatch. The eggs were then fertilized and placed in hatching jars. The Stone Laboratory housed nine hatching batteries. Each battery held 228 hatching jars (for a total of 2,052). Water was constantly run through the jars to keep the eggs rolling. Once the fry hatched, the current forced them up and out of the jars into the troughs and then down to collection basins. The fry were then moved to the raceways to grow another six inches before they were released back into Lake Erie.

More than any other president, Roosevelt focused on the essential value of fish hatcheries for the nation. In 1936, Dr. Gabrielson hired Robert Rucker, an expert on fish pathology, to grapple with depletion issues in California and the Pacific Northwest. Rucker helped build the Western Fish Disease Laboratory in Seattle.[65] He would spend thirty-seven years in his laboratory (or in the field), finding answers for those who relied on a stable and healthy fish population. As his career went on, he devoted as much time to research on the effect of pollution on fish as naturally caused diseases.

IV

Throughout early 1939, FDR worked to give the CCC a public relations boost. Roosevelt emphasized that enrollees had planted 1.7 billion trees in its first six years in operation, saving millions of acres of farmland from soil erosion and helping communities recover from flooding and hurricanes. Robert Fechner, the organization's director, was on the cover of *Time* that year; the accompanying article reeled off CCC statistics: 104,000 miles of truck trails built; 71,692 miles of telephone lines laid; forty thousand bridges completed; forty-five thousand (plus) buildings added; dams of various sizes and types finished; and so on. Buried in all this was an analysis of the CCC's positive and negative environmental impacts.[66]

Because 1940 was an election year, Roosevelt wanted to tout his CCC-related accomplishments—2.5 million men put back to work and more than $500 million making its way back to the families of enrollees—to help secure continued congressional funding for his program.[67] There wasn't a newspaper editor in America who didn't respect those figures. As the *Pittsburgh Press* noted in August 1939, "No other relief agency has been so popular with the American people as the Civilian Conservation Corps."[68]

With the notable exception of Representative Oscar Stanton De-Priest of Illinois, who opposed it because it discriminated against racial minorities, the CCC seemed to be beyond congressional reproach. Thomas Beck of *Collier's* helped celebrate it by publishing an anniversary article, "The CCC—Indispensable." As Rexford Tugwell, a member of FDR's brain trust, later recalled, "the CCC quickly became too popular for criticism."[69]

As part of the intensified public relations, Fechner also contributed a piece to the *Washington Evening Star*, explaining that the CCC had taught about seventy-five thousand illiterate enrollees how to read and write. Another 700,000 had used the CCC as a springboard to finish high school or complete vocational training. For Fechner, the former labor organizer, the CCC was always about training young men; there seemed to be nothing of Muir in him. "Virtually every enrollee has been improved in health," he wrote. "All have been taught to work."[70]

When the CCC was first established in 1933, its primary detractors had been labor unions and socialists. However, by 1939, in a surprising turn of events, it was visionary environmentalists who were leading the charge against Roosevelt's pet program. That June, the administration took an unexpected hit to its conservation strategy (though the sting wasn't immediately felt) when Aldo Leopold delivered an address in Milwaukee to a joint meeting of the Society of American Foresters and the Ecological Society of America. Leopold was highly critical of the New Deal approach to land management and was especially critical of the CCC. He accused the CCC of purging beech, white silver, and tamarack trees in the Midwest in the name of "timber stand improvement" and creating unwelcome pollens by planting the wrong trees in the wrong parts of the country. The farmland around Bismarck became a severe allergy zone. The buckthorn the CCC planted was now considered an undesirable "invasive species" in North Dakota.[71]

Leopold became first chair of the Department of Wildlife Management at the University of Wisconsin that year. For five years, since 1934, his innovative course, Wildlife Ecology 118, had been growing in renown, even exerting influence on the operational protocols of both Patuxent and the North Dakota Limited-Interest Program. An argument could be made that FDR's two announcements of 1939 regarding additional refuges in North Dakota echoed the recommendations in Leopold's highly influential *Game Management*. That book—along with Leopold's articles on the value of farm ponds, farmer-sportsmen alliances, game warden training, evergreen planting, wildflower maintenance, bird banding, and prairie chicken restoration—helped persuade USDA to help farmers in the upper Midwest revitalize their worn-out lands. Even though Leopold was challenging Roosevelt's aggressive conservationist vision as damaging certain terrains, he nevertheless admired the president.[72]

Patuxent was too new for Leopold to grasp the great leap the administration was making in 1939 to enhance the human ability to protect ecosystems and wildlife. His problem was really with the National Wildlife Research Center in Denver, which specialized in predator control. First founded in 1886 as the Division of Economic Ornithology, NWRC-Denver was famous for eradicating predator species throughout the West. NWRC-Denver fought the proliferation of animal predators (wolves and wolverines) and bird menaces (the blackbird and grackle) in hopes of protecting domesticated animals, minimizing damage to forest resources, reducing wildlife hazards to aviation, and curtailing rodent damage to crops and rangeland. If Henry Wallace had come to Denver instead of Patuxent, he would have been photographed not cuddling a fox kit, but grabbing a dead coyote by the scruff of the neck.

Leopold looked on FDR's policies with a mixture of opinions. What made the president so impressive to Leopold was that he didn't abandon the migratory bird refuge campaign during his second term, even though the specter of war threatened the United States. After Germany's invasion of Poland in September 1939, it seemed clear to most Americans that the nation would eventually get involved in the European war. And the threat posed to world peace by Japan was no less severe. But even with such a full plate, FDR regularly inquired about Arizona desert bighorn, North Dakota grasslands, birdlife in Maryland, and the first forestry projects in the Ozark highlands of Missouri.

While Leopold had constructively criticized the CCC at its weak spots, he recognized the value of many of the New Deal's land rehabilitation projects executed by the Forest Service, including the initiative in the Ozarks. During FDR's first term, the Forest Service purchased more than 3.3 million acres of fallow farmlands and clear-cut woodlands in southern Missouri. In this spirit, the president wanted to turn the area, blessed with numerous "first magnitude" natural springs and hundreds of caves for exploration, into a bustling recreational hub. (It didn't hurt that the scenic Eleven Point River ran through the region.) CCCers helped reforest the acreage, which had been stripped for timber. After the Ozarks were rehabilitated, FDR, on September 11, 1939, only ten days after Germany invaded Poland, established two huge national forests in Missouri: Mark Twain and Clark.[73] Leopold might not have liked the picnic tables and boating facilities erected by the CCC, but then, he himself sought solace—and prime hunting—in a refuge he called "the shack" along the Wisconsin River. FDR only wanted the same privilege for all Americans. Designating the acreage in Missouri as national forestland would at least keep the extraction industries at bay.

On November 19, 1939, President Roosevelt laid the cornerstone in Hyde Park for his presidential library—the first in American history. Two bills the president had signed during his first term—the Reorganization Act of 1933 and the Historic Sites Act of 1935—enabled the National Park Service to preserve the heritage of American presidents. FDR would later donate most of the Springwood estate's forests, hills, gardens, and spectacular views of the Hudson River to the Department of the Interior.[74] Under the watchful eye of Harold Ickes, this first presidential library, a repository for FDR's public and private papers, was soon completed on the Springwood grounds.

At the press event that November at Springwood, FDR was the beneficiary of his own invention. He spoke of his lifelong kinship with Dutchess County. Members of his Hyde Park Home Club told the press that Franklin hadn't changed much since his first run for public office in 1910; his blue eyes hadn't lost any of their trademark brightness. And his penchant for forestry was stronger than ever. Years and decades would pass, the president told the assembled people, while his sycamores, maples, poplars, and white pines continued to grow. Speaking in the third person about his own past, he said, "Half a century ago a small boy took especial delight in climbing an old tree, now unhappily gone, to pick and eat ripe sickle pears. That was just about one hundred feet to the west of where I am standing now. And just to the north he used to lie flat between the strawberry rows and eat sun-warmed strawberries—the best in the world." Roosevelt went on to explain how he used to dig into woodchuck holes with his dogs. "Some of you are standing on top of those holes at this minute. Indeed, the descendants of those same woodchucks still inhabit this field and I hope that, under the auspices of the national archivist, they will continue to do so for all time."[75]

Roosevelt's Springwood had gone through many renovations since he was a bird-obsessed boy during the Gilded Age. Seen across a broad green, the flat rooflines of the estate's main house were bracketed by wings, and the long facade was lined with a covered veranda. When FDR returned to the estate as an adult, he seemed most proud of the mounted birds he had collected in his adolescent years, which Sara had long since placed in the front hall cabinets for all to see. This bird collection was a happy reminder of his cherished youth, when he had traipsed all over Dutchess County looking for new entries to record in his American Ornithologists' Union diaries.

Because the president was always eager to escape the nerve-taxing confines of the White House, Springwood had become, as Eleanor Roosevelt recalled, more of an "official residence." As president, FDR made two hundred trips to Hyde Park. At times the "first couple" had to bring in extra help to assist "his mother's employees because of the large number of guests who followed the president and the extra staff that must come with him; but the manner of life changed very little."[76] All sorts of structural accommodations, including the construction of a tiny elevator, were made to allow the president's wheelchair access to every room. Whenever suggestions were made that perhaps the first-floor library should be transformed into a bedroom, FDR brushed them aside. A phone next to his bed, wired directly to the White House, allowed him to conduct official business while at Hyde Park. But true privacy was hard to come by at Springwood, so he frequently disappeared to Top Cottage.

"I found that on my trips to Hyde Park from Washington it was almost impossible to have any time to myself in the big house," Roosevelt lamented. "The trips were intended primarily for a holiday—a chance to read, to sort my books, and to make plans for roads, tree plantings, etc. This was seemingly impossible because of (a) visitors in the house; (b) telephone calls; (c) visits from Dutchess County neighbors; (d) visits from various people who, knowing I was going to be in Hyde Park, thought it an opportune time to seek some interview."[77]

During the Christmas season of 1939, FDR took stock of his presidency. Since late 1937 the economy had hit the road bump known as the "Roosevelt recession." Blaming "economic royalists" who hoarded fortunes but refused to invest in the stock market or expand businesses, Roosevelt began downsizing the Public Works Administration. The projects in North Dakota were some of the PWA's last hurrahs.

But the president reflected with pride on the PWA's accomplishments. It had financed nine thousand highways and streets; seven thousand educational buildings; eight hundred health care facilities; six hundred city halls and courthouses; 350 airports; and fifty housing projects.[78] And those statistics do not include the thousands of public playgrounds the PWA built using discretionary funds. Among FDR's favorite PWA projects was the Central Park Zoo. Ickes, who had grown up in Altoona, Pennsylvania, was himself proud that the New Deal had electrified the train line of the Pennsylvania Railroad into New York and Washington, D.C. Always thinking about how PWA construction work would hold up a century later, Ickes was described by historian William E. Leuchtenberg as "a builder to rival Cheops."[79] And he was also a preservationist to rival John Muir. "Ickes," biographer Jeanne Nienaber Clarke wrote, "saw potential national parks almost everywhere he looked."[80]

Though Ickes's leadership of the PWA was considerably less problematic than Harry Hopkins's of the Works Progress Administration (which revolved around smaller projects and employed unskilled laborers), the agency was disbanded by Congress in 1939. After the German invasion of Poland in September 1939, the New Deal continued to prioritize affordable public housing for the poor, but with the looming threat of Nazi expansionism on the horizon, FDR knew the economy had to shift toward war production in 1940. By the end, more than ten million Americans had benefited from the PWA's public works as an alternative to direct relief.[81] "What the PWA sought to do," Ickes wrote in his book *Back to Work*, "was to get honest work at honest wages on honest projects, which was a great deal more difficult a task than giving away money."[82]

But a telltale sign that the PWA had run its course in 1939 was that Roosevelt no longer ballyhooed new hydroelectric dams or public transmission systems. That July, PWA counsel Ben Cohen told the president that Congressman Lyndon Johnson was proving to be a master of Texas's version of the TVA: the Colorado River Authority. An appreciative Roosevelt brought LBJ officially into the New Deal family, asking him to serve as director of the Rural Electrification Administration. Johnson declined the offer, telling Roosevelt—one of his heroes—that his job wasn't promoting "big Washington" projects, but developing a stronger "contract with the people of Texas."[83] The president would have to do the same with citizens in the entire country in 1940 if he wanted to win a third term in the White House.

As 1939 came to a close, a tragedy struck the New Deal conservation movement: Bob Marshall, the Forest Service's chief of recreation and one of the country's greatest wilderness advocates, died on a nighttime train ride from New York to Washington. He was only thirty-eight years old. As a mountaineer, adventurer, and writer, Marshall was the closest thing the Great Depression generation had to the long-gone John Muir.[84] It was Marshall who had led the Wilderness Society; called for the nationalization of 80 percent of American forests; and overtly opposed racial discrimination in the Department of Agriculture. It's been argued that Marshall, unencumbered by ego, was "personally responsible" for preserving more wilderness than any other non-president in U.S. history.[85] "He brought with him a real appreciation of the problems of the underprivileged and fought their battles so that they might share in the recreational and spiritual benefits of our land and our resources," Ickes said. "His conception of conservation made him a leader among conservationists. The wilderness areas he worked so hard to perpetuate remain as his monuments." (The Bob Marshall Wilderness in Montana and Mount Marshall in the Adirondacks were named in his honor.)[86]

That same year the conservation movement had taken another blow: Fechner, director of the CCC, died. He was only sixty-three years old. During Fechner's last days, Roosevelt sat at his bedside at Walter Reed Hospital, cheering him on, but both Fechner's heart and lungs collapsed. At Fechner's instructions, six CCC boys from Rock Creek Park and Fort Dupont Park camps served as his pallbearers. Because Fechner was a Spanish-American War veteran, he was buried at Arlington National Cemetery.[87] All over America at CCC camps, flags were flown at half-mast in honor of the symbol of both human and land reclamation. To lead the corps, Roosevelt hired James McEntee, a close friend of Fechner's, and for three decades his loyal assistant director.[88]

PART FOUR

WORLD WAR II AND GLOBAL CONSERVATION, 1940–1945

CHAPTER EIGHTEEN

"AN ABUNDANCE OF WILD THINGS"

I

One evening in early 1940, Secretary of Labor Frances Perkins escorted Daniel Tobin, president of the Teamsters union, to the Oval Office for a meal with Roosevelt. "Mr. President, you have to run for a third term," Tobin implored. "Don't talk to me about your fishing trips next winter—you are going to be right here in the White House." "No, Dan," Roosevelt replied, "I just can't do it. I have been here for a long time. . . . I want to go home to Hyde Park to take care of my trees. I have a big planting there, Dan."[1]

Tobin was stunned. That spring he followed European affairs closely and grew ever more adamantly opposed to the aggressive spread of fascism there. He felt that the world situation demanded strong leadership. Was Roosevelt really going to slide off the world stage—with Hitler's Third Reich, Mussolini's Italian empire, and Hirohito's imperial Japan threatening global democracy—to become a tree farmer?

In early 1940, speculation was rampant over whether or not Roosevelt would defy the two-term tradition to make an unprecedented third play for the White House. The president's private remarks about retiring to Dutchess County, and his agreement to write articles for *Collier's*, with editor Thomas Beck purportedly paying him $75,000 per year for his post-presidential thoughts, indicated to many that indeed FDR was ready to retire.[2] Even Eleanor Roosevelt was initially opposed to the idea of a third term, as was labor leader John Lewis. And so was his vice president, John Nance Garner of Texas. Many Southern Democrats thought that "Cactus Jack," having dutifully served for two terms, deserved the Oval Office in his own right. Other potential Democratic Party replacements for FDR included

Secretary of State Cordell Hull, Agriculture's Henry Wallace, and Postmaster General James Farley. Ickes, however, felt, as Tobin had, that a third FDR term was essential, endorsing his boss in a July 1939 article titled "Why I Want Roosevelt to Run Again." According to Ickes, to deny the president was to "deny democracy itself."[3]

In the midst of the overall political swirl, a freighted rumor circulated that FDR, in his last year, would transfer the U.S. Forest Service from USDA to Interior, thereby recasting it as the "Department of Conservation," something Ickes had long coveted. Senator John H. Bankhead of Alabama was disturbed by a report in *Forestry* venturing that FDR was indeed poised for the controversial transfer. Bankhead, a Democrat who loathed Ickes, threatened that if the transfer occurred, then all of FDR's favored agriculture bills would stall out in Congress.[4]

Gifford Pinchot also implored Roosevelt to keep the Forest Service out of Interior's ever-expanding portfolio. Pinchot's public thrust at the time was his belief that "permanent peace" relied directly on international cooperation in conservation and on the equitable distribution of natural resources. Since he believed firmly that resources, meaning "forests, waters, lands, and minerals," were for only one thing, "the lasting good of men," he trusted that fight with the Forest Service. Angry with Ickes, Pinchot sent a disapproving letter to the White House on January 13, and he took the unusual step of including 139 letters of complaint signed by prominent academics at American forestry colleges.[5]

Pinchot's power play backfired. Roosevelt began his reply by taking his former mentor to the woodshed, comparing this "group drive" tactic to that of "Father Coughlin or the United States Chamber of Commerce or the cattlemen's associations, or, for that matter, horrid things like the K.K.K. itself."[6] That Pinchot, a longtime friend, was trying to strong-arm him was outrageous. FDR then sidestepped the issue of transferring the Forest Service in favor of a friendly, but not inconsequential discussion of the two departments in question. "And incidentally," he ended his letter to Pinchot, "the days have passed when any human being can say that the Department of Agriculture is wholly pure and honest and the Department of the Interior is utterly black and crooked."[7] Pinchot immediately shot back another letter, squarely on topic, insisting, "To uproot the Service from its lifelong surroundings would do great injury."[8]

Pinchot's and Bankhead's railing against the reorganization scheme may have hit a nerve with the president, but the knockout blow came from George W. Norris of Nebraska, one of Roosevelt's favorite senators, the impresario of the TVA. Over lunch at the White House, Norris told the president that transferring the Forest Service to Interior would be detrimental politically and a disastrous blow to New Deal conservationism. Norris well understood the rising backlash against progressivism; he had switched party affiliation from Republican to Independent in 1936 and had concerns with his own political future. According to Norris, the transfer of the Forest Service would be likely to split conservationist friends into two bitter factions.[9] Furthermore, Norris expressed his thoughtful concern that Ickes, who was known as a curmudgeon, had certain "peculiar qualities" that hampered his ability to properly lead Interior. Too often Ickes sided with ECC activists Rosalie Edge and Irving Brant. Cognizant that Ickes was personally very close to the President, Norris made the offbeat suggestion that "Honest Harold" was suited to spur the War Department forward. "I appreciate the fact that Mr. Ickes' real interest is in conservation," Norris wrote to FDR after their lunch. "In the War Department he would have the opportunity to do a real service in carrying out ideas in conservation."[10] Later in the month, Bankhead informed the president that the plan to transfer the Forest Service had almost no chance of approval in the Senate. Pushing the reorganization plan, doomed to fail, simply wouldn't be worth the fight.

The Pinchot-Bankhead-Norris "group drive" had effectively quashed Ickes's dream of a Department of Conservation. The outlook for a transfer of the Forest Service in 1940 looked remote. Lobbying on his own behalf, Ickes met privately with Norris, to no avail. Norris backpedaled a little, telling Ickes that he didn't object to the transfer per se; instead, his primary concern was with its timing during a presidential election year. If FDR had done it in 1939, Norris said, he wouldn't have squawked. But to Norris an open fight over the transfer in the spring of 1940 would adversely affect FDR's chances for reelection come fall in the South and West.

Irritated, sensing defeat, a fuming Ickes offered Roosevelt his resignation in early February. "Forestry has become a symbol to me," Ickes dejectedly explained to the president. "I have had one consistent ambition since I have been Secretary of Interior, and that has been to be head of a Department of Conservation, of which, necessarily, Forestry would be the keystone. I have not wanted merely to be a Secretary of the Interior. I have wanted to leave office with the satisfaction that I had accomplished something real and fundamental. I have told you frankly that, as this Department is now set up, it does not interest me. So I have come to the reluctant conclusion that, as matters now stand, I cannot be true to myself nor measure up to the high standards you have a right to expect of a man whom you have honored by making him a member of your Cabinet. Accordingly, I am resigning as Secretary of Interior and, at your pleasure, I would like my resignation to take effect not later than the 29th of February."[11] When Harry Hopkins heard that Ickes had submitted a letter of resignation, he rolled his eyes. "He is stubborn and righteous which is a hard combination," Hopkins wrote in his diary. "He is the 'great resigner'—anything that doesn't go his way, he threatens to quit. He bores me."[12]

Ickes wasn't merely grandstanding. He was sickened to see the Forest Service allowing the overcutting of woodlands in the North Cascades, the Porcupine Mountains of Michigan, and the North Woods of Maine for short-term gains. Speaking for many, Republican congresswoman Clare Boothe Luce once caustically remarked that Ickes had "the soul of a meat axe and the mind of a commissar."[13] Perhaps there was truth in her barb, but America's national parks and national monuments had no more steadfast friend than Ickes. It was he, more so than FDR, who moved beyond the wise-use confines of conservation and became a genuine environmental warrior in the tradition of John Muir. Roosevelt leaned away from commercial interests more than other presidents, before or since, but he was indeed a tree farmer and saw some room for compromise. Ickes didn't. Like FDR, he hoped future generations would be able to wander among the New England hills, Utah canyonlands, Missouri bottoms, Georgia pine woods, and Dakota grasslands and see stretches of America just as it had been when the *Mayflower* arrived in the New World. That was part of Ickes's soul that gave a reason to the meat ax.

When Roosevelt received Ickes's resignation letter, however, he grew discomfited. He hated to lose such an able cabinet officer, one who stood up to special interests. Three FDR loyalists—William O. Douglas, Pa Watson, and Dr. McIntire—advised the president to disregard Pinchot and the others, listen to Ickes, and strip the Forest Service away from the USDA. But Roosevelt knew that Bankhead had been right: the political timing was off. Picking up the telephone, he vented his frustration on Ickes. "I could tell from his voice that he was highly excited and troubled," Ickes wrote in his diary. "He shouted at me that I was making life miserable for him. I told him that this was not my desire at all but that I had no option since it was now too late to transfer Forestry, in view of the time that had been given to the opposition to be built up."[14] Once calmed down, FDR asked Ickes to meet in the Oval Office in two days' time to discuss the matter.

For two days Ickes was full of hope and the foolish optimism that headway was being made. Roosevelt, he imagined, had come to his senses, realizing that national forests would receive stronger protections under Interior leadership. When the meeting occurred, Roosevelt, as if a therapist, listened attentively to everything Ickes said about the public lands. There was a lot of general agreement. As Ickes was leaving the Oval Office, however, the president handed him a sealed envelope. Inside was a handwritten note:

> We—you and I—were married, "for better, for worse"—and it's too late to get a divorce and too late for you to walk out of the home—anyway. I need you! Nuff said.
>
> Affec. FDR.[15]

Ickes was crestfallen. With every fiber he believed a Department of Conservation would have been a great boon for wild America. But in the end, he was compelled to rescind his resignation. "It is pretty difficult to do anything with a man who can write such a letter," Ickes wrote of FDR in his diary. "It really left me no option except to go along."[16]

II

On April 3, 1940, the dream of Roosevelt and Ickes for Isle Royale National Park became a reality. The war raged across the oceans and FDR wrestled with his own political future, but for the distant northern preserve, events already in motion finally came to fruition. Five years earlier, Roosevelt had signed an executive order allocating federal funds for the northern Lake Superior archipelago, which belonged to the state of Michigan and was located fifteen miles from the Minnesota-Canadian border. The new national park and its accompanying islands amounted to forty square miles of primeval forests and lakes. Roosevelt and Ickes had been opposed to a luxury hotel being built on Isle Royale or any attempt to turn it into "the Bermuda of the North" for the Great Lakes cruise industry.[17] Determined to keep Isle Royale roadless, Roosevelt asked the CCC to establish camps at Senter Point on the western tip to clear the debris left behind by shuttered lumbering operations. The NPS designation ensured that industry would never return. Because of the intense snowy conditions in Michigan, Isle Royale was the only national park that Roosevelt would order closed during the winter months. During the summer, though, outdoor aficionados could hike the white spruce and balsam fir forests or canoe around the inland lakes, enjoying the abundance of wildlife. "Isle Royale," the *New York Times* reported, "is of much interest to the nature lover, sportsman, geologist, and archaeologist."[18]

The idea for Isle Royale National Park had been in the works since 1931, when Congress passed the authorizing legislation, but it took the New Deal to prompt the federal government into purchasing necessary private acreage and allocating government subsidies to the effort. The CCC prepared the island for National Park designation, replanting the forests and building structures. In 1936, after a wildfire burned through a quarter of the Isle Royale forestlands, the CCC restored lost habitats. After the disaster, Roosevelt lobbied Congress again for national park status.[19]

People of Michigan had mixed feelings about the park. The *De-*

troit News promoted the Isle Royale with editorial vigor while the *Detroit Free Press* countered that the Roosevelt ecology project was a boondoggle comparable to a "railway to the moon."[20] The copper lobby of the Keweenaw region derided Roosevelt's Lake Superior wilderness park as an absurd waste of taxpayers' money ($700,000 had already been allocated during FDR's first term). But Roosevelt, who had a powerful ally in Republican senator Arthur H. Vandenberg, remained stalwart. Critics groused that there was no single defining feature that Isle Royale was, so to speak, just a wilderness. But that was exactly the point. Wildlife expert Adolph Murie, who had studied the populations of moose and wolves on the main island, wrote the definitive report on the "worthiness" of the national park in 1929, delineating the archipelago's breathtaking shoreline, high interior ridges, inland lakes, and ancient beaches. "True wilderness," Murie asserted, "is more marvelous (and harder to retain) than the grandiose spectacular features of our outstanding parks."[21]

Ickes was gratified that the Roosevelt administration had forbidden any new roads or automobile tourism in Isle Royale. "I am not in favor of building any more roads in the National Parks," Ickes declared. "This is an automobile age, but I do not have much patience with people whose idea of enjoying nature is dashing along a hard road at fifty or sixty miles an hour. I am not willing that our beautiful areas ought to be opened up to people who are either too old to walk, as I am, or too lazy to walk, as a great many young people are who ought to be ashamed of themselves. I do not happen to favor the scarring of a wonderful mountainside just so that we can say that we have a skyline drive. It sounds poetical, but it may be an atrocity."[22] Deeming developers "parasites," Ickes also lectured his NPS superintendents to save other primeval spots in the existing national parks from pandering. "I do not want any Coney Island," Ickes said. "I want as much wilderness, as much nature preserved and maintained as possible. . . . I think parks ought to be for people who love to camp and love to hike and who . . . [want] a renewed communion with Nature."[23]

While "wilderness" was Ickes's personal mandate, the president kept promoting the NPS's historic site initiatives in 1940. During his second term, Roosevelt had added numerous heritage landmarks to the National Park Service through the authority of the National Historic Sites Act of 1935. Among them were Fort Laramie (Wyoming), an essential nineteenth-century trading post located in the Platte River Valley; Hopewell Furnace (Pennsylvania), an "iron plantation" complete with blast furnaces, the ironmaster's house, and a company store; the Old Customs House (Pennsylvania); the Jefferson National Expansion Memorial in St. Louis (where the Gateway Arch would be completed in 1965); Cumberland Gap (Kentucky, Tennessee, Virginia), the key pioneer passageway through the lower central Appalachians; and Federal Hall Memorial (New York), the first capitol building of the United States under the Constitution and the site of George Washington's first inauguration. Having brought Mount Rushmore into the National Park Service portfolio in 1933, when only Washington's face existed there, the Roosevelt administration subsequently oversaw the dedication of the faces of Jefferson (1936), Lincoln (1937), and TR (1939).[24]

Eleanor Roosevelt brought attention to a site important in Hispanic history, and one that reflected FDR's far-reaching ambitions for the National Park Service. Whenever he could, he encouraged states to develop their own parks, and so it was that NPS architects supported Texas in its challenging attempt to re-create the 1749 Mission Nuestra Señora Espiritu Santo de Zuniga in southeast Texas. The mission had largely disappeared by 1820, but CCC workers rebuilt it under the auspices of the NPS, with its meticulous historical research. Eleanor Roosevelt, visiting the region in 1940, enthusiastically took a side trip to the site. "Even the administration building is in keeping with everything else," she marveled, "hand hewn beams, hand wrought nails, all made on the spot. I do not think this could have been accomplished unless the director had been an artist with a real feeling for the work he is doing."[25]

The team of Franklin and Eleanor intervened to ensure that the National Park Service acquired the 211-acre Vanderbilt Mansion, just a few miles north of Hyde Park. The federal acquisition of the estate was proof that the Gilded Age of his youth had passed into history. Built in the 1890s, the estate was willed by Frederick Vanderbilt to his niece, Margaret Van Alen. She received it on his passing in 1938. After trying to sell it, she was glad to take FDR's friendly suggestion that she donate the treasure house to the nation. In doing so, she expressed her wish to "keep my place as it is—a memorial to Uncle Fred and a national monument."[26] FDR gladly accepted the gift on behalf of the Department of the Interior. To help improve the property, the president instructed the CCC to spruce up the grounds. Eleanor Roosevelt became personally involved in the Vanderbilt Mansion preservation effort. "When these places are taken over by the Park Service, it takes some time to put them in order," she explained in a "My Day" column from August 1940. "In the case of this estate, no one has lived there for the past few years and the gardens and greenhouses which require constant keeping up had, of course, greatly deteriorated. I imagine that gradually they will come back to their former beauty. Then visitors will not only see the house itself, surrounded by a collection of variegated trees which cannot be equaled anywhere else up or down the Hudson River Valley, but they will enjoy the beautiful gardens which have been developed by three different generations of owners."[27] Years later, the National Park Service decided to manage the Vanderbilt Mansion, Springwood, and Val-Kill jointly, referring to the three Hyde Park locations simply as the Roosevelt-Vanderbilt National Historic Sites.[28] The Roosevelts were probably aware of that possibility when their neighbor's home was opened to the public.

Meanwhile, Pinchot, pleased that Roosevelt had heeded his advice about keeping the Forest Service in the Agriculture Department, announced that he would "actively support" the president's reelection in 1940 as a "nonpartisan."[29] To placate Ickes, Roosevelt, without much media fanfare, would soon consolidate the Biological Survey and Bureau of Fisheries—arms of the USDA and Department of Commerce, respectively—to form the new Fish and Wildlife Service (FWS).[30] He then placed it in the Department of the Interior. The FWS reorganization during the summer of 1940 was interpreted by Washington insiders as a ringing endorsement of Ickes as the central intelligence behind America's natural resources management policies.[31] It very nearly transformed Interior into the Department of Conservation that Ickes coveted (the only major public land agency missing from his control was the Forest Service).

As mid-1940 drew near, the nation was still uncertain about FDR's personal plans. The women in Roosevelt's life all insisted that he shouldn't run for a third term. They felt that, having worked under pressure for eight years, having suffered a slight heart attack in February, and having been savaged with sinus infections, he needed to rest. The president stayed elusive on the subject of a third term in early 1940, keeping people guessing. But by May, with Hitler having invaded France, Belgium, Luxembourg, Denmark, the Netherlands, and Norway, Roosevelt, as his speechwriter Sam Rosenman noted, "became determined to stay in the White House until the Nazis were defeated."[32] And so he did. Once France and Norway fell to the Nazis in June, Hitler's spring offensive reached a pinnacle of success. Roosevelt delivered fireside radio chats denouncing isolationists and unveiling new plans for building up American defenses. His hat was in the ring.

The German blitzkrieg in Europe meant that American conservation became a lesser concern for the president. The 1940 presidential election, like the one in 1916, played out against the backdrop of the European war. With Garner tacitly running against the president, the choice of Roosevelt's running mate in the 1940 election was a major topic for discussion in his inner circle. A leading contender was Henry Wallace. Four of FDR's closest associates on public lands issues—Ickes, Brant, Morgenthau, and Delano—thought that Wallace was a brilliant bureaucrat but a boor. Frances Perkins, among others, urged the president to select Wallace. Determined to spike Wallace, Ickes worked back channels to get William O. Douglas the VP nod.

As the year wore on, the presidential election continued to play out in a climate of global crisis. Even though Japanese troops had captured vast territory in China, and European democracies across the Continent had fallen to Hitler—with Britain being pummeled by German bombing—the GOP isolationist wing urged neutrality. Politicians from both parties, though, relentlessly spoke of "war preparedness" and "industrial mobilization." Factories were beginning to ramp up the effort to manufacture ships, planes, and munitions at a rate that could help rebuild America's military.[33] In July, Roosevelt, in a bipartisan gesture, appointed Republicans Henry L. Stimson and Frank Knox as the heads of the War and Navy Departments, respectively.

During the summer of 1940, Roosevelt reconfigured the CCC as a national defense measure. Young men would be trained in operating and repairing mechanized equipment on behalf of the armed forces. This modification was a far cry from reforestation and wildlife rehabilitation, but it was intended to help extend the life of the corps. It didn't mean the CCC was giving up its ecological mission entirely, however. "The work projects proposed for our fields, forests, wildlife refuges, and parks," the new CCC director James J. McEntee wrote in *American Forestry*, "would keep the Corps of the present size occupied to from thirty to fifty years. . . . Until such time as these youths can be absorbed in private industry, business and agriculture, I believe there is justification for continuance of the CCC. I believe it should be made permanent."[34]

In late June, after supervising the planting of ten sequoia seedlings that had been shipped to Springwood from the High Sierra, FDR left Hyde Park for the White House, ready for the grueling campaign. The Democrats were gathering in Chicago for their convention. Eleanor Roosevelt—having spent much of the year traveling—stayed behind in Dutchess County. Every morning the *New York Times* carried grim headlines about European refugees and wartime taxes being imposed around the globe, while the sunshine danced on the rippling Hudson River. This juxtaposition of American pastoral light and European darkness made the first lady ponder the fairness of life. "I drove through the woods just as the sun was setting last night, a most mysterious magic hour," she wrote. "There was a little soft light on the deep green leaves. A fat woodchuck scuttled across the road ahead of me. A little white-tailed rabbit ran along the road, too frightened to get out of the way, until I stopped the car and let him run to cover. How can one think of these woods converted into a battlefield? Peace seems to be in the heart of them and yet, I remember some just like them outside of Paris and in the forests of Germany and England."[35]

On July 19, the first lady listened by radio from Val-Kill as her husband accepted the Democratic Party's presidential nomination for a third time. The convention in Chicago had chosen Roosevelt without hesitation. Once off the airwaves, Franklin telephoned his wife with a request: Could she fly to Chicago as his surrogate? "I think he hoped I might be able to give the delegates a personal sense of the appreciation he feels for their confidence in him," ER surmised, "even though the service required is such a heavy responsibility."[36]

The big speculation in Washington was over who would be Roosevelt's vice-presidential running mate. In a last-ditch effort, Ickes, Delano, and Watson pleaded with Roosevelt again not to choose Wallace, but Roosevelt rejected their belief that Wallace was weird, especially in his spiritual inquiry.[37] Postmaster General James Farley cast aspersions on Wallace as a mystic; FDR lit into him. "He's not a mystic, he's a philosopher," he scolded him. "He's got ideas. He thinks right. He'll help people think."[38]

Southern Democrats also had doubts about Wallace. When Governor Eurith D. Rivers of Georgia asked Governor Leon C. Phillips of Oklahoma what he made of the push for Wallace as vice president, the latter said, "Henry's my second choice." "Who's your first choice?" asked Rivers. "Any son of a bitch," replied Phillips, "red, white, black, or yellow, that can get the nomination."[39]

Roosevelt knew Wallace was the most accomplished Agriculture secretary in American history. During his nearly eight years as secretary, he turned the department into an effective engine for agricultural betterment, by removing acreage from production, paying farmers not to produce, and bringing the federal government into the lives of farmers across America. When Wallace first came to work for Roosevelt, the USDA had 40,000 employees; he left that summer of 1940 with 146,000 workers. USDA's expenditures more than quadrupled from $280 million in 1932 to $1.5 billion in 1940. Wallace deserved credit for co-creating the Agricultural Adjustment Administration, the Soil Conservation Service, and the Farm Security Administration and for waging noble fights against diseases that devastated plants and animals alike, from locust plagues to bark beetles to brucellosis to Dutch elm disease. "For Wallace there was no such thing as too much science," biographers John C. Culver and John Hyde concluded, "nor could there ever be an end to it."[40]

And so FDR went with Wallace. It was the first time that a candidate took the prerogative of choosing his own vice president; before 1940, conventions made the selection. Ickes's consolation with FDR's choice was that the obstinate Agriculture secretary would no longer run the Forest Service.[41]

That July, Roosevelt acted to organize America's protected ecosystems into a durable and comprehensive system of sanctuaries called the National Wildlife Refuge System.[42] No fireworks or front-page newspaper stories accompanied the system's birth—there was just a taut 255-word policy statement followed by a list of the refuges, called NWRs, which were being grouped within the Interior bureaucracy, while they received new names. For example, under the presidential proclamation, the Pelican Island Bird Refuge in Florida, the first federal bird preserve (established by Theodore Roosevelt in 1903) officially became Pelican Island National Wildlife Refuge.[43] Over ten million acres of wildlife habitat, formerly in USDA, were moved into the Department of the Interior's portfolio.[44]

The expansion of the NWR system remains Franklin Roosevelt's most enduring accomplishment in environmental conservation. In 1933, Roosevelt had inherited sixty-seven loosely affiliated holdings with an array of confusing names; over time he had streamlined them into a coherent system of marshes, prairie potholes, deserts, mountains, and coastal areas that numbered 252 units by the summer of 1940. Prodded and challenged by Ickes, Roosevelt willed the system into existence. His executive orders and presidential proclamations protected 700 species of birds, 220 species of mammals, 250 varieties of reptiles and amphibians, more than 1,000 types of fish, and an uncounted number of invertebrates and plants.[45] "Nobody, it seems," Irving Brant correctly asserted in his 1989 memoir *Adventures in Conservation with Franklin D. Roosevelt*, "thought to congratulate Franklin Roosevelt, who took time out from a herculean economic recovery task to grasp and perform a job in conservation by action that antagonized most of his friends and enemies."[46]

At the same time, Roosevelt's new Fish and Wildlife Service was emerging with its own identity. Dr. Gabrielson, known for his sustained concentration on species survival, was appointed by Roosevelt as the first head of the newly configured FWS. One of Gabrielson's initial actions was to fly to Alaska to inspect the primary nesting colonies of seabirds; he later wrote about this trip in his 1959 book *The Birds of Alaska*.[47] That October, Gabrielson reached a worldwide milestone in conservation: the Inter-American Convention on Nature Protection and Wildlife Preservation (signed by thirteen North American, Central American, and South American nations at the Pan-American Building in Washington, D.C.). This was a major step forward in protecting migratory birds in the Western Hemisphere.[48] And under Gabrielson's brilliant leadership FWS would "plant" approximately 200 million game fish—salmon, trout, steelhead, bass, pike, catfish, and perch—annually in federal refuge waters.[49]

Looking for an issue to distinguish itself, the FWS immediately sought scientific and political ways to protect the American bald eagle (*Haliaeetus leucocephalus*) in 1940. That June, after a long fight on Capitol Hill, the president had finally secured the passage of the Bald Eagle Protection Act. Unfortunately, the Territory of Alaska was exempt from the new law.[50] Emboldened by Ickes, FSW agents now traveled to southeastern Alaska to collect bald eagle carcasses and transport them to Patuxent Wildlife Research Center or Denver Wildlife Research Laboratory, where the contents of their stomachs could be examined in order to determine what the species ate in the wild. Ickes had also directed Dr. Ira Gabrielson to make an inventory of bald eagles on the thirty-seven NWRs in the lower forty-eight. FWS biologists, reporting to Gabrielson, expected the numbers to be robust in the Pacific Northwest and along the Mississippi River, but the big surprise was that St. Marks NWR in Florida—a sanctuary created by Herbert Hoover, which the CCC had restored starting in March 1933—had seen the largest increase of bald eagles.[51]

III

In June, at the Republican National Convention in Philadelphia, Wendell Willkie, a whipsmart New York lawyer and businessman who had been critical of the TVA, PWA, WPA, and CCC, emerged as the party's presidential nominee. The fact that the Indiana-raised Willkie had never been elected to public office of any kind before was anomalous—and a sign of how desperate the GOP was for a fresh face. So unknown was Willkie in GOP circles outside of New York and Washington that in May 1940 a Gallup poll placed him far behind Thomas Dewey, Arthur Vandenberg, and Robert Taft as the likely nominee. Willkie had risen from modest circumstances to great success as chief of the Commonwealth and Southern Corporation, the country's largest electric utility. As a former Democrat—and a delegate for FDR in 1932—Willkie had never liked Roosevelt's extension of government into areas served by private business, particularly those regarding the Taylor Grazing Act and TVA. In the realm of foreign policy, Willkie was a levelheaded interventionist who constantly warned the public that the American "way of life" was in competition with "Hitler's sway."[52] Deriding Roosevelt as "the third-term candidate," Willkie traveled over 34,000 miles and visited thirty-four states, hoping to gain momentum.

That both Roosevelt and Willkie were internationalists was indisputable. Both favored Great Britain over Germany, bringing an almost bipartisan cast to the election. Sculptor Albert Christensen, influenced by Gutzon Borglum, decided to carve the faces of Roosevelt and Willkie into a sandstone wall outside Moab, Utah, as a tribute to American exceptionalism. Called *Unity*, the public sculpture demonstrated that Republicans and Democrats were of one mind about defeating Hitler. But, according to the federal government, Christensen had committed an act of vandalism. To his chagrin, U.S. federal agents soon chiseled down the faces. Christensen, who lived in a series of unusual cave-like rooms called Hole N" the Rock, which he had carved out of solid rock near *Unity*, wasn't beaten, though. He decided to create another memorial on his own property. On his second attempt, he honored only Roosevelt. (This kitschy version can still be seen today near Moab.)[53]

On the campaign trail, the president painted Willkie as just another mouthpiece for special interests. Roosevelt proudly showcased his own New Deal record with its impressive roster of work-relief, public power, and conservation accomplishments. Ohio, for example, had been tragically deforested by the time FDR moved into the White House. When Ohio was first opened for settlement, in 1783, about 95 percent of its more than 26 million acres were blanketed with forests. As Roosevelt took to the platform to speak at the Cleveland Public Auditorium in early November 1940, only three million forested acres remained. The CCC reforested hilly sections of the Cuyahoga Valley, built bridges, and dammed Salt Run to create Kendall Lake (all or part of what became Cuyahoga National Park in 1974). But even with the CCC's intense efforts, Ohio's existing forests were being cut over at a rate three times higher than the rate at which they were growing. "I see an America," Roosevelt told a cheering crowd in Cleveland just before Election Day, "whose rivers and valleys and lakes—hills and streams and plains—the mountains over our land and nature's wealth deep under the earth—are protected as the rightful heritage of all the people."[54]

In 1940, one particularly onerous problem the CCC faced in Ohio was trying to save the Cuyahoga River, which had caught fire multiple times. The hundred-mile-long river, which flows through Akron and Cleveland before emptying into Lake Erie, had long since been ruined by industry. Ohioans had dumped so much effluent and debris into the Cuyahoga that the river emanated a rank odor even when it was frozen in the winter. Roosevelt might have preferred the passage of the Lonergan-Mundt-Clark bill, which would have made clean streams a public right "in the name of sound economics, health, recreation, and common decency," according to conservationist Philip G. Platt, former president of the Pennsylvania Division of the Izaak Walton League of America.[55] The President ultimately backed instead Senator Alben Barkley's somewhat similar bill, giving business more leeway. The time seemed ripe for legislation on water pollution, but in the end neither bill was enacted; the Barkley bill passed both houses of Congress, but stalled in conference.[56]

Courting sportsmen was a Roosevelt political tactic in 1940. His new National Park Service director, Newton E. Drury, who replaced Cammerer in August, was a genius at connecting conservation to American values. World War I veterans were trotted out to talk about how the very thought of America's pristine wilderness had helped fortify them in battle overseas. Ready to reenlist in the army, one Great War veteran sought guarantees that the Blue Ridge Mountains wouldn't be timbered while he was fighting overseas. Ickes circulated his testimony all around Interior. "Loving the outdoors," the veteran was quoted as saying in *Virginia Wildlife* magazine, "I should hate to find that our conservation program had been junked—not only because we should conserve the resources which have made America great, but because I want again to fish clear streams and tramp through unspoiled fields and forests with a dog and a gun. I want to come back to the America I have always known—an America of freedom, of opportunity, and of happy living. . . . Clear waters, green fields and forests, fertile soils, an abundance of wild things, and freedom to use and enjoy these resources properly—these I hope America will always have."[57]

The new boss at Agriculture, as of early September, was Wallace's former undersecretary, Claude R. Wickard. An Indianan, Wickard was a graduate of Purdue and had been a farmer before entering government service. Sympathetic to conservationist issues, he was expert in the area most needed during the war years: food production. Ickes, watching his old nemesis Wallace on the campaign trail with FDR, staked out territory with his new nemesis, Wickard. After reading a *Mining World* article in the summer of 1940 called "Conservation—Should It Serve—or Only Save?" Ickes turned troublemaker. Deeming the essay an "obvious biased attack" against the National Park Service, he fired off a letter to *Mining World* and accused the Forest Service (and the extraction industries) of turning a blind eye to destructive logging operations in the Pacific Northwest. Convinced that treasured ecosystems such as the North Cascades and the Olympics—government-owned tracts in the region of Olympic National Park—should be stripped from Forest Service jurisdiction and given over to Interior, Ickes lambasted Gifford Pinchot's long-standing "wise-use" philosophy as rank philistinism. Some priceless American forestlands, such as California's redwood kingdom and the Upper Peninsula of Michigan, Ickes argued, needed to be left in "pristine condition" without human interference. "If you had taken the trouble to analyze for your readers the meaning of the term 'multiple use,' you would have found that it is a meaningless expression," Ickes wrote to Miller Freeman, publisher of *Mining World*. "It is definitive of nothing. Its conflicting premises are as subject to question as are the promises of patent medicine that claims to cure all ills."[58]

Offended by Ickes's ill-tempered letter, a scripture-toting Freeman fired back that Jesus Christ had been against "hoarding" (Interior) and for "utilization" (Agriculture). It was an ineffective riposte. Freeman then accused Ickes of hurling "Chicagoanese" invective at his detractors and resorting to "roguery" against those "who have the temerity to honestly disagree with you."[59] For Ickes, Freeman's squawk offered even greater incentive to make sure the Pacific Northwest and Great Lakes "timber swine" were stopped dead in their tracks. But while rapacious timbermen and mine-owners were the enemy for Ickes, it was Secretary Wickard who had the national forests he wanted. Considering Wickard a political novice, Ickes sought creative ways to undermine the Forest Service every chance he got.

IV

On September 2, 1940, FDR delivered a speech in perhaps the most majestic setting of his career: the Laura Spellman Rockefeller Memorial in Newfound Gap at the heart of the Great Smoky Mountains National Park. The effort to create the park had spanned decades. "We are at last definitely engaged in the task of conserving the bounties of nature," FDR said that day, "thinking in the terms of the whole of nature."

In early September 1940, Franklin and Eleanor journeyed to Great Smoky Mountains National Park in western North Carolina and eastern Tennessee. In a formal dedication ceremony, Roosevelt claimed both the park and the trans-mountain road—an excellent paved highway—as New Deal success stories. The park, created in a partial way in 1930, was finally completed after being twice postponed. "The drive goes through the most beautiful scenery," Eleanor wrote. "Once in the park, I think you are impressed by the wonderful care which is being given the area. I saw no signs of forest fires, or of blights which have killed so many of our trees in other parts of the country. There is much virgin timber in these woods, but you have to go a little off the main road to see it. A policy of careful wildlife conservation will probably bring back much of the game which has disappeared."[60]

Flanked by three southern governors, the president delivered an impassioned defense of public lands. The Great Smoky Mountains National Park, chartered in 1926, was the first to have its land and building costs subsidized with federal funds; previous parks had been carved out of existing government property or had been purchased with private donations. The park, a hiking enthusiast's wonderland, ran along seventy-one miles of the Great Smoky Mountains, the highest of the Appalachian Mountains, clad in autumnal foliage and dotted with creeks plunging over falls, pristine lakes, open meadows, and forested peaks. It didn't escape Roosevelt's notice that more than 150 species of trees grew in the Smokies. He was particularly proud of his administration's role in saving 200,000 acres of primeval forest, the largest contiguous tracts of virgin red spruce and unspoiled hardwoods to be found in the United States. "Around us here, there are trees, trees that stood before our forefathers ever came to this continent; there are brooks that still run as clear as on the day the first pioneer cupped his hand and drank from them," Roosevelt declared. "In this Park, we shall conserve these trees, the pine, the red-bud, the dogwood, the azalea, and the rhododendron, we shall conserve the trout and the thrush for the happiness of the American people."

Most American presidents would have simply dedicated the Great Smokies delineating the region's wonderful features, letting it go at that. But Roosevelt had come to preach the old-time gospel of conservation. His words aimed to undercut reckless exploiters of natural resources: "We used up or destroyed much of our natural heritage just because that heritage was so bountiful," he sermonized. "We slashed our forests, we used our soils, we encouraged floods . . . all of this so greatly that we were brought rather suddenly to face the fact that unless we gave thought to the lives of our children and grandchildren, they would no longer be able to live and to improve upon our American way of life."[61]

Following the Great Smokies ceremony, Eleanor Roosevelt traveled all over New York State, celebrating the glories of the Taconic State Parkway and the beauty of the Adirondacks countryside. While Eleanor worked the eastern states, the president headed west. It was partly a campaign trip and partly a victory lap. From 1933 to 1940 the West beat out all other regions in America for per capita New Deal payments for work relief and loans. According to historian Richard White, the Rocky Mountain states received $716 per capita and the Pacific Coast states $424. For midwesterners, by comparison, the amount was $380. Roosevelt garnered 20 percent more votes in 1932 and 1936 in the West than other statewide Democrats up for election during the Great Depression. The president brought the West into modern America with a combination of federal exemptions, executive orders, public power enterprises, legislation, and conservation. The new West routinely sought federal assistance, and the WPA, PWA, CCC, and SCS were only too glad to oblige. New Deal bureaucracies modernized the region for better and worse. "When compared to the West a half-century before," White wrote, "the scope of the changes was staggering."[62]

Continuing to use Idaho as a model for New Deal work-relief projects, Roosevelt had CCC Company 297 at Priest River institute training programs in radio operation and FBI-style fingerprinting skills that the U.S. military anticipated for entry into World War II.[63] In October, the United States initiated a peacetime draft, in which Secretary of War Stimson drew the first number out of a bowl containing 7,836 numbers to decide which young men would be called first into military training. Before long, approximately 800,000 men were drafted, most as enlisted men or NCOs.[64]

Realizing that millions of voters didn't want U.S. intervention in World War II, Roosevelt professed a measure of neutrality. At a speech in Boston on October 30, he reassured Americans that he wasn't going to involve their nation in the European war. "I have said this before, but I shall say it again and again and again," Roosevelt insisted. "You boys are not going to be sent into any foreign wars."[65]

Those who considered themselves environmentalists or ecologists in 1940, and the numbers were cultish, voted for FDR. Some bemoaned Roosevelt's damming of the Columbia River but applauded his heroic rescue of the Olympic Forest. When a reporter asked Rosalie Edge of the ECC to name a Republican whom Wendell Willkie should tap as Interior secretary if he were to win in November, she went blank. When Irving Brant was asked the same question, he too drew a blank: "I don't know anybody in that damn party to which I once belonged who would do it." These remarks by Edge and Brant reflected how completely FDR had brought the TR-Pinchot Bull Moose wing of the Republican Party into the Democratic fold, and for many, conservation was the draw.[66]

On November 5, 1940, election day, Franklin, Eleanor, and Sara Roosevelt all voted together in Hyde Park's quaint town hall. Because Dutchess County was so staunchly Republican, an old joke was that only "the trees" considered him a great president. After casting his ballot, FDR was wheeled outside, where he heard a photographer shout, "Will you wave to the trees, Mr. President?" Roosevelt good-naturedly shot back, "Go climb a tree!" before waving at the stately oaks and elms. Having voted against Roosevelt in 1932 and 1936, the county was about to do so again in 1940. Only the trees of Hyde Park thought FDR was a great president.[67]

Roosevelt had nothing to worry about. He crushed Willkie by a margin of 449 electoral votes to 82, which amounted to 85 percent of the electoral votes, while also winning 55 percent of the popular vote. Voters evidently felt that, with World War II threatening to involve the United States, changing presidents in favor of a novice like Willkie was too risky. Although Willkie had many attributes, including foreign affairs acumen, it is unlikely that any Republican could have unseated Roosevelt.[68]

V

The president left for a Caribbean fishing trip aboard the USS *Tuscaloosa* to celebrate his reelection. Of his key advisers, only Harry Hopkins was invited along. Roosevelt doodled sailfish and swordfish he dreamed of catching, and novelist Ernest Hemingway tried to help in the effort, radioing Roosevelt on the *Tuscaloosa* from his home outside Havana with tips about where the best fish could be found. While baiting rod-and-reel, FDR received an urgent letter from Churchill requesting American aid—in the form of war materiel—for British forces around the world.[69] The answer Roosevelt formulated would soon become the Lend-Lease Act.

Relieved that the campaign was over, Eleanor Roosevelt allied herself with the National Audubon Society to ban the use of feathers in fashion and to expose the millinery industry as a kind of mafia.[70] "More than thirty years before, they led the fight to stop the slaughter of wild birds for their plumage," Eleanor wrote in praise of the Audubon Society's efforts. "It appears, we ladies in those days used too many pretty feathers from wild birds on our hats and in other decorative ways. Now the National Audubon Society has conducted an investigation and finds that they must start a new campaign. They ask the women of the United States to help them. We ladies are guilty, of course. If we realized that we were stamping out so many beautiful wild birds and destroying the species for all time, we would not be very happy, no matter how becoming our headdress might be. But most of us buy such things with little thought as to what lies behind the product." She continued, "I hope, therefore, that the Audubon Society's crusade will be very successful, and that all of us who like to think we are well dressed, will shun the use of feathers obtained by killing wild birds. We should look askance at anyone who cannot say: 'I bought this before 1940,' and hope that if such a lady buys feathers of the banned variety we can at least say of her that fashions are against her."[71]

At the end of 1940, in mid-December, Roosevelt spoke glowingly about his impressive year in conservation and historic preservation. Approximately 750,000 acres were added to the National Park Service system including the 454,000-acre Kings Canyon National Park (California); Isle Royale National Park (Michigan); and over 200,000 acres of forestlands, hot springs, and waterfalls were added to Olympic National Park (Washington).[72] Mammoth Cave and Big Bend were pending. In terms of historical sites, Roosevelt protected Arizona's Tuzigoot archaeological site; the McLean House, where Robert E. Lee surrendered to Ulysses S. Grant at Appomattox Courthouse, Virginia; and the Custer Battlefield Cemetery in Montana, to name a few.

On January 6, 1941, Roosevelt delivered his State of the Union address, recommending lend-lease aid for Great Britain and enumerating the famous "Four Freedoms" (freedom of speech and worship, from want and fear). Not included on the list was Frederic Delano's "freedom to enjoy outdoors recreation," which Delano believed should be an American birthright. Because of New Dealers like Delano, Wirth, and Ickes, Americans had grown accustomed to having state parks and national parks available to enjoy—no matter where they lived.

On January 20, Roosevelt, happy and relaxed, delivered his third inaugural address with his eighty-six-year-old mother, Sara, in attendance. "Perhaps she moved a bit more slowly than she did in 1933," the *Washington Star* reported, "perhaps she relied a little more on her polished cane as she passed from room to room in the White House. However, she'd had a grand time throughout the inaugural weekend, and hers was an honor that had come to no other American mother."[73]

As CCC boys from the West proudly marched in the Inauguration Day parade down Pennsylvania Avenue, a very different event was unfolding for CCCers far to the West. At a camp near Morrison, Colorado, corpsmen had built the most wondrous open-air amphitheater in North America: Red Rocks. Located fifteen miles from Denver and six thousand feet above sea level in the Rockies, it was originally known as the Garden of the Gods (because of its otherworldly formations, which were embedded with dinosaur fossils from the Jurassic period). The amphitheater is surrounded by sandstone rocks taller than Niagara Falls. A kaleidoscope of different rust-red colors appeared in the rocks, depending on the weather. Working in tandem with the WPA, which built the parking lots, CCC crews blasted away tons of rock, fought off rattlesnakes, and succeeded in constructing the 9,525-seat theater.

The Red Rocks Amphitheater near Denver under construction in the late 1930s. A natural stage seemed to grow there from the outcropping of sandstone and conglomerate imbued with iron oxide. It looked more natural than it was, however, requiring six years of hard work by CCC workers to create Red Rocks as a viable open-air theater. Owned by the city of Denver, it opened on June 15, 1941.

Red Rocks Amphitheater is a monument to careful planning and high-quality construction, and the architecture blends elegantly into the natural environment. The master architect was Burnham Hoyt, who participated in the design of Radio City Music Hall and also designed many prominent public buildings in Denver.[74] The Red Rocks Amphitheater would officially open on June 15, 1941. At the dedication concert, all of the CCC boys reunited to take a bow at one of the greatest public works projects of the era. The Morrison CCC camp in Colorado was preserved intact to serve as a historical reminder of the rustic craftsmanship of the New Deal.

Another CCC unit that Roosevelt admired was CCC Company 1837 of Phoenix, Arizona. These CCCers, devoted to Roosevelt's conservation vision, planted more than 7.4 million trees and built

512,093 erosion-control check dams (low ridge, made of gravel to slow runoff). In Phoenix, the CCC helped create South Mountain Park, the world's largest city park, at 16,000 acres. Thousands of CCC enrollees constructed more than forty miles of hiking and equestrian trails, eighteen buildings, 734 fire pits, thirty water faucets, an erosion-control structure, several lookout shelters, and other outdoor recreation features.[75]

At the time of Roosevelt's third inaugural, 5 percent of America's total male population was then serving or had participated in the CCC nationwide. In Arizona, that number reached 20 percent: every month, it seemed, a new CCC camp opened there. Senator Carl Hayden, an Arizona Democrat, thought the agency was so beneficial that he arduously lobbied for it to become permanent.[76]

Even with the CCC successes in the Grand Canyon and Colorado Rockies, FDR knew that securing congressional appropriations for his pet project for a ninth year would be difficult.[77] After all, Congress was already pressuring him to abolish the Farm Security Administration (it would be officially closed in 1942). Making adjustments, Roosevelt had CCC companies working on military installations around America, building roads, clearing U.S. Army maneuvering areas, grading and draining landing fields, and constructing rifle ranges.[78] He was focusing on the war in a far grander way, as well, working on "lend-lease" as a means of sending supplies and materials to Britain. Congress passed FDR's Lend-Lease Bill and, starting March 11, 1941, the Arsenal of Democracy was officially in business.[79]

Ominous reports from Europe and the Pacific didn't diminish FDR's appetite for new national parks. On July 1, 1941, Kentucky's Mammoth Cave, the longest cave system in the world, entered the NPS portfolio. The behemoth cave's stalactites, stalagmites, gypsum flowers, and ribbonlike flowstone had been a top-draw tourist attraction for years. After the CCC built comfortable trails for tourists, attendance rose. With more than 390 miles of passages—and such features as Grand Avenue, Frozen Niagara, Jennie Lind's Armchair, and Fat Man's Misery—there was much for curiosity seekers, spelunkers, and amateur adventurers alike to explore. An Echo River tour gave visitors a boat ride along an underground river. It was mindboggling to contemplate that hundreds of miles of subterranean passageways had never even been explored. The provision for the park was authorized by Congress in 1926 on the understanding that 45,306 acres had to be secured—most from a private landowner—before development could begin. With the goal finally met, Mammoth Cave National Park opened on July 1. As a lagniappe, National Park status meant that the cave system's rare bats, northern cavefish, and albino shrimp would receive federal protection.[80]

Another place in Kentucky that President Roosevelt helped preserve was ornithologist John James Audubon's former home in Henderson, Kentucky. Built in the Norman style reminiscent of the great painter's childhood home in France, the Audubon Museum linked natural history to national history. In 1933, Roosevelt had encouraged Emma Guy Cromwell, president of Kentucky's burgeoning state park system, to build a museum to honor the life and legacy of Audubon, who had lived in Henderson for several years, starting in 1810. Kentucky soon acquired three hundred acres in the Wolf Hills, east of Henderson, for John James Audubon State Park. The widow of Audubon's great-grandson donated a trove of memorabilia to the New Deal–inspired project, including an original four-volume set of *The Birds of America*. By 1941 not only had the house been turned into the museum, but the CCC and WPA also erected stone picnic shelters and dug scenic Wilderness Lake. Roosevelt was proud that the CCC had likewise built museums at other state parks to help tell the history—human, geological, and biological—of countless areas around the United States.

In August 1941, six weeks after Mammoth Cave had been dedicated, Roosevelt met with Winston Churchill at the Atlantic Conference in Newfoundland's Placentia Bay. The Nazis had invaded the Soviet Union in June, and America was clearly heading toward war. Because the summit meeting was top secret, Roosevelt spent a day fishing in Nonquitt, Massachusetts, in full view of the press and then, undetected, boarded the USS *Augusta* for the cold waters off Newfoundland for the conference. After arriving at Placentia Bay, Roosevelt fished for halibut until Churchill arrived. "It is a really beautiful harbor," he wrote Daisy Suckley, "high mountains, deep water, & fjord-like arms of the sea."[81]

Winston Churchill had traveled to Newfoundland to convince Roosevelt to bring the United States actively into the war as an ally against Germany. Britain had been at war for two torturous years and was running out of steam. A sympathetic Roosevelt, however, wasn't prepared to make such a great leap, not with the isolationists in Congress threatening impeachment if he dared. And a minor dispute occurred with Churchill over Britain's colonial policy.[82] The postwar world Roosevelt envisioned was anti-colonial, one where big countries didn't use the war to grasp new territory and strip developing countries of their natural resources. Eventually Roosevelt and Churchill cobbled together a joint declaration, issued on August 14, promoting the ideal goals of winning World War II: no territorial aggrandizement; self-determination for all the people; and freedom of the seas. The Atlantic Charter—a declaration in the noble tradition of the U.S. Bill of Rights, guaranteeing the rights of *all* nations—was also a step in positioning America to join Great Britain in fighting the tyranny of Germany's Third Reich.

When Ding Darling heard about the Atlantic Charter—and the secretive circumstances in which the visionary document had been drafted—he drew a cartoon that perfectly encapsulated the modus operandi of America's angler in chief. It showed a huge shark (Germany) that had been hung upside down with a rope as Roosevelt and Churchill, smoking away, stood nearby with their fishing rods in hand.[83] With America still hobbled by isolationists hoping to avoid war with Hitler, the president, living up to his 1940 campaign pledge of neutrality, wasn't tickled by the cartoon.

The Atlantic Charter meeting had been conducted in such strict confidentiality that FDR sailed right past Campobello without stopping to see his ailing mother, who was summering on the island.[84] Early in September, feeling listless, Sara Roosevelt returned to Springwood. Startled by her mother-in-law's declining health, Eleanor called Franklin at the White House to urge him to come home; he did. Sara, unable to greet her son on the Springwood portico as she had planned, instead tied a blue ribbon in her hair and reclined on a chaise longue, waiting for her only child to arrive. When FDR entered the room, the president and his mother gossiped about the Hudson River valley and the world at large.

That evening, September 7, Sara slipped into a coma and died in the same bed in which she had given birth to Franklin. What happened next might have been a scene from a Hollywood movie: just moments after Sara stopped breathing, a thunderous boom swept across Springwood. The Secret Service thought a bomb had exploded. Quickly investigating, they discovered that the tallest deciduous tree at Springwood, an oak, had fallen. "Although geologists would later say that it was not unheard of, that the Hudson Valley had only a shallow layer of earth to support heavy trees, no one on the estate that afternoon doubted that the tumbling oak was a sign," historian Jan Pottker wrote. "Franklin left his mother's still body and went out to where the tree had fallen. He sat there staring at the roots torn from the earth and thick trunk heavy on the ground."[85]

CHAPTER NINETEEN

"THE ARMY MUST FIND A DIFFERENT NESTING PLACE!"

I

Trumpeter swans, native to North America, are the largest species of waterfowl in the world, and President Roosevelt wanted them permanently protected. With blinding white plumage and a vocalization like a French horn, trumpeters are the royalty of the wetlands. Five feet long, weighing twenty to thirty pounds, graceful in manner and majestic in flight, *Cygnus buccinator* were huge creatures with a wingspan averaging eight feet. Their sheer magnificence, sadly, worked against them. For trumpeters, easy to spot, were hunted for their meat and feathers, and annihilated to the brink of extinction. During the early Dust Bowl, ornithologists counted fewer than seventy trumpeters alive in all of North America.

Addressing the tragic situation, Roosevelt, like the good American Ornithological Union member he was, established Red Rocks Lake National Wildlife Refuge (in eastern Montana's Centennial Valley), as a trumpeter swan safe haven. Almost half of the world's trumpeters lived here because the hot springs provided year-round open waters. As a public service, Roosevelt also had the USDA distribute posters of trumpeters throughout the intermountain West with the government warning "EXTINCTION? THINK Before You Shoot!"[1] At the historic North American Wildlife Conference in 1936, an entire session had been devoted to trumpeter restoration.[2] Due to the Roosevelt administration's diligence, the number of wild trumpeters rose to over two hundred by 1941. The species had a fighting chance, but it was still a slim one, as biologists ramped up relocation and protection efforts in Ruby Lake, Nevada, and Malheur, Oregon.

In late November 1941, with World War II engulfing Europe, a bureaucratic tussle ensued between the U.S. Army and the Department of the Interior over the fate of a trumpeter wintering ground at Henry's Lake, Idaho. The president was predisposed to align himself with the conservationists. The U.S. Army's 10th Mountain Division had been conducting military exercises around Henry's Lake, only fifteen miles from Yellowstone. Secretary of War Stimson had ordered soldiers to train at the high-altitude base there, ideal terrain for ski jumps, cross-country runs, downhill skiing maneuvers, and slaloming in winter snowpack. The $20 million facility, on which construction started that October, would soon accommodate thirty-five thousand soldiers and encompass one hundred thousand acres of the Caribou-Targhee National Forest.

Enter the indomitable Rosalie Edge of the Emergency Conservation Committee (ECC), who insisted in a letter that the War Department relocate the entire army base at Henry's Lake.[3] She was outraged that the military facility, only a few miles east of the Red Rocks NWR, and in close proximity to Yellowstone Park, was on the swans' migration route between the two winter havens. If the 10th Mountain Division activated the proposed artillery range, the trumpeters would disappear. Nothing came of her dissent letters except a couple disheartening War Department courtesy replies. Instead of giving up, Edge lobbied higher up the food chain, reaching out directly to Ickes through Irving Brant. Once aware of the situation in Idaho, Ickes was fully on board with Edge to save the swans.

Word leaked out that the army might be on course to inadvertently exterminate the species. Spurred on by Ickes and Edge, the entire eastern establishment conservationist community grew outraged that the trumpeters were being disturbed by the army. With public relations in his favor, Ickes hit hard. "I beg of you that this Department be at least consulted before the Army takes unto itself anymore lands within the jurisdiction of this Department, especially if they are lands within national parks, national monuments, or wildlife refuges," Ickes wrote Roosevelt in late November. "It is utterly discouraging to have a body of men who don't care about the sort of thing that this Department is charged with fostering and protecting, who are marching in and taking possession just as Hitler marched in and took possession of the small democracies of Europe."[4]

Comparing anyone, let alone the Army's ski troopers, to Adolf Hitler was shocking. There was no worse epithet to hurl at a fellow American in late 1941, but Ickes was furious about the impending desecration of Henry's Lake. He demanded a War Department investigation aimed at halting the construction of the 10th Mountain base.[5] Irving Brant, meanwhile, explained to Stimson and General Emory S. Adams, adjutant general of the army, that Henry's Lake was "the solitary unprotected point" on the short flyway of the trumpeters between Yellowstone and Red Rock Lakes and therefore a key habitat for survival of the species.[6]

With militaristic Germany and Japan constantly in his sights, Roosevelt nevertheless found time to weigh in on the matter, decisively, and with humor. "Considering the size of the United States, I think that Irving Brant is correct," Roosevelt wrote to Stimson. "Please tell Major General Adams or whoever is in charge of this business that Henry Lake, Utah [*sic*], must be struck from the Army planning list for any purposes. The verdict is for the Trumpeter swan and against the Army. The Army must find a different nesting place!"[7]

Stimson didn't want the army's ski troopers to abandon Henry's Lake. Nor did he respect Ickes's penchant for igniting interdepartmental squabbles. That the relocation directive came from the commander in chief by way of a re-nesting joke probably didn't sit well with him, either. Nonetheless, Stimson followed orders, and found a suitable location for the 10th Mountain Division at Camp Hale in Colorado.[8] Gracefully conceding defeat over Henry's Lake, Major General Adams wrote to Edge that "appropriate steps [were] being taken to discontinue all planning activities in connection with that site."[9]

Roosevelt's verdict sent a broad message: protected public lands and species weren't open to the military, not even in the time of national emergency. In early December 1941, Roosevelt worried that munitions companies, gearing up for the onrush of wartime production, were using America's waterways for industrial dumping. FDR's instinct was to push for tighter regulation of extraction industries and chemical companies. To accomplish this, the president partnered with the Izaak Walton League—which had fought for a uniform national pollution policy since 1934. The League's goal was to prevent industrial waste and untreated sewage from indiscriminately being dumped into American waterways.[10]

This was no slight issue. The presence of raw sewage, toxic chemicals, oil, hospital refuse, acids from mines, cyanide, animal waste, and agricultural runoff was all too common in major rivers such as the Mississippi, Ohio, Susquehanna, and Missouri. Along the Colorado River, uranium was being mined haphazardly in late 1941 without proper federal oversight—although this was an activity that demanded careful regulation. The last of the Hudson River's famous oyster beds were dying. Boston Harbor, Newport News, and San Diego Bay stank of gasoline, fertilizer, and industrial by-products. Concerned about polluted waterways, Roosevelt wanted to establish an oversight authority for every principal watershed in the country, one that would have the power to fully enforce federal anti-pollution regulations. FDR felt strongly that Congress needed to appropriate funds for such an agency so that sewage-treatment facilities could be built in municipalities that were unable to finance construction on their own.

"I am sure you are aware of my continued interest in the protection of our streams and lakes from the harmful effects of undue amounts of polluting materials," Roosevelt wrote Kenneth Reid of the Izaak Walton League's Chicago chapter on December 3. "I appreciate the need for the application of corrective measures and to this end Federal assistance is being given to the construction of sewage treatment plants where added contributions of polluting material are resulting from the defense program."[11]

Four days later, on December 7, the Japanese attacked Pearl Harbor on Oahu, Hawaii. By the time the bombardment ended, after two hours and twenty minutes, more than 2,400 Americans were dead and 1,200 wounded. Nearly two hundred American aircraft were destroyed, and all eight of the Pacific Fleet's battleships were damaged or sunk. "They caught our ships like lame ducks!" Roosevelt fumed to William "Wild Bill" Donovan, chief of U.S. foreign intelligence. "Lame ducks, Bill."[12]

Speaking to Congress the following day, peering into the crowded chamber, Roosevelt explained, "Yesterday, December 7, 1941—a date which will live in infamy—the United States of America was suddenly and deliberately attacked" by Japan on the Hawaiian Islands and Guam, the Philippines, Wake Island, and Midway Island.[13] A few sentences later he asked Congress to declare war so that "this form of treachery shall never endanger us again." The president received a standing ovation. Both the House and Senate immediately voted in favor of war.[14]

When Hitler declared war on the United States four days later, Roosevelt was suddenly engaged in a two-theater conflagration: one against Japan (in the Pacific) and one against Germany and Italy (across Europe). Nevertheless, even after the Pearl Harbor attack, Roosevelt, battening down the hatches, remained vigilant about protecting America's forests, public lands, and wildlife.

One key Roosevelt land-conservation project, long in the bureaucratic pipeline, came to fruition only nine days after Pearl Harbor. On December 16, 1941, FDR issued Executive Order 8979, designating 1.92 million acres on the Kenai Peninsula—including the twenty-five-mile-long, six-mile-wide Tustumena Lake—as the Kenai National Moose Range.[15] Roosevelt's grand Alaska Territory sanctuary, similar in size to the Desert Game Range in Nevada, would protect "the natural breeding and feeding range of the giant Kenai moose (*Alces gigas*)."[16] The Kenai Peninsula, a biological crossroads, was teeming with bald eagles, elk, loons, and warblers, and more than twenty species of fish, including chinook, coho, sockeye, and pink salmon.[17] Roosevelt's order wasn't very popular in Alaska. "Get out! We don't allow any game wardens around here," became a popular anti–federal government refrain in the territory.[18] (In August 1944 a war-weary FDR would go fishing in Alaskan waters.)

In some ways, Roosevelt's conservation policies were deprioritized with America at war. National parks—like Olympic—would be used as a staging area for U.S. troops. Nevertheless, regulations in some situations may have been eased, but FDR's historic conservation gains weren't forsaken during the war years. And that was itself an accomplishment. To keep an eye on land gougers and petroleum profiteers, Roosevelt appointed Ickes petroleum coordinator for National

Defense in late May 1941 to lord over the production and distribution of oil reserves during World War II. This guaranteed that America's public lands wouldn't be assaulted by extraction companies; Ickes loved being the watchdog and prosecutor of despoilers.[19]

II

After a recuperative Christmas at Hyde Park, Franklin and Eleanor returned to the White House in early January 1942 for what they knew would be a difficult year. ER managed to retain her characteristic equanimity by taking solitary walks along the Tidal Basin and reflecting on America's storied past. "The longer I live here, the more the Washington Monument grows on me," the First Lady wrote after a walk in the rain. "It changes in color with the atmosphere and it is beautiful at all times. Yesterday evening, the tracery of the bare trees near it stood out against its white background. It had a misty soft outline, which was entirely different than the clear-cut look it had against a blue sky."[20]

On January 6, 1942, President Roosevelt delivered his first wartime State of the Union address, calling for national unity in the face of global fascism and for the mass mobilization of manufacturing to ensure the Allies' victory. He pushed for the production of sixty thousand planes, forty-five thousand tanks, twenty thousand antiaircraft guns, and six million deadweight tons of merchant ships. A War Production Board was founded later that month to supervise the distribution of critical wartime resources such as timber, ore, copper, tin, zinc, and gasoline. The pressure to open up public lands to satisfy war demand was intense. Roosevelt and Ickes were determined to continue their emphasis on long-term environmental stewardship, but Americans were thinking in terms of board feet, not new national parks.

In early 1942, victory over Germany and Japan seemed far away. But Roosevelt, unlike Churchill, had determined that global conservation was a casus belli of winning the war, an integral part of his Four Freedoms. Just days after Pearl Harbor, Eleanor Roosevelt had tried to remind Americans that long-term conservation values were part and parcel of democratic thinking. "I wish I could say that whenever I see magnificent trees cut down, I could also see plantations of new trees, but I have not noticed that as yet," Eleanor Roosevelt wrote. "One important lesson we still must learn is that we cannot ask anything which comes from our soil and not return something to the soil for the use of generations to come."[21]

Of particular concern to Ickes were the vulnerable forests of northern Michigan. In the war frenzy for airplanes, jeeps, tanks, and myriad other pieces of military equipment for the U.S. government, Detroit-based companies like Ford Motor and General Motors, he feared, would pillage the pristine Porcupine Mountains. The "Porkies" contained one of the last stands of virgin hardwoods in North America and were on the docket for national park designation as a kind of Great Lakes bookend with Isle Royale. The forests of the Porcupine Mountains—so named by the Ojibwa because the tree line resembled the quilled rodent—were replete with sugar maple (*Acer saccharum*), American basswood (*Tilia americana*), eastern hemlock (*Tsuga canadensis*), and yellow birch (*Betula alleghaniensis*). The ecosystem was also thought to be rich in copper. It was a cornucopia of natural wonders and resources. And Ickes knew it was under attack by industrialists. By 1942, trees there were being lumbered at the staggering rate of 100,000 board feet per day. The increase in logging production was facilitated by the completion of a railroad line through the most densely forested section of the Porcupines.

Because Ickes considered Chicago his hometown, protecting Great Lakes ecosystems was of keen interest. Reporting to him was Dr. John Van Oosten, the director of the Great Lakes Biological Laboratory and a preeminent fishery scientist.[22] Having persuaded Congress in 1940 to designate Isle Royale as a "biologically intact" national park, Ickes pined for another, in spite of the outbreak of war. Struck by the fierce beauty of the Porcupine Mountains, whose long basalt and conglomerate escarpments ran parallel to Lake Superior, he began to prepare for the fight over federal designation. "In these times few people have the vision to think of the future even if a small effort will result in mighty gains for latter generations," Ickes wrote to Roosevelt. "You have the vision . . . you are the only one who can do anything to save the wonderful Porcupine Mountain area."[23] With that, he recommended a new report from Brant, who had recently traveled to Michigan to gather information. "The lumbering railroad, just built, runs directly into the most beautiful area which will therefore be almost the first cut," Brant wrote Ickes. "In this region as a whole, the nature of the soil, loose sand, means that this kind of lumbering will produce a desert."[24]

Determined to save the primeval forest, Brant cleverly suggested that General Motors, which owned about one-fourth of the forestland in the Porkies, donate it to the federal government. "I think General Motors should be asked by the president to give this land to the government for a national park," Brant wrote to Ickes, "as a goodwill token return for what it is getting out of the government in defense contracts. It is valuable land. When a stand of hardwoods is so fine that it is worth being made into a national park it has high commercial value. But it is an invisible drop in the bucket of General Motors' war supply profits."[25] Although Roosevelt allowed Ickes to float the Porkies proposal with General Motors CEO Alfred P. Sloan Jr., the company, claiming that it did not want to set a precedent, declined to cede the acreage.

Meanwhile, another strategy reflected by a bill in Congress, involved having the Reconstruction Finance Corporation lend $30 million to the USDA to make the Porcupine Mountains a national forest. It stalled, in part, because conservationists weren't comfortable with a provision calling for the loan to be repaid through revenues derived from lumber contracts. Eventually the president found a solution that was more than acceptable: working with the Michigan legislature in Lansing to establish the Porcupine Mountain Wilderness State Park. Thanks to FDR's intervention, the "last area of virgin northern hardwoods in the country" was saved for posterity. Just as Ickes and Brant had predicted, the Porcupine Mountains became a popular postwar tourist destination, visited by outdoors enthusiasts hoping to encounter and photograph gray wolves, wolverines, and peregrine falcons.[26]

Conservative Democrats and Republicans in Congress tried in 1942 to "frustrate" what remained of Roosevelt's conservation agenda and "even dismantle earlier achievements."[27] Legislators defunded the National Resources Planning Board, whose chair had been Frederic Delano, and moved headstrong to abolish both the CCC and the National Youth Administration because of the war. The two programs together cost $400 million per year. It no longer made sense, to many on Capitol Hill, with global democracy on the line, to have CCC men in fighting trim build a giant red rock amphitheater in Colorado, dig a reservoir in North Dakota, or save a manatee herd in Florida. Roosevelt tried to remind Congress that his work-relief agencies did more than just plant trees and save wildlife; they had trained over two million young men to become effective soldiers. "I cannot agree with those who take the position that these agencies should be terminated," Roosevelt said in March. "I feel the youth agencies have a definite place in the all-out war effort."[28]

Determined to keep the CCC funded, Roosevelt had his U.S. Army generals defend the work-relief achievements in national parks and monument areas. Douglas MacArthur, in his annual report to the secretary of war in 1933, had argued that the army's involvement in the CCC provided "renewed evidence of the value of systematic preparation for emergency, including the maintenance of trained personnel and suitable supplies and the development of plans and policies applicable to a mobilization."[29] Omar Bradley soon declared that if not for the CCC boys, the army would have been critically short of cooks, truck drivers, and machine operators after Pearl Harbor. Bradley was grateful for the training those thousands and thousands of former CCCers had received prior to enlistment—it prepared them with the discipline needed to fight Hitler and Tojo.[30] *Time* magazine declared that "more continuously" than any other New Deal project, the CCC had "respect of the foes as well as friends of Franklin Roosevelt."[31] Humphrey Bogart, the actor, was even enlisted to praise Roosevelt's pet program, describing the agency as "a 14-karat opportunity for young men."[32]

Nevertheless, Congress wanted to defund FDR's Tree Army. With America at war, there really was no choice.[33] Roosevelt didn't have the votes to stop it. "There was no doubt then that we were through with the CCC," Conrad Wirth of the Department of the Interior remembered of the days following Pearl Harbor, "it was no longer a question of reorganizing it, but rather of disbanding it."[34] In late March 1942 the *Washington Post* reported that the total number of CCC camps would be reduced from fifteen hundred the previous year to only six hundred. And most non-defense work by the CCC was ordered ended.[35] A few New Dealers kept up the fight, though. Out of desperation, James J. McEntee, director of the CCC, tried to persuade the public that his agency was a "physical conditioner for youths."[36] Likewise, Ickes claimed that the CCC had directly assisted the U.S. military by helping build convalescence centers for soldiers on army reservations and erecting communication lines essential for military operations.[37] The cause was slipping, though.

To justify keeping the CCC alive, Roosevelt needed a dramatic alibi to shop on Capitol Hill. Senator Elbert Thomas of Utah, a sympathizer, had written the White House a cogent letter about his fear that Japanese arsonists would torch the West Coast's wild forests. The threat of sabotage, Thomas believed, was FDR's best gambit. "You are dead right about the danger of forest fires on the Pacific Coast," the president replied to Thomas. "It is obvious that many of them will be deliberately set on fire if the Japs attack there—and even if they do not there I don't think people realize that both the CCC and NYA are doing essential war work and that if they are abolished some machinery will have to be started to take their places—and probably at a net increased cost."[38]

Longtime enemies of the New Deal jumped at the opportunity to publicly ridicule the CCC. Governor Leon Phillips of Oklahoma, a Democrat, testified before the Senate Labor Committee that the CCC was "poison to our boys," teaching the art of laziness. No wording was too strong for Phillips, who charged that a "great majority" of first-time prisoners at Oklahoma's state reformatory were ex-CCCers. McEntee quickly pounced on Phillips's "dastardly insult to the young people of Oklahoma" as libelous.[39]

The *Washington Post* reported that Roosevelt was furious because his CCC boys were being defamed.[40] After years of amazing press, his domestic pet project was now taking hits. A common refrain in Washington was that the CCC was now a *waste* of money and manpower. Bad press ensued. One story told of a fictitious "ghost" CCC camp that a clerk from the Department of the Interior had kept "running" for four years with bogus records he "kept in perfect order," faking promotions, sick leaves, and disciplinary orders.[41] Another unpleasant story was that CCC officials burned equipment because of vermin infestations in the camps. Representative Edward Creal, a Democrat from Kentucky, defended the CCC in this case, stating,

"There is a lot of vermin in this war program, and a lot in the departments, and there is a lot of vermin in the arguments on this floor, and if you would strike out all the vermin that is in the *Congressional Record* here, then, instead of looking like a Sears, Roebuck catalog, the *Congressional Record* would look like a postcard, and we would be able to effect a good deal of economy."[42]

III

On February 19, 1942, Roosevelt signed Executive Order 9066, which authorized the internment of more than 120,000 Japanese Americans, 62 percent of whom were American born. Ickes was outraged by Roosevelt's incarceration of his fellow citizens, calling it un-American. To drive his dissent home, Ickes had four Japanese men and three women transferred from a relocation camp in Arizona to his home in Olney, Maryland, to help raise chickens. (It was legal for Japanese Americans to live at liberty outside of designated military areas.) "I do not like the idea of loyal citizens," Ickes told the *New York Times*, "whatever their race or color being kept in relocation centers."[43] The government also herded up German and Italian nationals, though that was to be expected in time of war. It didn't put German Americans or Italian Americans behind fences, except in a very few cases. By contrast, Japanese Americans were deemed "dangerous" to the public peace and safety of the United States.[44]

Roosevelt's decision to round up American citizens was a flagrant violation of human rights and morally reprehensible: a product of American prejudice toward the Japanese following Pearl Harbor. The fear of Japanese sabotage on American soil was widespread. Four days after Roosevelt issued EO 9066, however, the Japanese navy indeed bombed the town of Goleta, California, near Santa Barbara, firing on the Ellwood Oil Field from a submarine in the Pacific. While little damage was inflicted—an oil well was temporarily decommissioned and only an orange grove caught fire—the attack raised alarms throughout the West.[45] If Japan could shell Goleta, then wasn't the entire Pacific Coast at risk? The Los Padres National Forest, approximately one hundred miles from Los Angeles, seemed especially vulnerable to a similar attack. "[We] must guard against Japanese incendiary bombs and incendiary fires during the dry season," the president wrote to the head of the Bureau of the Budget. "This is essential for our national future."[46]

The president, following the attack on Goleta, ordered Los Padres closed to the public as a preventative measure against potential arsonists or even slovenly campers. An arsonist's match could threaten the Los Angeles Basin or San Diego. Ironically, closing Los Padres would have a side benefit for an imperiled flock of sixty to eighty California condors (*Gymnogyps californianus*), a symbol of the state's wilderness. This endangered species, which had telescopic vision and a wingspan of nearly ten feet, had been making a last stand along the forest's boundaries. By closing Los Padres to protect it from arsonists, Roosevelt was also helping this charismatic species survive. He knew that the condor—especially when nesting—was easily disturbed by machinery, campfires, and other types of human interference. (He had previously curtailed logging in the forest so as not to frighten the condors.)[47] Wildlife biologists had two unique difficulties in trying to protect the condors circa 1942: the species took five years to reach sexual maturity and females laid only one egg per nesting cycle. The condor—like the whooping cranes of Gulf Texas and trumpeter swans around Yellowstone—would survive only if the Roosevelt administration oversaw its protection in an unprecedented way.

Reduced to begging for congressional appropriations from opponents of the New Deal, Roosevelt focused on the CCC and invited Senator Kenneth Douglas McKellar of Tennessee, a Democrat, to the White House for lunch. McKellar had been friendly to the New Deal earlier in the 1930s, but he stiffened into an incorrigible obstacle to the continuance of its signature programs. His opposition to the CCC may have stemmed from a reasonable concern about expenditures in time of war, though it was also said that McKellar, and some of his colleagues, were racist segregationists who could not abide the inclusion of African Americans in the CCC. They were especially disgusted by the evidence that it—and other New Deal programs—helped to pull black workers into the mainstream of American employment. Turning on his famous charm, Roosevelt argued that the pending legislation to defund the CCC should be scrapped for the sake of national security.

However, the threat of Japanese arsonists wasn't strong enough to dissuade McKellar, whose bill (Senate Bill S. 2295, introduced February 23, 1942) gained momentum on Capitol Hill. Grasping at straws, FDR offered to integrate the National Youth Administration into the CCC as a sort of supra-National Guard. McKellar wasn't biting. The advent of World War II, he argued, made New Deal conservation irrelevant.

On May 4, 1942, Roosevelt made a last-ditch plea to retain at least a skeleton network of CCC camps through the fiscal year 1942–43. The House Committee on Appropriations denied his request. A few weeks later, FDR instructed his budget director, Harold Smith, to make clear to the Senate that defunding the CCC wouldn't save taxpayers a nickel. According to Smith, the CCC currently had 158 camps on military reservations, 42 more camps doing similar work, and 150 camps tasked with protection against forest fires; the administration calculated that the CCC cost taxpayers $80 million annually. Abolishing it and hiring other laborers to do its work would cost $125 million, wasting $45 million of taxpayer revenue. "The elimination of the CCC will call for a wholly separate appropriation to take its place in two of the CCC activities," a defiant Roosevelt explained in a letter to Smith. "The first is the need for forest fire protection, especially on the Pacific Coast and back as far as the Rockies, where we must guard against Japanese incendiary bombs and incendiary fires during the dry season. This is essential for our national future. The second is the building of roads and other facilities for [military] camps. These have to be built by someone and I shall have to ask for a special appropriation to let the work out by contract instead of having it done by the CCC."[48]

On June 5 the House voted to defund the CCC, but approved $500,000 to liquidate the program properly. The jig was up. Virginia's Camp Roosevelt, the very first CCC camp in a national forest, began dismantling itself, trucking out iron beds, cooking utensils, encyclopedias, and Ping-Pong tables.[49] Thousands of CCC boys, already comfortable in uniform, were encouraged to enlist in the U.S. Armed Forces to help fight World War II. Only a few hundred CCCers found ways to continue working in national and state parks. Just two weeks after FDR's final appeal to the federal budget director, the CCC had only 136 enrollees left.[50]

Supreme Court Justice William O. Douglas tried to rescue the president, suggesting that the West Coast develop "civilian irregulars" and "wilderness patrols" under federal military supervision to patrol forests to keep an eye out for wildfires.[51] Douglas's security-meets-conservation program, however, couldn't get funded. But Roosevelt did launch an effective public awareness campaign against wildfires. The U.S. Forest Service adopted fire prevention slogans like "Don't Aid the Enemy" and "Our Carelessness, Their Secret Weapon."[52] The president soon brought his fire protection message to the nation in an impassioned radio address. "Uncontrolled fire, even in normal times, is a national menace," he proclaimed. "Today, when every machine is being taxed to its fullest productive capacity . . . when agents of our enemies are seeking to hinder us by every possible means, it is essential that destructive fire be brought under stricter control in order that victory may be achieved at the earliest date."[53]

Never one to shy away from a new federal agency, FDR established the War Advertising Council (WAC) to muster the know-how of Madison Avenue firms for the war effort. Roosevelt cleverly enlisted the council to mount a nationwide fire prevention campaign, which would center on the slogan: "Careless Matches Aid the Axis—Prevent Forest Fires." Not only did Secretary of Agriculture Claude Wickard take to the radio to publicize the slogan, but the phrase also appeared in twelve million mail inserts, two million leaflets, and more than fifteen thousand billboards. On a number of these ads a heinous (and racist) caricature of General Hideki Tojo, the Japanese prime minister, could be seen with a match in hand, ready to burn a national forest to the ground. Another poster depicted Tojo and Hitler floating around a wind-whipped blaze under the banner: "Our Carelessness: Their Secret Weapon."[54]

On August 11, 1942, the last CCC boys, eighty-two members of Veterans' Company 3822 in Goliad, Texas—a camp the first lady had once visited—were dismissed.[55] A phenomenal era in conservation had ended. From 1933 to 1942, the CCC had enrolled more than 3.4 million men to work in thousands of camps across America. Roosevelt had used the CCC as an instrument for both environmentalism and economic revitalization. Its erosion-control programs alone benefited forty million acres of farmland. The success the agency had in building up American infrastructure is impossible to deny: forty-six thousand bridges; twenty-seven thousand miles of fencing; ten thousand miles of roads and trails; five thousand miles of water-supply lines; and three thousand fire-lookout towers. Credited with establishing 711 state parks, the CCC also restored close to four thousand historic structures and rehabilitated 3,400 beaches.[56]

Nobody could deny the CCC's enduring legacy from 1933 to 1942: combating deforestation, dust storms, overhunting, water pollution, and flooding. In this way, the New Deal conservation revolution had already made a difference. Even while American troops were fighting in Europe and the Pacific, back home American lands brimmed with native grasses and cottonwoods, desert oases and high-country evergreens. The American land was healing and, in some regions, thriving. Around three *billion* trees had been planted by "the boys." The CCC was the single best land-rehabilitation idea ever adopted by a U.S. president and it rescued more than natural resources. As the *Alaskan*, a CCC publication in Juneau, praised the president's conservation force for having taken "baffled, furtive, tough, city youngsters in and transforming them into bronzed, clear-eyed, well-muscled soldiers in waiting."[57]

In its nine years of existence, the CCC introduced young American men to the rigors of outdoors living. By and large they had comported themselves well. It wasn't intended as a form of military preparation, as was the Hitler Youth in Germany, but a generation of toughened CCC enrollees indeed became a wave of GIs during the war. Pick any CCC company roster from 1933 to 1942, and you will find alumni who went on to win Purple Hearts, Bronze Stars, and Silver Stars during World War II. Approximately one out of every six men drafted to fight during World War II was an alumnus of the CCC.[58] Sadly, there was also a long list of heroic CCC alumni who were killed in action at Midway, Okinawa, Normandy, and Luzon, among other far-flung locales. As Major General James A. Ulio later explained, "Hardly a large unit in the Army is without key men who owe their important assignments either in whole or in part to the training they had in the CCC."[59]

On disbandment, many CCCers joined the industrial mobilization effort at home in such cities as Detroit, New Orleans, and Norfolk.

Take Ethric Brown of Iowa, who left the CCC to help build 1,500 B-26 bombers. Men like Brown who enlisted in the CCC were just as much a part of the "greatest generation" as the U.S. Army's Second Rangers, who stormed the cliffs of Pointe du Hoc, and James Doolittle, who took the air war to the Japanese mainland.

Once defunded, CCC camps closed with little fanfare. Equipment and facilities were used by the U.S. Army for maneuvers or special projects. Other camps were disassembled, leaving only foundations and forlorn remnants from their boom days. The temporary nature of camp structures such as barracks and mess halls meant that limited CCC camp "architecture" survived. Yet their infrastructure work in public forests and parklands remained as proud monuments to American can-doism, intrinsic parts of the U.S. national landscape heritage. A few CCC camps—including one in the Great Smokies—housed conscientious objectors in World War II. They engaged in a wide spectrum of conservation work.

Strangely, a few of the shuttered CCC camps were repurposed as POW internment facilities, including those in Kennedy, Texas; Kooskia, Idaho; Fort Lincoln, North Dakota; Fort Missoula, Montana; and Fort Stanton, New Mexico. Instead of tree planting and fish stocking, Japanese and German prisoners worked in slaughterhouses, constructed highways, broke rock, and fought forest fires. No longer was conservation education the prevailing sentiment in these camps; all was drudgery.

There was something heartbreaking about seeing CCC camps that were once on the cutting edge of a new environmentalism consigned to prisons encircled by barbed-wire fences.[60] The Fort Hunt, Virginia, camp became a top-secret POW center known as P.O. Box 1142. A number of prominent members of the Third Reich were questioned there, including rocket-technology pioneer Wernher von Braun, engineer Heinz Schlicke, and intelligence officer Reinhard Gehlen. German scientists spilled secrets about rocket technology on the very same land that George Washington had once farmed and where FDR's Tree Army had planted oaks.[61]

The CCC wasn't the only New Deal relief agency to lose its funding as a result of World War II. The WPA also folded in 1942 after having created 8.5 million jobs across nearly 1.4 million projects. The agency's impressive list of accomplishments included building, repairing, or refurbishing more than 600,000 miles of roads and streets; more than 120,000 public buildings (including schools and courthouses); and about 80,000 recreational facilities (including parks, pools, playgrounds, gyms, golf courses, ski areas, and skating rinks). San Antonio's River Walk and the presidential retreat Shangri-La (Camp David) also owe their existence to the WPA. The WPA's contributions to the cultural infrastructure of America include putting on more than 200,000 concerts and producing approximately 475,000 works of art.[62]

With the CCC and WPA gone, the president turned instinctively to the Boy Scouts of America, whose campaign "Every Scout to Save a Soldier" was in the midst of raising $355 million from bond sales for the war effort.[63] Roosevelt anointed the Boy Scouts as "messengers of the United States Government" and had them distribute millions of educational posters in towns across America. The president also tasked the Scouts with collecting scrap metal and cans—409.3 million pounds—for recycling. On a smaller scale, the Scouts also picked up where the CCC left off in forestry. Between 1942 and 1945, the Scouts planted almost two million trees.[64]

No one in Congress complained about Roosevelt's leaning on the nonpartisan Boy Scouts to get vital conservation work done. When the USDA needed able bodies for the 1943 harvest season, more than 704,000 Boy Scouts volunteered to help keep American agriculture strong. The organization was also responsible for planting an impressive 204,000 victory gardens.[65]

The victory gardens, or "food gardens for defense," were a clever way for the White House to encourage backyard farming and ease the pain of rationing food. When it came to winning a war, food was just as important as planes, ammunition, and tanks. Transportation and labor shortages, as well as the need for food for the military, induced FDR to urge citizens to grow their own produce. Approximately 20 million Americans, many of whom had never farmed before, answered his call. World records for food production were set in 1940 (and then beaten in 1941, 1942, and again in 1943). Leaflets filled with helpful growing and canning tips were disseminated free of charge.[66]

FDR wasn't the first to advocate for planting victory gardens. During World War I, President Wilson had launched a small but similar agricultural initiative. But FDR was a far better salesman. Franklin and Eleanor planted victory gardens at the White House, Hyde Park, on a vacant lot in New York City's Riverside Park, and in San Francisco's Golden Gate Park. Thrilled that America was breaking food-production records during his third term, the president called for the growing of corn on community plots. On a single-acre test plot, a family could produce more than a dozen bushels of corn by using two hundred pounds of superphosphate fertilizer. For city dwellers with few farming options, vacant industrial properties became fertile ground for tomato and lettuce patches. Using the slogan "Victory Gardens Are War Work," FDR spoke regularly about what it meant to grow fruit at home, make pickles and relishes from homegrown produce, ward off insects, store food for winter, and discover disease-resistant vegetables. The Roosevelts even promoted rooftop vegetable gardens.[67]

IV

By 1942 Roosevelt no longer had the luxury of visiting far-flung places like Hawaii Volcanoes, Glacier, or even the Great Smokies. All he could do was approve the last round of WPA silk-screen promotional posters, which became iconic, extolling America's national parks. His idea of a nature getaway during wartime was driving from Hyde Park to Red Hook on the Hudson, where he could look out with his binoculars at nearby Cruger Island, counting raptors. Driving his specially equipped automobiles, with hand controls instead of pedals, was liberating for him. Going down a steep embankment straight toward the Hudson like a daredevil gave him a vicarious thrill. One afternoon Franklin took Eleanor and Daisy to Red Hook to watch the bird life in memory of Maunsell Crosby. In a telling journal entry, Suckley reported that they "watched thousands of birds collecting for the night" along the railroad line. "The president is awfully interested, birds being one of his many hobbies," Suckley wrote. "Mrs. Roosevelt was frankly not specially interested."[68]

Once during the war, FDR had a lengthy talk with William D. Hassett, his correspondence secretary, about his long-standing admiration of essayist John Burroughs. As a boy, Roosevelt had been enthralled with Burroughs's book *Wake-Robin*. A winter walk across the Poughkeepsie-Highland Railroad Bridge brought him to the West Park front door of the "Sage of Slabsides." But in the 1940s, with the world spinning out of control, Roosevelt noted that Burroughs's "leisurely, discursive style" had seemingly fallen out of "taste."[69]

All three members of FDR's 1934 Committee on Wildlife Restoration—Darling, Leopold, and Beck—fought to protect the natural world during World War II. In 1942, Darling began negotiations with the Fish and Wildlife Service to limit development on Sanibel Island (where Theodore Roosevelt liked to spearfish). This barrier island off the southwest coast of Florida was known for its white-sand beaches, hardwood hammocks, and estuarine bays. The shallow gulf waters around Sanibel, the "seashell capital of America," made for excellent fishing and beachcombing. The island virtually screamed for federal protection. Fortunately, the Roosevelt administration had a vocal champion in the retired Darling. He was not only the former head of the U.S. Biological Survey (reorganized as the Fish and Wildlife Service), but he also owned a winter home on Captiva Island, just north of Sanibel. Awed by the many bird species that visited the mudflats on the islands from mid-October to April, Darling sought a new wildlife refuge and increased FWS protection for an assortment of egrets, herons, storks, and roseate spoonbills.[70]

By 1942, photographs replaced cartoons in the conservationists' call for action, in part because Ansel Adams's work during the Kings Canyon campaign had proved so vitally effective. Darling sent prints of Sanibel's famed seashells and roseate spoonbills to Ickes and Gabrielson. These photographs did the trick. Thanks to Darling's persistence, Sanibel National Wildlife Refuge (later enlarged and renamed the J. N. "Ding" Darling National Wildlife Refuge) was established in 1945 to protect one watery slice of Sanibel's wetlands, beaches, shell, silt, dry ridges, sloughs, and mangrove trees. Approximately 2,800 acres of the subtropical refuge were later dedicated as a wilderness area by Congress.[71]

During World War II, a consortium of farmers in the Midwest were bitter because federally protected ducks and geese had destroyed their crops. A sympathetic Roosevelt allowed Gabrielson to issue a slew of individual permits to help thin the flocks that were menacing crops. On the whole, the president viewed what he called "duck complaints" from midwesterners as a backhanded compliment; their gripes meant that his National Wildlife Refuge system had been successful in restoring populations. To the consternation of Rosalie Edge, he even took the woodcock (*Scolopax minor*) off the endangered species list.[72] Stories of waterfowl abundance were always music to Roosevelt's ears.

In October 1942, 700,000 mallards congregated in northeastern Colorado, devouring unharvested grain due to a wartime shortage of farm labor, principally in the cornfields bordering the Platte River. Due to the skyrocketing number of ducks, FDR instructed Secretary of Agriculture Wickard to issue a "community permit" authorizing any properly licensed hunter living along the Platte River to shoot mallards in the fields at all hours of the day during the period of enforcement. "This first community permit seemed to work quite well," Assistant Director of Fish and Wildlife Albert Day recalled. "At the time few birds per hunter were killed and the dispersal caused by the taking of a hunting sabbatical afforded considerable relief from damage."[73]

Roosevelt took a risk, amending his strict waterfowl protection policy of 1934 to suit particular circumstances in the Midwest and Colorado in 1942. Creating exceptions to the rules agonized FWS biologists. And watering down the law didn't make game wardens' jobs any easier. The FWS scientists had a point; Coloradans living along the Platte River soon begged for yet another round of special permits because the mallard population had swollen from 700,000 to 815,000. Unable to find migrant labor to harvest their crops, 1,140 Colorado farmers were provided exemption permits. The USDA allowed each farmer to kill nine ducks. There is no documentary evidence that anyone tried to exceed the bag limit. Word of FDR's generous "community permits" quickly spread throughout the West. Citizens in Washington State soon demanded special permits to shoot widgeons that were causing damage to the vegetable seed industry of the Skagit Flats near Puget Sound; Roosevelt allowed the permits to be issued.

Yet hunting and fishing licenses fell off in the seasons of 1942 and 1943 due to shortages of ammunition and tackle. Restrictions on gas and tires made it more difficult for sportsmen to visit remote places. But due to Duck Stamp and Pittman-Robertson revenues, New Deal fishing hatcheries and wildlife preserves flourished during World War II.

V

Throughout World War II, President Roosevelt's whereabouts were often shrouded in secrecy. A sizable security detail accompanied him everywhere. Members of the press who were assigned to the FDR beat filed their stories under a voluntary censorship agreement. The hideaway that Roosevelt frequented most was the former Recreation Demonstration Area (RDA) in the Catoctin Mountains of Maryland, named "Shangri-La" by Roosevelt—an allusion to the fictional monastery in James Hilton's 1933 novel *Lost Horizon.* It is known today as Camp David (after Dwight Eisenhower's grandson).

Roosevelt *(right)* took Prime Minister Winston Churchill of Britain on a fishing foray during their wartime conference at Shangri-La (since then renamed Camp David) in the Maryland mountains in May 1943. As Churchill wrote of FDR and their outing in his memoirs, "No fish were caught, but he seemed to enjoy it very much, and was in great spirits for the rest of the day."

How the president settled on the name Shangri-La is part of wartime lore. On April 21, 1942, Roosevelt was aglow with good news for the press. Lieutenant Colonel Jimmy Doolittle, a former stunt flier with a science degree from MIT, had led a bombing raid on Tokyo. The invasion was of questionable strategic importance, but it boosted morale. When asked from where Doolittle's planes had originated, a smiling FDR answered, "Yes, I think the time has come now to tell you. They came from our secret base at Shangri-La." (Actually, Doolittle and his fellow pilots took off from an aircraft carrier in the Pacific.)

The next day, escaping the notice of the press, the president slipped out of Washington to his mountain retreat at Shangri-La.[74] The Secret Service refused to let Roosevelt unwind at sea any longer (Germany sank more than 130 ships in the Atlantic during the first months of the war). In fact, his agents didn't want him on any boat. Roosevelt pleaded to take the presidential yacht on the Potomac River for starlight cruises past Mount Vernon, but the Secret Service remained firm. So Roosevelt opted for a secret retreat, sixty miles north of Washington in the hills of Maryland, as his weekend White House.

Ever since Ickes purchased his Head Waters farm in Olney, Maryland, Roosevelt thought about carving out a retreat of his own in the Old Line State. With a flash of enthusiasm he directed aides to scout the Catoctins in Frederick County for available land. Ickes thought that the president would particularly enjoy a CCC-improved RDA lot then known as Camp Hi-Catoctin, which was at an elevation of 1,800 feet. When FDR arrived at the site, a workman named William Renner helped him out of the car. Roosevelt took in the twilight vista, smiled, and pronounced to Renner, "This is a Shangri-La!"[75]

Roosevelt had looked for—and found—secluded sites before, like the Little White House in Warm Springs and Top Cottage in Hyde Park. Ever an eager rustic architect, FDR penciled designs on the NPS's cost estimate sheet. As was Roosevelt's wont, to construct the "Bear's Den"—large quarters with four bedrooms and two bathrooms—he specified local stone and timber be used as the primary building materials. One afternoon, Roosevelt told Conrad Wirth that he wanted a screened porch and terrace added to the "Bear's Den."[76] Walkways through the forest were also established. FDR demanded that the dogwood trees, azaleas, spicewood, hazelnut, witch hazel, and other wild shrubs remain, "just as God made them."[77]

Reshaping the old camp was great fun to Roosevelt. Within the old CCC barracks on the property, which he renamed "221B Baker Street" after the address of Sherlock Holmes, Roosevelt joked that the Shangri-La compound was his newfangled, land-locked yacht. In early July, the renovated buildings, including guest quarters, were completed at a cost of $130,000. Theatrically, Roosevelt inspected Shangri-La as if he were a naval officer christening a new ship. His first journal entry revealed his puckish humor; he wrote on the thick, cream-colored entry book for the compound, "U.S.S. Shangri-La—Launched at Catoctin July 5, 1942" before signing his name.[78]

Roosevelt's "trial run" at Shangri-La lasted from July 18 to 20. Enjoying the cool morning air, he worked on his stamp collection, did a little reading, and made some pressing phone calls. Early overnight guests included "Wild Bill" Donovan and Jim Byrnes of the Supreme Court. Although Roosevelt was disappointed not to be in Hyde Park or on the *Potomac*, he took quite nicely to this woodsy Maryland retreat, far removed from Washington. Top-secret intelligence reports were often read here. It wasn't a coincidence that the Office of Strategic Services was running a counterintelligence training center just down the road from the president's wartime retreat at Chopawamsic RDA (now Prince William Forest Park). For all of Roosevelt's humor, his hidden mountain camp was the backdrop for major decisions throughout the remainder of Roosevelt's administration.

Social life at Shangri-La was enlivened by guests such as Archibald and Ada MacLeish, Sam Rosenman, and Daisy Suckley.[79] The ineffable Winston Churchill considered Shangri-La "in principle a log cabin, with all modern improvements."[80] William O. Douglas remembered that his Catoctin visits to Shangri-La were always accompanied by the ritualistic shaking of dry martinis. "I had a rather lonely time at Shangri-La because FDR was holed up doing homework on endless problems—he would read, then doze, then dictate, then read, then doze," Douglas recalled. "I was company-in-waiting, ready to mix his favorite cocktail or to join him in idle chitchat."[81]

Everything at the mountain hideaway bore evidence of the president's distinctive personal touch. There was even a dress code: casual, no ties. Knowing that his secretary, Grace Tully, enjoyed a nightcap and card games, the president had a sign posted that read, visitors will beware of gamblers (especially female) on this ship.[82] Also on the walls were some of his favorite nautical prints. A telescope on a tripod was installed so that the president could gaze at constellations when the sky was clear. Guidebooks about the birds of Maryland were always within easy reach. Fireplaces were added to each bedroom, as the lodge proved itself and remained in use, even during the cold winter months. A chef was brought in to prepare popular Maryland dishes such as blue crab cakes, she-crab soup, and peach cobbler.

While the president didn't like being trailed by the Secret Service, he enjoyed the cloak-and-dagger procedures required to maintain secrecy. Throughout World War II Roosevelt and Harry Hopkins learned to talk in code in case their conversations were bugged. So at Shangri-La, Roosevelt spoke in a cryptic code that was nearly indecipherable. At the president's request, key people were assigned code names derived from figures at Hyde Park. Winston Churchill was referred to as "Moses Smith"; George Marshall as "Plog" (the longtime Springwood groundskeeper); Dwight Eisenhower as "Kueren" (a worker on the estate under Plog); Sir Stafford Cripps as "Mrs. Johannesen" (a weaver at Val-Kill); General Carl Spaatz as "Depew" (Sara Roosevelt's chauffeur); General Mark Clark as "Robert" (after Robert McGaughey, the Springwood butler); and so on. Interestingly, FDR chose code names from only people on his payroll, not from any of his Hyde Park neighbors.[83]

VI

Bird-watching remained a Roosevelt hobby during World War II. Daisy Suckley had been a member of the Rhinebeck Bird Club during her formative years living on the Wildenstein estate. In that spring of 1942, Suckley asked the president to participate in the Dutchess County May Census of birds. Since Maunsell Crosby conducted the first one in 1919, FDR had retained a "keen interest" in the annual bird-counting tradition.[84] Many species of birds FDR shot in the 1890s for his Springwood collection—sparrows, hawks, jays—were still thriving at Thompson Pond, only twenty miles from Hyde Park.

FDR decided that the pond, which had a higher concentration of bird species in its seventy-five acres than could be found in some of the larger western national parks, would be the perfect locale for a bird-counting collaborative. He and Suckley were accompanied by ornithologist Ludlow Griscom, the research curator at Harvard's Museum of Comparative Zoology, who had traveled with Crosby on an ornithological expedition to Panama in 1927.[85]

On May 10, 1942, an eager Roosevelt awoke at 3 a.m. for his outing with Griscom, Suckley, and a few others. After a quick breakfast at Springwood, the president grabbed his "bins"—birder lingo for binoculars—and ordered the top of his navy-blue phaeton rolled down for the predawn drive to Thompson Pond. Secret Service agents assigned to the president weren't happy about the outing but didn't dare complain aloud. Although the agents liked the president immensely, they were often exasperated by his childish antics, especially his fondness for playing "vehicular hide-and-seek" with his security detail. Time after time when driving around Hyde Park, he

FDR left his house in Hyde Park before dawn on the morning of May 10, 1942, for a "birding party." Those in the group were old friends and fellow bird-lovers. *Left to right*: Raymond Guernsey, Allen Frost, Margaret "Daisy" Suckley, and Ludlow Griscom.

often took abrupt U-turns, demonstrating the distinct advantage of a light car and longtime knowledge of rural Dutchess County. On a number of occasions, Roosevelt successfully slipped away from the agents for hours-long spells to get lost in rural New York.[86]

As Griscom sat in the backseat of the phaeton, the windshield splattered with crushed bugs that formed starburst patterns, the president held court on warblers, wrens, and Dutchess County history. "We call this car," Roosevelt joked to Griscom, "the Queen Mary."[87] When Griscom learned there were grenades on the floor of the car, fear washed over him. While Roosevelt happily chain-smoked, Griscom, quite sensibly, worried about an explosion. The president, utterly unfazed, acted as though it were the most normal thing in the world to have explosives rolling around on the floor of a fast-moving vehicle. That wasn't all. Griscom remembered thinking that such a noisy, crowded automobile convoy was no way to attract the American wigeon (*Anas americana*) and the black-crowned night heron (*Nycticorax nycticorax*) in the solitude of Thompson Pond.

Worried the outing might be a bust, Griscom asked Roosevelt to order the Secret Service agents to turn off their searchlights. "There is one person that the president of the United States can not tell what to do and that is a Secret Service man," Roosevelt told Griscom. "If you want to go back and plead with them, go ahead."[88] To the ornithologist's happy surprise, they obliged. In the hush of dawn, Roosevelt's party then patiently waited without speaking, hearing only the marginal sounds of insects and the "plop" of a turtle. Once Griscom began his series of birdcalls, marsh birds—Virginia rails, soras, and bitterns—began appearing. Dozens of avian species were observed that morning. "One incident included my father standing beside the car doing his usual chirping to call the birds in close," Andrew Griscom recalled. "At that moment a chickadee landed on his hat and explored it for a few seconds to the president's delight."[89]

To FDR, patiently looking for birds and planting trees were the God-ordained practices of a civilized democracy at its best. The

Germans were bombing Britain, American soldiers were dying on flyspeck islands in the Pacific Ocean, and factories in Detroit and Pittsburgh were pumping out munitions to satisfy the demand for war material. But the sun kept rising and setting, and Roosevelt thought it was important to tend his pines and continue the traditional bird counts. The president often attended St. James' Episcopal Church in Hyde Park when he was in town. But whatever spiritual solace he found during the services was sometimes marred by the pressing demand for handshakes, backslaps, and autographs afterward. Therefore, Roosevelt's nature-infused getaways to Top Cottage, Thompson Pond, and Shangri-La, away from the prying eyes of neighbors, helped him escape the ungodly pressures of being commander in chief.[90]

When Winston Churchill visited Hyde Park on June 19 and 20, 1942, Roosevelt treated him to a guided tour of Springwood's wooded roads and the best vantage points along the river. He was sharing one of his own deep-felt requirements. "I love this country of ours—every inch of it," Roosevelt told a B&O train official during the war. "And I want to take the time to really see it."[91] With his "map mind" at work, studying the waters and forests of America—from the Hudson River valley to the Olympic Peninsula—was Roosevelt's lifeblood. "The president drove me all over the estate, showing me its splendid views," Churchill warmly wrote in *The Hinge of Fate*. "I confess that when on several occasions the car poised and backed on the grass verges of the precipices of the Hudson I hoped the mechanical devices and brakes would show no defects. All the time we talked business, and though I was careful not to take his attention off the driving we made more progress than we might have done in formal conference."[92]

VII

In September 1942 the Army Corps of Engineers put the recently established Manhattan Project under the command of General Leslie R. Groves, tasking him with building the industrial capacity to process plutonium and uranium. A few months later, Hanford,

Washington—located at the confluence of the Yakima, Snake, and Columbia rivers—was deemed the perfect site. Grand Coulee Dam generated electricity for Hanford's "Site W," acquired by the federal government through eminent domain. Even though FDR had saved the Olympics, the one-two punch of the Grand Coulee and Site W has prevented many from ever venerating him as a true environmental hero.[93] This is unfortunate, for when Congress denied appropriations for new national wildlife refuges in 1943, the president nevertheless used his executive authority to establish a long list of them. It included Chassahowitzka in Florida; Great Meadows and Parker River in Massachusetts; Mingo in Missouri; Hailstone, Halfbreed Lake, and Lamesteer in Montana; the Missisquoi River Delta in Vermont; Slade in North Dakota; and Columbia in Washington State. Unfortunately, these new wartime refuges lacked congressional funding and illegal poaching persisted. Nevertheless, Roosevelt believed that in a postwar world they would receive proper maintenance.

The grandest of Roosevelt's unfunded 1943 NWRs was Chincoteague National Wildlife Refuge, on the Virginia end of Assateague Island. Thanks to field reports written by John Clark Salyer, Roosevelt understood that this vital saltwater marsh was a premier nesting and feeding spot for migratory birds on the Atlantic Flyway. The president's principal rationale for saving Chincoteague was that the greater snow goose (*Chen caerulescens atlanticus*) needed a sanctuary in the Mid-Atlantic. The refuge was the setting for Marguerite Henry's 1947 children's book *Misty of Chincoteague*, which made the island's wild ponies its most famous residents. The fourteen thousand acres of Chincoteague NWR, which the government purchased from the S. B. Fields family, were only a short drive from major metropolitan areas like Washington, D.C., and Richmond. Rachel Carson would write her first "Conservation in Action" booklet about this wildlife-rich ecosystem.[94]

Another wartime sanctuary created by the Roosevelt administration was the Susquehanna River NWR in northeast Maryland. For years FDR had asked the Department of Agriculture for reports on the canvasback and redhead duck situation along the river. He worried about the 250,000 American wigeon (*Anas americana*) that were feeding on wild celery, weeds, and grasses around the Susquehanna flats. His intervention blocked hunting and other activities across some thirteen thousand acres on the water and a small island. The survival of the upper Chesapeake islands, around Blackwater NWR, was essential for the health of migratory birds, including the wigeon. (Due to broad changes in the river, the Susquehanna NWR was reduced in size thirty-five years after its inception.)

From his years of sailing the North Atlantic, the president knew that migratory birds nested and fed in huge numbers on Monomoy Island, located eight miles off the southeastern tip of Cape Cod, Massachusetts, during their epic journeys from the Arctic and eastern Canada to Florida. Furthermore, Monomoy, unlike Martha's Vineyard, had almost no residents. There were only a couple of shacks and a smattering of cabins situated along the marsh ponds. In 1944, Roosevelt established a 7,604-acre NWR at Monomoy to protect the island's sandy beaches, shoal waters, migratory birds, and gray seals.[95] An abandoned Coast Guard station was repurposed as the new refuge's headquarters. With only 1 percent of the Atlantic coastline owned by the government, Roosevelt saw his protection of Chincoteague and Monomoy as critical acquisitions.

Eleanor Roosevelt believed that the surest way for environmental activists to persuade her husband to federally protect a marsh, or seashore, or woodlands was to schedule a one-on-one meeting with him, but the big challenge was finding a way to chat with a president consumed by global war. After being shown a map or series of photos of the ecosystem in question, FDR would as often as not be moved to action. Harold Ickes, Ira Gabrielson, Irving Brant, William Finley, Frances and Jean Harper, and Minerva Hamilton Hoyt had all seen this technique bear fruit. It was a photo album containing images of white pelican (*Pelecanus onocrotalus*) colonies in eastern Montana that persuaded FDR to establish the Lamesteer NWR.[96]

Recognizing that the Pacific Flyway also needed federal help—even during a war—Roosevelt established Colusa NWR and Sutter NWR in California. These sites became essential migratory bird havens within the large Sacramento National Wildlife Refuge complex pieced together by conservationists in the 1940s. In Texas, more than two thousand acres along the Rio Grande River were designated as Santa Ana NWR. Located in the southeast corner of Texas, it is also at the northern range of many Central American birds. As a naturally occurring tropical habitat, Santa Ana is a concentration zone for the black-bellied whistling duck (*Dendrocygna autumnalis*), fulvous whistling duck (*Dendrocygna bicolor*), mottled duck (*Anas fulvigula*), blue-winged teal (*Anas discors*), green-winged teal (*Anas carolinensis*), cinnamon teal (*Anas cyanoptera*), least grebe (*Tachybaptus dominicus*), anhinga (*Anhinga anhinga*), tricolored heron (*Egretta tricolor*), and other species. In Roosevelt's mind, Santa Ana NWR was the perfect complement to Big Bend National Park.

Even as Roosevelt established new NWRs, some that were already in existence deteriorated, owing to inadequate maintenance budgets and poaching. Dikes, water controls, roads, buildings, and other facilities were moved into the "deferred maintenance" category. Without the CCC, invasive species like the pinewood nematode (*Bursaphelenchus xylophilus*) and sickleweed (*Falcaria vulgaris*) spread far and wide in federal marshlands. Many NWR headquarters had no paid employees left. Local "friends" groups kept many NWRs in reasonably good shape with volunteer programs. But the very fact that the NWRs weren't targeted for dissolution during World War II gave Roosevelt's system historical permanence.[97]

If sacrifices were made on preserved lands, they were a constant part of life for citizens, too—though on a minuscule scale compared to the lot of those in the war zone. Americans were being asked to observe blackout drills; recycle metals, oil, and paper; and use ration cards to purchase commodities such as sugar, coffee, meat, and gasoline. FWS augmented the domestic war effort by promoting sources of protein other than the usual favorites of beef, lamb, and pork. The FWS, through its *Wildlife Leaflet* series offered wild game recipes, which encouraged the use of muskrat, turtle, and rabbit as main courses in rural areas. *Wildlife Leaflet* 218, "Domestic Rabbits in the Food for Freedom Purchases," the authors offered techniques for killing, dressing, and broiling domestic rabbits. In *Wildlife Leaflet* 229, muskrat was promoted as more delicious than traditional hamburger meat; recipes were included for muskrat à la terrapin, wine-fried muskrat, and muskrat salads.

During World War II, the Forest Service posted signs at the entrances of national forests and parks which read, "Land of Many Uses." This notion, the heart of Pinchot's conservation philosophy, was that recreation, wildlife, timbering and mining could all coexist. But Roosevelt knew that not all "uses" could work together in harmony. Cecil Andrus, a longtime governor of Idaho who served as secretary of the interior under Jimmy Carter, described the tension: "I've never seen people enjoying a picnic in an open-pit mine."[98] It's hard to allot credit in history for *preventing* something, but Roosevelt's insistence that national forests and national wildlife refuges not be pillaged for natural resources during the war was indeed proof of a brave conservation policy.

Refusing to be put on the defensive by the timber lobby, Roosevelt asked Nelson Brown, whom he had earlier asked to oversee lumber production in the Northeast, to describe in the *Journal of Forestry* his family's longtime commitment to responsible timbering. Brown's February 1943 article, "The President Practices Forestry," related how, for three hundred years, the Roosevelts had been pioneers in wise-use forestry. As president, Brown argued, FDR had made forest conservation a national undertaking. "His leadership in planting trees and in selectively cutting his woods has had an important influence on many other private timberland owners to 'Go thou and do likewise,' " Brown wrote. "The resulting examples of forest management are landmarks in the progress of forestry and are doing their part to offset the wolf cries of a long-heralded timber famine and the much publicized stories of denudation and depletion of our forest resources."[99]

Brown wanted other American tree farmers to understand that the president viewed his Hyde Park estate as living proof of his personal commitment to forestry. He wove statistics about FDR's silviculture successes into his *Journal of Forestry* article. The average number of trees Roosevelt cut per acre was still 16.7, which equaled 11 percent of the total number at Hyde Park that were ten inches or more in diameter. Brown noted that the volume of timber removed varied from 23 to 60 percent of the stand and averaged 3,588 board feet per acre (or 43 percent of the "merchantable stand"). What made these facts intriguing was that FDR kept his own detailed records about forestry on his farm—there were no corresponding financial records, for example. The commander in chief who didn't keep a diary of his historic meetings with Winston Churchill in Newfoundland or at the Casablanca Conference in Morocco documented the growth patterns of trees at Hyde Park and Warm Springs with amazing exactitude. According to Brown, only in late 1941, after Pearl Harbor, did FDR feel the need to fell a few overmature oaks on the Springwood grounds to contribute to the war effort against Germany and Japan. A naval vessel somewhere in the Atlantic or Pacific had received FDR's Dutchess County board feet, proving that the president was both a forester and a patriot.

VIII

On March 12, 1943, Harold Ickes lunched with FDR at the White House and discussed the proposed Jackson Hole National Monument.[100] (In this case, "hole" referred to a big valley ringed by high peaks.) The debate over whether this beautiful part of western Wyoming should be incorporated into Grand Teton National Park or left in the hands of ranchers and developers had been raging since the early 1920s. Residents of the Jackson Hole settlements—Jackson, Wilson, Kelly, and Moran—wanted the region free for cattle ranching. A smaller group favored a national monument designation for the valley, believing the protected land would encourage the creation of "dude ranches" for East Coasters or Europeans hungering for a fanciful taste of the cowboy experience portrayed in Hollywood movies.

Undeterred by the ranchers, Horace Albright, superintendent of

Yellowstone National Park from 1919 to 1929, began an effort to bring Jackson Hole under the purview of the National Park Service. Albright covertly arranged to take philanthropist John D. Rockefeller Jr. on a guided tour of the Tetons. Rockefeller was overwhelmed. In his estimation, it was perhaps the "most majestic" mountain vista in America. Collaborating with Albright, Rockefeller started quietly buying up all the ranchlands in the region in order to protect them

John D. Rockefeller Jr. and his wife, Abby, surveyed the Grand Teton Mountains during a boat ride on Jenny Lake, Wyoming, in 1931. Without the stealthy intervention of the powerful couple, the glories of the range might have been lost. Quietly buying land amid the chaotic development that prevailed there, they were eventually able to donate to the nation thirty-five thousand acres, crucial in the creation of the Grand Teton National Park, including the former Jackson Hole National Monument.

from cattle. There was one imperative: the real estate had to be purchased quietly. If Wyoming ranchers heard that "a Rockefeller was on the loose," they would instantly increase the prices of their land fivefold.[101]

After about thirty-five thousand acres had been purchased from ranchers for approximately $1 million, Rockefeller attempted to deed the land over to the federal government for the expansion of

Grand Teton National Park.[102] But with Albright having retired from the Department of the Interior in 1933, the movement to enlarge Grand Teton National Park hit a sizable roadblock. A consortium of Rocky Mountain stockmen and miners successfully lobbied Congress to refuse Rockefeller's generous land gift to the federal government.

Rockefeller had paid taxes on the Jackson Hole land for fifteen years, awaiting the moment when the political tensions would cool down enough to allow Congress to accept his donation. But with the onset of World War II Rockefeller, worried about losing momentum, began to play hardball. Late in 1942, he told his friend Harold Ickes that the government needed to make a decision about his gift one way or another. On February 10, 1943, Rockefeller wrote a strongly worded letter to FDR, which carried a backhanded compliment combined with an ultimatum. "I have now determined to dispose of the property, selling it, if necessary, in the market to any satisfactory buyer," Rockefeller wrote. "Because of your interest in the national parks and in the conservation of great areas for public use, I have preferred to advise you in advance of my intention, rather than to have you hear of it first as an established fact."[103]

Determined to bring Jackson Hole under Interior's purview, Ickes had an executive order drafted and began testing the waters for national monument designation at the White House. "My own view is that the President ought to set up a national monument before we lose an offer that will never be made again," Ickes wrote on February 27, 1943, to Pa Watson (with the intention that the back-channel message would be forwarded to FDR). "We need this land, not only to round out our park holdings in the Grand Teton area, including Jackson Lake, but also because much of this land would afford winter feed for elk and deer, thousands of which are without sufficient forage during the winter."[104] Hence Ickes's March 12 luncheon with Roosevelt.

Not long after that lunch, Roosevelt cheerfully agreed to use the Antiquities Act to bring 222,000 acres of Jackson Hole land, including the Rockefeller tract, under the National Park Service as a national monument. Presidential Proclamation No. 2578 saved Jackson Hole from ecological harm. An elated Rockefeller congratulated the president, telling him that his executive action had "made possible the preservation for all time of the most uniquely beautiful and dramatic of all the areas set aside for national park purposes."[105]

New Deal conservationists were also ecstatic about FDR's decision, but there was backlash from westerners who believed the federal government already owned too much of their region. Wyoming's sole representative in the House, Republican Frank Barrett, cited FDR's decree as proof that Hitler's fascist tactics had taken root in American soil, introducing a bill to immediately rescind the action. It passed both houses but Roosevelt remained firm, vetoing the measure. (Barnett would introduce this same bill every legislative session from 1943 until he left Congress in 1959, aside from when he served as Wyoming's governor from 1951 to 1953.)[106]

GOP senators were likewise angry. Milward Simpson of Wyoming led the charge against Roosevelt's proclamation, arguing that it was a federal land grab, that it would destroy the economy of Jackson Hole, and that it would cost ranchers their livelihoods. Roosevelt wasn't going to be flummoxed by anti-conservationists who mocked the New Deal for putting nature ahead of Stetson-hatted cowboys. Moreover, he loved the scenic Yellowstone-Tetons area, and relished the fact that the blowhards had been fooled by Rockefeller's dummy corporation.

There was, however, a last card to be played against the national monument. On May 2, 1943, the Jackson Hole controversy took an odd turn. A group of heavily armed Wyoming ranchers, led by Hollywood actor Wallace Beery, drove 550 calves across the national monument to a summer grazing range without securing the proper federal permit. This stunt was essentially a challenge to Charles Smith, the monument's superintendent, to arrest them. Instead of giving Beery free press, Smith treated him and his gang like gum on the bottom of a boot, sticky but not threatening to the designation of the national monument. From his Washington office, Ickes had a field day lampooning Beery's "mock heroics" and "mail-order regalia" as a pathetic cry for attention from a Hollywood has-been. And the Department of the Interior reported to the press that Beery, a Californian, hardly qualified as a Wyoming rancher. He had a Forest

Service permit for only a tiny cabin on a half acre of land bordering Jackson Lake and a "milk cow that had recently died."[107]

Beery nevertheless found a few gullible reporters willing to cover his cattle drive as a news story. Both *Time* and the *Saturday Evening Post* printed reports, while syndicated columnist Westbrook Pegler compared FDR's executive order to Adolf Hitler's absorption of Austria. "They anschlussed a tract of 221,610 acres," Pegler fumed about the national monument, as "Ickes's domain."[108] (On September 14, 1950, the bulk of Jackson Hole National Monument finally would be incorporated into a greatly expanded Grand Teton National Park.)

One downside of the Jackson Hole battle was that future administrations were hesitant to use the Antiquities Act to acquire land for the Interior Department; it wasn't until the 1970s that any further national monuments of significant size would be established.[109] But Roosevelt did use his executive privilege to name two other historic national monuments after Jackson Hole. In 1943, he designated George Washington Carver (Missouri), to honor the great Tuskegee agronomist; and in 1944 he established Harpers Ferry (West Virginia, Maryland, Virginia), to protect the site along the confluence of the Potomac and Shenandoah rivers where John Brown was captured in 1859.[110]

CHAPTER TWENTY

"CONSERVATION IS A BASIS OF PERMANENT PEACE"

I

Roosevelt studies a newly received globe in December 1942. After America's entry into World War II, two American generals steeped in planning noticed that it was somewhat difficult for the two key Allied leaders to communicate accurately on geographical matters. General Dwight Eisenhower suggested to General George C. Marshall that Roosevelt and Churchill should have identical globes. In December 1942, after intensive work by a team of cartographers and a Chicago globe-maker, Marshall was able to present Roosevelt with his massive globe. An identical one was shipped to Churchill.

Pastoralizing postwar Germany in the names of *denazification* and *demilitarization* was very much on President Roosevelt's mind even before World War II ended. It was on his own agenda at the Quebec Conference with Prime Minister Winston Churchill, held in mid-

September 1944. Prodded by Secretary of the Treasury Henry Morgenthau, Roosevelt sought to turn postwar Germany into a land of yeoman farmers. The Morgenthau Plan was immediately ridiculed by the State and War departments as global Jeffersonian agrarianism run amok. Changing an industrial powerhouse like Germany into a farm economy might take away the factories, but not necessarily the "innate Hun" militarism. It was assumed that because Morgenthau was Jewish, this thinking was guided by his hunger to punish the Germans for their Holocaust sins. There was truth in that assessment, but something else was at play. Morgenthau rightly claimed that his pastoralizing plan was similar to the New Deal's Rural Resettlement and Farm Security Administration programs. "The Morgenthau Plan for Germany assumed that the Germans deserved harsh punishment," historian Warren F. Kimball wrote, "but its ultimate thrust was an extension of the New Deal."[1]

Due to State and War Department blowback, Roosevelt repudiated his own Morgenthau Plan shortly after the Quebec Conference. Many critics of the plan—including top U.S. government strategists—mistakenly thought Roosevelt wanted to ban only heavy industry in the Ruhr Valley of Germany to prohibit the armament industry. Developing a new Germany where the population partook in small-scale farming, getting intimately acquainted with the land and forests, was Roosevelt's idea of the ideal peaceful European country. Fearful that the Nazi mentality had been inculcated into the mentality of the German people, Roosevelt intimated that after the war some members of the Third Reich should be transplanted to Central Africa to partake in TVA-style conservation work.[2]

Always the grand experimenter, still impressed with Germany's forestry program, Roosevelt thought that postwar Germany could join Norway and Denmark as model conservation-minded nations. To Roosevelt, pastoralism—"ploughshares over swords"—wasn't a punishment, but a way to bring Germany back to the cultured humanity of Goethe, Bach, and the Brothers Grimm. But in agriculture, Germany had long relied on huge estates led by a conservative aristocracy; this led, in his mind, to a dangerous consolidation of power. Breaking up these huge entities into small farms would prevent Germany from rearmament. However, with the Morgenthau Plan spiked, he looked for another course.

Throughout World War II Roosevelt continued raising trees at Hyde Park and Warm Springs. Using his own funds, he bought advertisements in the *New York Times* and *Washington Post* to help him sell "fine fresh cut trees" for the Christmas holiday. These ads announced that the seedlings he had planted in 1938 were now, in 1943, ready to be harvested. A seven-foot tree grown by the president cost $1.75, while fifteen-foot-high trees sold for $1.95, an "exceptional holiday bargain," the ads stated. The trees—sold at three Manhattan locations—came with Roosevelt's guarantee as well: it was "A NEW DEAL in Christmas Trees." They each carried a Roosevelt Val-Kill Farms tag that read, "NOT CUT BEFORE DECEMBER 1st."[3]

Archived at the FDR Presidential Library in Hyde Park, New York, is a fascinating "Christmas tree" folder with letters Roosevelt wrote to his friends to accompany a gift of a Christmas tree. In October 1943, he instructed Springwood's groundskeeper, William Plog, to give away evergreens on his behalf: "One tree to be delivered to Miss Suckley at the Library; one to be delivered to Miss Delano; one to be shipped to the Crown Princess of Norway at Bethesda, Maryland; and one for Prime Minister Winston Churchill, on which I will give you shipping directions."[4] Roosevelt's private secretary, Grace Tully, devised an elaborate method to deliver a tree to Churchill at 10 Downing Street: Plog wrapped it in burlap, crated it, and had it personally delivered by the army to Churchill. It was a touching gift.[5] "I frequently visited the President at Hyde Park, and was with him there when he completed the plans for the Roosevelt Library," William O. Douglas recalled. "I would tour the estate with him, and he would show me his latest project—the young Christmas trees which were turning into a good business venture for him."[6] Roosevelt was still running his forest plantations at Hyde Park and Warm Springs, carefully and purposefully selecting species, pruning the lower limbs of growing trees, making scientific soil preparations, overseeing weed and fungal control, and experimenting with seed enhancement by herbicide use.

Because the U.S. Navy needed redwood boards—which resisted warp and rot—Roosevelt suggested timbering species in Latin America so that America's giants could be left alone. When the navy told him it had located some albarco, a wood found along the Magdalena River in Colombia, he was ecstatic. It was even tougher than mahogany. The President even looked into establishing an agency called the Forest Products Service to help oversee the output of lumber for the war, without gouging public lands, but he was never able to get it up and running.[7]

Redwood lumber was in high demand during the war because it not only did not warp but also had insulation properties, soundproofing capabilities, and resistance to fire; so Roosevelt and Nelson Brown experimented with growing redwoods and sequoias on the East Coast. Throughout the war years, Roosevelt raised both species of these trees with Daisy Suckley at Wilderstein in Rhinebeck, New York. After trial and error, the president determined that neither species would ever grow in Dutchess County. So his hopes turned toward Tennessee and North Carolina. "I do not think it is possible to make them grow in the East—with one possible exception," Roosevelt wrote to Vice President Henry Wallace. "As you know, the rainfall in the Great Smoky Mountains National Park, or a little south thereof, is the highest in the East and I am going to get the Park Service to try planting them there in several correct locations."[8]

Ickes had grown accustomed to Roosevelt's quirky silviculture needs and meteorological concerns. Why not let the president continue his forestry whims in the Great Smokies as a fun release from the wartime White House pressure cooker? "The location should be protected from winds and, at the same time, the soil should be as rich and deep as possible," FDR wrote Ickes that March. "There are many such places in the Smokies."[9] The redwoods and sequoias didn't grow in the Smokies as Roosevelt had hoped, and although a couple took root at the White House and the Washington institution, Dumbarton Oaks, they eventually died. But Roosevelt *was* successful in saving the Calaveras Sequoia Grove in Northern California (the largest and finest grove remaining outside of a national or state park).[10]

By the fall of 1943, the Secret Service had grown fearful that the president was vulnerable to attack at Shangri-La. While the wooded buildings of the Maryland compound could hardly be seen by air, single-engine pilots were occasionally spotted flying over the camp. False rumors were planted with the press that Roosevelt's new secret retreat was Hoover's old camp on the Rapidan River (the one he had rejected in favor of the Catoctin Mountains). About 130 Marines kept around-the-clock watch over Shangri-La. Still, the Secret Service worried that too many reporters knew that Roosevelt's weekend White House was in the Catoctins. On the recommendation of the Secret Service, he chose as his new Shangri-La financier Bernard Baruch's plantation "Hobcaw Barony," near Georgetown, South Carolina. "Moss waves from the branches of the trees," Eleanor Roosevelt wrote about Baruch's plantations. "Those through which we approached the house and immediately around it, are some of the most beautiful old trees I have seen. Even though we drove through some swampy land, this place seems to me a friendly cheerful place, lacking that eerie quality which I find often prevalent in Southern landscape."[11]

Even though the president's health was problematic, with high blood pressure, migraines, and trembling hands, he kept a rigorous traveling schedule. On November 27, Roosevelt met with Churchill and Chiang Kai-shek in Cairo to discuss postwar Pacific strategy. Then he traveled to Tehran for a summit with Churchill and Stalin to decide on the timing of a cross-channel invasion of France.[12] Flying at low altitude over the arid Iranian countryside, a hazardous enterprise, in December 1943, President Roosevelt—en route to Tehran for a much-anticipated tripartite summit with British prime minister Winston Churchill and Soviet premier Joseph Stalin—saw the telltale signs of widespread soil erosion and deforestation.[13] During World War II, Roosevelt had made seven foreign trips, traveling 306,265 miles, but never before had seen such poor land stewardship as in Egypt and Iran. Grass and trees were as scarce as bodies of water; there weren't even clumps of bushes along the dry riverbeds. No one had thought to construct modern aqueducts and reservoirs. Aware that Iran was not a naturally harsh desert, he was appalled by the ecological abuse, envisioning a Shelterbelt for the Middle East.

"Of course, I do not pretend to know Iran well on account of the shortness of my visit, but may I write you about one of the impressions which I received on my air trip to Tehran?" Roosevelt wrote to twenty-four-year-old Shah Mohammed Reza Pahlavi of Iran, the conference's host, from the White House. "It relates to the lack of trees on the mountain slopes and the general aridity of the country which lies above the plains. All my life I have been very much interested in reforestation and the increase of the water supply that goes with it. May I express a hope that your Government will set aside a small amount for a few years to test out the possibility of growing trees or even shrubs on a few selected areas [and] to test out the possibility of trees which would hold the soil with their roots and, at the same time, hold back floods? We are doing something along this line in our western dry areas and, though it is a new experiment, it seems to be going well. It is my thought that if your Government would try similar small experiments along this line it would be worthwhile for the future of Iran."[14]

The shah was touched that Roosevelt, in the midst of war, cared about Iranian conservation (or the lack thereof). Convinced that the landscape of any country was the people's portrait of themselves and that love of nature was universal, Roosevelt likewise wrote to King Ibn Saud of Saudi Arabia in February 1944 on the same topic. Having formed an alliance with Saudi Arabia that was anchored on petroleum strategy, he hoped to convince the Saudi leaders to think and act more sensitively toward the land, embracing their responsibility as environmental stewards. "My avocation, as you probably know, is the increase in water supply and in reforesting vacant land," Roosevelt wrote to Ibn Saud. "I feel sure that the Kingdom of Saudi Arabia has a great future before it if more agriculture land can be provided through irrigation and through the growing of trees to hold the soil and increase the water supply."[15] Ibn Saud's reply—a polite invitation for FDR to visit Saudi Arabia—contained no mention of forest conservation.[16]

In the spring of 1944, with D-Day approaching, Roosevelt spent occasional afternoons at Top Cottage with Daisy Suckley, marveling at the rhododendrons and azaleas and the springtime birdlife, feeling that this green patch of Dutchess County was his own self-expression. At Top Cottage, he kept over one hundred natural history books, his most dog-eared being Chief of Forestry N. H. Egleston's *Handbook of Tree Plants*.[17] On May 19, Suckley wrote in her diary that the president "just talked quietly about the view, the dogwood, a little about the coming invasion of Europe." That D-Day, in other words, received equal billing with dogwoods was typical of both Roosevelt (the arborist) and Daisy Suckley (the pastoralist). The Hudson River Valley helped the president put world events in perspective that May. In stressful times, he turned to Ray Bergman's *Trout* and J. Fletcher Street's *Brief Bird Biographies* as forms of diversion and relaxation.[18]

Early in June, just days before the planned invasion, the president made a short trip to Charlottesville, Virginia, to stay at Kenwood, the secluded seventy-eight-acre home of Pa Watson. Although there was a guest cottage at Kenwood, Roosevelt preferred to sleep in the main house's front bedroom, where he could better hear the birds chatter. Escaping from the Washington pressure-cooker and playing cards with Pa Watson helped Roosevelt recharge his strength.[19]

On June 5, 1944, after four days in Charlottesville, Roosevelt returned to the White House. To deflect reporters' questions about the war, he bantered amiably about forestry and the pink tulips blooming at the White House, never hinting that an English cross-channel invasion was imminent. Just after midnight on June 6, the president retired to the Lincoln Bedroom. Within the hour the first Allied troops would be landing on the beaches at Normandy. He knew the day was going to be long, no matter the outcome. The First Lady, who suffered bouts of insomnia, was too anxious to even try to sleep. She paced around the White House, awaiting General George C. Marshall's report. While the lights stayed on in the White House, Allied forces were storming the five battlefield beaches of Normandy: Omaha and Utah (Americans), Gold and Sword (British), and Juno (Canadians). At 3 a.m., Eleanor woke up Franklin, who put on his favorite gray sweater and sipped coffee before starting a round of telephone calls, which lasted over five hours.

That afternoon FDR visisted Andrew Jackson's magnolia on the White House grounds, planted in 1835, for a moment of respite. Later, in an Oval Office press conference, he described the transformative events of D-Day. The armada that had crossed the turbulent waters of the English Channel from Dover that morning was the largest in the history of the world—the ships carried approximately fifty-five thousand American, British, and Canadian soldiers. On the radio later that evening he offered a prayer: "With thy blessing we shall prevail over the unholy forces of our enemy."[20]

Roosevelt didn't forget about America's national parks, even on such a red-letter day. Bills establishing national parks had taken, on average, anywhere from six to nine years to get through Congress. Big Bend National Park in Texas, in purgatory since 1935, was finalized on arguably the most momentous day of the twentieth century. Just before 1 p.m. on June 6, 1944, the president met briefly with four steadfast Texan promoters of Big Bend: Speaker of the House Sam Rayburn, Senator Tom Connally, Congressman R. Ewing Thomason, and Fort Worth businessman Amon Carter. A White House photographer captured Carter handing FDR the deed to the 708,000-acre Big Bend, a cross-section of geological eras.[21] (A blown-up photograph of the hand-off adorns the wall of the Panther Junction visitor center at Big Bend National Park, attesting to the act of D-Day preservationism.)

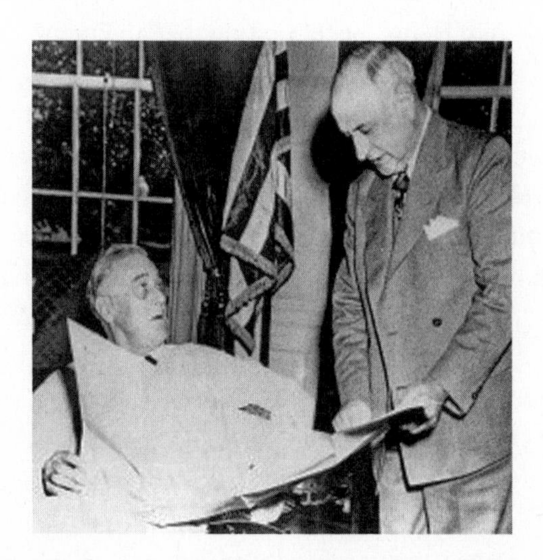

On June 6, 1944, newspaper publisher Amon Carter (*standing*) presented FDR with the deed for a 708,000-acre tract that would form the basis of Big Bend National Park in south Texas. Carter, who started out in poverty, became the owner of the *Fort Worth Star-Telegram*, as well as the city's first radio station, a stake in American Airlines, and many other ventures, mostly within the state he loved. Carter was one of several undauntable Texans who confronted ranching interests in order to preserve the Big Bend region.

Big Bend had been a Roosevelt priority for years. Whole mountain ranges, hundreds of square miles of Chihuahua Desert, and a complex of largely undisturbed canyons made by the Rio Grande River were now preserved for the ages. Five separate life zones were found in the region that was home to over 450 bird species. "I have heard so much of the wildness and the beauty of this still inaccessible corner of the United States and also of its important archeological remains that I very much hope that some day I shall be able to travel through it myself," Roosevelt had once written Congressman Thomason. "Furthermore, I feel sure that it will do much to strengthen the friendship and good neighborliness of the people of Mexico and the people of the United States."[22]

Just six days later, on June 12, with America in a celebratory mood over the offensive in France, the NPS took over Big Bend National Park. Because the United States and Mexico shared watersheds and ecosystems around the Rio Grande canyons of Santa Elena, Mariscal, and Boquillas, Roosevelt thought that Big Bend—which William O. Douglas described as a "geological potpourri" of "painted skylines, painted cliffs, painted alcoves"—could demonstrate how critical joint conservation was in maintaining future world peace.[23] To Roosevelt, Big Bend wouldn't be complete until both sides of the Rio Grande ecosystem were off limits to developers. No longer would forests be cut for timber to fuel the quicksilver mines of Terlingua and Study Butte, two mining camps turned into ghost towns.[24] Later that year, Roosevelt appealed directly to his Mexican counterpart to help usher in a new era of environmental diplomacy along the Rio Grande. "In the United States we think of the Big Bend region in terms of its international significance and hope that the Mexican people look forward in the same spirit to the establishment of an adjoining national park in the States of Chihuahua and Coahuila," Roosevelt wrote to President Manuel Ávila Camacho. "These adjoining parks would form an area which would be a meeting ground for the people of both countries, exemplifying their cultural resources and advancement, and inspiring further mutually beneficial progress in recreation and science and the industries related thereto."[25] That international park has yet to be completed, although Barack Obama signed a statement with Mexico's president, Felipe Calderon, in 2012, expressing interest in realizing FDR's dream.

II

In July 1944, not without some reservations, the president accepted the Democratic nomination for a fourth term. In good times Roosevelt heartily enjoyed delivering stump speeches and shaking hands, but in 1944, the thought of campaigning against Thomas Dewey of New York bored him. Also, consumed with fatigue, Roosevelt yearned for Springwood and Top Cottage. He even nursed a dream that the United Nations—his grand plan for a body of global governance—might be headquartered in Dutchess County.[26] "All that is within me," the president wrote just before accepting the nomination, "cries out to go back to my home on the Hudson River, to avoid public responsibilities, and to avoid the publicity which in our democracy follows every step of the Nation's Chief Executive."[27] But Roosevelt, knowing his work wasn't finished, that his voice resonated democracy, allowed the nomination to proceed. Although he didn't officially commit to running for reelection until mid-summer, he played no cat-and-mouse games. "If the convention should nominate me, I shall accept," Roosevelt said. "If the people elect me, I will serve."[28]

The controversy at the convention pertained to FDR's vice-presidential running mate for the November election. The problem for Roosevelt was that he liked Wallace and admired his work ethic, as did most Americans. Wallace's problem, however, was that he was a liberal and several powerful Democratic bosses—who had already seen the party move to the left under FDR—wanted him off the ticket. Although one would think that FDR could have designated his dog, Fala, as the vice-presidential candidate and still won the election in 1944, he bowed to the bosses. For the sake of party unity, he dismissed Wallace and opened the question of a replacement. But if Wallace was purged, who would be number two? It was not a slight question with Roosevelt's health rapidly deteriorating. The man FDR wanted to be his running mate was Supreme

Court justice William O. Douglas. Of the leading vice-presidential candidates, only Douglas, the leading high-profile preservationist for the Cascades of Washington and the White Mountains of New Hampshire, shared Roosevelt's deep love of conservation. For this reason alone, Ickes thought there was no smarter New Dealer alive in America from a political and legalistic perspective than Douglas.

One the eve of the Democratic Convention in Chicago, the president had dictated to Grace Tully that he would be happy to run "with either Bill Douglas or Harry Truman" on the ticket; Douglas's name was put *before* Truman's. But the Democratic National Committee chairman, Robert E. Hannegan of Missouri, made a switch, putting Truman first. Whether Roosevelt was party to this remains open to debate. Truman was nominated, while Douglas remained on the Supreme Court, and remained a vociferous champion of the environment well into the 1970s.

"We have lost much with our environment," Douglas would bemoan in coming years. "We allow engineers and scientists to convert nature into dollars and into goodies. A river is a thing to be exploited, not treasured. A lake is better as a repository of sewage than as a fishery or canoe-way. We are replacing a natural environment with a symbolic one."[29]

On the campaign trail in 1944, Roosevelt would excitedly tell rural voters that yields could increase exponentially by dutifully regulating the timing of harvest. He'd talk about the benefits of healthy forests at every stop, how they absorbed rainfall, refilled groundwater aquifers, and slowed runoff from storms—the same gospel he'd been preaching since 1910. A favorite biblical verse of FDR's—"When thou shalt besiege a city a long time, in making war against it to take it, thou shall not destroy the trees thereof by forking an axe against them" (Deuteronomy 20:19)—became a maxim for him.[30] At one whistle stop, a voter remarked to Roosevelt that he spoke about trees, soil, and water more than the war in Europe and the Pacific. "I fear," FDR replied, "that I must plead guilty to that charge."[31]

It seemed to George C. Marshall that the president saw the world in terms of seaports, rivers, and forests rather than cities. Even during military strategy meetings in the White House's map room, Roosevelt invoked postwar conservation of natural resources. War consumed the president's every waking hour. But his interest in geography and conservation aided his strategic thinking in dozens of ways. With U.S. forces moving deliberately across France, marching to Berlin, Roosevelt started planning the global peace. His grand hope was to succeed where Woodrow Wilson's League of Nations had foundered—world peace forever.[32]

Three weeks after D-Day, President Roosevelt met with Gifford Pinchot at the White House to discuss holding a global conservation conference at the war's end. The best hope of the postwar world, Pinchot believed, lay in the care and restoration of the world's forests, deserts, wildlife, and waterways ravaged by years of bombings and misuse. "Roosevelt," Pinchot wrote, "unlike Taft, Wilson, and Hoover, grasped the full implications of the idea at once, received it with immense enthusiasm, and expressed the desire for rapid action, even to the possibility of launching the movement in the autumn."[33]

During World War II, Pinchot, a survivor of three heart attacks, remained indefatigable in his conservation efforts, sitting on various commissions charged with boosting civilian morale. Although he had officially retired from government in 1935 and spent his free time fishing in Pennsylvania's streams, he was still a vocal advocate for conservation reforms, vehemently opposing any hint of a transfer of the Forest Service to the Department of the Interior. Congressmen consulted him on conservation-related issues. Agreeing that "all Americans must do their part," Pinchot invented fishing tackle that could be used by U.S. military personnel who found themselves marooned on lifeboats. Likewise he developed a "water substitute" derived from the juices of raw fish.[34] And his speech delivered at the Eighth Pan American Scientific Conference, held in early May 1940, motivated the president to look seriously at holding an international conservation conference. Pinchot argued forcefully that "conserving, utilizing, and distributing natural resources to the mutual advantage of all nations might well remove one of the most dangerous of all obstacles to a just and permanent world peace." Pinchot's clarion call for global environmental standards evolved with Roosevelt's help throughout the early 1940s before being distilled into a single phrase: "Conservation is a basis of permanent peace."[35]

With this enduring slogan, Roosevelt asked his White House staff to help Pinchot draft a working proposal for a summit in just three weeks' time.[36] As Election Day drew closer, Roosevelt sent Pinchot an encouraging note about the finished proposal. "Remember that I have not forgotten that conservation is a basis of permanent peace, and I have sent the enclosed to Cordell Hull," he wrote on October 24. "I think something will happen soon. You must, of course, be on the American Delegation!"[37] If Theodore Roosevelt brought the conservationist revolution to America, FDR, in the fall of 1944, was going to bring it to the wider world. At the Dumbarton Oaks Conference, held in Washington, D.C., from August 21 to October 2, 1944, the United States and its allies outlined the contours of a new postwar international organization: the United Nations.[38] In addition to all else, this new organization would be the appropriate body to establish international ground rules for timbering, mining, and drilling.

The courtly, white-haired Hull, the longest-serving secretary of state in U.S. history, had worked doggedly at Dumbarton Oaks to establish the United Nations. With the outline in place, he retired from public life. But before he did, FDR had ordered him explicitly to ensure that global conservation was a top priority in any postwar diplomatic organization:

> In our meetings with other nations I have a feeling that too little attention is being paid to the subject of the conservation and use of natural resources.
>
> I am surprised that the world knows so little about itself.
>
> Conservation is the basis of permanent peace. Many different kinds of natural resources are being wasted; other kinds are being ignored; still other kinds can be put to more practical use for humanity if more is known about them. Some nations are deeply interested in the subject of conservation and use and other nations are not at all interested.

It occurs to me, therefore, that even before the United Nations meet for the comprehensive program which has been proposed, it could do no harm—and it might do much good—for us to hold a meeting in the United States of all of the united and associated nations for what is really the first step toward gathering for the purpose of a world-wide study of the whole subject.

The machinery could at least be put into effect to carry it through.

I repeat again that I am more and more convinced that conservation is a basis of permanent peace.

Will you let me have your thoughts on this?

I think the time is ripe.[39]

III

The Republicans chose as their nominee Thomas E. Dewey, a former Manhattan prosecutor who had been elected governor of New York in 1942. A national hero as a crime-buster, Dewey struggled to find a meaningful wedge issue for the GOP. Realizing that it would be undignified to question Roosevelt's health, Dewey attempted, particularly in the West, to use the president's enthusiasm for the Antiquities Act against him. Insinuating that the Roosevelt administration was filled with communist sympathizers, Dewey focused his disapproval on FDR's controversial presidential proclamation 2578 of 1943, which had designated Jackson Hole a national monument. It was in the news, because incensed lawmakers from Wyoming had introduced congressional legislation to rescind it.

As a counterattack, Roosevelt asked Speaker of the House Sam Rayburn of Texas to kill any such bill. "Jackson Hole is one of the most scenic areas in the country and a wintering ground for the southern Yellowstone elk herd, the largest elk herd in the United States," Roosevelt wrote to Rayburn. "Last year to avoid losing Mr. Rockefeller's gift to carry out the commitment, made during the administrations of Presidents Coolidge and Hoover, I established Jackson Hole National Monument, under the authority of . . . the Antiquities Act. This Act had been used for this purpose many times before. Five Republican Presidents and one Democratic President before me have established a total of seventy-one national monuments under the authority of this Act, many of them larger than the Jackson Hole NM."[40] In the end, Rayburn couldn't stop the bill, which made it through both houses of Congress; Roosevelt exercised a pocket veto.[41]

The battle over Jackson Hole National Monument intensified during the 1944 presidential election. Dewey delivered blistering speeches about New Deal "land grabs" that made westerners "wards" of Uncle Sam; case in point, 51 percent of Wyoming had been given over to public lands. In Sheridan, from a rostrum at the historic Sheridan Inn, a favorite haunt of Buffalo Bill, Dewey argued that designating Jackson Hole as a national monument had deprived Teton County communities of tax revenue for its public schools and had shown a complete "lack of respect for the rights and opinions of the people affected."[42] The designation, the GOP nominee asserted, was anti-American collectivism; the president simply didn't understand rugged western frontier values. After the speech at the Sheridan Inn, Roosevelt's so-called land grabs became a prominent talking point for Dewey at nationwide campaign stops as Election Day approached. "The Government takes the land under various guises," Dewey charged in one stump speech. "Sometimes in the guise of conservation, sometimes under the guise of monuments, thousands of acres for a monument. The amount taken (in the West) for military purposes is infinitesimal."[43]

Charges that Roosevelt was "land grabbing" in Wyoming were ludicrous; accepting Rockefeller's generous gift of 33,000 acres only made sense. Roosevelt was rather cautious in accepting lands under other conditions. As a case in point, there were more than 400,000 acres of land that the federal government rehabilitated during the New Deal as Recreation Demonstration Areas. The NPS could have seized these RDAs, but Roosevelt sought Congressional authorization in 1942 to deed twenty-six of them back to the states.[44]

To counter Dewey, Roosevelt unleashed Ickes, who charged that Dewey, "in his beagle-like snuffing about for votes," had maligned the three Republicans most responsible for establishing Jackson Hole: Calvin Coolidge, Herbert Hoover, and John D. Rockefeller Jr. After ticking off a few sizable western national monuments Hoover had established by executive orders, Ickes reminded Dewey (and the electorate) that "no ranch owner or cattleman lost a thing" as a result of FDR's decision to designate Jackson Hole as a national monument: "We have not even interfered with the trespassers who have grazed their cattle on the public domain without paying the customary fee," he said. "All that this Administration did was to carry out, in good faith, the obligation entered into by Presidents Coolidge and Hoover."[45] Congress agreed with Dewey, passing H.B. 2241, abolishing Jackson Hole National Monument. Roosevelt promptly vetoed it, to the surprise of no one.[46] "It is disturbing," Eleanor Roosevelt would later write about the Jackson Hole fight of the 1940s, "to find how little real enthusiasm there seems to be among our people for the preservation of our national parks."[47]

Throughout the 1944 campaign, Roosevelt proudly invoked the conservation accomplishments of the New Deal. One-third of America was covered in forestlands. Over 180 million acres of woodlands in forty states were part of the national forest system: enough commercial forests left to maintain maximum sustained yield to win World War II. Furthermore, in 1935, there had been fewer than 30 million waterfowl in America; now there were over 140 million from the Cascades to the Cumberland Plateau, to the Pennsylvania Wilds. In this spirit of accomplishment, on October 29, ten days before the election, Roosevelt delivered a spellbinding speech in Clarksburg, West Virginia, centered on the theme "only God can make a tree."

In 1933, Roosevelt had been rocked by the sight of clear-cut hilltops in West Virginia. Now, more than a decade later, he returned to praise West Virginians for reforestation efforts, though he criticized them for not doing enough: "It doesn't amount to very much, this cost of planting trees," he told Clarksburg residents, "and yet the hillsides of West Virginia of our grandparents' day were much more wonderful than they are now. It's largely a deforested State. And I believe that from the point of view of the beauties of nature, from the point of view of all that trees can be, and from the point of view of your own grandchildren's pocketbooks, the small number of cents, the small number of dollars that go into a reforestation, are going to come back a thousandfold."[48]

Around the same time that the president was delivering his West Virginia tree sermon, Eleanor Roosevelt visited the Audubon Nature Center in Greenwich, Connecticut, where she expressed the conviction that young people needed education about the natural world to better understand the "interdependence of human kind—the animals, the oceans, the Earth, and human beings." In her "My Day" column, she described how hundreds of acres of Connecticut woodlands had become an outdoor learning center. "The whole thing, of course, is 'conservation,' but it is so easy to understand because it is done so simply," she wrote. "It shows clearly the whole set-up of nature, going from the underlying rock, through plant and animal life, to human beings at the peak."[49] At the White House, noticing that the squirrels on the ground looked scrawny, she called the National Zoological Park to have them caught, put on a special diet, and rereleased.[50]

Beginning in mid-1944, FDR had several visits with Lucy Mercer Rutherfurd, a former social secretary to Eleanor Roosevelt with whom he'd had an affair in the late 1910s. They hadn't seen each other, as far as is known, since the mid-1920s, and Roosevelt's daughter Anna arranged their rendezvous that summer. When Anna saw how these meetings lifted her father's spirits, she continued setting them up behind her mother's back. At the end of the year, Rutherfurd came to Warm Springs to see FDR, and he took her on a sightseeing tour of Meriwether County. Later, Lucy told Anna Roosevelt, the president's daughter, about his unquenchable enthusiasm for nature. Even when discussing Hitler and Hirohito during a private picnic at Dowdell's Knob, the president would stop and talk about the land: "I just couldn't get over thinking of what I was listening to," Rutherfurd told Anna, "and then he would stop and say, 'You see that knoll over there? That's where I did this-or-that' or 'you see that bunch of trees?' "[51]

IV

On November 7, 1944, Roosevelt trounced Thomas Dewey by 432 electoral votes to 99. Harry Truman was the new vice president. Although FDR had met Truman only a handful of times, he knew Truman was a hawk on waste and government inefficiency, and that the Truman Committee of 1941 had saved the U.S. government $10 billion to $15 billion in military spending, as well as countless American lives.[52] But Truman had evinced no serious interest in conservation.

Roosevelt appointed Edward Stettinius to succeed Cordell Hull as Secretary of State. Handsome, silver-haired with a perpetual tan, Stettinius, had been an executive at U.S. Steel before entering public service, was also a former head of the War Resources Board, and had worked at the Office of Lend-Lease Administration. Distrustful of ecology, he thought that natural resources were to be *used*. Even though the president—working in conjunction with Pinchot—had decided that the global conservation of natural resources would be a major component of postwar diplomacy, Stettinius wasn't convinced. In the meantime, war diplomacy and relationships with America's allies—most notably Stalin's Soviet Union—ate up Stettinius's working days.

Exhausted from the 1944 campaign, looking forward to reviving his spirits in Warm Springs, Roosevelt grew frustrated with bureaucratic inaction on the UN conservation conference. The problem was that the State Department despised Ickes and Pinchot and felt that the president was thinking too much about forestry and irrigation. Stettinius had the audacity to distribute a State Department memo proposing to delay the conference.[53] "I am not satisfied with the Department's attitude on a Conservation Conference," an irritated Roosevelt wrote to Stettinius around Thanksgiving. "Whoever wrote the memorandum for you has failed to grasp the real need of finding out more about the world's resources and what we can do to improve them."[54]

Roosevelt knew that the postwar world would be beset by ecological devastation and lack of clean water. Because the United States would be the world's premier exporter, the nation's future would be melded with that of the rest of the world. Global environmental standards would be a prerequisite for democracies to flourish and for developing countries to avoid a Dust Bowl of their own. In Roosevelt's mind, this task belonged to the State Department, not to the Departments of Interior or Agriculture. "Just for example, take the case of Persia [Iran]," he continued. "The greater part of it, i.e., the North, used to be a forested country. Today it is utterly bare with a few cattle and a few very poor crops in the small valleys. The people are destitute. Anyone who knows forestry would say that an immediate program of tree planting is the only hope for the Persia of the future."[55] Stettinius was learning the hard way that FDR didn't take global conservation lightly.[56]

After the Christmas holiday Stettinius worked to get back into Roosevelt's good graces, writing to him about a sudden desire on the part of the State Department to establish a "national tribute grove" honoring the U.S. armed forces fighting in World War II. The secretary of state wanted the president's advice about whether Joseph Grew—who had been a staunch advocate for protecting and preserving California's redwoods and had been ambassador to Japan at the time of Pearl Harbor—would be an appropriate choice as chairman of the committee to establish the grove. Roosevelt used the overture to prod Stettinius to prioritize global conservation. "It is perfectly all right for Joe Grew to act as President of 'The Save-the-Redwoods League' but I think it is important to get the idea abroad that some long-time conservationists are interested [in the UN conference]," Roosevelt brusquely replied. "For example, Gifford Pinchot, who is undoubtedly our No. 1 conservationist, should be in this thing."[57]

Even though Japan was losing the war, Roosevelt continued worrying about protecting the forests of the West Coast. As the president had predicted, the Japanese military employed a wildfire strategy against the United States, launching approximately six thousand paper hydrogen balloons carrying thirty thousand incendiary devices into the jet stream. The first balloon had been launched on November 3, 1944, and approximately three hundred of those bombs actually reached American soil, but the vast majority of them hadn't detonated properly.[58] In early May 1945, one such firebomb landed at the Fremont National Forest near Bly, Oregon, killing schoolteacher Elsye Mitchell and wounding five Sunday school students—the only American casualties on continental U.S. soil.[59]

SMOKEY SAYS—
Care will prevent
9 out of 10 forest fires!

The first appearance of Smokey the Bear was on a poster drawn by Albert Staehle in August 1944. Before Smokey, Walt Disney lent rights to his character Bambi, the deer fawn, to the U.S. Forest Service in its effort to stop man-made wildfires. After the term of the loan was over, the Forest Service sought its own spokes-animal, and Staehle, working with the Ad Council, introduced Smokey to the world.

Around this time the U.S. Ad Council began using a cartoon bear named "Smokey" in its posters as part of a larger fire prevention campaign. Artist Albert Staehle drew the first image of Smokey, outfitted in a park ranger's hat, pouring a bucket of water on a campfire; it carried the caption "Smokey says—Care *will* prevent 9 out of 10 forest fires!" While Roosevelt remained concerned that Japanese arsonists might succeed in burning down the lush forests of the Pacific Northwest, he knew that the cause of accidental forest fires was most likely to be careless citizens (generally by leaving a campfire lit, tossing matches, or operating machinery in arid counties).[60] It wasn't long before Smokey the Bear began appearing in brochures on forest fire prevention, teaching Americans about proper forest stewardship. It was yet another example of the administration trying to educate young people about how to be good stewards of the great American outdoors.[61]

V

As Roosevelt worked on his fourth inaugural address in January 1945, he kept in regular touch with Pinchot about their United Nations conservation summit.[62] Whatever animus Ickes had once felt toward Pinchot had dissipated; among other things, they commiserated with each other about surviving their heart attacks. "Glad to be on friendly terms with him again," Ickes wrote in his diary. "I have always liked him very much indeed in spite of our temporary breaking away from each other."[63] Both Ickes and Pinchot now consolidated their influence to push forward the dream of global conservation. "When you told me at luncheon on [the day before his inauguration] that you are going to take up the proposed conference on conservation as a basis of permanent peace with Churchill and Stalin," Pinchot wrote to Roosevelt that January, "I saw great things ahead and was more delighted than I can easily say."[64]

Before leaving for Crimea in the Soviet Union to meet with Churchill and Stalin at the Yalta Conference, the president used the fortieth anniversary of the U.S. Forest Service to promote his global conservation crusade. Roosevelt also wanted a Tree Farm System for private forests, supported by the Weyerhaeuser Timber Company, to take root across America. The first Tree Farm was designated in Montesano in Washington State. The Texas Forest Service, in conjunction with the Texas Long Leaf Lumber Company, helped the federal government inaugurate Tree Farms in that state. Sixty-five tree farmers were certified, tending a combined 846,000 acres. Three national forests that Roosevelt had established—Davy Crockett, Sabine, and Sam Houston—were also succeeding beyond the president's wildest dreams.[65]

Eleanor Roosevelt also commemorated the Forest Service anniversary. "When we think what the Forestry Service has accomplished in the last 40 years, we should all be celebrating," she wrote in "My Day" that February, "because their work is helping to preserve one of our great national assets. We do not begin to reforest sufficiently anywhere in our country, but year by year we are learning more about trees, their care and their value, and eventually we may discover that each one of us owning any land has a responsibility to the nation to keep some of it in trees. Perhaps I am particularly conscious of this because so much of my husband's land at Hyde Park is tree land."[66]

Although Roosevelt was an ecologically minded president, his conservation legacy is marred by a few major blind spots. One major concern of the employees of Roosevelt's Fish and Wildlife Service during World War II was the increased use of pesticides. The widespread use of the most pervasive toxin—a molecule of chlorinated hydrocarbon called dichloro-diphenyl-trichloroethane (DDT)—became a cause for concern throughout the scientific conservation community. DDT had first been synthesized in 1874, but it wasn't until 1939 that it was found to have a practical use: killing insects. It had been sprayed directly on troops because the chemical controlled mosquitoes, lice, and ticks. As a delousing agent, DDT worked wonders.[67]

As a field experiment, Dr. Gabrielson of the Patuxent Wildlife Research Center ordered a FWS plane to spray DDT on 117 acres of Maryland marsh and woodlands. The result was horrifying. Just ten hours after the spraying, fish in the Patuxent River died off in large numbers. More tests were conducted in Patuxent's artificial ponds, where the outcome was even more grim. Two of Gabrielson's top scientists, Joseph Linduska and Clarence Cottam, fed different animals food laced with DDT. All of the test subjects demonstrated "excessive nervousness, loss of appetite, tremors, muscular twitching, and persistent rigidity of the leg muscles, the last continuing through death."[68]

It is unreasonable to criticize Roosevelt for allowing DDT to be used during World War II. To GIs, it was understandably preferable to be hosed down with the chemical than to catch tropical diseases such as malaria. Luckily, the Patuxent Wildlife Research Center was equipped to look into the effects of such synthetic chemicals. Rachel Carson, of FWS, wrote an alarming article about the deleterious effects of DDT on wildlife. "We have heard a lot about what DDT will soon do for us by wiping out insect pests," she said. "The experiments at Patuxent have been planned to show what other effects DDT may have if applied to wide areas; what it will do to insects that are beneficial or even essential; how it may affect waterfowl, or birds that depend on insect food; whether it may upset the whole delicate balance of nature if unwisely used."[69]

In early 1945, Roosevelt was also overseeing the Manhattan Project, based in New Mexico, the effort led by Major General Leslie Groves to build an atomic bomb. The radioactivity released from the bombs, when detonated, would make DDT poisoning seem mild in comparison. Both DDT and nuclear proliferation were scourges of the early Cold War era, and in both cases Roosevelt received both credit and blame.[70]

Roosevelt's penchant for creating dams was yet another area where he fell short as a conservationist president. Public power projects such as the TVA and Grand Coulee were holy causes to the president. Determined to undermine the greed of private power companies, he took every opportunity to champion public power projects. History proved Roosevelt was shortsighted about the ecological damage these dams caused. No matter how Roosevelt spun a dam project, building a reservoir like Lake Mead for recreation wasn't the same as protecting the finest salmon and trout runs in America. Overly impressed with engineering and experts, Roosevelt did not consider that the loss of a river bottom was not "progress" at all. While FDR deserves high marks for forestry, wildlife protection, state and national parks management, and soil conservation, his dams in the name of the "public interest" devastated numerous riverine ecosystems.

VI

Roosevelt's fourth inaugural took place on January 20, 1945, on the South Portico of the White House. Haggard, with black circles ringing his eyes, he was too weary to be sworn in at the Capitol as tradition dictated. A huge throng of spectators, most with black umbrellas, braved a cold, steady rain to witness the quickest inauguration in American history. Gifford and Cornelia Pinchot were among those shivering nearby to cheer him on. "We have learned that we must live as men and not as ostriches," Roosevelt said that day. "We can gain no lasting peace if we approach it with suspicion and mistrust—or with fear."[71]

After the ceremony, FDR briefly mingled with family and friends at a reception before excusing himself to go to bed. The next afternoon, under a veil of secrecy, he left Washington for Newport News to board the USS *Quincy*. His companions and confidants for the trip to Malta, and then Yalta, were Pa Watson and FDR's daughter Anna, who marveled at her father's knowledge of Virginia's shorebirds.[72] He had asked Pinchot, Ickes, and Brant to work on the global conservation initiative while he was away. The president's major stipulation for the conservation conference was that each participating nation should receive a seat for only a single, top-flight representative, ideally, the head of state.

Four strategic imperatives dominated the agenda at Yalta: the voting rules and membership criteria for the United Nations; the future of Eastern Europe, particularly Poland; the disposition of a defeated Germany; and the entry of the Soviet Union into the war in the Pacific. "If Tehran was in many ways a rehearsal for Yalta," David M. Kennedy wrote in *Freedom from Fear*, "then Yalta in turn set the stage for the dawning international regime that became known as the Cold War."[73]

Because Roosevelt kept no formal notes from his meetings at Yalta, it's impossible to ascertain how forcefully he pressed for global conservation. Churchill and Stalin weren't likely to be enthusiastic about his global conservation push, however; both men had more desperate matters on their minds. What was decided at Yalta was the partitioning of postwar Europe, with guidelines for the governance of Germany after its inevitable surrender, and an agreement that the Soviet Union would enter the war against Japan once Hitler had lost.

King Abdul Aziz ibn Saud of Saudi Arabia and President Roosevelt *(right)* met on board the USS *Quincy* in Great Bitter Lake in Egypt on February 14, 1945. Colonel William Eddy *(kneeling)* served as translator, while Admiral Wiliam D. Leahy looked on *(at left)*. Roosevelt, who was on the way home from the Yalta Conference, charmed the king, while laying serious groundwork for the postwar relationship between the two nations. "Every now and then," Eddy later recalled, "I could catch him off guard and see his face in repose. It was ashen in color; the lines were deep; the eyes would fade in helpless fatigue. He was living on his nerve."

On February 11, Roosevelt left Yalta for the *Quincy*, which had been moored in Great Bitter Lake in the Suez Canal. To the surprise of Churchill, FDR held secret talks with three prominent African and Middle Eastern leaders: King Ibn Saud of Saudi Arabia, King Farouk I of Egypt, and Emperor Haile Selassie of Ethiopia. Global conservation and the plight of Jewish refugees from Europe were both discussed. FDR then finally boarded the USS *Quincy* for the trip home. On February 23, he held a press conference on the ship. Connecting the nascent United Nations with global conservation, Roosevelt drove home the point that "reforestation is the best hope" for the prosperity of civilization:

"Wouldn't that be a long-time proposition?" one reporter asked.

"Growing trees is a long-time proposition," Roosevelt retorted.

"Do you mean that the conference looked ahead a great many years?" another reporter wanted to know.

"Sure," Roosevelt agreed, "we are looking at the human race, which we hope won't end in fifty years."[74]

A few days later, while crossing the Atlantic, Pa Watson suffered a cerebral hemorrhage and died. "One moment he was breathing," Anna Roosevelt Boettiger recorded in her diary, "and the next his pulse had stopped."[75] Roosevelt was beside himself with grief. Losing Pa was a terrible blow to him. Although Watson had spent his adult life in the U.S. Army, it was the navy that honored him when the *Quincy* finally docked in Newport News. Days later, Watson's remains were interred at Arlington National Cemetery, not far from the grave of General John Joseph "Blackjack" Pershing. For the burial Roosevelt, his face anguished and drawn, had his limousine park a few feet from the open grave. His relationship with Watson far transcended official interactions; he was perhaps his closest male friend. Now, the grim reality of his own human mortality—along with numerous other worries and cares—hung over the president like a dark cloud that wouldn't lift.

On March 28, 1945, Pinchot wrote to the president about his "rough plan" for the grandiose conservation summit. Not an inkling of the strategy was shared with Stettinius. "Here it is," he wrote, "T. R. introduced conservation to America. Nothing could be more fitting than that you, who have already done so much for conservation on this continent, should crown your good work by rendering the same great service to the rest of mankind."[76] Drained from the combination of the Yalta Conference and Pa Watson's death and suffering from dangerously high blood pressure, Roosevelt never had a chance to respond to Pinchot's letter. But he did see William O. Douglas to talk about conservation and was very alert, remembering out of thin air the names of Justice Douglas's two horses, Thunder and Lightning.[77]

In late March, Roosevelt returned to Warm Springs, his face looking like something death brought with it in a suitcase. Joining him in the Georgia countryside were his daughter, Anna; Daisy Suckley; Lucy Mercer Rutherfurd; and other relatives and colleagues. Swimming in the thermal pools, going for country drives, and resting among the pines trees allowed his overworked analytical powers to resuscitate themselves. On April 11, Henry Morgenthau arrived in Warm Springs and dined with the president, discussing methods for the re-pastoralization of postwar Germany, still eager to transform the war-torn country into Europe's breadbasket. Morgenthau later noted that though FDR seemed to have lost some of his coordination, the president seemed "happy and enjoying himself."[78] Roosevelt planned to work on his opening speech for the United Nations organizational conference on April 25.

On the morning of April 12, Roosevelt dressed smartly. Elizabeth Shoumatoff, a friend of Rutherfurd, was to paint a portrait of the president that day. As Roosevelt posed for the portrait, a sharp pain seized him, and his eyes rolled back in his head. A terrified Shoumatoff called out, "Lucy, something has happened!" Roosevelt had suffered a major cerebral hemorrhage. A few hours later, he died at the Little White House. Word spread rapidly across the land. It seemed impossible that the fearless commander in chief, was among the American war dead.[79] The last presidential order Roosevelt signed directed the U.S. Postal Service to issue a stamp celebrating the United Nations to coincide with the global summit in San Francisco later that month.[80]

Americans were stunned by the news of FDR's death. Declaring the "story is over," Eleanor Roosevelt tried to help the nation heal, putting her husband's personality in perspective. For public consumption she spoke of recently listening to Franklin hold forth at dinner about helping Saudi Arabia, Iran, and North Africa jumpstart a postwar reforestation project. She scolded him, saying that once Hitler and Tojo were defeated, he needed to let other people solve the world's problems. "With very characteristic emphasis," she recalled, "he turned to me and said, 'I like to be where things are growing.' "[81]

For Harry Truman, sworn in as America's thirty-third president less than four hours later in the White House, the moment felt as though a whole galaxy had crashed down on top of him. Pinchot was also in Washington, D.C., trying to organize the global conservation summit when his wife called him with the grim news. "At first I couldn't understand what she said," Pinchot wrote in his diary.

"Then came the dreadful news of the president's death. At first I didn't believe it. But it was true."[82]

Ralph McGill, a reporter for the *Atlanta Constitution*, was abroad when he heard the news of Roosevelt's passing. "To a Georgian far from home there was a sudden and bitter nostalgia for home at the news of the President's passing in Warm Springs," he wrote. "I could see the dogwood in bloom and the green of the trees. I knew that the peach blossoms were out and that the warm Georgia sun had been like a benediction to the tired body of the ailing president. And I wanted to be home with my own fellow Georgians as they mourned him. It was said of Abraham Lincoln when death claimed him that a tree is measured best when it is down. So it will be with Franklin D. Roosevelt. The tree is down and the historians will begin to measure and will find what the hearts of millions of Americans and peoples of the world already knew, that here was the tallest man America has ever given the world."[83]

EPILOGUE

"WHERE THE SUNDIAL STANDS"

I

On Friday morning, April 13, 1945, a throng of Georgian farmers and townspeople crowded the Warm Springs depot to see Franklin D. Roosevelt's flag-draped coffin carried on the *Ferdinand Magellan* to begin an eight-hundred-mile journey to Union Station in Washington, D.C. Eleanor Roosevelt had wisely tapped George Marshall, the master logician of both the CCC and the U.S. Army, to oversee the transport arrangements. As the eleven-car funeral train headed northward, with the first lady aboard, over two million citizens stood along the tracks in fields, hamlets, towns, and cities in five states to bid farewell to their leader. "The President would have enjoyed the ride," wrote Merriman Smith in the *New York Herald Tribune*. "He loved to sit beside the broad window of his private car and comment on the condition of the soil and forests."[1]

On Saturday, April 14, at 4 p.m., a funeral service was held for Roosevelt at the White House. Some two hundred family members and friends joined together in the East Room to pay their respects. Even though Eleanor Roosevelt had requested no flowers, the White House was overflowing with fragrant bouquets from thousands of mourners. Pinchot and Ickes stood next to each other, a New Deal united front at the dawn of the age of Truman. After the ceremony, Roosevelt's casket was placed on a caisson drawn by six white horses, accompanied by battalions of blue jackets, field artillery, air forces, and women's auxiliary forces, to the railroad depot filled with mourners. Not since 1865, when Abraham Lincoln was assassinated by John Wilkes Booth, had such collective shock and grief engulfed the nation's capital.

Once the tributes in Washington were over, Roosevelt's casket traveled north by train, along the Atlantic line through Baltimore and Philadelphia, and passed New York City and eventually Poughkeepsie, until verdant Hyde Park revealed itself in springtime glory. All five of Roosevelt's children, as well as Eleanor, were aboard. Apple trees and lilacs were happily in bloom. The April sky was cerulean blue. As Ickes reported in his diary, "The air was clear and cool. Everything was dignified and in good taste."[2] As a cannon fired twenty-one times, the coffin was moved from the railcar to a caisson drawn by six handsome horses. A seventh horse, hooded, was brought into the procession symbolizing a lost warrior, boots reversed in the stirrups. Drummers played a dirge as the mourners climbed up the steep hill toward the house, and FDR's gravesite nearby. Five years earlier Roosevelt had shown his longtime estate superintendent William Plog his chosen burial site ("where the sundial stands in the garden").[3] In coming months, a simple marble monument marking

The funeral for Franklin Roosevelt took place on April 15, 1945, in the rose garden at Springwood in Hyde Park. Three days before, the president had died of a cerebral hemorrhage in his "Little White House" at Warm Springs, Georgia. The news was a shock like few that America has ever had.

the burial site of Franklin and Eleanor was crafted to FDR's exact specifications and placed on the grave.

President Truman and hundreds of cadets from West Point were gathered at the rose garden of the Springwood estate. Many top echelon New Dealers weren't quite convinced that Truman would be up to the job as president. Ickes feared that Truman would be too easily intimidated by stockmen associations, the oil lobby, and timber corporations and would fail to establish new national monuments (or enlarge existing ones) in the Colorado Plateau or wildlife refuges in the Gulf South. As Ickes would sneer in 1948, Truman was the kind of pro-business leader who allowed "the oil companies to get away with murder."[4] Ickes, who left office in 1946, was right. The only national monument that Truman established was Effigy Mounds in Iowa. But in 1947 Truman did bring Roosevelt's beloved Everglades fully into the National Park Service.[5]

Well into the twenty-first century, Springwood remained the only place in the United States where a president had been born, grown up, and laid to rest. In a very specific provision in his will, Roosevelt had requested that the over half million trees he planted between 1912 and 1945 be protected in perpetuity. If one of his trees died, then another was to be planted in its place. The National Park Service was tasked with overseeing this program. The pond where Roosevelt swam in his efforts to recover from polio and the hemlock hedge were also carefully preserved. Even the river bluffs across the Hudson would be preserved as a memorial. "My husband's spirit will live in this house, in the library, and in the quiet garden where he wished his body to lie," Eleanor Roosevelt said. "It is his life and his character and his personality which will live with us and which will endure and be imparted to those who come to see the surroundings in which he grew. . . . He would want them to enjoy themselves in these surroundings and to draw from them rest and peace and strength as he did all the days of his life."[6]

The state of Georgia designated the Little White House and its grounds a state historic site. Most of the trees Roosevelt had planted in Warm Springs were protected, and an elegant walkway—now known as the Walk of Flags and Stones—was erected as a memorial.

Eventually each of the states donated its own block of indigenous stone to the historic site. Gifts from states where the CCC had excelled were especially innovative. Arizona donated a cross section of richly colored petrified logs from its Petrified Forest National Monument. The Iowa State Conservation Commission provided a rough piece of Sioux County quartzite from Welch Lake. William D. Hassett, Roosevelt's correspondence secretary, presented the Little White House memorial with a slab of blue granite from Vermont.

Immediately after the funeral, FDR's associates and friends boarded the train back to Washington. Everybody was glum—except Ickes. Roosevelt usually had a stiff drink after a funeral, so Ickes, in the same spirit, went hunting for whiskey, and procured a flask. He tossed back a drink and passed the bottle around. Within a few moments the mood of the mourners—including Postmaster General Frank Walker (the supplier of the hooch); Labor Secretary Frances Perkins; and Press Secretary Jonathan Daniels and his father, Josephus Daniels, the former naval secretary to Woodrow Wilson—turned giddy with nostalgia. Ickes's wife, Jane, insisted that her husband be a gentleman and wander to the president's private Pullman car to wish Truman good luck. A somewhat reluctant Ickes obliged.

That's when the trouble occurred. Ickes encountered Democratic National Committee chairman Edwin Pauley, acting as gatekeeper to the president. Pauley, who had made a vast fortune as a California oilman, nonchalantly asked Ickes if the Departments of Justice and Interior could make sure that the California tidelands could be opened to the private sector for drilling. To Ickes, these tidelands belonged to the American people to be collectively enjoyed in perpetuity. Back in 1938, the Roosevelt administration had established the Channel Islands National Monument off the coast of Santa Barbara, home to the largest seal and sea lion breeding colonies in America.[7] At that time Ickes had wanted the entire coast of California, from Camp Pendleton to the Oregon line, protected as a public trust. So he was deeply offended when Pauley intimated that he would donate $300,000 to $400,000 to the California Democratic Party if Ickes would cooperate. Deeming this overture a "treacherous" bribe, a frosty exchange ensued between Ickes and Pauley. Luckily Truman woke before Ickes got too heated. But the matter would come to a boil in 1946, when Ickes was fired as secretary of the interior by Truman. He died in 1952, leaving behind an eighty-thousand-word diary, deposited at the Library of Congress, which constitutes a primary-source treasure trove of conservation ideas and policy initiatives in the age of FDR.[8]

Ickes wasn't the only New Deal conservationist to leave the new administration. With his dear friend gone, Henry Morgenthau resigned to promote Christmas tree farms, pastoral living, and the new nation of Israel. (After Morgenthau died in 1967, all 840 volumes of his diary were donated to the Franklin D. Roosevelt Presidential Library in Hyde Park, New York.) Gifford Pinchot died in October 1946, shortly before his autobiography, *Breaking New Ground*, was published; a national forest was named after him in Washington State. Suffering with bad health, Ding Darling stopped drawing cartoons in the late 1940s, spending his remaining years creating what would become the J. N. "Ding" Darling National Wildlife Refuge in Florida. And Frederic Delano died in 1953, working behind the scenes for the beautification of America until his last breath.

As historians look back over the twentieth century, they discover that it was the Roosevelt-Ickes united front that turned the National Park Service into perhaps the most beloved agency in the U.S. government. The key moment was in 1933, when Roosevelt, under the authority of the Reorganization Act approved by Congress, transferred America's national military parks and other areas of military importance previously administered by the War Department, and the national monuments lying within the national forests, which had been in the portfolio of the Forest Service of the Department of Agriculture, to the jurisdiction of the NPS. Roosevelt's "master plan" made it clear that he wanted the NPS to represent both America's historical and natural history and heritage. These Roosevelt transfers, early in the New Deal, as well as the dozens of new sites established during his long presidency, gave the NPS more than five times as many areas as on August 25, 1916, when the Organic Act was signed by President Wilson.[9]

Roosevelt's legacy in conservation was large and contained multitudes. Between 1932 and 1945, the president had also established numerous national forests, many through the acquisition of abandoned or cut-over land. All of these native forests, as Pinchot believed, were living testimonials to Roosevelt's unshakable conviction that no landscape, not even the Mojave Desert, should be bereft of robust trees. "It is an error to say that we have 'conquered Nature,' " Roosevelt told Congress more than a decade earlier. "We must, rather, start to shape our lives in a more harmonious relationship with Nature. This is a milestone in our progress toward that end. The future of every American family everywhere will be affected by the action we take."[10]

Just how important all the New Deal wildlife refuges were became abundantly clear after World War II. With soldiers coming home, trained to shoot guns, there was a marked increase in hunting and fishing. Many veterans wanted to take a well-earned month or two for recreation. In 1945, eight million hunting licenses were issued, an increase of one million licenses since the start of the war. That same year the issuance of fishing licenses rose by a half million. But because of Roosevelt's new National Wildlife Refuges, the increased hunting was absorbed without depleting wildlife populations.

Even though the CCC had been dissolved in 1942, the work of the "boys" remained visible from coast to coast. Durable structures ran the gamut from rustic Bascom Lodge near the summit of Mount Greylock in Massachusetts to the stonework observation tower atop Mount Tamalpais in California; from the Art Moderne visitor center at Ocmulgee National Monument in Georgia to the handsome picnic enclosures in Cuyahoga County, Ohio.[11] Over the course of its nine-year existence, the CCC conserved more than 118 million acres of national resources throughout America, more acreage than all of California.[12] During the Cold War, statewide conservation corps continued to involve young people in their late teens and early twenties—though these organizations never caught the public's imagination as Roosevelt's Tree Army did. Beginning in the 1960s, women were recruited for such programs. In the twenty-first century, projects in California, Colorado, and Missouri similar to those of the CCC have become unqualified successes.

In 2007, two national CCC alumni groups merged to form the CCC Legacy Foundation, setting up headquarters in St. Louis and Edinburg, Virginia. Although the organization collected oral histories and helped museums acquire memorabilia, perhaps its greatest achievement was the unveiling of sixty-two bronze statues honoring CCC veterans in state and federal parks. The first statue was placed in North Higgins Lake State Park in Roscommon, Michigan. The most moving, however, can be found in Highlands Hammock State Park in Sebring, Florida. The artwork was dedicated to Emil Billitz, who was paralyzed as a result of a truck accident that occurred during his CCC service. Some of the statues have plaques accompanying them. The plaque on the statue at the School of Conservation in Branchville, New Jersey, reads, "These men participated in the world's most famous conservation program. America will never be able to repay them. All that is great and good about conservation we owe to the CCC."[13]

Perpetuating a permanent CCC might have been the best tribute to FDR's visionary commitment to the environment. Veteran environmentalists—including Eleanor Roosevelt—were disappointed by the Truman and Eisenhower administrations. Following Franklin's death, Eleanor retired to her serene Val-Kill cottage, doting on her children and grandchildren more than ever before. In 1946, however, President Truman asked her to be a delegate to the United Nations General Assembly. In this role she became the most respected human rights activist in the world. Upon resigning in 1952, Eleanor traveled the world, sometimes promoting global reforestation projects, assuming the role of elder stateswoman of the Democratic Party. She died on November 7, 1962.

During the 1960s, the CCC reentered the public discourse when President John F. Kennedy urged Congress to establish the Peace Corps, and President Lyndon B. Johnson incorporated the Job Corps and his domestic answer to the Peace Corps, Volunteers in Service to America (VISTA) as part of the Economic Opportunity Act of 1964. But LBJ wasn't happy with these programs. One can hear LBJ on a tape (August 7, 1964) complaining to White House aide Bill Moyers about the administration's poverty programs. "I thought," he says, "we were going to have CCC camps."[14]

On May 30, 1959, Lyndon B. Johnson symbolically turned a shovel on a recently planted tree in front of the Franklin Roosevelt Library on the Springwood grounds. Mrs. Roosevelt looked on approvingly. Johnson, then a Senate majority leader with his eye on a run for the White House, regarded Roosevelt as his political hero.

President Kennedy, impressed with FDR's work-relief conservation record and national seashore push, identified his own "New Frontier" within the legacy of the New Deal. His secretary of the interior, Stewart Udall, of Arizona, was as brilliantly effective in the role as his hero Ickes had been. Justice William O. Douglas—the direct link between the New Deal and the Great Society in the conservation realm—argued that the "preservation of values technology will destroy . . . is indeed the New Frontier."[15] Running for president in 1960, Kennedy advocated using Roosevelt's Cape Hatteras model for saving seashores as wildlife refuges and recreational areas. During his brief presidency, he established three national seashores: Cape Cod, Massachusetts; South Padre Island, Texas; and Point Reyes, California. In 1961, speaking at the dedication of Gifford Pinchot's home, Grey Towers, to the National Park Service, JFK sounded like an echo of FDR. "In the field of [natural] resources," he warned, "the opportunities which are lost now can never be won back."[16]

By the time U.S. Fish and Wildlife alumna Rachel Carson published *Silent Spring* in 1962—the turning point in American environmentalism—FDR was widely praised for having saved such landscapes as the Kenai Peninsula, the Okefenokee, the Olympics, the Great Smokies, Isle Royale, Joshua Tree, Capitol Reef, Jackson Hole, Mammoth Cave, Kings Canyon, the Everglades, Big Bend, and the Desert Game Range of Nevada. Irving Brant, the activist of the Emergency Conservation Committee, once finished with his multivolume James Madison biography project, published *Adventures in Conservation with FDR*, in 1988, giving Roosevelt vast credit for championing the ecological well-being of American landscapes. Justice Douglas, in two *My Wilderness* books, also enshrined the New Deal as being the incubator for the environmental justice movement of the 1960s and beyond.[17] Through political know-how, legislative muscle, and fearlessness in using executive authority, FDR had crafted a conservation legacy to match or even surpass that of Theodore Roosevelt.

How shortsighted Congress had been for not making the CCC a permanent government agency became glaringly apparent during the Cold War and beyond. Once World War II got the American economy over the hump of the Great Depression, his Tree Army was seen as a relic of a bygone era. Many CCC records were merely thrown out. The closest the nation got to resuscitating the CCC was with President Bill Clinton's AmeriCorps NCCC (National Civilian Community Corps), a full-time, team-based residential program for men and women ages eighteen to twenty-four. As in the CCC, these AmeriCorps volunteers were assigned for ten months to one of five campuses located in Denver; Sacramento; Perry Point, Maryland; Vicksburg, Mississippi; and Vinton, Iowa. AmericaCorps, in its mission to complete "service projects" throughout the United States, as director Kate Raftery explained, "was truly based on the CCC model even in terms of the military aspects that the CCC had."[18] A big difference was that whereas the CCC had approximately 250,000 young men working on reforestation projects, AmeriCorps averaged around 1,200 men and women annually. During the mid-1990s a marvelous Environmental Corps (E-Corps) program took root in Texas aimed at restoring the state park system FDR helped to build.

President Barack Obama, working with his secretaries of Interior, Ken Salazar and Sally Jewell, sought to bring back Roosevelt's idea of the CCC by establishing a national council, among eight federal departments, to put America's youth and returning veterans to work protecting and restoring the great American outdoors. The 21st Century Conservation Service Corps (21CSC), built on the legacy of the CCC, aimed to train a new generation of what Salazar called "environmental stewards" committed to improving the conditions of America's public lands.[19] The 21CSC program absorbed the Texas corps and other state versions of a youth-oriented conservation effort.

Once the fear of climate change became a twenty-first-century reality, there were calls from Al Gore to Pope Francis to start weaning humans off fossil fuels, replacing the natural resources with renewable sources of energy. Some environmentalists called for a huge federal undertaking on the scale of the Apollo space program or the Marshall Plan to discover new energy technologies. But Bill McKibben, the founder of 350.org, a cutting-edge environmental nonprofit, thought a new CCC-inspired tree-planting army was what was needed. "The CCC planted 3 billion trees (which would be no small help with global warming)," wrote McKibben. "Imagine an army of similar size trained to insulate American homes and stick solar photovoltaic panels on their roofs. They could achieve, within a year or two, easily noticeable effects on our energy consumption; our output of carbon dioxide might actually begin to level off. And imagine them laying trolley lines back down in our main cities or helping erect windmills across the plains. All this work would have real payoff—and none of it can be outsourced."[20]

II

Back at the Department of the Interior, following the FDR's funeral, Ickes brainstormed about how to properly honor his late boss. A number of options were considered, including planting a grove of tulip poplars at Springwood; establishing an FDR monument on the National Mall in Washington D.C.; carving his face on Mount Rushmore; and renaming Patuxent Wildlife Research Center in his honor. Brazil's foreign minister, Pedro Leão Velloso, suggested that a memorial service be held for FDR in Muir Woods National Monument during the upcoming United Nations conference in San Francisco. Logistically this made sense to Ickes because delegates from forty-six nations would be in attendance. Owen A. Tomlinson, regional director of the National Park Service, wrote the custodian of Muir Woods in early May informing him the FDR memorial service would be held in Cathedral Grove as a "tribute to the late President Roosevelt's leadership in conservation."[21]

On May 19, 1945, five hundred delegates helping to organize the United Nations attended a memorial service for the late president at the Muir Woods National Monument north of San Francisco. In humble tribute, they sat among the towering trees, listening to a navy band and eloquent speeches. Even in death, Roosevelt brought attention to trees and the beauty they bring to even the saddest of days.

The National Park Service issued a press release on May 12 declaring Muir Woods the most fitting location for honoring the late president. "The site in the monument chosen for the meeting is aptly named—Cathedral Grove, it was pointed out. In this quiet grove is the impressiveness of a temple. Massive fluted columns, the trunks of the great coast redwoods, support a ceiling of green and the sunlight filters in as through a church window. It is a place designed by nature to engender a feeling of peace and reverence, in keeping with the humanitarian ideals responsible for the United Nations conference."[22]

In the late afternoon of May 19, 1945, United Nations delegates boarded buses at the Fairmont Hotel to trek the sixteen miles across the Golden Gate Bridge to Muir Woods National Monument for the 5 p.m. service at "Cathedral Grove," on the western slope of Mount Tamalpais. What the delegates didn't realize while driving on Highway 101 was that outside their bus windows was the handiwork of Roosevelt's CCC, including the Mountain Theatre at Mount Tamalpais (a natural amphitheater with stone seating for four thousand).[23]

At Cathedral Grove the combination of the majestic trees and the solemnity of the occasion caused the congregation of five hundred to speak in hushed, reverential tones. That serenity was shattered when the U.S. Navy band struck up "America the Beautiful" and "The Star-Spangled Banner," behind a makeshift stage. Four wounded veterans of World War II formed an honor guard.[24] Tomlinson of the National Park Service spoke first. He recounted how President Roosevelt believed that the redwood groves, painted deserts, rain forests, and bird-breeding grounds were all part of the rightful heritage of the American people. To Roosevelt, public lands were the heart and soul of the nation. That he wanted to bring his New Deal beautification revolution to the rest of the world—"Conservation Is a Basis of Permanent Peace"—was at the center of the memorial ceremony at Muir Woods.

Chairman Leão Velloso spoke eloquently about Roosevelt's courage and audacity in the face of adversity: "Much has been said, and it is certain much more will be said through the ages, of Franklin Roosevelt, but never enough will be said."[25] More than any other country, Brazil would go on to emulate the restoration example of the New Deal, initiating a one-billion-tree planting program to revitalize the Amazon forest.[26]

Likewise, Field Marshal Jan Christian Smuts, prime minister of the Union of South Africa, spoke about how Roosevelt drew strength from the natural world. "Here among the great redwoods, this great man will find fitting and congenial company," Smuts said as a plaque was laid next to an ancient tree. "Here, henceforth, will be the company of the giants."[27]

The final eulogy was that of Edward Stettinius, who had telephoned Pinchot shortly after Roosevelt's death in an attempt to extinguish any hard feelings. "Stettinius called about 4:30 to talk about proposed World Conference," Pinchot wrote in his diary on April 19. "He is for it on basis of conservation and said repeatedly he was heartily for it as a step toward permanent peace. That is a great surprise and a great satisfaction."[28]

At Cathedral Grove, surrounded by the UN delegates, looking around him at the old-growth trees, and sunbeams dancing down on the forest floor, Stettinius spoke from the heart. "I often heard him talk of the trees he planted and grew at Hyde Park," he said of FDR. "He rests for all time in hallowed grounds surrounded by these and older trees that held for him such cherished memories." Then Stettinius, pointing around him at the trees encircling them, connected Roosevelt to the sublime Cathedral Grove, and to the ebb and flow of American history. "These great redwoods at Muir Woods National Monument are the most enduring of all trees," he said. "Many of them stood here centuries before Christopher Columbus landed in the New World. They will be here centuries after every man now living is dead. They are as timeless and as strong as the ideals and faith of Franklin D. Roosevelt."[29]

Fifty years later, on May 19, 1995, the United Nations and the National Park Service gathered at Cathedral Grove to observe the fiftieth anniversary of the memorial service and the legacy of the United Nations. The following year, on July 8, 1996, an eight-hundred-year-old redwood tree crashed down in Cathedral Grove next to the plaque honoring FDR. Many visitors wondered why the debris from the two-hundred-foot giant wasn't removed; it was obstructing the Roosevelt memorial. The official National Park Service response spoke legions about the conservation ideals that FDR had imprinted on the nation: "The tree, which toppled gracefully up-slope, caused no damage and required no cleanup. The tree will remain where it fell, providing nutrients to the soil, nesting for birds, bedding for plants, and water for everything."[30]

Roosevelt would have been proud.

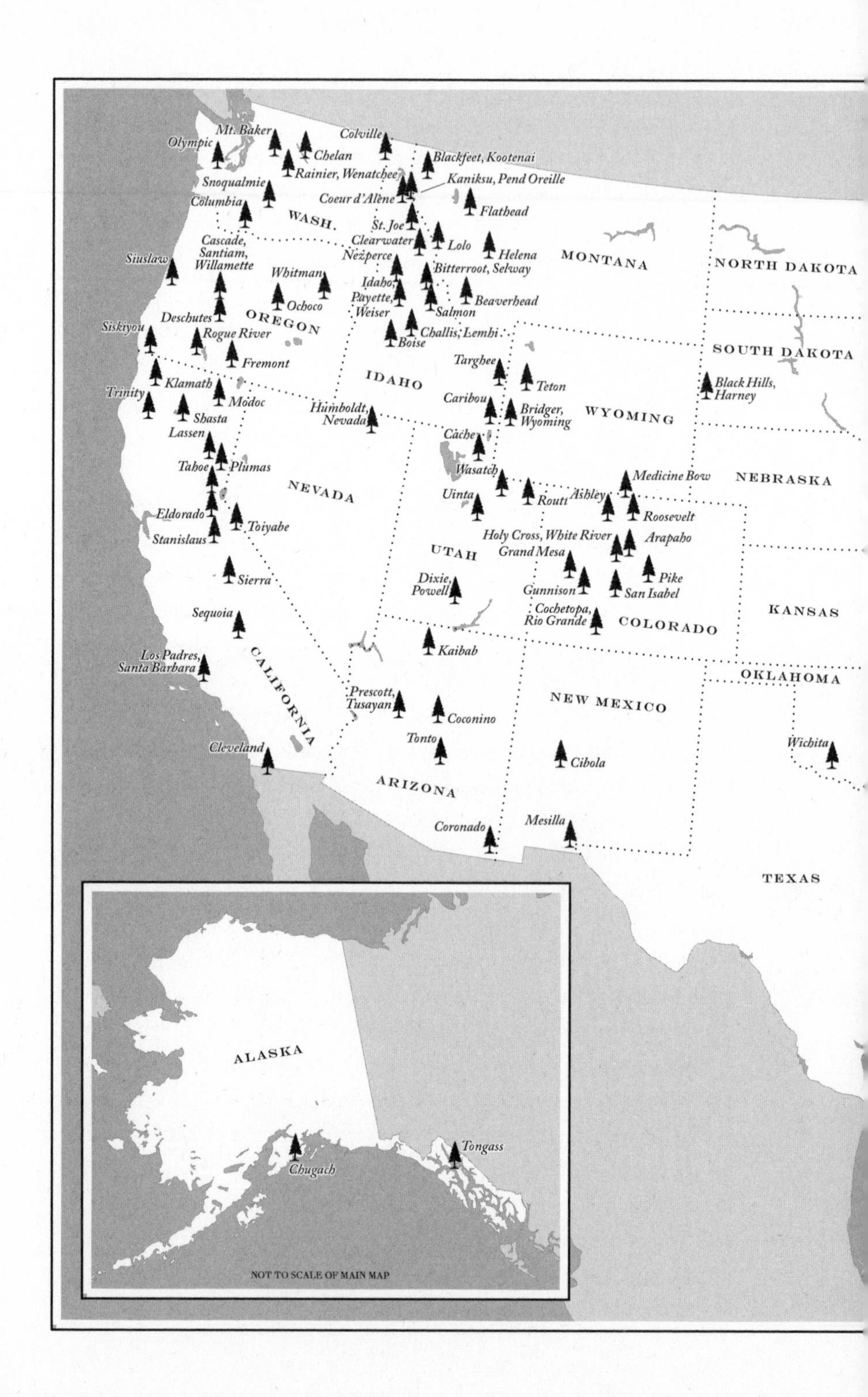

Olympic
Mt. Baker
Chelan
Colville
Rainier, Wenatchee
Blackfeet, Kootenai
Kaniksu, Pend Oreille
Snoqualmie
Coeur d'Alene
Flathead
Columbia
WASH.
St. Joe
Clearwater
Lolo
Helena
Cascade, Santiam, Willamette
Nezperce
Bitterroot, Selway
MONTANA
NORTH DAKOTA
Siuslaw
Whitman
Idaho, Payette, Weiser
Ochoco
Beaverhead
Salmon
Deschutes
OREGON
Challis, Lembi
Siskiyou
Rogue River
Boise
Fremont
SOUTH DAKOTA
Targhee
IDAHO
Teton
Black Hills, Harney
Klamath
Trinity
Modoc
Caribou
Shasta
Humboldt, Nevada
Bridger, Wyoming
WYOMING
Lassen
Cache
Tahoe
Plumas
Wasatch
Medicine Bow
NEBRASKA
NEVADA
Uinta
Routt
Ashley
Roosevelt
Eldorado
Toiyabe
Stanislaus
Holy Cross, White River
Arapaho
UTAH
Grand Mesa
Sierra
Dixie, Powell
Pike
Gunnison
San Isabel
KANSAS
Sequoia
Cochetopa, Rio Grande
COLORADO
Kaibab
Los Padres, Santa Barbara
CALIFORNIA
OKLAHOMA
Prescott, Tusayan
NEW MEXICO
Coconino
Tonto
Cleveland
Wichita
Cibola
ARIZONA
Coronado
Mesilla
TEXAS
ALASKA
Tongass
Chugach
NOT TO SCALE OF MAIN MAP

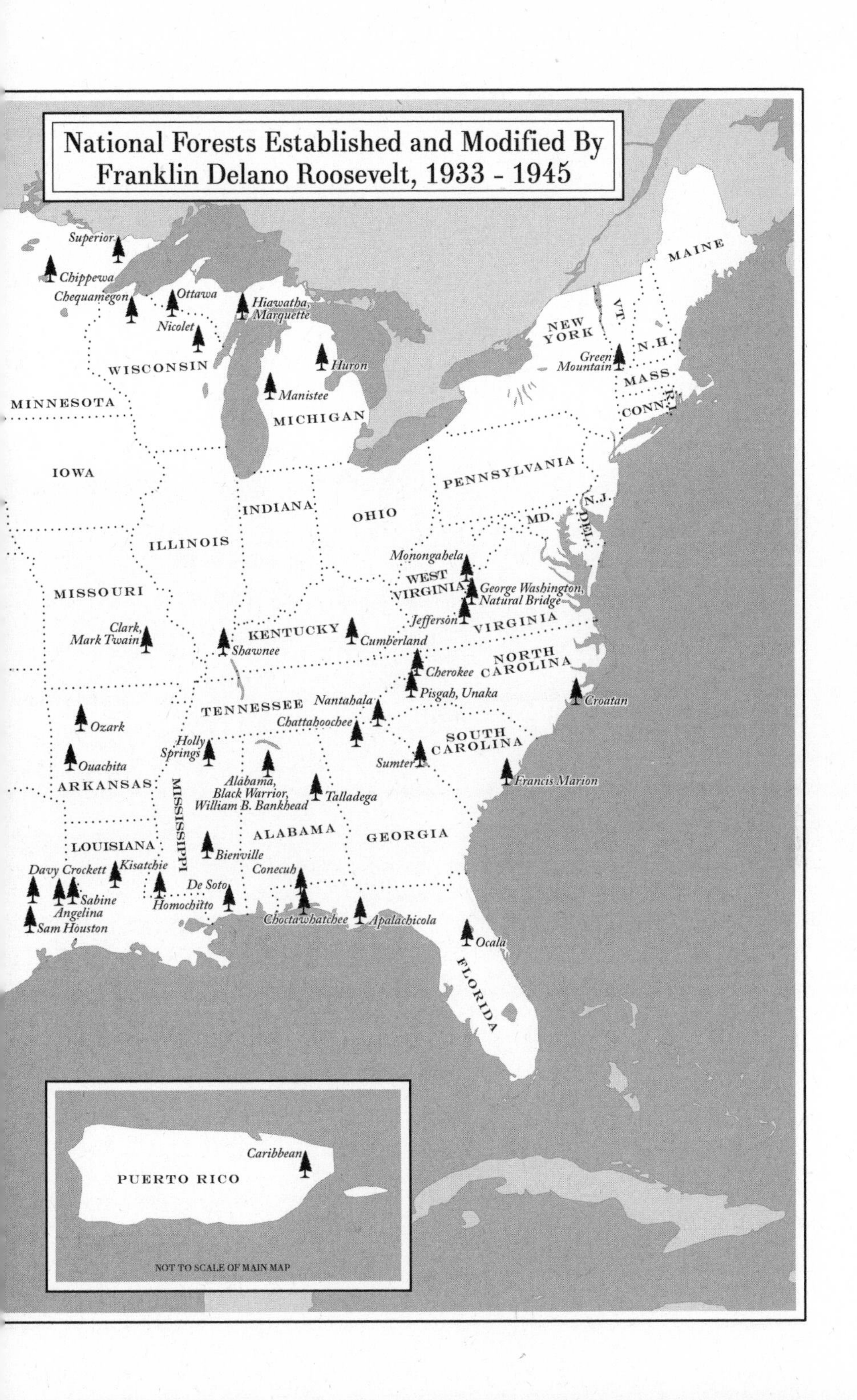
National Forests Established and Modified By Franklin Delano Roosevelt, 1933 - 1945
Superior
Chippewa
Chequamegon
Ottawa
Nicolet
Hiawatha, Marquette
Huron
Manistee
WISCONSIN
MINNESOTA
MICHIGAN
MAINE
NEW YORK
VT.
N.H.
Green Mountain
MASS.
CONN.
R.I.
IOWA
PENNSYLVANIA
INDIANA
OHIO
N.J.
MD.
DEL.
ILLINOIS
Monongahela
WEST VIRGINIA
George Washington, Natural Bridge
Jefferson
VIRGINIA
MISSOURI
Clark, Mark Twain
KENTUCKY
Shawnee
Cumberland
Cherokee
NORTH CAROLINA
Pisgah, Unaka
Croatan
TENNESSEE
Nantahala
Chattahoochee
Ozark
Holly Springs
SOUTH CAROLINA
Sumter
Ouachita
Francis Marion
ARKANSAS
MISSISSIPPI
Alabama, Black Warrior, William B. Bankhead
Talladega
ALABAMA
GEORGIA
LOUISIANA
Bienville
Kisatchie
Davy Crockett
Conecuh
De Soto
Sabine
Angelina
Sam Houston
Homochitto
Choctawhatchee
Apalachicola
Ocala
FLORIDA
Caribbean
PUERTO RICO
NOT TO SCALE OF MAIN MAP

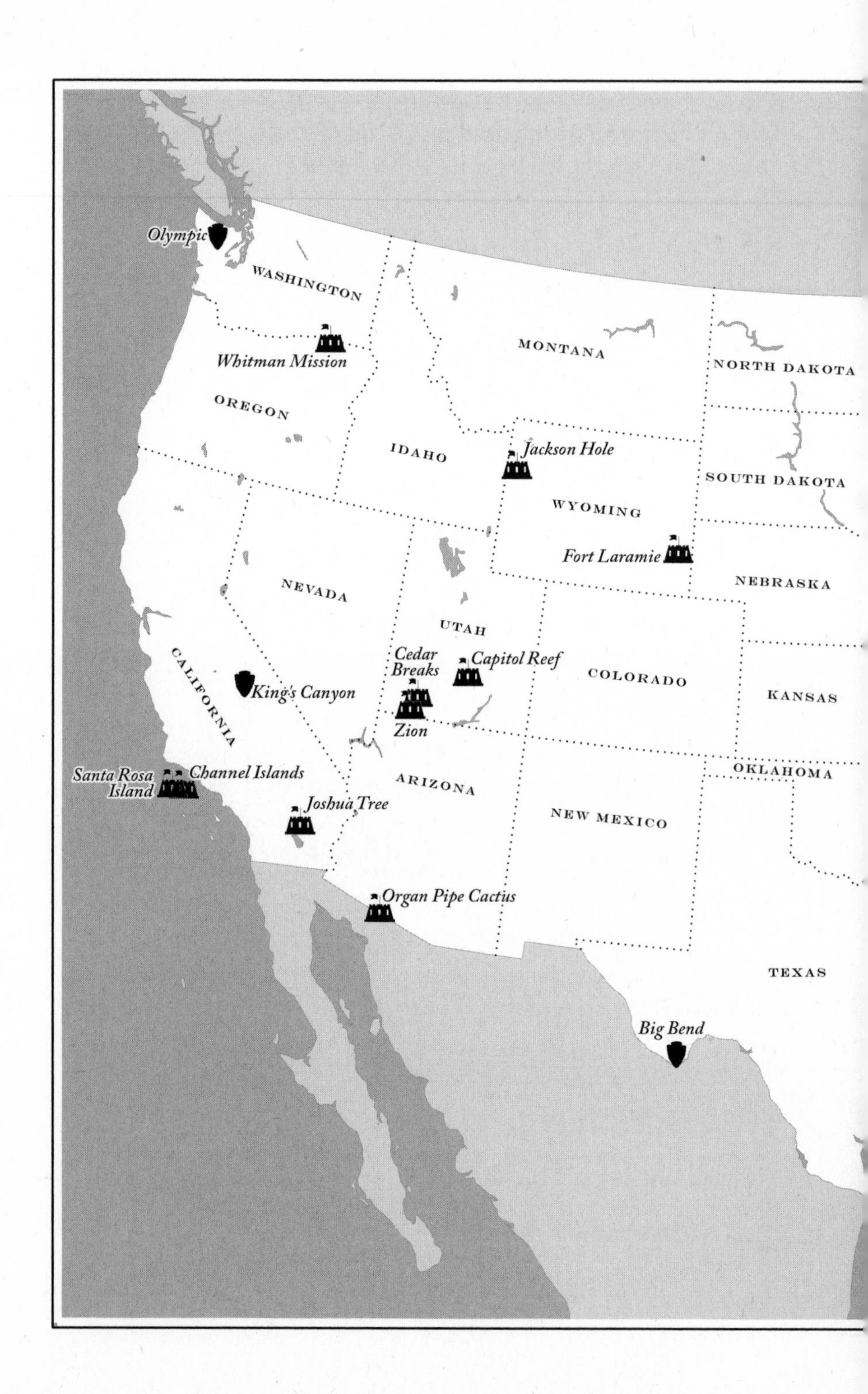
Olympic
WASHINGTON
Whitman Mission
OREGON
MONTANA
NORTH DAKOTA
IDAHO
Jackson Hole
SOUTH DAKOTA
WYOMING
Fort Laramie
NEBRASKA
NEVADA
UTAH
Cedar Breaks
Capitol Reef
COLORADO
KANSAS
CALIFORNIA
King's Canyon
Zion
Santa Rosa Island
Channel Islands
ARIZONA
OKLAHOMA
NEW MEXICO
Joshua Tree
Organ Pipe Cactus
TEXAS
Big Bend

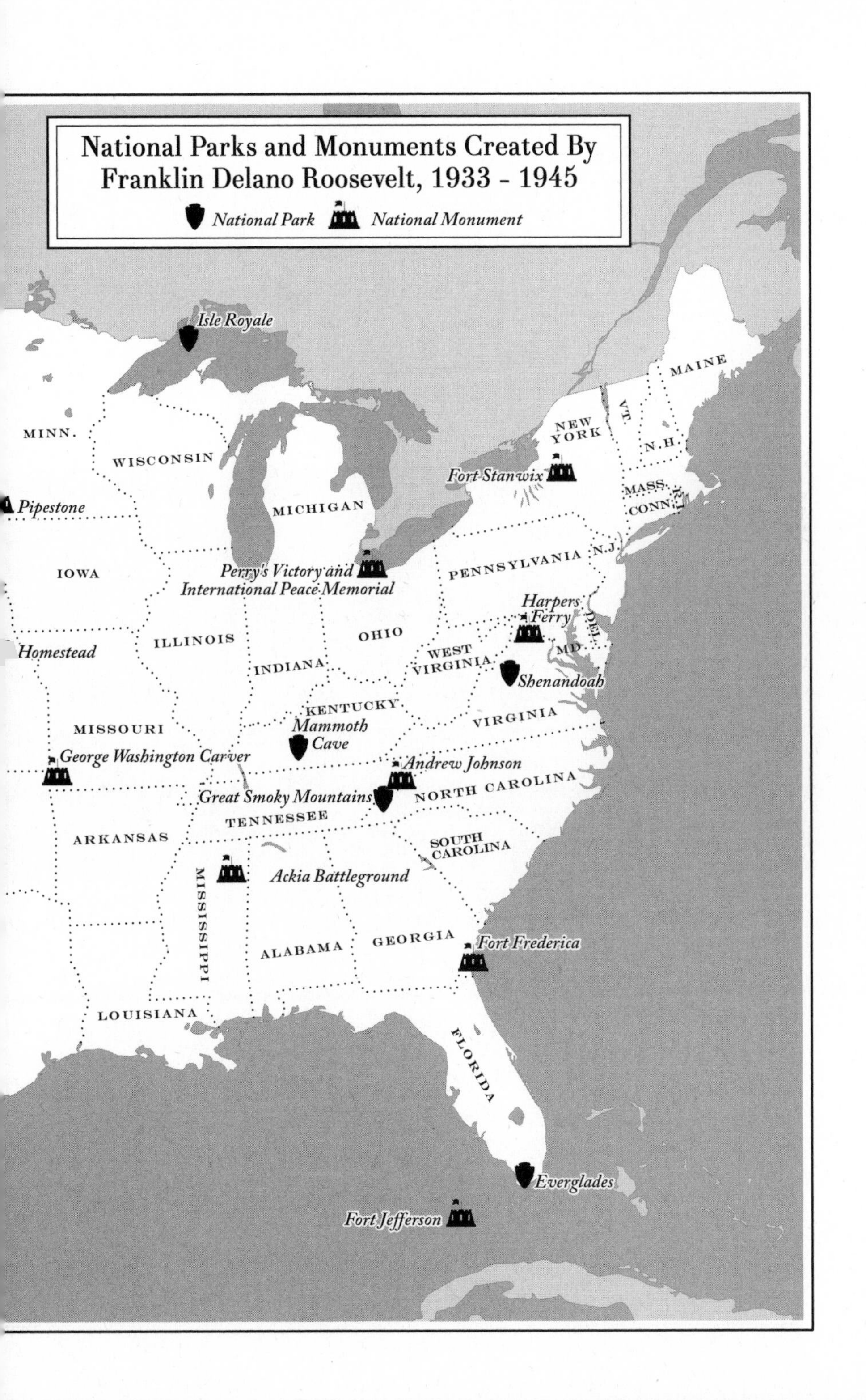
National Parks and Monuments Created By
Franklin Delano Roosevelt, 1933 - 1945
National Park
National Monument
Isle Royale
MINN.
WISCONSIN
MICHIGAN
Pipestone
IOWA
Perry's Victory and
International Peace Memorial
ILLINOIS
Homestead
INDIANA
OHIO
MISSOURI
George Washington Carver
KENTUCKY
Mammoth
Cave
Andrew Johnson
Great Smoky Mountains
TENNESSEE
ARKANSAS
MISSISSIPPI
Ackia Battleground
ALABAMA
GEORGIA
LOUISIANA
Fort Frederica
FLORIDA
Everglades
Fort Jefferson
SOUTH
CAROLINA
NORTH CAROLINA
VIRGINIA
WEST
VIRGINIA
Shenandoah
MD.
DEL.
Harpers
Ferry
PENNSYLVANIA
N.J.
NEW
YORK
Fort Stanwix
MASS.
CONN.
R.I.
VT.
N.H.
MAINE

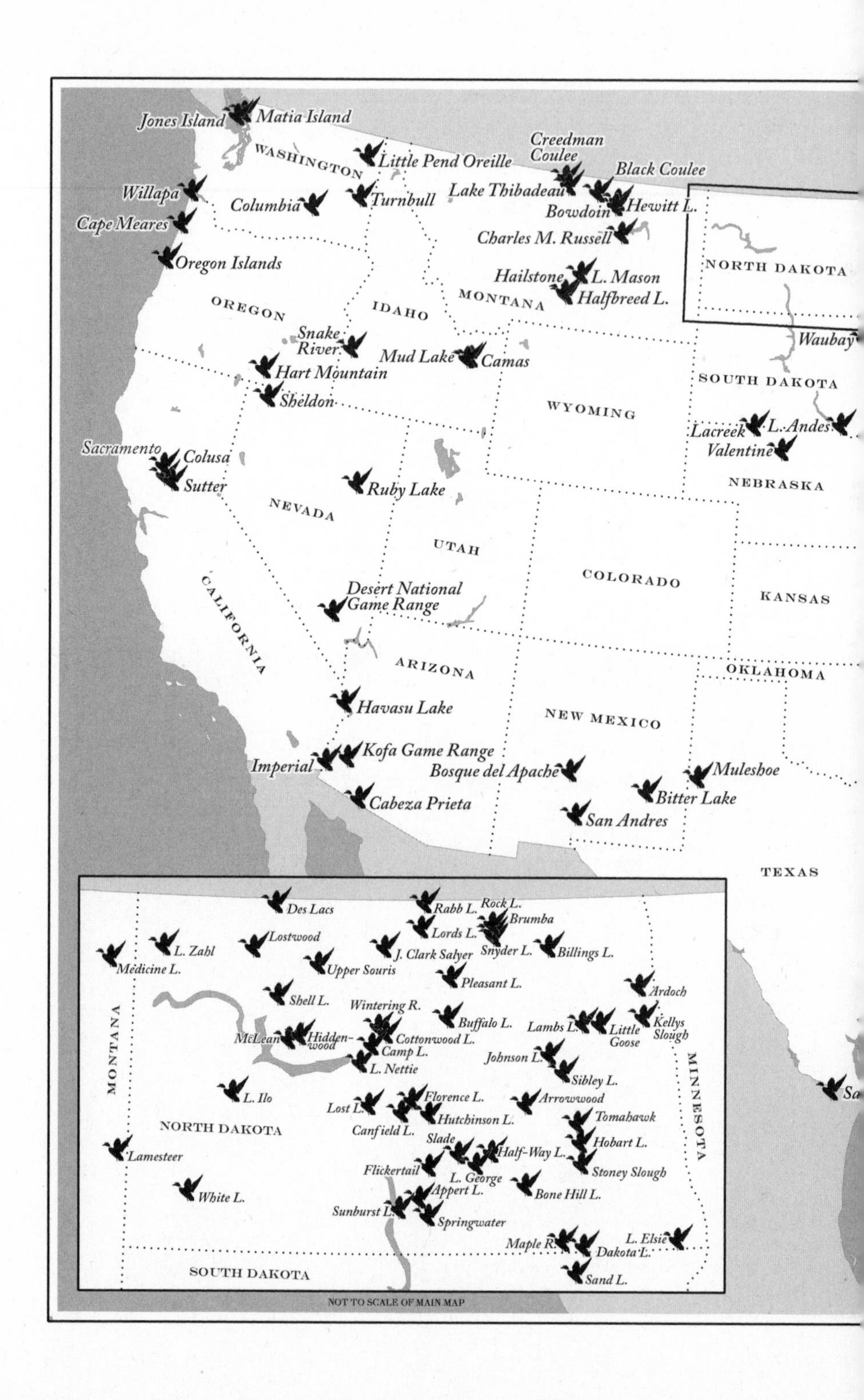
Jones Island
Matia Island
WASHINGTON
Little Pend Oreille
Creedman Coulee
Black Coulee
Willapa
Columbia
Turnbull
Lake Thibadeau
Bowdoin
Hewitt L.
Cape Meares
Charles M. Russell
Oregon Islands
Hailstone
L. Mason
Halfbreed L.
NORTH DAKOTA
OREGON
IDAHO
MONTANA
Snake River
Mud Lake
Camas
Waubay
Hart Mountain
SOUTH DAKOTA
Sheldon
WYOMING
Lacreek
L. Andes
Valentine
Sacramento
Colusa
Sutter
NEBRASKA
Ruby Lake
NEVADA
UTAH
COLORADO
KANSAS
CALIFORNIA
Desert National Game Range
ARIZONA
OKLAHOMA
Havasu Lake
NEW MEXICO
Kofa Game Range
Imperial
Bosque del Apache
Muleshoe
Bitter Lake
Cabeza Prieta
San Andres
TEXAS
Des Lacs
Rabb L.
Rock L.
Brumba
Lords L.
L. Zahl
Lostwood
J. Clark Salyer
Snyder L.
Billings L.
Medicine L.
Upper Souris
Pleasant L.
Ardoch
Shell L.
Wintering R.
Buffalo L.
Lambs L.
Little Goose
Kellys Slough
McLean
Hidden-wood
Cottonwood L.
Camp L.
L. Nettie
Johnson L.
Sibley L.
MONTANA
MINNESOTA
L. Ilo
Lost L.
Florence L.
Arrowwood
Hutchinson L.
Tomahawk
NORTH DAKOTA
Canfield L.
Hobart L.
Slade
Half-Way L.
Lamesteer
Flickertail
L. George
Stoney Slough
Appert L.
Bone Hill L.
White L.
Sunburst L.
Springwater
Maple R.
L. Elsie
Dakota L.
SOUTH DAKOTA
Sand L.
NOT TO SCALE OF MAIN MAP

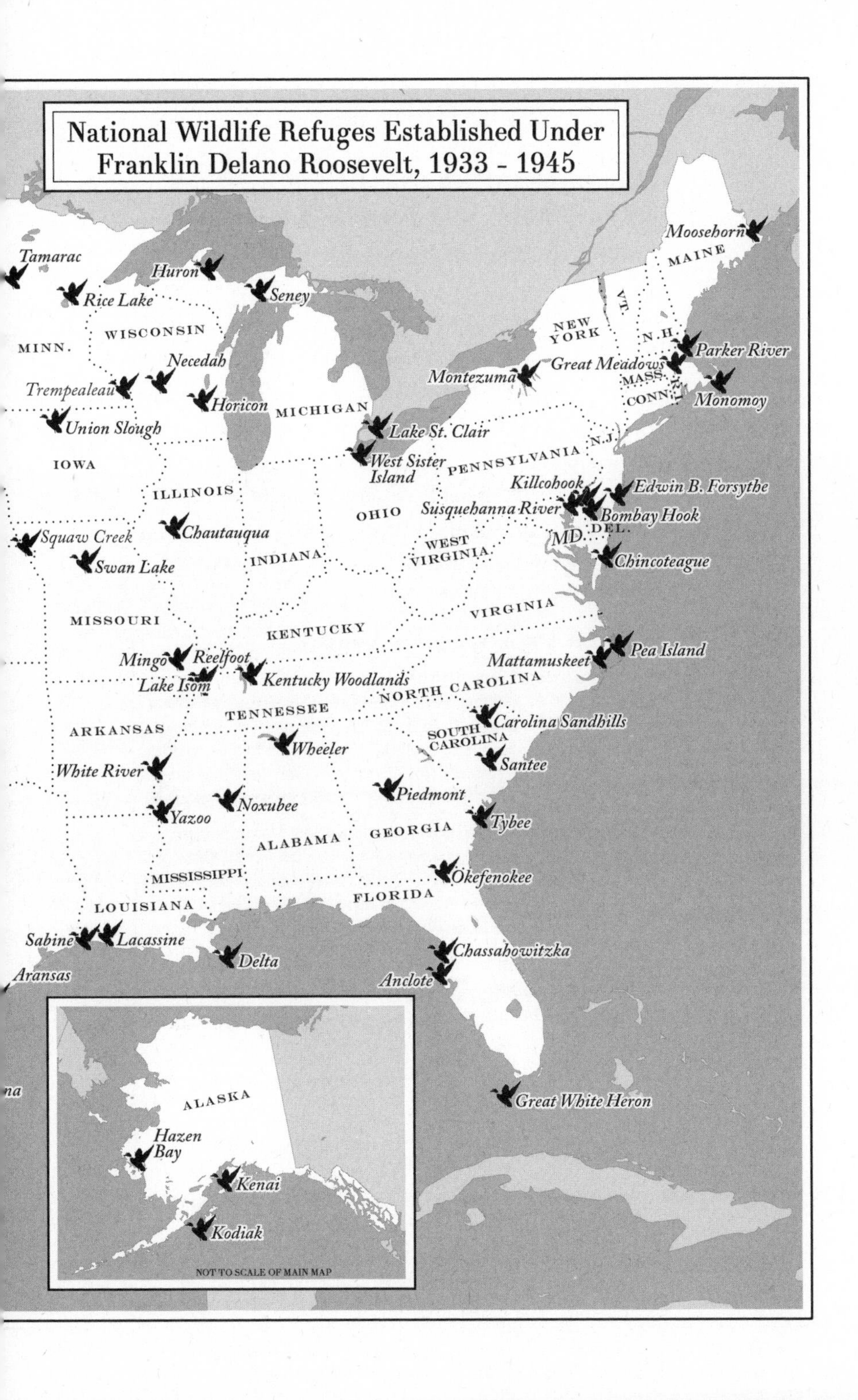
National Wildlife Refuges Established Under Franklin Delano Roosevelt, 1933 - 1945
Moosehorn
MAINE
Tamarac
Huron
Rice Lake
Seney
WISCONSIN
MINN.
NEW YORK
VT.
N.H.
Parker River
Great Meadows
MASS.
Monomoy
CONN.
Necedah
Montezuma
Trempealeau
Horicon
MICHIGAN
Union Slough
Lake St. Clair
N.J.
IOWA
West Sister Island
PENNSYLVANIA
Killcohook
Edwin B. Forsythe
ILLINOIS
OHIO
Susquehanna River
Bombay Hook
Chautauqua
MD.
DEL.
Squaw Creek
WEST VIRGINIA
Swan Lake
INDIANA
Chincoteague
MISSOURI
VIRGINIA
KENTUCKY
Pea Island
Mingo
Reelfoot
Mattamuskeet
Lake Isom
Kentucky Woodlands
NORTH CAROLINA
TENNESSEE
Carolina Sandhills
ARKANSAS
SOUTH CAROLINA
Wheeler
Santee
White River
Piedmont
Noxubee
Yazoo
Tybee
GEORGIA
ALABAMA
Okefenokee
MISSISSIPPI
FLORIDA
LOUISIANA
Sabine
Lacassine
Chassahowitzka
Delta
Aransas
Anclote
ALASKA
Great White Heron
Hazen Bay
Kenai
Kodiak
NOT TO SCALE OF MAIN MAP

APPENDIX A

NATIONAL PARK SYSTEM AREAS AFFECTED UNDER THE REORGANIZATION OF AUGUST 10, 1933

According to the *Annual Report of the Department of the Interior, 1933* (Washington, D.C.: U.S. Government Printing Office, 1933), Franklin D. Roosevelt's June 10, 1933, executive order gave the National Park Service jurisdiction over all the Civil War battlefields; all of the national monuments; most of America's national cemeteries; both the Fine Arts Commission and the National Capital Park and Planning Commission; and all parks, monuments, and public buildings in the District of Columbia. The Reorganization went into effect on August 10, 1933.

NUMBER	PARK, STATE	DATE, DESCRIPTION
1	National Capital Parks/National Mall and Memorial Parks, DC	Originated as Office of National Parks, Buildings and Reservations in the Department of Interior on August 10, 1933, under Executive Order 6166, transferring functions of Office of Public Buildings and Public Parks of the National Capital. Name changed to National Park Service on March 2, 1934. "National Capital Parks" first used in DC appropriation act of June 4, 1934; "National Capital Parks (Central)"; name changed to "National Mall and Memorial Parks" in April 2005, Reorganization on August 10, 1933.
2	Ford's Theatre National Historic Site, DC	Redesignated Lincoln Museum on February 12, 1932; transferred from Office of Public Buildings and Public Parks of the National Capital August 10, 1933, as Lincoln Museum; redesignated Ford's Theatre, on April 14, 1965; combined with House Where Lincoln Died (also transferred on August 10, 1933) on June 23, 1970, as Ford's Theatre National Historic Site.
3	Lincoln Memorial, DC	Transferred from Office of Public Buildings and Public Parks of the National Capital on August 10, 1933 (authorized on February 9, 1911; dedicated in 1922).
4	Washington Monument, DC	Transferred from Office of Public Buildings and Public Parks of the National Capital on August 10, 1933 (accepted on August 2, 1876; dedicated 1885).
5	White House, DC	Transferred from Office of Public Buildings and Public Parks of the National Capital on August 10, 1933.
6	George Washington Memorial Parkway, VA–DC	Transferred from Office of Public Buildings and Public Parks of the National Capital on August 10, 1933 (established on May 29, 1930).
7	Chickamauga and Chattanooga National Military Park, GA and TN	Transferred from War Department on August 10, 1933 (established on August 19, 1890).
8	Fort Donelson National Battlefield, TN	Established as National Military Park on March 26, 1928; transferred from War Department on August 10, 1933; redesignated National Battlefield on August 9, 1985.

NUMBER	PARK, STATE	DATE, DESCRIPTION
9	Fredericksburg and Spotsylvania County Battlefields Memorial National Military Park, VA	Transferred from War Department on August 10, 1933 (established on February 14, 1927).
10	Gettysburg National Military Park, PA	Transferred from War Department on August 10, 1933 (established on February 11, 1895).
11	Guilford Courthouse National Military Park, NC	Transferred from War Department on August 10, 1933 (established on March 2, 1917).
12	Kings Mountain National Military Park, SC	Transferred from War Department on August 10, 1933 (established on March 3, 1931).
13	Moores Creek National Battlefield, NC	Transferred from War Department on August 10, 1933, as National Military Park (established on June 2, 1926); redesignated on September 8, 1980.
14	Petersburg National Battlefield, VA	Transferred from War Department on August 10, 1933 as National Military Park (established on July 3, 1926); redesignated as National Battlefield on August 24, 1962.
15	Shiloh National Military Park, TN	Transferred from War Department on August 10, 1933 (established on December 27, 1894.
16	Stones River National Battlefield, TN	Transferred from War Department on August 10, 1933, as National Military Park (established on March 3, 1927); redesignated as National Battlefield on April 22, 1960.
17	Vicksburg National Military Park, MS	Transferred from War Department on August 10, 1933 (established on February 21, 1899).
18	Abraham Lincoln Birthplace National Historic Site, KY	Transferred from War Department on August 10, 1933, as National Park; established on July 17, 1916, as National Park; redesignated National Historical Park on August 11, 1939; redesignated on September 8, 1959.
19	Fort McHenry National Monument and Historic Shrine, MD	Transferred from War Department on August 10, 1933, as National Park (authorized on March 3, 1925); redesignated on August 11, 1939.
20	Antietam National Battlefield, MD	Transferred from War Department as National Battlefield Site on August 10, 1933 (established on August 30, 1890, as Antietam National Battlefield Site); redesignated on November 10, 1978.
21	Appomattox Court House National Historical Park, VA	Transferred from War Department as Appomattox Battlefield Site on August 10, 1933 (authorized on June 18, 1930); changed to Appomattox Court House National Historical Monument on August 13, 1935; redesignated National Historical Park on April 15, 1954.
22	Brices Cross Roads National Battlefield Site, MS	Transferred from War Department as National Battlefield Site on August 10, 1933 (established on February 21, 1929).
23	Jean Lafitte National Historical Park and Preserve, LA	Transferred from War Department as Chalmette Monument and Grounds on August 10, 1933 (established on March 4, 1907); changed to National Historical Park in 1939; Chalmette absorbed into Jean Lafitte National Historical Park and Preserve November 10, 1978.
24	Cowpens National Battlefield, SC	Transferred from War Department as National Battlefield Site on August 10, 1933 (established March 4, 1929); redesignated as National Battlefield on April 11, 1972.
25	Fort Necessity National Battlefield, PA	Transferred from War Department as National Battlefield Site on August 10, 1933 (established on March 4, 1931); redesignated as National Battlefield on August 10, 1961.
26	Kennesaw Mountain National Battlefield Park, GA	Transferred from War Department as National Battlefield Site on August 10, 1933; authorized on February 18, 1917, as National Battlefield Site; redesignated National Battlefield Park on June 26, 1935.

NUMBER	PARK, STATE	DATE, DESCRIPTION
27	Tupelo National Battlefield, MS	Transferred from War Department as National Battlefield Site on August 10, 1933 (established on February 21, 1929); changed to National Battlefield on August 10, 1961.
28	White Plains National Battlefield, NY	Transferred from War Department as Battlefield Site on August 10, 1933; removed from NPS on May 20, 1956.
29	Big Hole National Battlefield, MT	Transferred from War Department as National Monument on August 10, 1933 (established as Big Hole Battlefield National Monument on June 23, 1910); changed to Big Hole National Battlefield on May 17, 1963.
30	Cabrillo National Monument, CA	Transferred from War Department on August 10, 1933 (proclaimed on October 14, 1913).
31	Castle Pinckney National Monument, NY	Transferred from War Department on August 10, 1933; abolished on March 29, 1956.
32	Father Millet Cross National Monument, NY	Transferred from War Department on August 10, 1933; turned over to state on September 7, 1949.
33	Castillo de San Marcos National Monument, FL	Transferred from War Department on August 10, 1933, as Fort Marion National Monument (proclaimed on October 15, 1924); redesignated on June 5, 1942.
34	Fort Matanzas National Monument, FL	Transferred from War Department on August 10, 1933 (proclaimed on October 15, 1924).
35	Fort Pulaski National Monument, GA	Transferred from War Department on August 10, 1933 (proclaimed on October 15, 1924).
36	Meriwether Lewis National Monument, TN	Transferred from War Department on August 10, 1933; absorbed by Natchez Trace Parkway, TN, AL, MS, on September 10, 1961.
37	Hopewell Culture National Historical Park, OH	Transferred from War Department on August 10, 1933, as Mound City Group National Monument; incorporated in Hopewell Culture National Historical Park on May 27, 1992.
38	Statue of Liberty National Monument, NY	Transferred from War Department on August 10, 1933 (accepted on March 3, 1877; dedicated in 1886; designated a National Monument on October 15, 1924.
39	Wright Brothers National Memorial, NC	Transferred from War Department on August 10, 1933, as Kill Devil Hill Monument National Memorial (authorized on March 2, 1927); redesignated on December 1, 1953.
40	New Echota Marker National Memorial, GA	Transferred from War Department on August 10, 1933—not activated by the NPS.
41	Arlington House, VA (the Robert E. Lee Memorial)	Transferred from War Department on August 10, 1933, as Custis-Lee Mansion (authorized on March 4, 1925); redesignated on June 30, 1972.
42	Battleground National Cemetery, DC	Transferred from War Department on August 10, 1933; unit of NCP.
43	Antietam National Cemetery, MD	Transferred from War Department on August 10, 1933; deleted as separate area in 1974; is part of its park.
44	Vicksburg National Cemetery, MS	Transferred from War Department on August 10, 1933; deleted as separate area in 1974; is part of its park.
45	Gettysburg National Cemetery, PA	Transferred from War Department on August 10, 1933; deleted as separate area in 1974; is part of its park.
46	Chattanooga National Cemetery, TN	Transferred from War Department on August 10, 1933; deleted as separate area in 1974; is part of its park.

NUMBER	PARK, STATE	DATE, DESCRIPTION
47	Fort Donelson National Cemetery, TN	Transferred from War Department on August 10, 1933; deleted as separate area in 1974; is part of its park.
48	Shiloh National Cemetery, TN	Transferred from War Department on August 10, 1933; deleted as separate area in 1974; is part of its park.
49	Fredericksburg National Cemetery, VA	Transferred from War Department on August 10, 1933; deleted as separate area in 1974; is part of its park.
50	Poplar Grove National Cemetery, VA	Transferred from War Department on August 10, 1933; deleted as separate area in 1974; is part of Petersburg National Military Park, VA.
51	Yorktown National Cemetery, VA	Transferred from War Department on August 10, 1933; deleted as separate area in 1974; is part of Colonial National Historical Park, VA.
52	Gila Cliff Dwellings National Monument, NM	Transferred from Forest Service, USDA, on August 10, 1933 (proclaimed on November 16, 1907).
53	Tonto National Monument, AZ	Transferred from Forest Service, USDA, on August 10, 1933; (proclaimed on December 19, 1907).
54	Jewel Cave National Monument, SD	Transferred from Forest Service, USDA, on August 10, 1933; (proclaimed on February 7, 1908).
55	Oregon Caves National Monument, OR	Transferred from Forest Service, USDA, on August 10, 1933; (proclaimed on July 12, 1909).
56	Devils Postpile National Monument, CA	Transferred from Forest Service, USDA, on August 10, 1933; (proclaimed on July 6, 1911).
57	Walnut Canyon National Monument, AZ	Transferred from Forest Service, USDA, on August 10, 1933; (proclaimed on November 30, 1915).
58	Great Basin National Park, NV	Transferred from Forest Service on August 10, 1933 (proclaimed as Lehman Caves National Monument on January 24, 1922); incorporated into Great Basin National Park on October 27, 1986.
59	Timpanogos Cave National Monument, UT	Transferred from Forest Service, USDA, on August 10, 1933 (proclaimed on October 14, 1922).
60	Chiricahua National Monument, AZ	Transferred from Forest Service, USDA, on August 10, 1933 (proclaimed on April 18, 1924).
61	Lava Beds National Monument, CA	Transferred from Forest Service, USDA, on August 10, 1933 (proclaimed on November 21, 1925).
62	Olympic National Park, WA	Transferred on August 10, 1933, from Forest Service as Mount Olympus National Monument (proclaimed on March 2, 1909, as National Monument); redesignated as National Park on June 29, 1938.
63	Sunset Crater Volcano National Monument, AZ	Transferred from Forest Service, USDA, on August 10, 1933 (proclaimed on May 26, 1930, as Sunset Crater National Monument); redesignated on November 16, 1990.
64	Saguaro National Park, AZ	Transferred from Forest Service, USDA, on August 10, 1933 (proclaimed on March 1, 1933); redesignated as National Monument to National Park October 14, 1994.

APPENDIX B

NATIONAL WILDLIFE REFUGES ESTABLISHED UNDER FRANKLIN D. ROOSEVELT, 1933–1945

I compiled this list from U.S. Fish and Wildlife Service, *Directory: National Wildlife Refuges* (1960); from *Annual Report of Lands Under Control of the U.S. Fish and Wildlife Service* (2012); and with the help of USFWS historian Dr. Mark Madison. The acreages listed below are the most current estimates available for each refuge unless otherwise noted. After World War II, some of Roosevelt's refuges changed names or turned into state parks. But this listing is how the system looked in 1945.

LOCATION	REFUGE	DATE ESTABLISHED	ACREAGE	PRIMARY PURPOSE
Alabama	Wheeler	July 7, 1938	34,430	Geese, ducks, quail, mourning doves.
Alaska	Hazen Bay[a]	December 14, 1937	15,948	Steller's eiders, cackling geese, pintails, brant, emperor geese.
	Kenai[b]	December 16, 1941	1,912,425	Kenai moose, brown bears, Dall's sheep, mountain goats.
	Kodiak	August 19, 1941	1,990,417	Kodiak bears, black-tailed deer, waterfowl.
Arizona	Cabeza Prieta[c]	January 25, 1939	860,000	Desert bighorn sheep, Mexican pronghorns, peccaries, Gambel's quail, white-winged doves.
	Havasu Lake	January 22, 1941	37,515	Geese, ducks, Gambel's quail, white-winged doves, herons, desert bighorn sheep, muskrats.
	Imperial	February 14, 1941	26,000	Ducks, geese, herons, Gambel's quail, desert bighorn sheep.

LOCATION	REFUGE	DATE ESTABLISHED	ACREAGE	PRIMARY PURPOSE
	Kofa Game Range	January 25, 1939	660,000	Desert bighorn sheep, white-winged doves, Gambel's quail.
California	Colusa (also AZ)	March 31, 1945	4,686	Waterfowl.
	Havasu Lake	January 22, 1941	37,515	Geese, ducks, Gambel's quail, white-winged doves, herons, desert bighorn sheep, muskrats.
	Imperial (also AZ)	January 14, 1941	26,000	Ducks, geese, herons, Gambel's quail, desert bighorn sheep.
	Sacramento	February 27, 1937	10,819	Snow geese, cackling geese, Ross's geese, whistling swans, ducks, pheasants, shorebirds.
	Sutter	March 30, 1945	2,600	Waterfowl.
Delaware	Bombay Hook	June 22, 1937	16,251	Greater snow goose, black ducks, blue-winged teal, gadwall, shorebirds.
	Killcohook	February 3, 1934	580[d]	Waterfowl.
Florida	Anclote	April 5, 1939	208[e]	Herons, egrets, cormorants, terns, black skimmers.
	Chassahowitzka	June 15, 1943	31,000	Waterfowl, Florida cranes, limpkins, white ibises.
	Great White Heron	October 27, 1938	200,000	Great white herons, roseate spoonbills, brown pelicans, white-crowned pigeons, Key deer.
Georgia	Okefenokee	March 30, 1937	403,119	Florida cranes, herons, white ibises, ducks, limpkins, alligators, otters, black bears.
	Piedmont	January 18, 1939	35,000	Waterfowl, quail, mourning doves, wild turkeys, beavers.
Idaho	Camas	October 8, 1937	10,578	Canada geese, whistling swans, ducks, sage grouse, pheasants, long-billed curlews, pronghorns.

LOCATION	REFUGE	DATE ESTABLISHED	ACREAGE	PRIMARY PURPOSE
	Snake River[f]	August 17, 1937	376	Canada geese, ducks.
Illinois	Chautauqua	December 23, 1936	4,488	Mallards, wood ducks, bald eagles.
Iowa	Union Slough	September 19, 1938	3,334	Waterfowl, upland game birds.
Kentucky	Kentucky Woodlands	August 30, 1938	65,759[g]	Ducks, geese, wild turkeys, deer.
Louisiana	Delta	November 19, 1935	48,000	Blue geese, snow geese, ducks, egrets, shorebirds, alligators.
	Lacassine	December 30, 1937	35,000	Blue geese, snow geese, mottled ducks, roseate fulvous tree ducks, herons, rails.
	Sabine	December 7, 1937	124,511	Blue geese, snow geese, mottled ducks, roseate spoonbills, glossy ibises, herons, muskrats, raccoons, alligators.
Maine	Moosehorn	July 1, 1937	28,751	Woodcock, black ducks, grouse, deer.
Maryland	Susquehanna River	August 24, 1939	1.5	Whistling swans, canvasbacks, ruddy ducks.
Massachusetts	Great Meadows	May 3, 1944	250	Waterfowl.
	Monomoy	June 1, 1944	7,604	Black ducks, eiders, scoters, shorebirds.
	Parker River	December 30, 1942	4,662	Waterfowl.
Michigan	Huron	January 21, 1938	147	Cormorants, ducks, terns.
	Lake Saint Clair	September 21, 1943	880	Canada geese, ducks.
	Seney	December 10, 1937	95,212	Sandhill cranes, ducks, Canada geese, deer, otters.
Minnesota	Mud Lake	March 23, 1937	60,772[h]	Geese, ducks, sharp-tailed grouse, moose.
	Rice Lake	October 31, 1935	18,208	Waterfowl, muskrats, beavers.
	Tamarac	May 31, 1938	42,724	Waterfowl, deer, ruffed grouse, muskrats, beavers.

LOCATION	REFUGE	DATE ESTABLISHED	ACREAGE	PRIMARY PURPOSE
Mississippi	Noxubee	June 14, 1940	48,000	Wild turkeys, deer, quail, wood ducks, mourning doves.
	Yazoo	December 7, 1936	12,941	Canada geese, ducks, herons.
Missouri	Mingo	June 7, 1944	21,676	Geese, ducks, herons, raccoons.
	Squaw Creek	August 23, 1935	7,350	Blue geese, snow geese, ducks, white pelicans.
	Swan Lake	February 27, 1937	10,975	Geese, ducks, prairie chickens.
Montana	Black Coulee	January 18, 1938	1,308	Waterfowl, sage grouse.
	Bowdoin	February 14, 1936	15,551	Canada geese, ducks, white pelicans, herons, cormorants, sage grouse, pronghorns.
	Charles M. Russell[1]	December 11, 1936	915,814	Sharp-tailed grouse, sage grouse, pronghorn, deer, mountain sheep, elk, waterfowl.
	Creedman Coulee	October 25, 1941	2,728	Waterfowl, prairie chickens, sage grouse, shorebirds.
	Hailstone	December 31, 1942	2,700	Waterfowl, sage grouse.
	Halfbreed Lake	May 19, 1942	4,318	Waterfowl, sage grouse.
	Hewitt Lake	March 7, 1938	1,360	Waterfowl, sage grouse.
	Lake Mason	June 2, 1941	16,814	Ducks, sharp-tailed grouse, sage grouse, muskrats.
	Lake Thibadeau	September 23, 1937	3,868	Ducks, sharp-tailed grouse, sage grouse.
	Lamesteer	May 19, 1942	800	Waterfowl, sharp-tailed grouse.
	Medicine Lake	August 19, 1935	31,533	Geese, ducks, sandhill cranes, sharp-tailed grouse, shorebirds, gulls, terns, muskrats.
Nebraska	Valentine	August 14, 1935	71,516	Ducks, geese, sharp-tailed grouse, pheasants, shorebirds, muskrats.

LOCATION	REFUGE	DATE ESTABLISHED	ACREAGE	PRIMARY PURPOSE
Nevada	Desert National Game Range[j]	May 20, 1936	1,600,000	Bighorn sheep, mule deer, Gambel's quail, mourning doves.
	Ruby Lake	July 2, 1938	37,632	Canada geese, ducks, sage grouse, sandhill cranes, shorebirds.
	Sheldon	December 21, 1936[k]	573,504	Pronghorns, mule deer, sage grouse, waterfowl.
New Jersey	Edwin B. Forsythe[l]	October 5, 1939	40,000	Brant, ducks, shorebirds, gulls, terns, rails.
New Mexico	Bitter Lake	October 8, 1937	24,536	Waterfowl, little brown cranes, scaled quail.
	Bosque del Apache	November 22, 1939	57,331	Ducks, geese, Mexican ducks, Gamel's quail, scaled quail.
	San Andres	January 22, 1941	57,215	Bighorn sheep, mule deer, white-winged doves, harlequin quail, Gambel's quail, scaled quail.
New York	Montezuma	September 12, 1938	7,000	Geese, ducks, muskrats.
North Carolina	Mattamuskeet	December 18, 1934	50,173	Canada geese, whistling swans, ducks, herons, deer, raccoons.
	Pea Island	April 8, 1938	31,534	Great snow geese, brant, Canada geese, ducks, shorebirds, gulls, terns, otters.
North Dakota	Des Lacs	August 22, 1935	19,500	Ducks, geese, grebes, sharp-tailed grouse, deer, muskrats.
	Upper Souris	August 27, 1935	32,092	Geese, ducks, sharp-tailed grouse, pheasants, sandhill cranes, shorebirds, muskrats, deer.
	Arrowwood	September 4, 1935	50,173	Whistling swans, geese, ducks, prairie chickens, sharp-tailed grouse, pheasants.
	J. Clark Salyer[m]	September 4, 1935	58,593	Geese, ducks, grebes, white pelicans, prairie chickens, deer, sharp-tailed grouse, muskrats.

LOCATION	REFUGE	DATE ESTABLISHED	ACREAGE	PRIMARY PURPOSE
	Kellys Slough	March 19, 1936	1,270	Geese, ducks, sharp-tailed grouse, prairie chickens, gray partridges, muskrats.
	Lostwood	September 4, 1935	26,747	Waterfowl, sharp-tailed grouse, sandhill cranes, shorebirds, muskrats.
	Appert Lake	May 10, 1939	1,160[n]	Waterfowl.
	Billings Lake	May 10, 1939	760[o]	Waterfowl, upland game birds.
	Bone Hill Creek	May 10, 1939	640	Waterfowl.
	Buffalo Lake	May 10, 1939	2,096[p]	Ducks, sharp-tailed grouse, prairie chickens, pheasants.
	Camp Lake	May 10, 1939	535	Waterfowl.
	Canfield Lake	May 10, 1939	453[q]	Waterfowl.
	Dakota Lake	May 10, 1939	2,784	Waterfowl.
	Flickertail	May 10, 1939	640[r]	Waterfowl.
	Florence Lake	May 10, 1939	1,888	Waterfowl.
	Half-Way Lake	May 10, 1939	160	Waterfowl, pheasants.
	Hutchinson Lake	May 10, 1939	479[s]	Waterfowl.
	Johnson Lake	May 10, 1939	2,002	Geese, ducks, sharp-tailed grouse, prairie chickens.
	Little Goose	May 10, 1939	359[t]	Waterfowl, gray partridges, pheasants, shorebirds.
	Lords Lake	May 10, 1939	1,915[u]	Waterfowl.
	Lost Lake	May 10, 1939	960	Waterfowl.
	Ardoch	June 12, 1939	2,676[v]	Whistling swans, geese, ducks, white pelicans, muskrats.
	Brumba	June 12, 1939	1,977[w]	Waterfowl.
	Cottonwood Lake	June 12, 1939	1,013[x]	Waterfowl.
	Hiddenwood	June 12, 1939	568	Waterfowl, pheasants.
	Hobart Lake	June 12, 1939	2,077	Waterfowl, pheasants.
	Lake Elsie	June 12, 1939	634[y]	Waterfowl.
	Lake George	June 12, 1939	3,118[z]	Waterfowl, grouse.

LOCATION	REFUGE	DATE ESTABLISHED	ACREAGE	PRIMARY PURPOSE
	Lake Ilo	June 12, 1939	4,034	Ducks, sharp-tailed grouse, gray partridges, pheasants.
	Lake Nettie	June 12, 1939	3,055	Waterfowl, gulls.
	Lake Zahl	June 12, 1939	3,823	Waterfowl, upland game birds, shorebirds, muskrats.
	Lambs Lake	June 12, 1939	1,286	Waterfowl, sharp-tailed grouse, gray partridges.
	Maple River	June 12, 1939	1,120	Waterfowl, prairie chickens, pheasants.
	McLean	June 12, 1939	760	Waterfowl.
	Pleasant Lake	June 12, 1939	1,001	Waterfowl, upland game birds.
	Rock Lake	June 12, 1939	5,506[aa]	Waterfowl, muskrats.
	Shell Lake	June 12, 1939	1,835	Waterfowl.
	Sibley Lake	June 12, 1939	1,077	Geese, ducks, sharp-tailed grouse, prairie chickens, deer.
	Rabb Lake	February 3, 1941	800[bb]	Waterfowl, upland game birds.
	Snyder Lake	February 3, 1941	1,550[cc]	Waterfowl.
	Springwater	February 3, 1941	640[dd]	Waterfowl.
	Stoney Slough	February 3, 1941	2,000	Ducks, gray partridges, pheasants.
	Sunburst Lake	February 3, 1941	495	Waterfowl.
	Tomahawk	February 3, 1941	440	Waterfowl.
	White Lake	February 3, 1941	1,040	Waterfowl.
	Wintering River	February 3, 1941	399[ee]	Waterfowl.
	Slade	October 10, 1944	3,000	Waterfowl, sharp-tailed grouse, prairie chickens, deer.
Ohio	West Sister Island	August 2, 1938	77	Geese, ducks, white pelicans, quail, Franklin's gulls, herons, shorebirds.
Oregon	Cape Meares	August 19, 1938	138.51	Shorebirds, band-tailed pigeons, black-tailed deer.

LOCATION	REFUGE	DATE ESTABLISHED	ACREAGE	PRIMARY PURPOSE
	Hart Mountain	December 12, 1936	270,608	Pronghorns, mule deer, sage grouse, valley quail, waterfowl.
	Oregon Islands	May 6, 1935	1,083	Cormorants, gulls, murres, puffins.
South Carolina	Carolina Sandhills	March 17, 1939	45,348	Quail, wild turkeys, wood ducks.
	Santee	July 31, 1941	15,000	Waterfowl, herons, raccoons, otters.
	Tybee	May 9, 1938	100[ff]	Shorebirds.
South Dakota	Lacreek	August 26, 1935	16,410	Geese, ducks, sandhill cranes, pheasants, sharp-tailed grouse, shorebirds, muskrats.
	Lake Andes	February 14, 1936	5,638	Geese, ducks, shorebirds.
	Sand Lake	September 4, 1935	21,498	Geese, ducks, pheasants, prairie chickens, Franklin's gulls, gray partridges, white pelicans, cormorants, muskrats.
	Waubay	December 10, 1935	4,650	Ducks, geese, pheasants, Franklin's gulls, shorebirds, muskrats, deer.
Tennessee	Lake Isom	August 12, 1938	1,846	Ducks, herons, mourning doves.
	Reelfoot[gg]	August 28, 1941	10,428	Ducks, geese, herons, quail, mourning doves, muskrats, raccoons, mink.
Texas	Aransas	December 31, 1937	114,657	Whooping cranes, sandhill cranes, roseate spoonbills, egrets, herons, geese, ducks, wild turkeys, shorebirds, hawks, deer, peccaries.
	Muleshoe	October 24, 1935	5,809	Canada geese, ducks, sandhill cranes, scaled quail, shorebirds.

LOCATION	REFUGE	DATE ESTABLISHED	ACREAGE	PRIMARY PURPOSE
	Santa Ana	September 1, 1943	2,088	Tree ducks, mottled ducks, herons, egrets, ibises, least grebes, doves, red-billed pigeons, chachalacas, green jays.
Virginia	Chincoteague[hh]	May 13, 1943	14,000	Greater snow geese, shorebirds, gulls, terns, brant.
Washington	Columbia	September 6, 1944	29,596	Waterfowl.
	Jones Island	March 30, 1937	179[ii]	White-winged scoters, band-tailed pigeons, cormorants, gulls.
	Little Pend Oreille	May 2, 1939	42,593	Franklin's grouse, blue grouse, ruffed grouse, deer, black bears, waterfowl.
	Matia Island	March 30, 1937	145[jj]	Cormorants, guillemots, harlequin ducks, scoters, band-tailed pigeons.
	Turnbull	July 30, 1937	18,654	Ducks, geese, ruffed grouse, valley quail, shorebirds.
	Willapa	January 22, 1937	11,000	Black brant, Canada geese, ducks, shorebirds, blue grouse, black bears, black-tailed deer, raccoons.
Wisconsin	Horicon	July 16, 1941	21,400	Canada geese, ducks, whistling swans, muskrats.
	Long Tail Point	October 23, 1936	103[kk]	Waterfowl.
	Necedah	March 14, 1939	43,696	Canada geese, ducks, deer, ruffed grouse, sharp-tailed grouse, prairie chickens, beavers, muskrats, mink.
	Trempealeau	August 22, 1936	6,226	Waterfowl, American egrets.

NOTES

[a] Hazen Bay is now part of Yukon Delta NWR.

[b] Originally established as Kenai National Moose Range.

[c] Originally established as Cabeza Prieta Game Range.

[d] Acreage upon establishment. Its NWR status was revoked in 1998; the site is currently under the purview of the Army Corps of Engineers.

[e] Acreage upon establishment. No longer an NPS unit.

[f] Integrated with Deer Flat National Wildlife Refuge in 1963.

[g] Acreage upon establishment. No longer an NPS unit.

[h] Acreage upon establishment. No longer an NPS unit.

[i] Established as Fort Peck Game Range in 1936, renamed in 1963 after artist Charles M. Russell, and reclassified as a national wildlife refuge in 1976.

[j] Originally established as Desert Game Range.

[k] Originally established in 1931 as Charles Sheldon Wild Life Refuge, the unit, under FDR, was enlarged in May 1936 and subsequently redesignated in December 1936 as Charles Sheldon Antelope Range. A small portion of the refuge exists within the boundaries of Oregon. On December 12, 1936, Hart Mountain National Antelope Refuge, a companion unit to Charles Sheldon National Wildlife Refuge, was also established.

[l] Edwin B. Forsythe NWR was formed in 1984 when two existing refuge parcels—the Brigantine Division, established by FDR in 1939; and the Barnegat Division, established in 1967 by LBJ—were combined.

[m] Originally established as Lower Souris Migratory Waterfowl Refuge.

[n] Acreage upon establishment

[o] Acreage upon establishment. Established as Billings Lake Migratory Waterfowl Refuge, this unit is no longer part of the NPS.

[p] Acreage upon establishment.

[q] Acreage upon establishment.

[r] Acreage upon establishment. No longer an NPS unit.

[s] Acreage upon establishment.

[t] Acreage upon establishment.

[u] Acreage upon establishment.

[v] Acreage upon establishment.

[w] Acreage upon establishment.

[x] Acreage upon establishment.

[y] Acreage upon establishment. No longer an NPS unit.

[z] Acreage upon establishment.

[aa] Acreage upon establishment.

[bb] Acreage upon establishment.

[cc] Acreage upon establishment.

[dd] Acreage upon establishment.

[ee] Acreage upon establishment.

[ff] Acreage upon establishment.

[gg] Portions of this refuge are located within the state boundaries of Kentucky.

[hh] A very small portion of Chincoteague NWR is contained within the boundaries of Maryland.

[ii] Acreage upon establishment. No longer an NPS unit.

[jj] Acreage upon establishment. Now managed as part of San Juan Islands National Wildlife Refuge.

[kk] Acreage upon establishment. No longer an NPS unit.

APPENDIX C

NATIONAL PARKS AND NATIONAL MONUMENTS CREATED BY FRANKLIN D. ROOSEVELT FOLLOWING THE REORGANIZATION OF AUGUST 10, 1933

Cedar Breaks National Monument (UT)	August 22, 1933
Everglades National Park (FL)	May 30, 1934
Great Smoky Mountains National Park (NC/TN)	June 15, 1934
Fort Jefferson National Monument[a] (FL)	January 4, 1935
Big Bend National Park (TX)	June 20, 1935
Fort Stanwix National Monument (NY)	August 21, 1935
Ackia Battleground National Monument (MS)	August 27, 1935
Andrew Johnson National Monument (TN)	August 29, 1935
Shenandoah National Park (VA)	December 26, 1935
Homestead National Monument of America (NE)	March 19, 1936
Fort Frederica National Monument (GA)	May 26, 1936
Perry's Victory and International Peace Memorial National Monument (OH)	June 2, 1936
Whitman National Monument (WA)	June 29, 1936
Joshua Tree National Monument[b] (CA)	August 16, 1936
Zion National Monument[c] (UT)	January 22, 1937
Organ Pipe Cactus National Monument (AZ)	April 13, 1937
Capitol Reef National Monument[d] (UT)	August 2, 1937
Pipestone National Monument (MN)	August 25, 1937
Channel Islands National Monument[e] (CA)	April 26, 1938
Olympic National Park (WA)	June 29, 1938
Fort Laramie National Monument (WY)	July 16, 1938
Santa Rosa Island National Monument (FL)	May 17, 1939
Kings Canyon National Park (CA)	March 4, 1940
Isle Royale National Park (MI)	April 3, 1940
Mammoth Cave National Park (KY)	July 1, 1941
Jackson Hole National Monument (WY)	March 15, 1943
George Washington Carver National Monument (MO)	July 14, 1943
Harpers Ferry National Monument (VA/WV)	June 30, 1944

[a] Became Dry Tortugas National Park in 1992.

[b] Became Tree National Park in 1994.

[c] Became Zion National Park in 1956.

[d] Became Capitol Reef National Park in 1971.

[e] Became Channel Islands National Park in 1980.

APPENDIX D

ESTABLISHMENT AND MODIFICATION OF NATIONAL FOREST BOUNDARIES BY FRANKLIN D. ROOSEVELT, MARCH 1933 TO APRIL 1945

Information in this appendix is from United States Department of Agriculture Forest Service Lands and Realty Management Staff, *Establishment and Modification of National Forest Boundaries and National Grasslands: A Chronological Record, 1891–2012*, FS-612, 2012. I've included only acreage *added* by FDR.

NAME	STATE	DOC TYPE	DATE APPROVED	CITATION	EFFECT
1933					
Modoc	CA	PL 432	MAR 04, 1933	47 Stat. 1563	Land added.
Gunnison	CO	PL 437	MAR 04, 1933	47 Stat. 1569	Land added.
Cascade	OR	EO 6104	APR 06, 1933		Land combined with Santiam and established Willamette.
Santiam	OR	EO 6104	APR 06, 1933		Land combined with Cascade and established Willamette.
Willamette	OR	EO 6104	APR 06, 1933		Established.
Harney	SD	EO 6117	MAY 02, 1933		Land added.
Cache	ID/UT	EO 6172	JUN 15, 1933		Land added.
Natural Bridge	VA	EO 6210	JUL 22, 1933		Land transferred to George Washington. Name discontinued.
George Washington	VA/WV	EO 6210	JUL 22, 1933		Land added from Natural Bridge.
Dixie	AZ/UT	Proc. 2054	AUG 22, 1933	48 Stat. 1705	Land transferred to Cedar Breaks National Monument.
Coeur d' Alene	ID	EO 6303	SEP 30, 1933		Smaller portion added from Pend Oreille.

NAME	STATE	DOC TYPE	DATE APPROVED	CITATION	EFFECT
Kaniksu	ID/MT/WA	EO 6303	SEP 30, 1933		Larger portion added from Pend Oreille.
Pend Oreille	ID/MT/WA	EO 6303	SEP 30, 1933		Consolidated larger part with Kaniksu. Remainder transferred to Coeur d'Alene.
Columbia	WA	EO 6333	OCT 13, 1933		Land added from Rainier.
Snoqualmie	WA	EO 6334	OCT 13, 1933		Land added from Rainier.
Wenatchee	WA	EO 6335	OCT 13, 1933		Land added from Rainier.
Rainier	WA	EO 6333, EO 6334, EO 6335	OCT 13, 1933		Land divided among Columbia, Snoqualmie, and Wenatchee.
Snoqualmie	WA	EO 6336	OCT 13, 1933		Land transferred to Mt. Baker.
Mt. Baker	WA	EO 6336	OCT 13, 1933		Land added from Snoqualmie.
Ashley	UT	EO 6409	NOV 07, 1933		Land transferred to and from Wasatch.
Wasatch	UT/WY	EO 6409	NOV 07, 1933		Land transferred to and from Ashley.
Nicolet	WI	Proc. 2060	NOV 13, 1933	48 Stat. 1715	Mondeaux and Oconto purchase. Units added. Moquah and Flambeau units transferred to Chequamegon.
Chequamegon	WI	Proc. 2061	NOV 13, 1933	48 Stat. 1716	Established 1934.
1934					
Saint Joe	ID	PL 73-182	APR 30, 1934	48 Stat. 649	Land added.
Pike	CO	PL 73-194	MAY 03, 1934	48 Stat. 657	Land added.
Cochetopa	CO	PL 73-195	MAY 03, 1934	48 Stat. 658	Land added.
Ochoco	OR	PL 73-218	MAY 11, 1934	48 Stat. 772	Land added.
Boise	ID	PL 73-228	MAY 17, 1934	48 Stat. 779	Land added.
Klamath	CA/OR	EO 6786*	JUN 30, 1934		Land transferred to Shasta.
Shasta	CA	EO 6786*	JUN 30, 1934		Land added from Klamath.
Uinta	UT/WY	EO 6801-A	JUL 27, 1934		Land added.
Wasatch	UT	EO 6801-B	JUL 27, 1934		Land added.
Kaibab	AZ	EO 6806	AUG 04, 1934		Land added from Tusayan.
Tusayan	AZ	EO 6806	AUG 04, 1934		Land transferred to Kaibab.

NAME	STATE	DOC TYPE	DATE APPROVED	CITATION	EFFECT
Black Hills	SD/WY	EO 6809	AUG 04, 1934		Land added.
Cochetopa	CO	EO 6881	OCT 22, 1934		Boundary adjusted between Cochetopa and Gunnison.
Gunnison	CO	EO 6881	OCT 22, 1934		Boundary adjusted between Gunnison and Cochetopa.
Tusayan	AZ	EO 6882	OCT 22, 1934		Land transferred to Prescott. Name discontinued.
Prescott	AZ	EO 6882	OCT 22, 1934		Land added from Tusayan. Land transferred to Tonto.
Tonto	AZ	EO 6882	OCT 22, 1934		Land added from Prescott.
Nicolet	WI	EO 6886	OCT 27, 1934		Mondeaux unit transferred to Chequamegon.
Chequamegon	WI	EO 6886	OCT 27, 1934		Mondeaux unit of Nicolet added.
Selway	ID	EO 6889	OCT 29, 1934		Land divided among Bitterroot, Clearwater, Lolo, and Nezperce.
Bitterroot	ID/MT	EO 6889	OCT 29, 1934		Land added from Selway.
Clearwater	ID	EO 6889	OCT 29, 1934		Land added from Selway.
Lolo	ID/MT	EO 6889	OCT 29, 1934		Land added from Selway.
Nezperce	ID	EO 6889	OCT 29, 1934		Land added from Selway.
1935					
Uinta	UT/WY	EO 6944	JAN 09, 1935		Land transferred to Wasatch.
Wasatch	UT	EO 6944	JAN 09, 1935		Land added from Uinta.
Marquette	MI	PL 81	MAY 28, 1935	48 Stat. 307	Land added.
Tongass	AK	EO 7059	MAY 31, 1935		Land added.
Luquillo	PR	EO 7059-A	JUN 04, 1935		Name change to Caribbean.
Caribbean	PR	EO 7059-A	JUN 04, 1935		Established.
Siskiyou	CA/OR	PL 131	JUN 13, 1935	49 Stat. 338	Land added.
Cibola	NM	PL 156	JUN 20, 1935	49 Stat. 393	Land transferred to Indian Reservation.
Blackfeet	MT	EO 7082	JUN 22, 1935		Land divided between Flathead and Kootenai.
Flathead	MT	EO 7082	JUN 22, 1935		Land added from Blackfeet.
Kootenai	ID/MT	EO 7082	JUN 22, 1935		Land added from Blackfeet.

NAME	STATE	DOC TYPE	DATE APPROVED	CITATION	EFFECT
Clearwater	ID	EO 7087	JUN 27, 1935		Land transferred to Lolo.
Lolo	ID/MT	EO 7087	JUN 27, 1935		Land from Clearwater.
Tongass	AK	EO 7088	JUN 27, 1935		Land added.
Flathead	MT	EO 7118	JUL 29, 1935		Exec. Order 7082 of JUN 22, 1935, corrected.
Medicine Bow	WY	PL 288	AUG 20, 1935	49 Stat. 662	Land added.
Pisgah	NC	PL 328	AUG 26, 1935	49 Stat. 800	Land added from Oteen Hospital.
Fremont	OR	Proc. 2143	OCT 14, 1935	49 Stat. 3477	Land added.
Willamette	OR	Proc. 2151	DEC 07, 1935	49 Stat. 3486	Land added.
1936					
Unaka	NC/TN/VA	Proc. 2165	APR 21, 1936	49 Stat. 3506	Part combined w/other land to establish Jefferson. 1 FR 227.
George Washington	VA/WV	Proc. 2165	APR 21, 1936	49 Stat. 3506	Part combined w/other land to establish Jefferson. 1 FR 227.
Jefferson	VA	Proc. 2165	APR 21, 1936	49 Stat. 3506	Established.
Monongahela	VA/WV	Proc. 2166	APR 28, 1936	49 Stat. 3510	Land added. Land transferred to George Washington. 1 FR 295.
George Washington	VA/WV	Proc. 2167	APR 28, 1936	49 Stat. 3513	Portion of Monongahela 1 FR 297 and other land added.
Nicolet	WI	EO 7359	MAY 05, 1936	1 FR 342	Land added.
Ozark	AR	Proc. 2168	MAY 13, 1936	49 Stat. 3516 1 FR 408	Land added.
Apalachicola	FL	Proc. 2169	MAY 13, 1936	49 Stat. 3516 1 FR 408	Established.
Cache	ID/UT	EO 7378	MAY 22, 1936	1 FR 446	Land added.
Kisatchie	LA	Proc. 2173	JUN 03, 1936	49 Stat. 3520	Established. 1 FR 544.
Rogue River	CA/OR	PL 642	JUN 04, 1936	49 Stat. 1460	Added Oregon and California lands.
Bienville	MS	Proc. 2175	JUN 15, 1936	49 Stat. 3521	Established. 1 FR 606.
Holly Springs	MS	Proc. 2176	JUN 15, 1936	49 Stat. 3522	Established. 1 FR 606.
DeSoto	MS	Proc. 2174	JUN 17, 1936	49 Stat. 3524	Established. 1 FR 609.
Alabama	AL	Proc. 2178	JUN 19, 1936	49 Stat. 3526	Name changed to Black Warrior. Proc. 1423 of Jan 15, 1918, modified. 1 FR 645.
Black Warrior	AL	Proc. 2178	JUN 19, 1936	49 Stat. 3526	Established. 1 FR 645.

NAME	STATE	DOC TYPE	DATE APPROVED	CITATION	EFFECT
Pisgah	NC	Proc. 2183	JUL 8, 1936	50 Stat. 1735	Land transferred to Cherokee. 1 FR 763.
Unaka	NC/TN/VA	Proc. 2183	JUL 08, 1936	50 Stat. 1735	Land transferred to Cherokee. 1 FR 763.
Cherokee	GA/NC/TN	Proc. 2183	JUL 08, 1936	50 Stat, 1735	Portion of Pisgah and Unaka added. 1 FR 763.
Cherokee	GA/NC/TN	Proc. 2184	JUL 09, 1936	50 Stat. 1739	Portion combined w/ other land to establish Chattahoochee. 1 FR 776.
Nantahala	GA/NC	Proc. 2184	JUL 09, 1936	50 Stat. 1739	Combined w/other land to establish Chattahoochee
Chattahoochee	GA	Proc. 2184	JUL 09, 1936	50 Stat. 1739	Established. 1 FR 776.
Nantahala	NC	Proc. 2185	JUL 09, 1936	50 Stat. 1742	Boundary redescribed. 1 FR 778.
Francis Marion	SC	Proc. 2186	JUL 10, 1936	50 Stat. 1744	Established. 1 FR 792.
Unaka	NC	Proc. 2187	JUL 10, 1936	50 Stat. 1744	1 FR 792. Land transferred to Pisgah. Name discontinued.
Pisgah	NC	Proc. 2187	JUL 10, 1936	50 Stat. 1745	Unaka and other land added. 1 FR 792.
Sumter	SC	Proc. 2188	JUL 13, 1936	50 Stat. 1750	Established. 1 FR 799.
Conecuh	AL	Proc. 2189	JUL 17, 1936	50 Stat. 1754	Established. 1 FR 859.
Talladega	AL	Proc. 2190	JUL 17, 1936	50 Stat. 1755	Established. 1 FR 860.
Homochitto	MS	Proc. 2191	JUL 20, 1936	50 Stat. 1758	Established. 1 FR 872.
Croatan	NC	Proc. 2192	JUL 29, 1936	50 Stat. 1759	Established. 1 FR 909.
Bienville	MS	EO 7412	AUG 10, 1936	1 FR 781	Proc. 2175 of JUN 15, 1936, corrected..
Uinta	UT/WY	EO 7429	AUG 17, 1936	1 FR 1125	Land added..
Talladega	AL	EO 7443	AUG 31, 1936	1 FR 1312.	Proc. 2190 of JUL 17, 1936, modified.
Ouachita	AR	Proc. 2201	OCT 12, 1936	50 Stat. 1777	Land added. 1 FR 1600.
Angelina	TX	Proc. 2202	OCT 13, 1936	50 Stat. 1780	Established. 1 FR 1601.
Davy Crockett	TX	Proc. 2203	OCT 13, 1936	50 Stat. 1782	Established. 1 FR1603.
Sabine	TX	Proc. 2204	OCT 13, 1936	50 Stat. 1787	Established. 1 FR 1606.
Sam Houston	TX	Proc. 2205	OCT 13, 1936	50 Stat. 1789	Established. 1 FR 1608.
Wichita	OK	Proc. 2211	NOV 27, 1936	50 Stat. 1797	Abolished. 1 FR 2148.
Santa Barbara	CA	EO 7501	DEC 03, 1936	1 FR 2141	Name changed to Los Padres.

NAME	STATE	DOC TYPE	DATE APPROVED	CITATION	EFFECT
Los Padres	CA	EO 7501	DEC 03, 1936	1 FR 2141	Established.
Pike	CO	EO 7513	DEC 16, 1936*	1 FR 2159	Land transferred to Arapaho.
Roosevelt	CO	EO 7513	DEC 16, 1936*	1 FR 2159	Land transferred to Arapaho.
Arapaho	CO	EO 7513	DEC 16, 1936*	1 FR. 2159	Land added from Pike and Roosevelt.
Superior	MN	Proc. 2213	DEC 28, 1936	50 Stat. 1799	Land added. 1 FR 2250.
Chippewa	MN	Proc. 2216	DEC 29, 1936	50 Stat. 1803	Land added. 2 FR 2.
Chequamegon	WI	Proc. 2218	DEC 31, 1936	50 Stat. 1806	Land added. 2 FR 22.
Nicolet	WI	Proc. 2219	DEC 31, 1936	50 Stat. 1807	Land added. 2 FR 23,
1937					
Ottawa	MI	Proc. 2220	JAN 11, 1937	50 Stat. 1808	Land added. 2 FR 63.
Green Mtn.	VT	Proc. 2225	FEB 08, 1937	50 Stat. 1813	Land added. 2 FR 284.
Coconino	AZ	Proc. 2226	FEB 23, 1937	50 Stat. 1817	Land transferred to Montezuma Castle National Monument. 2 FR 360.
Cumberland	KY	Proc. 2227	FEB 23, 1937	50 Stat. 1818	Established. 2 FR 361.
Arapaho	CO	EO 7572	MAR 09, 1937	2 FR 520	EO7513 of DEC 16, 1936, amended.
Pike	CO	EO 7572	MAR 09, 1937	2 FR 520	EO 7513 of DEC 16, 1936, amended.
Rio Grande	CO	SIO	MAR 27, 1937	2 FR 647	Boundary adjusted (Re surveyed).
Tonto	AZ	Proc. 2230	APR 01, 1937	50 Stat. 1825	Land transferred to Tonto National Monument. 2 FR 644.
Dixie	AZ/UT	EO 7607	APR 19, 1937*	2 FR 720	Transferred Moapa division to Nevada.
Nevada	NV	EO 7607	APR 19, 1937*	2 FR 720	Land added to Moapa division of Dixie.
Chugach	AK	EO 7610	APR 23, 1937	2 FR 749	EO 6966 of FEB 8, 1935, amended.
Tongass	AK	EO 7624	MAY 29, 1937	2 FR 938	Land added.
Hiawatha	MI	SAO	JUN 07, 1937	2 FR 979	Land added.
Ocala	FL	SAO	JUN 30, 1937	2 FR 1136–1361	Land added.
Harney	SD	Proc. 2244	JUL 12, 1937	50 Stat. 1842	Land added.
Uinta	UT/WY	EO 7663	JUL 17, 1937	2 FR 1251	Land added.

NAME	STATE	DOC TYPE	DATE APPROVED	CITATION	EFFECT
Rouge River	OR	PL 214	JUL 27, 1937	50 Stat. 534	Added Oregon and California (O&G) lands.
Ocala	FL	SAO	AUG 03, 1937	2 FR 1625	Sec. of Agriculture Order of JUN 30, 1937 corrected.
Columbia	WA	PL 257	AUG 12, 1937	50 Stat. 622	Land added.
Snoqualmie	WA	PL 333	AUG 21, 1937	50 Stat. 739	Land added.
Ouachita	AR	EO 7719	OCT 08, 1937	2 FR 2110	Land added.
Tongass	AK	EO 7742	NOV 19, 1937	2 FR 2524	Land added.
Chattahoochee	GA	Proc. 2263	DEC 07, 1937	51 Stat. 404	Land added. 2 FR 2758.
Chugach	AK	EO 7781	DEC 30, 1937	3 FR 1	EO 5517 of DEC 17, 1930 amended.
1938					
Tahoe	NV	PL 428	FEB 12, 1938	52 Stat. 28	Land added.
Humboldt	NV	EO 7884	MAY 09, 1938	3 FR 913	Santa Rosa division transferred to Toiyabe.
Nevada	NV	EO 7884	MAY 09, 1938	3 FR 913	Land transferred to Toiyabe.
Toiyabe	NV	EO 7884	MAY 09, 1938	3 FR 913	Reestablished.
Talladega	AL	Proc. 2285	MAY 11, 1938	52 Stat. 1548	Land added. 3 FR 928.
Cache	UT	PL 505	MAY 11, 1938	52 Stat. 347	Land added.
Kaniksu	ID/WA	PL 546	MAY 26, 1938	52 Stat. 443	Land added.
Coronado	AZ/NM	Proc. 2288	JUN 10, 1938	52 Stat. 1551	Land transferred to Chiricahua National Monument. 3 FR 1399.
Black Hills	SDAVY	PL 615	JUN 15, 1938	52 Stat. 686	Land added.
Ochoco	OR	PL 624	JUN 15, 1938	52 Stat. 692	Land added.
Rio Grande	CO	PL 679	JUN 20, 1938	52 Stat. 781	Land added.
Trinity	CA	PL 683	JUN 20, 1938	52 Stat. 797	Land added.
Apalachicola	FL	Proc. 2289	JUN 21, 1938	53 Stat. 2453	Land added. 3 FR 1505.
Modoc	CA	PL 692	JUN 22, 1938	52 Stat. 835	Land added.
Shasta	CA	PL 692	JUN 22, 1938	52 Stat. 835	Land added.
Lassen	CA	PL 692	JUN 22, 1938	52 Stat. 835	Land added.
Shasta	CA	PL 693	JUN 22, 1938	52 Stat. 836	Land added.
Klamath	GA	PL 693	JUN 22, 1938	52 Stat. 836	Land added.
Plumas	CA	PL 694	JUN 22, 1938	52 Stat. 838	Land added.
Tahoe	CA	PL 694	JUN 22, 1938	52 Stat. 838	Land added.
Lassen	CA	PL 694	JUN 22, 1938	52 Stat. 838	Land added.

NAME	STATE	DOC TYPE	DATE APPROVED	CITATION	EFFECT
Olympic	WA	PL 778	JUN 29, 1938	52 Stat. 1241	Land transferred to Olympic National Park.
Ocala	FL	Proc. 2293	JUL 16, 1938*	53 Stat. 2462	Land added. 3 FR 1769.
Chattahoochee	GA	Proc. 2294	AUG 02, 1938	53 Stat. 2463	Land added. 3 FR 1961.
Ouachita	AR	Proc. 2296	AUG 30, 1938	53 Stat. 2465	Land added. 3 FR 2169.
Coconino	AZ	Proc. 2300	SEP 24, 1938	53 Stat. 2469	Land transferred to Walnut Canyon National Monument. 3 FR 2321.
Arapaho	CO	SIO	SEP 24, 1938	3 FR 2445	Boundary adjusted to conform to survey.
Grand Mesa	CO	SIO	SEP 24, 1938	3 FR 2445	Boundary adjusted to conform to survey.
Holy Cross	CO	SIO	SEP 24, 1938	3 FR 2445	Boundary adjusted to conform to survey.
Routt	CO	SIO	SEP 24, 1938	3 FR 2446	Boundary adjusted to conform to survey.
Lemhi	ID	EO 7986	OCT 08, 1938	3 FR 2435	Land divided between. Challis and Salmon. Name discontinued.
Targhee	ID/WY	EO 7986	OCT 08, 1938	3 FR 2435	Land transferred to Challis and Salmon.
Challis	ID	EO 7986	OCT 08, 1938	3 FR 2435	Land added from Lemhi and Targhee.
Salmon	ID	EO 7986	OCT 08, 1938	3 FR 2435	Land added from Lemhi and Targhee.
Manistee	MI	Proc. 2306	OCT 25, 1938	53 Stat. 2492	Established. 3 FR 2577.
George Washington	VA/WV	Proc. 2311	NOV 23, 1938	53 Stat. 2499	Boundary redescribed. 3 FR 2779.
Marquette	MI	Proc. 2313	NOV 25, 1938	53 Stat. 2505	Land added. 3 FR 2799.
Deschutes	OR	Proc. 2316	DEC 05, 1938	53 Stat. 2511	Land added. 3 FR 2881.
Deschutes	OR	FR Notice	DEC 09, 1938	3 FR 2885	Proc. 2316 of DEC 5, 1938, corrected.
Cochetopa	CO	EO 8030	DEC 29, 1938	3 FR 3187	Land transferred to Rio Grande.
Rio Grande	CO	EO 8030	DEC 29, 1938	3 FR 3187	Land added from Cochetopa.
1939					
Hiawatha	MI	Proc. 2318	JAN 3, 1939	53 Stat. 2518	Land added. 4 FR 75.
Marquette	MI	Proc. 2319	JAN 11, 1939	53 Stat. 2520	Proc. 2313 of NOV 25, 1938, corrected. 4 FR 2025.

NAME	STATE	DOC TYPE	DATE APPROVED	CITATION	EFFECT
Tongass	AK	Proc. 2330	APR 18, 1939	53 Stat. 2534	Land transferred to Glacier Bay National Monument. 4 FR 1661.
Whitman	OR	Proc. 2332	APR 26, 1939	53 Stat. 2536	Land added. 4 FR 1716.
Cache	ID/UT	Proc. 2333	APR 28, 1939	53 Stat. 2537	Land added. 4 PR 1763.
Homochitto	MS	EO 8100	APR 28, 1939	4 FR 1725	Land added.
Cache	ID/UT	EO 8130	MAY 11, 1939	4 FR 2017	Land transferred to Caribou.
Caribou	ID/WY	EO 8130	MAY 11, 1939	4 FR 2017	Land added from Cache.
Marquette	MI	Proc. 2336	MAY 11, 1939	53 Stat. 2541	Proc. 2313 of NOV 25, 1938, corrected. 4 FR 2044.
Mt. Baker	WA	Asst. Sec.	MAY 15, 1939		Land added. Of Comm, ltr.
Beaverhead	ID/MT	Proc. 2339	JUN 29, 1939	53 Stat. 2544	Land transferred to Big Hole National Monument. 4 FR 2747.
Kaniksu	ID/WA	PL 374	AUG 10, 1939	53 Stat. 1347	Land added.
Wenatchee	WA	PL 394	AUG 11, 1939	53 Stat. 1412	Land added.
Black Hills	SD	EO 8240	SEP 06, 1939	4 FR 3836	EO 4244 of JUN 5, 1925, amended.
Chattahoochee	GA	Proc. 2355	SEP 06, 1939	54 Stat. 2645	Land added. 4 FR 3859.
Cache	ID/UT	Proc. 2356	SEP 06, 1939	54 Stat. 2648	Land added. 4 FR 3860.
Shawnee	IB	Proc. 2357	SEP 06, 1939	54 Stat. 2649	Established. 4 FR 3860.
Mark Twain	MO	Proc. 2362	SEP 11, 1939	54 Stat. 2655	Established. 4 FR 3907.
Clark	MO	Proc. 2363	SEP 11, 1939	54 Stat. 2657	Established. 4 FR 3908.
Arapaho	CO	SIO	NOV 28, 1939	4 FR 4864	Boundary adjusted to conform to survey.
Holy Cross	CO	SIO	NOV 28, 1939	4 FR 4864	Boundary adjusted to conform to survey.
1940					
Olympic	WA	Proc. 2380	JAN 02, 1940	54 Stat. 2678	Land transferred to Olympic National Park. 5 FR 57.
Siuslaw	OR	PL 402	JAN 17, 1940	54 Stat. 14	Land added.
Huron	MI	Proc. 2384	JAN 31, 1940	54 Stat. 2684	Land added. 5 FR 593.
Idaho	ID	EO 8355	FEB 25, 1940	5 FR 827	Land deleted and transferred to Salmon.
Salmon	ID	EO 8355	FEB 25, 1940	5 FR 827	Land added from Idaho.
Wasatch	UT	Proc. 2387	MAR 02, 1940	54 Stat. 2687	Land added. 5 FR 927.

NAME	STATE	DOC TYPE	DATE APPROVED	CITATION	EFFECT
Sequoia	CA	PL 424	MAR 04, 1940	54 Stat. 41	Land transferred to Kings Canyon National Park.
Sierra	CA	PL 424	MAR 04, 1940	54 Stat. 41	Land transferred to Kings Canyon National Park.
Harney	SD	PL 519	MAY 22, 1940	54 Stat. 218	PL 629 of JUN 15, 1938, corrected.
Sequoia	CA	Proc. 2411	JUN 21, 1940	54 Stat. 2710	Land transferred to Kings Canyon National Park. 5 FR 2379.
Choctawhatchee	FL	PL 668	JUN 27, 1940	54 Stat. 628	Land transferred to War Department. 54 Stat. 655.
Chattahoochee	GA	PL 2415	JUL 12, 1940	54 Stat. 2716	Added Bankhead-Jones Act lands within boundary to National Forest.
Talladega	AL	PL 2415	JUL 12, 1940	54 Stat. 2716	Added Bankhead-Jones Act lands within boundary to National Forest.
Ouachita	AR	PL 2415	JUL 12, 1940	54 Stat. 2716	Added Bankhead-Jones Act lands within boundary to National Forest.
Apalachicola	FL	PL 2415	JUL 12, 1940	54 Stat. 2716	Added Bankhead-Jones Act lands within boundary to National Forest.
Chequamegon	WI	PL 2415	JUL 12, 1940	54 Stat. 2716	Added Bankhead-Jones Act lands within boundary to National Forest.
Nicolet	WI	PL 2415	JUL 12, 1940	54 Stat. 2716	Added Bankhead-Jones Act lands within boundary to National Forest.
Talladega	AL	PL 2415	JUL 12, 1940	54 Stat. 2716	Added and reserved public lands within boundary to National Forest.
De Soto	MS	PL 754	JUL 19, 1940*	54 Stat. 773	Authorized land transferred to Secretary of War subject to reverter clause.
Cherokee	TN	TVA 56799	AUG 12, 1940	5 FR 4512	Land added from Tennessee Valley Authority (TVA).
Chattahoochee	TN	TVA 56798	NOV 15, 1940*	5 FR4515	Land added from Tennessee Valley Authority (TVA).
Ozark	AR	Proc. 2422	AUG 27, 1940	54 Stat. 2736	Land added.
Lolo	MT	EO 8544	SEP 19, 1940	5 FR 3761	Land transferred to Helena.

NAME	STATE	DOC TYPE	DATE APPROVED	CITATION	EFFECT
Oelena	MT	EO 8544	SEP 19, 1940	5 FR 3761	Land added from Lolo.
Kaibab	AZ	SIO	SEP 30, 1940	5 FR 4055	Boundary adjusted to conform to survey.
Routt	CO	SIO	SEP 30, 1940	5 FR 4055	Boundary adjusted to conform to survey.
Siuslaw	OR	PL 877	NOV 25, 1940	54 Stat. 1210	Land added.
De Soto	MS	SAO	DEC 04, 1940	Not Published	Land transferred to Secretary of War under PL 754 of JUL 19, 1940.
1941					
Shasta	CA	Proc. 2459	FEB 19, 1941	55 Stat. 1613	Land added. 6 FR 1097.
Wyoming	WY	EO 8709	MAR 10, 1941	6 FR 1400	Name changed to Bridger.
Bridger	WY	EO 8709	MAR 10, 1941	6 FR 1400	Reestablished.
Toiyabe	NV	SIO	MAR 18, 1941	6 FR 1636	Boundary adjusted to conform to survey.
Cache	ID/UT	Proc. 2484	MAY 12, 1941	55 Stat. 1641	Land added. 6 FR 2457.
Wenatchee	WA	Proc. 2490	JUN 03, 1941	55 Stat. 1651	Land added. 6 FR 2741.
Ouachita	AR	EO 8906	SEP 23, 1941	6 FR 4877	Land transferred to Ozark.
Ozark	AR	EO 8906	SEP 23, 1941	6 FR 4877	Land added from Ouachita.
Mt. Baker	WA	Asst. Sec.	DEC 23, 1941		Land added. Of Comm, ltr.
1942					
Fremont	OR	EO 9060	FEB 12, 1942	7 FR 1059	Land added.
Powell	UT	PL 485	MAR 07, 1942	56 Stat. 141	Proc. 1952 of MAY 4, 1931, corrected.
Dixie	AZ/UT	PL 486	MAR 07, 1942	56 Stat. 141	Boundary adjusted between Dixie and Cedar Breaks National Monument.
Stanislaus	CA	FR Notice	APR 04, 1942*	7 FR 2916	Land transferred to Yosemite National Park.
Cache	ID/UT	EO 9124	APR 07, 1942*	7 FR 2692	Land transferred to Caribou.
Caribou	ID/WY	EO 9124	APR 07, 1942*	7 FR 2692	Land added from Cache.
Boise	ID	PL 582	JUN 05, 1942	56 Stat. 320	Land added.
Salmon	ID	PL 582	JUN 05, 1942	56 Stat. 320	Land added.
Targhee	ID/WY	PL 582	JUN 05, 1942	56 Stat. 320	Land added.
Black Warrior	AL	PL 595	JUN 06, 1942	56 Stat. 327	Name changed to William B. Bankhead.

NAME	STATE	DOC TYPE	DATE APPROVED	CITATION	EFFECT
William B. Bankhead	AL	PL 595	JUN 06, 1942	56 Stat. 327	Established.
Cleveland	CA	PL 606	JUN 11, 1942	56 Stat. 358	Land added.
Sequoia	CA	PL 795	DEC 09, 1942	56 Stat. 1044	Land added.
1943					
Teton	WY	Proc. 2578	MAR 15, 1943	57 Stat. 731	Land transferred to Jackson Hole National Monument. 8 FR 3277.
Olympic	WA	Proc. 2587	MAY 29, 1943	57 Stat. 741	Land transferred to Olympic National Park. 8 FR 7365.
Colville	WA	PLO 162	AUG 23, 1943*	8 FR 12349	Land transferred to Chelan.
Chelan	WA	PLO 162	AUG 23, 1943*	8 FR 12349	Land added from Colville.
Kaniksu	WA	PLO 163	AUG 23, 1943*	8 FR 12349	Land transferred to Colville.
Colville	WA	PLO 163	AUG 23, 1943*	8 FR 12349	Land added from Kaniksu.
1944					
Payette	ID	PLO 217	MAR 18, 1944	9 FR 3655	Land transferred to Boise.
Boise	ID	PLO 217	MAR 18, 1944*	9 FR 3655	Established. Payette (old) added.
Idaho	ID	PLO 218	MAR 18, 1944*	9 FR 3655	Forest combined w/Weiser and established Payette (new).
Weiser	ID	PLO 218	MAR 18, 1944*	9 FR 3655	Forest combined w/Idaho and established Payette (new).
Payette	ID	PLO 218	MAR 18, 1944*	9 FR 3655	Established (Payette, new).
Mesilla	NM/TX	PLO 220	APR 06, 1944	9 FR 4031-3	Established.
Angelina	TX	PL 539	DEC 22, 1944	58 Stat. 911	Authorized transfer of Nacogdoches County land to National Forest.
1945					
Plumas	CA	Proc. 2635	JAN 13, 1945	59 Stat 853	Land added. 10 FR 693.
Eldorado	CA/NV	Proc. 2636	JAN 13, 1945	59 Stat. 854	Land added. 10 FR 693.
San Isabel	CO	PLO 257	JAN 17, 1945*	10 FR 1243	Land transferred to Rio Grande.
Rio Grande	CO	PLO 257	JAN 17, 1945*	10 FR 1243	Land transferred from San Isabel.
Cochetopa	CO	PLO 258	JAN 17, 1945*	10 FR 1243	Land distributed among Gunnison, Rio Grande, and San Isabel.

NAME	STATE	DOC TYPE	DATE APPROVED	CITATION	EFFECT
Gunnison	CO	PLO 258	JAN 17, 1945*	10 FR 1243	Land added from Cochetopa.
Rio Grande	CO	PLO 258	JAN 17, 1945*	10 FR 1243	Land added from Cochetopa.
San Isabel	CO	PLO 258	JAN 17, 1945*	10 FR 1243	Land added from Cochetopa.
Powell	UT	PLO 260	JAN 19, 1945*	10 FR 1244	Abolished. Land transferred to Dixie.
Dixie	AZ/UT	PLO 260	JAN 19, 1945*	10 FR 1244	Abolished. Land added from Powell.
Holy Cross	CO	PLO 263	FEB 19, 1945*	10 FR 2251	Abolished. Land transferred to White River.
White River	CO	PLO 263	FEB 19, 1945*	10 FR 2251	Land added from Holy Cross.

* Effective date is different from date approved.

APPENDIX E

THE NINE CIVILIAN CONSERVATION CORPS AREAS

First: Connecticut, Maine, Massachusetts, New Hampshire, Rhode Island, Vermont

Second: Delaware, New Jersey, New York

Third: District of Columbia, Maryland, Pennsylvania, Virginia

Fourth: Alabama, Florida, Georgia, Louisiana, Mississippi, North Carolina, South Carolina, Tennessee

Fifth: Indiana, Kentucky, Ohio, West Virginia

Sixth: Illinois, Michigan, Wisconsin

Seventh: Arkansas, Iowa, Kansas, Minnesota, Missouri, Nebraska, North Dakota, South Dakota,

Eighth: Arizona, Colorado, New Mexico, Oklahoma, Texas, Wyoming (excluding western Wyoming and Yellowstone National Park)

Ninth: Alaska, California, Hawaii, Idaho, Montana, Nevada, Oregon, Utah, Washington, and western Wyoming and Yellowstone National Park

APPENDIX F

CIVILIAN CONSERVATION CORPS—BASIC FACTS (COURTESY OF THE CCC LEGACY FOUNDATION)

- Duration of the program: April 5, 1933, to June 30, 1942
- Nicknames: "Roosevelt's Tree Army," "Tree Troopers," "Soil Soldiers," "Cees," "3 Cs," "Colossal College of Calluses," "Woodpecker Warriors"
- Total men enrolled: 3,463,766
- Juniors, veterans, and Native American enrollees: 2,876,638
- Territorial enrollees: 50,000 (estimated)
- Non-enrolled personnel: 263,755
- Average enrollee: eighteen to nineteen years old, 147 pounds, five feet eight inches tall
- Average weight gain of enrollees in first three months: 11.5 pounds
- Number of illiterate enrollees taught to read: 40,000
- Average number of camp operating in the United States per year: 1,643
- Total number of different camps: 4,500
- Highest elevation of a CCC camp: 9,200 feet above sea level—Rocky Mountain Park, Colorado
- Lowest elevation of a CCC camp: 270 feet below sea level—Death Valley, California
- Camp locations: Every state in the Union, plus Alaska, Hawaii, Puerto Rico, the U.S. Virgin Islands

- Total cost: $3 billion
- Approximate cost per enrollee in 1940 for food, clothing, overhead, and allotments to dependents: $1,000
- Allotments to dependents: $662,895,000
- Number of people who directly benefited from enrollees' checks: 12 million to 15 million
- Value of work in 1942 dollars: $2 billion
- Miles of roads built: 125,000 miles
- Miles of telephone lines strung: 89,000 miles
- Miles of foot trails built: 13,000 miles
- Farmland that benefited from erosion-control protection: 154 million square miles
- Range revegetation: 814,000 acres
- Firefighting man-days: More than 6 million
- Number of enrollees who died fighting fires: 29
- Overall death rate: 2.25 per thousand
- State parks developed: 800
- Public campground development: 52,000 acres
- Mosquito control: 248,000 acres
- Number of fish stocked: 972 million
- Historic restoration: 3,980 structures
- Number of trees planted: between 2 billion and 3 billion
- Number of federal government agencies participating in some capacity: 25
- Unofficial motto: "We can take it!"

ACKNOWLEDGMENTS

Just after Franklin D. Roosevelt won the 1932 presidential election the *New York Times* published a series of biographical articles about him. When asked what differentiated him from his distant cousin Theodore Roosevelt, the president-elect turned instinctively to forestry. Franklin, unlike TR, disliked the lumberjack's ax. "I like to plant trees," FDR said, "not cut them down." And plant he did. From 1933 to 1945 Roosevelt's New Deal conservation programs planted around three billion trees from California to New Hampshire, Alaska to Puerto Rico. Planting trees was to FDR a God-affirming act. Determined to avoid the long-ballyhooed timber famine, sickened by much-publicized stories of soil erosion and depletion of the nation's forest resources, Roosevelt, during his two terms as governor of New York and his entire four terms in the White House (March 4, 1933, to April 12, 1945), became America's visionary forester in chief.

Essentially, this book is a sequel to my *The Wilderness Warrior: Theodore Roosevelt and the Crusade for America* (New York: HarperCollins, 2009). TR was brilliantly effective in preserving America's most visually stunning and wildlife rich terrain, but at the same time he exerted only light control over industries—notably timber extraction on public and private land—that FDR held to a higher standard of conduct and often outlawed completely from rich forestland. The two Roosevelts—TR and FDR—are undisputedly America's great revolutionary conservation presidents. Yet FDR's conservation heroics remained, up until this publication, overshadowed in history by TR. This is largely because TR wrote vivid travel articles about his outdoor adventures in places such as the Grand Canyon and Yellowstone, Africa and the Amazon. By contrast, the intrepid FDR never kept journals of his travels to Hawaii Volcanoes and the Olympics, the Everglades and Dry Tortugas. As documented in this book, FDR, for example, cruised to the Galápagos Islands in 1938 to lead a Smithsonian Institution–sponsored marine biology expedition. Certainly TR would have self-servingly written about discovering, in the Darwinian tradition, new species of flora and fauna. By contrast, all FDR managed after weeks of scientific collecting on Ecuador's islands were a couple of lighthearted letters to Eleanor.

The appendixes of this book inventory the mind-numbing number of national parks, national forests, and national historic sites FDR preserved. There was no room for a spreadsheet delineating the hundreds of state parks the New Deal built from scratch. While Thomas Jefferson, Abraham Lincoln, Benjamin Harrison, John F. Kennedy, Lyndon B. Johnson, Richard M. Nixon, Jimmy Carter, Bill Clinton, and Barack Obama were likewise outstanding environment-minded presidents, the Roosevelts made the management of natural resources the top domestic issue during their White House tenures.

My journey to write a biography of FDR from an environmental history perspective began in Hyde Park, New York. As a trustee of the Franklin D. Roosevelt Presidential Library (FDRL), administered by the National Archives and Records Administration (NARA), I regularly examine Great Depression and World War II literature. Whenever a new archival collection pertaining to Franklin and Eleanor is processed, I grow excited. The digitalization of many FDRL collections has enabled twenty-first-century scholars everywhere to conduct primary research without having to travel to Hyde Park. Nevertheless, for this book, living in the Hudson River Valley ecosystem was essential. Many weeks were spent studying documents in the Arthur M. Schlesinger Jr. Research Room at the FDRL. The most useful collections were the Civilian Conservation Corps (CCC) records and FDR speech files. But I also harvested great returns from the personal papers of FDR, Louis Howe, Henry Wallace, Nelson Brown, and Eleanor Roosevelt. At the FDR Library I'm proud to have worked with both directors Lynn A. Bassanese and Paul M. Sparrow. Archivists Bob Clark (especially), Virginia Lewick, Kristen Strigel Carter, Matthew Hanson, and Sarah Malcolm went beyond the call of duty.

While researching at FDRL I had the privilege of staying at the amazing Payne Estate, located on the west bank of the Hudson River in Ulster County, New York. Marist University owns this 42,000-square-foot Beaux Arts–style Mediterranean palazzo built in 1905. We called the place "the Mansion" and had free run of the sixty well-manicured acres for the summers of 2013 and 2014. My humble gratitude goes to the longtime president of Marist, Dennis J. Murray, for allowing the Brinkley family to experience the life of Hudson River Valley gentry. He has been a valued friend for many years. And special thanks are due to Dr. Murray's assistant, Eileen Sico, and estate manager, Tony Sill, for making our summers so bucolic. Marist University is an exciting institution.

I owe a profound debt of gratitude to Ambassador William J. vanden Heuvel. When I was a young assistant professor at Hofstra University, way back in 1992, Bill asked me to join the board of directors of the Franklin and Eleanor Roosevelt Institute (FERI). The primary objective of FERI was to promote the legacy and ideals of both Roosevelts. I was just thirty-two and had never sat on a board before. At the time, I was living on Bedford Street in Greenwich Village (across from Chumley's Bar) and doing the reverse train commute to Hempstead, Long Island, to teach at Hofstra. Every couple of weeks Bill invited me to lunch along with his closest friend, Arthur M. Schlesinger Jr. Before long, the three of us launched a book series with St. Martin's Press called The World of the Roosevelts. And under the auspices of FERI we organized academic conferences for the fiftieth anniversaries of Pearl Harbor, the Atlantic Charter, Casablanca, D-Day, and other seminal historical events. A former UN ambassador, Bill also spearheaded the successful effort to build the Franklin D. Roosevelt Four Freedoms Park on Roosevelt Island in the East River. With an unobstructed view of the United Nations, this New York City memorial is lined with rows of trees, just as Roosevelt, the inveterate park planner, would have liked.

FERI was fortunate to have John F. Sears as its executive director in the 1990s. John supervised the restoration of Top Cottage (FDR's Dutchess County getaway home) and established the Roosevelt Foundation for United States Studies at Moscow State University in the Soviet Union. In this all-things-Roosevelt spirit, FERI cohosted a first-rate conference in 2002 titled Recovering the Environmental Legacy of FDR. As the title intimates, the conference looked at the neglected legacy of the New Deal's conservation policies. Many topics covered in this book—the Soil Conservation Service, the Shelterbelt, Cape Hatteras National Seashore, the TVA, and the Taylor Grazing

Act—were discussed at this conference. The pioneering New Deal historian Neil M. Maher spoke on the CCC, while environmental historian Paul Sutter ably illuminated the "wilderness" movement during the New Deal. The end result was the publication of Harry L. Henderson and David B. Woolner, eds., *FDR and the Environment* (London: Palgrave Macmillan, 2009) for our World of the Roosevelts series. It served as a cornerstone for me in developing this book.

There is so much excellent scholarship on Franklin and Eleanor Roosevelt's legacy that if I included a proper bibliography, it would run around fifteen pages. Instead, I've chosen, for space purposes, to credit appropriate scholars in the endnotes. However, a few books and sources deserve special mention. A cache of primary documents relevant to environmental history can be found in Edgar B. Nixon, *Franklin D. Roosevelt and Conservation, 1911–1945*, 2 vols. (Hyde Park, NY: National Archives and Records Service, Franklin D. Roosevelt Library, 1957). All scholars find very useful Samuel I. Rosenman, ed., *The Public Papers and Addresses of Franklin D. Roosevelt*, 4 vols. (New York: Random House, 1937–1940), and Geoffrey C. Ward, ed., *Closest Companion: The Unknown Story of the Intimate Friendship Between Franklin Roosevelt and Margaret Suckley* (New York: Simon & Schuster, 1995), which offers extraordinary insights into FDR's thinking about the natural world.

I profited mightily from Arthur M. Schlesinger Jr.'s three-volume masterpiece, The Age of Roosevelt: *The Crisis of the Old Order: 1919–1933* (1957); *The Coming of the New Deal: 1933–1935* (1958); and *The Politics of Upheaval: 1935–1936* (1960). Before his death in 2007, Arthur told me that Frederic A. Delano was an unsung hero of the New Deal. Because Delano was FDR's uncle, he slipped into and out of the Oval Office without being marked on the White House logbooks. His influence on Franklin was large. Indeed, as I soon learned, he was FDR's crucial adviser on all land-planning–related issues. As chairman of the Board of the American Planning and Civic Association, Delano oversaw the publication of the organization's *Annual* from 1932 to 1945. These reports are the mother lode for anybody interested in U.S. preservation philosophy during the Great Depression and New Deal land-planning accomplishments.

While I was writing this book, the Olmsted Center for Landscape Preservation published a series of "Historic Resource" studies pertaining to Springwood, Val-Kill, and Top Cottage. Packed with photos, maps, and research, these Olmsted Center reports—written with the help of the State University of New York College of Environmental Science and Forestry in Syracuse—were of inestimable value to me, as were back issues of the *New York Times*, *Washington Post*, *New Yorker*, *Los Angeles Times*, *Living Wilderness*, *Journal of Environmental History*, *Scenic and Historic America*, *Forestry News Digest*, *Journal of Forestry*, *National Geographic*, and *American Forests*. The Forest History Society (Durham, North Carolina) publishes the elegant periodical *Forest History Today*. Every issue offers cutting-edge articles pertaining to U.S. conservation history. I find it must reading.

A special mention is due Curt Meine, author of the masterful *Aldo Leopold: His Life and Work* (Madison: University of Wisconsin Press, 2010). Curt diligently proofread the manuscript and offered numerous helpful suggestions. He is a true gentleman and scholar. And the noble Char Miller—author of *Gifford Pinchot and the Making of Modern Environmentalism* (Washington, DC: Island Press, 2001)—always answers my U.S. Forestry Service questions.

A key component of this book is the working relationship between FDR and Harold Ickes. The best books on Ickes are T. H. Watkins, *Righteous Pilgrim: The Life and Times of Harold L. Ickes, 1874–1952* (New York: Henry Holt & Company, 1990); Jeanne Nienaber Clarke, *Roosevelt's Warrior: Harold L. Ickes and the New Deal* (Baltimore: Johns Hopkins University Press, 1996); and Linda Lear, *Harold L Ickes: The Aggressive Progressive, 1874–1933* (New York: Garland Publishing, 1981). The diaries of Harold L. Ickes (published and unpublished) are a true treasure trove of information about New Deal conservation. The Library of Congress does an admirable job of making these diaries—as well as the Gifford Pinchot Papers—available to scholars.

The key CCC book is Neil Maher's landmark *Nature's New Deal* (New York: Oxford University Press, 2007). Others that were exceedingly helpful were Cynthia Brandimarte, *Texas State Parks and the CCC* (College Station: Texas A&M Press, 2013); Stan Cohen, *The Tree Army: A Pictorial History of the Civilian Conservation Corps, 1933–1942* (Missoula, MT: Pictorial Histories Publishing Co, 1980); Alfred E.

Cornebise, *The CCC Chronicles: Camp Newspapers of the Civilian Conservation Corps, 1933–1942* (Jefferson, NC: McFarland & Co., 2004); Diane Galusha, *Another Day, Another Dollar: The Civilian Conservation Corps in the Catskills* (Delmar, NY: Black Dome Press, 2008); Edwin G. Hill, *In the Shadow of the Mountain: The Spirit of the Civilian Conservation Corps* (Pullman: Washington State University Press, 1990); Perry H. Merrill, *Roosevelt's Forest Army: A History of the Civilian Conservation Corps, 1933–1942* (Montpelier: Perry H. Merrill Publisher, 1984); and John Salmond, *The Civilian Conservation Corps, 1933–1942: A New Deal Case Study* (Durham, NC: Duke University Press, 1967). Ren and Helen Davis, *Our Mark on This Land: A Guide to the Legacy of the Civilian Conservation Corps in America's Parks* (Granville, OH: McDonald & Woodward Publishing Company, 2011) is a wonderful guidebook and history of CCC sites across the nation. Special thanks to the Davises for proofreading my manuscript and offering cogent commentary.

Special thanks to Joan Sharpe, president of CCC Legacy (Edinburg, Virginia). Joan introduced me to surviving CCC veterans, now all in their nineties. Their heartfelt reminiscences were illuminating. The CCC museums in Chesterfield, Virginia, and Sebring, Florida, were also quite helpful. I've been equally impressed with the CCC legacy projects under way in California, Iowa, Arizona, Missouri, New York, Idaho, Wisconsin, Michigan, Texas, and Pennsylvania. The first CCC camp in a national park was Sequoia, and under the superintendent Woody Smeck they do a marvelous job of interpreting CCC history. Everywhere I travel around America, the CCC is fondly remembered. As FDR's assistant secretary of agriculture Rexford G. Tugwell once quipped, the CCC was "one thing in these troubled times of which not even the Republicans can complain."

At the University of California–Berkeley, I profited from the "Living New Deal" project (www.livingnewdeal.org), run by leading WPA scholar Gray Brechin, a landmark digital-historical repository for the accomplishments of FDR's New Deal across the nation. Nick Taylor's *American-Made* (New York: Random House, 2008) has become the must-read book on the WPA.

Spending time in Dutchess County is great fun because I have a coterie of friends living there. Fredrica and Jack Goodman of Poughkeepsie—proud Rooseveltians—have been like family to me for decades. And then there are David and Manuela Roosevelt, who now live just a stone's throw from the FDR Library. Their stories about FDR and ER are priceless. Finally, Joan Burroughs, the granddaughter of nature essayist John Burroughs, sponsored me as a guest lecturer at both Vassar College and the John Burroughs Sanctuary in Esopus, New York (a short walk from the Payne estate).

While most of my research was conducted at FDRL, the Library of Congress, and the National Conservation Training Center in Shepherdstown, West Virginia, other archival repositories were visited. These included the Smithsonian Institution (Waldo Schmidt Papers); the USGS–Patuxent Wildlife Research Center; the Groton Preparatory School; the Columbia University Oral History Project; the Cornell Lab of Ornithology, Cornell University; the Idaho Conservation League, Boise, Idaho; the Center for Biological Diversity, Tucson, Arizona; the Roosevelt House at Hunter College in New York City; and the Roosevelt-Vanderbilt National Historic Site Archives in Hyde Park.

Eleanor Roosevelt was also a true-blue conservationist. Her "My Day" columns, written from 1935 to 1962, are brimming with natural history observations. Scholars owe George Washington University—especially Allida Black—gratitude for putting them online.

Across the Atlantic, I visited on a number of occasions the Roosevelt Study Center (RSC), located in the twelfth-century Abbey of Middelburg, the Netherlands. The RSC is the most dynamic American Studies center in Europe. Great thanks go to director Cornelius A. van Minnen for teaching Europeans about the enduring legacies of Theodore, Franklin, and Eleanor Roosevelt.

My collaboration with the American Museum of Natural History in New York proved beneficial, and I'm very grateful to director Ellen Futter for her support of my work. Periodically, I guest-lecture at the museum on U.S. conservation history. Likewise the New-York Historical Society, under the inspired leadership of Louise Mirrer, has allowed me to be impresario of its ongoing forum on the U.S. presidency. Her able colleagues Jennifer Schantz, Margi Hofer, and Dale Marsha Gregory are consummate professionals.

Kennon Moody, author of *FDR and His Hudson Valley Neighbors*

(Poughkeepsie, NY: Hudson House Publishing, 2013), proofread the manuscript. His knowledge of the Hudson River Valley is unsurpassed. Professor Mark Carnes of Barnard College escorted me around Newburgh, New York (home of the Delanos). This proved very instructional in my imagining the role the river played in the lives of the Delano family. Mark, along with Jill Leopore, is the glue that keeps the Society of American Historians thriving. Being a member of this organization, getting to know other historically minded writers, is a great privilege.

For assistance large and small, my thanks to Wint Aldrich, John Auwaerter, Lowell Baier, Rocky Barker, David Beard, Shane Bernard, Dick Beahrs, Allida Black (an encyclopedia on all things ER), Reed Bohne, Talmadge Boston, Camille Bradford, Nate Brostrom, Doug Brown, Michael Brune, William J. Bryan, Clark Bunting, Tom Campion, Dorothy Canter, Ben Carter, Dan Chu, Richard A. Coon, William Cronon, Kate Damon, Ginny Davis, Harry Dennis, Patricia Duff, Bob Dumaine, Dave Foreman, Gail Friedman, Lynda Garrett, Gary and Elizabeth Goetzman, Doris Kearns Goodwin, Jeff Gronauer, Dale Hall, Bruce Hamilton, Patricia Hart, Martin Heinrich, David and Sarah Holbrooke, Vicky Hoover, Nancy Roosevelt Ireland, Bob Irvin, James M. Johnson, Emma Juniper, Michael Kellett, David Klinger, John A. Knox, Brad Knudsen, Chieu and Mike Komarek, Jill Krastner, Karen Bates Kress, Jane B. Kulow, Bob Kustra, Howard Labanara, Joette Langianese, Linda Lear, Douglas Leen, Jim Madsen, Tim Mahoney, Terri Martin, Mike Matz, Clayton Maxwell, Lisa Mighetto, Char Miller, Kristin Miller, Neil Mulholland, Jim Mullen, Patty Murray, Jay and Georganne Nixon, Michael Northrup, Matt Nye, Joy Oakes, Brian O'Donnell, Uri Perrin, John Podesta, Christopher Pryslopski, John F. Reiger, Allison Whipple Rockefeller, George Roderick, Jedediah Rogers, Anne Roosevelt, David Roosevelt, Simon Roosevelt, Winthrop Roosevelt, Alfred Runte, Ken Salazar, Alexandra Schlesinger, Steve Schlesinger, Larry Schweiger, Janet Seegmiller, Kabir Sehgal, Laurie M. Shaffer, Cindy Shogan, Mike Simpson, Jay Slack, Carter Smith, Gary Snyder, Melanie Spoo, Will Swift, Melissa Switzer, Mark Tercek, Paul Tritaik, Tom Udall, Robert M. Utley, Peter Van Tuyn, Melody Webb, Jann Wenner, Brooke Williams, Jamie

Williams, Terry Tempest Williams, Felecia Wong, David Yarnold, Suzanne Ybarra, Lee and Anne Yeakel, and David Zilberman.

I am deeply indebted to historian Mark Madison of U.S. Fish and Wildlife's National Conservation Training Center in Shepherdstown, West Virginia, for guiding me through the maze of archival materials in the collection. He also kindly proofread the manuscript. While I was writing this book, the U.S. Fish and Wildlife asked me to be a judge for the 2013 Federal Duck Stamp Contest, held in Maumee, Ohio. Since 1934, the sale of these FDR-inspired stamps has raised more than $850 million to protect more than six million acres of prime habitat on our nation's national wildlife refuges. Thanks to my fellow judges—John E. Cornley, John Ruthven, Charles "Chad" Snee, and Mamie Parker—all now friends.

All Americans should visit Roosevelt's Little White House State Historic Site in remote Warm Springs, Georgia, ably managed by Robin Glass. All the items in the modest cottage have been left exactly as they were when FDR died there. The thermal pools where Roosevelt once swam are now run by the Roosevelt Warm Springs Institute for Rehabilitation. Special thanks to David Burke, who discovered this book's rare cover photo of FDR *standing* at Warm Springs, inspecting a pine tree. Burke is the heart and soul of Little White House preservation.

Franklin Roosevelt's conservation legacy has enormous significance to the modern-day National Park Service. It was FDR—not Woodrow Wilson—who configured the agency that resembles today's NPS. Two bills passed during Roosevelt's first White House term—the Reorganization of 1933 and the Historic Sites Act of 1935—had a more profound impact on the National Park Service System than did any other legislation since the Organic Act of 1916. To write this book, I consulted with a wide range of NPS and Interior leaders, including John Jarvis, John Sprinkle, Jason Jurgena, Bob Sutton, and Tracy Baetz. Also David Nimkin and Suzanne Dixon of the National Parks Conservation Association (NPCA) helped direct me to a half dozen national park "friends" groups. Jamie Williams of the Wilderness Society and Michael Brune of the Sierra Club helped me better understand the roles their nonprofits played during the

New Deal era. On issues pertaining to American wilderness history, Doug Scott of Seattle is The Man.

Touring the Stewart Lee Udall Department of the Interior Building (aka Harold Ickes's Palace) is one of the great tourist-friendly things to do in Washington, D.C. The New Deal is alive in the corridors and offices of this grand edifice. Secretary of the Interior Sally Jewell facilitated my visits to the Udall Building and encouraged the staff to help me. Together with Janet Napolitano, we participated in a three-way public conversation at the University of California–Berkeley on the future of the national parks. In addition to Jewell, the NPS is very lucky to have Sarah Olson serving as superintendent of Roosevelt-Vanderbilt National Historic Sites. Nobody has more preservation savvy or works harder than Sarah to protect the integrity of Springwood. Together we explored FDR's tree farm, local woodlands, the Hudson River, and Top Cottage. I treasure her as a kindred spirit.

My dear friend Julie Fenster, author of many seminal books of American history, proofread and helped edit the manuscript. Her input is always smart and on point. All my kids adore Julie's stable of horses in upstate New York. She takes them riding. And Rob Fleder ably edited early chapter drafts.

At HarperCollins, my editor Jonathan Jao was a constant source of support. His standards are among the best in the business. Jonathan Burnham epitomizes an excellent publisher. We are good friends and have amazing shorthand. Michael Morrison, president and publisher, U.S. General Books and Canada at HarperCollins, has been a friend since the 1990s. I'm always grateful for his wise counsel. A new addition to the HarperCollins family—Sofia Groopman—did a dazzling job of shepherding this book through the production process. Competence should be her middle name. Thanks also to Trent Duffy (editor), David Koral (production editor), Leah Carlson-Stanisic (design), Kate D'Esmond (publicist extraordinaire), and Katie O'Callaghan (marketing). They're a marvelous team. Lisa Bankoff of ICM remains the best of agents and counselors.

My indispensable assistant on this book was Mark Davidson. A brilliant New Deal scholar in his own right, Mark earned his PhD in cultural musicology at the University of California–Santa Cruz while helping me from Austin. Nobody knows more about WPA folk music collecting than he does. An Internet wizard, he helped me track down academic articles and obscure facts. Likewise, in the early stages of this book, Virginia Northington, a University of Texas–Austin graduate, served as my research assistant. Although Virginia now lives in Washington, D.C., we stay in constant touch.

Every fall semester I teach a class on U.S. environmental history at Rice University. Over the years my students have stimulated my thinking about New Deal conservation. Special thanks to Kim Ricker, Jean Aroom, and the wonderful staff of the GIS/Data Center of Rice's Fondren Library for preparing the maps that appear in this book. I'm also the presidential historian for CNN. It's exhilarating to be part of a first-rate news team. While the producers and on-air talent didn't directly help with this book, their enthusiasm for my work helps fuel me forward.

Carolyn Merchant, professor of environmental history, philosophy, and ethics at the University of California–Berkeley, continues to be a source of inspiration to me. It was Merchant who made sure I was aware of female environmental warriors of the New Deal, such as Minerva Hamilton Hoyt and Jean Harper. Her knowledge of the Progressive Era, her mensch-like words of wisdom, her marvelous mind, and, above all, her openhearted generosity are deeply appreciated.

My wife Anne and our children—Benton, Johnny, and Cassady—went on this environmental history journey with me. Together we lived in the Hudson River Valley, hiked the Lost Mine Trail at Big Bend, beach-combed in Aransas, swam in the fresh spring water of Balmorhea, stargazed in Joshua Tree, photographed the sequoias of Kings Canyon, and visited many other FDR-established national and state parks. Getting to "work" in scenic towns such as Jackson Hole, Wyoming, and Front Royal, Virginia, is as good as it gets for a historian of Wild America. And all the love in the world to my mother and father—Anne and Edward Brinkley of Laguna Niguel, California—who brought the National Park Service into my life as a boy.

Austin, Texas
November 23, 2015

NOTES

CHAPTER 1: "ALL THAT IS IN ME GOES BACK TO THE HUDSON"

1. Eleanor Roosevelt, *Franklin D. Roosevelt and Hyde Park: Personal Reflections of Eleanor Roosevelt* (Washington, DC: Department of the Interior, 1977), p. 1.

2. Jan Pottker, *Sara and Eleanor: The Story of Sara Delano Roosevelt and Her Daughter-in-Law, Eleanor Roosevelt* (New York: St. Martin's, 2004), p. 1.

3. Hudson's log, quoted in Carl Carmer, *The Hudson* (New York: Fordham University Press, 1989), pp. 12–13.

4. Geoffrey C. Ward, *Before the Trumpet: Young Franklin Roosevelt, 1882–1905* (New York: Harper & Row, 1985), pp. 120–21.

5. Lisa Nowak, *Cultural Landscape Report for Eleanor Roosevelt National Historic Site* (Boston: Olmsted Center for Landscape Preservation, 2005), p. 5.

6. "The New Summer White House an Old Hudson River Estate," *New York Times*, March 12, 1933.

7. Harry T. Peters, *Currier and Ives: Printmakers to the American People* (New York: Doubleday, Doran, 1942).

8. Leo Marx, *The Machine in the Garden* (New York: Oxford University Press, 1964), p. 3.

9. Pottker, *Sara and Eleanor*, p. 59.

10. Franklin D. Roosevelt (hereafter, as author, FDR), "Address at the Cornerstone Laying of the Franklin D. Roosevelt Library in Hyde Park," November 19, 1939, Franklin D. Roosevelt Presidential Library, Hyde Park, NY (hereafter cited as FDRL).

11. Olin Dows, *Franklin Roosevelt at Hyde Park: Documented Drawings and Text* (New York: American Artists Group, 1949), pp. 23–29.

12. John F. Sears, "FDR and the Land," in *Historic Resource Study for the Roosevelt Estate* (draft report prepared for National Park Service, July 2004), p. 3.

13. John F. Sears, *Historic Resource Study for the Roosevelt Estate* (Boston: National Park Service, 2004), p. 3. This study was undertaken with help from the Olmsted Center for Landscape Preservation.

14. Gardner Bridge, "Sea Attractive to Roosevelt," *New York Times*, November 11, 1932.

15. Thomas Cole, "Essay on American Scenery," *American Monthly Magazine*, Vol. 1 (January 1836), pp. 1–12.

16. Article VII, Section 7, of New York Constitution, quoted in "History of the Adirondack Park," New York State Adirondack Park Agency, last updated 2003, http://apa.ny.gov/about_park/history.htm; see also James M. Glover, *A Wilderness Original: The Life of Bob Marshall* (Seattle: Mountaineers, 1986).

17. Charles Sprague Sargent, *Report on the Forests of North America* (Washington, DC: U.S. Government Printing Office, 1884).

18. Pottker, *Sara and Eleanor*, pp. 11–13.

19. Ruth Valenti, "Algonac Estate, Home for Delano Family," *Evening News* (Newburgh), May 15, 1987.

20. Catherine Sedgwick quoted in David E. Shi, *The Simple Life: Plain Living and High Thinking in American Culture* (Athens: University of Georgia Press, 2007), p. 106.

21. Clara Steeholm and Hardy Steeholm, *The House at Hyde Park* (New York: Viking, 1950), p. 10.

22. Andrew Jackson Downing, "On the Moral Influence of Good Houses," in Robert Twombly, ed., *Andrew Jackson Downing: Essential Texts* (New York: W. W. Norton, 2012), p. 117.

23. Angela Miller, *Empire of the Eye: Landscape Reforestation and American Cultural Politics, 1825–1875* (Ithaca, NY: Cornell University Press, 1993), p. 13.

24. Steeholm and Steeholm, *The House at Hyde Park*, p. 10.

25. Sara Delano Roosevelt Diary, November 11, 1882, FDRL.

26. Ward, *Before the Trumpet*, p. 121.

27. Kenneth S. Davis, *FDR: The New York Years, 1928–1933* (New York: Random House, 1994), p. 5.

28. Pottker, *Sara and Eleanor*, pp. 25–26.

29. Ibid., p. 63.

30. Elliott Roosevelt, ed., *F.D.R.: His Personal Letters, Early Years* (New York: Duell, Sloan and Pearce, 1947), p. 39.

31. Pottker, *Sara and Eleanor*, p. 12.

32. Ted Morgan, *FDR: A Biography* (New York: Simon and Schuster, 1985), p. 22.

33. Steeholm and Steeholm, *The House at Hyde Park*, p. 6.

34. James MacGregor Burns, *Roosevelt: The Lion and the Fox* (New York: Harcourt, Brace and World, 1956), p. 7.

35. Sara Roosevelt, "Foreword," *Crum Elbow Folks* (Philadelphia: J. P. Lippincott, 1938).

36. "Next Week," *Collier's*, June 18, 1932.

37. Douglas Brinkley, *The Wilderness Warrior: Theodore Roosevelt and the Crusade for America* (New York: HarperCollins, 2009), p. 91.

38. Robert Barnwell Roosevelt, *Game Fish of the Northern States of America and British Provinces* (New York: Carleton, 1862); *The Game-Birds of the Coasts and Lakes of the Northern States of America: A Full Account of the Sporting Along Our Sea-Shores and Inland Waters, with a Comparison of the Merits of Breech-Loaders and Muzzle-Loaders.* (New York: Carleton, 1866); and *Superior Fishing; Or, The Striped Bass, Trout, Black Bass, and Blue-Fish of the Northern States. Embracing Full Directions for Dressing Artificial Flies with the Feathers of American Birds; An Account of a Sporting Visit to Lake Superior, Etc., Etc.* (New York: Orange Judd, 1884).

39. Edwin G. Burrows and Mike Wallace, *Gotham: A History of New York City to 1898* (New York: Oxford University Press, 1998), p. 1196.

40. "An Engineering Marvel: An Urgent Need for Water," Friends of the Old Croton Aqueduct, accessed September 2, 2013, http://www.aqueduct.org/content/engineering-marvel.

41. Frances F. Dunwell, *The Hudson: America's River* (New York: Columbia University Press, 2008), p. 274.

42. James Tobin, *The Man He Became: How FDR Defeated Polio to Win the Presidency* (New York: Simon and Schuster, 2013), p. 41.

43. Stanley Weintraub, *Young Mr. Roosevelt: FDR's Introduction to War, Politics, and Life* (New York: Da Capo, 2013), p. 11.

44. Tobin, *The Man He Became*, p. 41.

45. Steeholm and Steeholm, *The House at Hyde Park*, pp. 85–86.

46. Ward, *Before the Trumpet*, p. 160.

47. Ibid., p. 121; and Conrad Black, *Franklin Delano Roosevelt: Champion of Freedom* (Washington, DC: PublicAffairs, 2003), p. 15.

48. Elliott Roosevelt, *F.D.R.: His Personal Letters, Early Years*, p. 293.

49. Stephen Cernek, "From Picturesque to Profane: A Cultural History of the Hudson River, Palmer Falls," *Hudson River Valley Review* (Spring 2013), pp. 52–70.

50. Six of FDR's first seven trips to Europe are recorded in the Sara Delano Roosevelt Diary, Roosevelt Family Papers, FDRL.

51. FDR, address at Clarksburg, West Virginia, October 29, 1944 (transcript), FDRL.

52. Geoffrey C. Ward, "Franklin D. Roosevelt, Birdwatcher," *Audubon* (January 1990).

53. FDR, "Birds of the Hudson River Valley," 1894, FDRL.

54. Geoffrey C. Ward, *American Originals: The Private Worlds of Some Singular Men and Women* (New York: HarperCollins, 1991), p. 245.

55. Mark V. Barrow Jr., *A Passion for Birds: American Ornithology After Audubon* (Princeton: Princeton University Press, 1998), p. 118.

56. Oliver H. Orr Jr., *Saving American Birds* (Gainesville: University Press of Florida, 1992), pp. 206–18.

57. Rita Halle Kleeman, *Gracious Lady: The Life of Sara Delano Roosevelt* (New York: D. Appleton-Century, 1935), p. 174.

58. FDR, "Bird Diary," April 16, 1896, FDRL.

59. Ward, *Before the Trumpet*, p. 164.

60. Mrs. James Roosevelt, *My Boy Franklin* (New York: Ray Long and Richard R. Smith, 1933), p. 15.

61. Ibid., p. 16.

62. Bridge, "Sea Attractive to Roosevelt."

63. Gilbert Schrank, Professor Emeritus at Nassau Community College, to Douglas Brinkley, October 26, 2013. Schrank spent a summer cataloging the papers of the Linnaean Society.

64. FDR, "Hawaii, the Paradise of the Pacific," February 1895, Linnaean Society of New York Archive, American Museum of Natural History; also Gilbert Schrank to Douglas Brinkley, November 5, 2013. This essay was discovered by Dr. Gilbert Schrank while I was writing this book. It's an exciting new document quoted here for the first time.

65. FDR, "Bird Diary," February 8, 1896, through February 11, 1896, FDRL.

66. William King Gregory, *Biographical Memoir of Frank Michler Chapman, 1864–1945* (Washington, DC: National Academy of Sciences, 1947).

67. "Uniting the Ornithologists," *New York Times*, September 27, 1883.

68. Robert B. Fisher, "Biological Survey Unit, USGS Patuxent Wildlife Research Center" (Washington, DC: USGS Patuxent Wildlife Research Center, National Museum of Natural History, 2011).

69. Edward H. Graham, *The Land and Wildlife* (New York: Oxford University Press, 1947), p. 32.

70. Gregory, *Biographical Memoir of Frank Michler Chapman.*

71. "Dr. Chapman Dies; Ornithologist, 81," *New York Times*, November 17, 1945.

72. "Big Old Houses: A Dutchess County Dowager," August 19, 2012, http://bigoldhouses.blogspot.com/2012/08/a-dutchess-county-dowager.html.

73. FDR, "Bird Diary," February 18, 1896, FDRL.

74. FDR, to Thomas M. Upp, July 16, 1913, in Edgar B. Nixon, ed., *Franklin D. Roosevelt and Conservation 1911–1945*, Vol. 1 (Hyde Park, NY: General Services Administration, National Archives and Records Service, FDRL, 1957).

75. FDR, "Spring Song," *Foursome* (May 11, 1896), FDRL.

76. FDR, "Bird Diary," February 19, 1896, FDRL.

77. Frank Freidel, *Franklin D. Roosevelt: The Apprenticeship* (Boston: Little, Brown, 1952), p. 33.

78. *New York at the World's Columbian Exposition* (Albany, NY: James B. Lyon, 1894), p. 8.

79. *Reading* (Pennsylvania) *Times*, November 11, 1932, p. 3, quoted in Ernest K. Lindley, *Franklin D. Roosevelt: A Career in Progressive Democracy* (New York: Bobbs-Merrill, 1931), p. 48.

80. Wallace Stegner, *Where the Bluebird Sings at the Lemonade Springs: Living and Writing in the West* (New York: Random House, 1992), p. 199.

81. David Schuyler, *Sanctified Landscape: Writers, Artists, and the Hudson River Valley, 1820–1909* (Ithaca, NY: Cornell University Press, 2012), p. 3.

82. Doris Kearns Goodwin, *No Ordinary Time: Franklin and Eleanor Roosevelt—The Home Front in World War II* (New York: Simon and Schuster, 1994), p. 74.

CHAPTER 2: "I JUST WISH I COULD BE AT HOME TO HELP MARK THE TREES"

1. Ward, *Before the Trumpet*, p. 163.

2. Blanche Wiesen Cook, *Eleanor Roosevelt*, Vol. 1, *1884–1933* (New York: Penguin, 1992), p. 148.

3. Classmate quoted in Freidel, *Roosevelt: The Apprenticeship*, p. 50.

4. FDR to James and Sara Roosevelt, March 13, 1898, FDRL.

5. FDR to James and Sara Roosevelt, February 1, 1898, FDRL.

6. FDR to James and Sara Roosevelt, November 30, 1898, FDRL.

7. FDR to James and Sara Roosevelt, May 14, 1899, FDRL.

8. FDR to James and Sara Roosevelt, May 12, 1899, FDRL.

9. FDR to James and Sara Roosevelt, May 16, 1899, FDRL.

10. "Birds Seen with Camera," *New York Times*, October 14, 1900.

11. Kenneth Davis, *FDR: The Beckoning of Destiny* (New York: Putnam, 1972), p. 121.

12. Ward, *Before the Trumpet*, p. 192.

13. Missouri v. Illinois, 26 *Supreme Court Reporter* (Eagan, MN: West, 1906), Google Books edition.

14. FDR to James and Sara Roosevelt, October 28, 1898, FDRL.

15. Geoffrey C. Ward and Ken Burns, *The Roosevelts: An Intimate History* (New York: Knopf, 2014), p. 477.

16. Ward, *Before the Trumpet*, p. 195.

17. Cook, *Eleanor Roosevelt*, Vol. 1, p. 148.

18. Ward, *Before the Trumpet*, p. 195.

19. Barry Yeoman, "From Billions to None," *Audubon* (May–June 2014), p. 32.

20. Diane Galusha, *Another Day, Another Dollar: The Civilian Conservation Corps in the Catskills* (Hendersonville, NY: Black Dome, 2008), pp. 6–7.

21. Elliott Roosevelt, ed., *F.D.R.: His Personal Letters* (New York: Duell, Sloan and Pearce, 1950), Vol. 1, p. 370.

22. FDR to James and Sara Roosevelt, December 3, 1900, FDRL.

23. "Death List of a Day: James Roosevelt," *New York Times*, December 9, 1900, p. 7.

24. Kleeman, *Gracious Lady*, p. 213.

25. Ward, *Before the Trumpet*, p. 226.

26. Lou Sebesta, "Balmville Tree—Living Landmark," *New York State Conservationist*, Vol. 53, no. 5 (April 1999), pp. 20–21.

27. Joseph Berger, "A Tree Survives; Foes Do Not; After 3 Centuries, a Town Shores Up a Symbol of Its Heart." *New York Times*, September 27, 1995, p. B7.

28. Margaret Logan Marquez, *Hyde Park on the Hudson* (Charleston, SC: Arcadia, 1996), p. 44.

29. Fred W. Herbert, *Careers in Natural Resource Conservation* (New York: Henry Z. Walck, 1965), p. 22.

30. FDR to Sara Roosevelt, January 12, 1901, FDRL.

31. Philip M. Boffey, "Franklin Delano Roosevelt at Harvard," *Harvard Crimson*, December 13, 1957, http://www.thecrimson.com/article/1957/12/13/franklin-delano-roosevelt-at-harvard-phistorians/.

32. Friedel, *The Apprenticeship*, pp. 56–65.

33. Gifford Pinchot, "The Birth of Conservation," in *Breaking New Ground* (New York: Harcourt, Brace, 1947), p. 324.

34. Char Miller, *Gifford Pinchot and the Making of Modern American Environmentalism* (Washington, DC: Island, 2004), p. 117.

35. Pinchot, *Breaking New Ground*, p. 10.

36. "History: Yale School of Forestry and Environmental Studies," https://environment.yale.edu/about/history/.

37. Kim Heacox, *An American Idea: The Making of the National Parks* (Washington, DC: National Geographic, 2009), p. 146; David Stradling, *The Nature of New York: An Environmental History of the Empire State* (Ithaca, NY: Cornell University Press, 2010), p. 103.

38. FDR, Speech to the Yale University School of Forestry, June 20, 1934, FDRL.

39. "What Is to Become of Us?" *Fortune* (December 19, 1933).

40. FDR, "The Roosevelt Family," Roosevelt Family, Business, and Personal Papers, Box 36, FDRL.

41. Davis, *The Beckoning of Destiny*, p. 165.

42. FDR, "Remarks at Fort Peck Dam, Montana," August 6, 1934. Online at American Presidency Project, http://www.presidency.ucsb.edu/ws/?pid=14735.

43. Jonathan Worth Daniels, *Washington Quadrille: The Dance Beside the Documents* (New York: Doubleday, 1968), p. 229.

44. Theodore Roosevelt to Franklin Delano Roosevelt, November 29, 1904, FDRL.

45. Theodore Roosevelt quoted in Edmund Morris, *The Rise of Theodore Roosevelt* (New York: Random House, 1979), p. 440.

46. Eleanor Roosevelt, "My Day" (column), January 14, 1936. Source of "My Day" citations is Eleanor Roosevelt Papers, George Washington University, Washington, DC, http://www.gwu.edu/~erpapers/myday/.

47. Eleanor Roosevelt, *This Is My Story* (New York: Harper & Brothers, 1937), p. 6.

48. Joseph P. Lash, *Eleanor and Franklin* (New York: W. W. Norton, 1971), pp. 105–6.

49. Eleanor Roosevelt, "My Day," March 24, 1942.

50. FDR to Sara Roosevelt, July 15, 1905, FDRL.

51. Geoffrey C. Ward, *A First-Class Temperament: The Emergence of Franklin Roosevelt* (New York: Harper & Row, 1989), p. 23.

52. FDR to Sara Roosevelt, August 7, 1905, in Elliott Roosevelt, ed., *Personal Letters of Franklin D. Roosevelt*, Vol. 2 (New York: Duell, Sloan and Pearce, 1971), p. 50.

53. "German Water Power," *Economist* (February 18, 2006).

54. David Blackbourn, *The Conquest of Nature: Water, Landscape, and the Making of Modern Germany* (New York: W. W. Norton, 2006).

55. FDR quoted in Nixon, *Franklin D. Roosevelt and Conservation*, Vol. 1, p. 18.

56. David Brower, *Let the Mountains Talk, Let the Rivers Run* (New York: Harper Collins, 1995), pp. 72–74.

57. Gifford Pinchot, *The Adirondack Spruce* (New York: Critic, 1898).

58. National Park Service, "National Register of Historic Places Inventory—Nomination Form: Home of Franklin Delano Roosevelt National Historic Site," May 1979, pdfhost.focus.nps.gov/docs/NRHP/Text/66000056.pdf.

59. Eleanor Roosevelt, "My Day," April 26, 1945.

CHAPTER 3: "HE KNEW EVERY TREE, EVERY ROCK, AND EVERY STREAM"

1. Norman J. Van Valkenburgh and Christopher W. Olney, *The Catskill Park: Inside the Blue Line* (Hendersonville, NY: Black Dome, 2004), p. 21.

2. J. Anthony Lukas, *Big Trouble: A Murder in a Small Western Town Sets Off a Struggle for the Soul of America* (New York: Simon and Schuster, 1997), p. 617.

3. "Hyde Park: The President's Estate," *Life*, Vol. 6, no. 22 (May 29, 1939), p. 61.

4. William Plog oral history interview with George A. Palmer, November 13, 1947, FDRL. Given to author by Anne E. Jordan of the National Park Service.

5. Having read so much Dutchess County lore, FDR tried to rechristen Springwood as "Crum Elbow," which is what his historic river bend was called when Henry Hudson traded with members of the Wappinger tribe in 1609. To back up his case, he pointed to old nautical maps showing the Roosevelt estate at the Crum Elbow bend in the Hudson River. However, the new designation failed to stick.

6. Nelson C. Brown, "Governor Roosevelt's Forest," *American Forests*, Vol. 37, no. 5 (May 1931), p. 273.

7. Elliott Roosevelt, ed., *F.D.R.: His Personal Letters*, 4 vols. (New York: Duell, Sloan and Pearce, 1948); James Roosevelt, *Affectionately, FDR: A Son's Story of a Lonely Man* (New York: Harcourt Brace, 1959), p. 47.

8. Eleanor Roosevelt, *Franklin D. Roosevelt and Hyde Park*, p. 8.

9. Christopher Gray, "For Eleanor and Franklin, a Built-In Mother-in-Law," *New York Times*, June 8, 1997. The house was sold to Hunter College in the early 1940s and renamed "Sara Roosevelt Memorial House."

10. Black, *Franklin Delano Roosevelt: Champion of Freedom*, p. 50.

11. Burns, *The Lion and the Fox*, p. 15.

12. Doris Greenberg, "Shah Lays Wreath at Roosevelt Tomb; Ruler of Iran Tours Hyde Park, Eats Turkey with Widow of President and Son Elliott," *New York Times*, November 25, 1949, p. 16.

13. Burns, *The Lion and the Fox*, p. 4.

14. Hudson-Fulton Celebration Commission, *Official Program of the Hudson-Fulton Celebration* (New York: Redfield Brothers, Authorized Publishers, 1909), https://archive.org/details/officialprogramh00huds.

15. FDR, Press Conference at the President's Cottage, November 23, 1932, Warm Springs, GA.

16. William O. Douglas, *A Wilderness Bill of Rights* (New York: Little, Brown, 1965), p. 168.

17. John Burroughs, *Under the Maples* (Boston: Houghton Mifflin, 1921), p. 111. Google Books.

18. FDR, speech before Troy, NY, People's Forum, March 3, 1912, FDRL.

19. Wint Aldrich to Douglas Brinkley, December 16, 2013. See also FDR to Maunsell Crosby, March 12, 1913, FDRL.

20. C. Stuart Gager, ed., *Brooklyn Botanic Garden Record* (Lancaster, PA: Brooklyn Institute of Arts and Sciences), p. 111.

21. F. Kennon Moody, *FDR and His Hudson Valley Neighbors* (Poughkeepsie, NY: Hudson House Publishing, 2013), p. 4.

22. Elliott Roosevelt, *FDR: His Personal Letters*, Vol. 2, p. 154.

23. John Solan, "Nursing Forests Back to Health," *Conservationist* (February 2003).

24. Black, *Franklin Delano Roosevelt: Champion of Freedom*, p. 53.

25. Brian Black, "The Complex Environmentalist: Franklin Roosevelt and the Ethos of New Deal Conservation," in Harry L. Henderson and David B. Woolner, eds., *FDR and the Environment* (New York: Palgrave, 2005), p. 30.

26. Quoted in Helen Meserve, "The House That Became a Second Home to FDR," *Hyde Park Townsman*, September 7–8, 1983.

27. Eleanor Roosevelt, *I Remember Hyde Park*, p. 72.

28. Nixon, *Franklin D. Roosevelt and Conservation*, Vol. 1, p. 4.

29. FDR, "Remarks at Clarksburg, West Virginia," October 29, 1944.

30. FDR to G. O. Shields, February 20, 1911, FDRL.

31. Robert Caro, *The Power Broker: Robert Moses and the Fall of New York* (New York: Knopf, 1974), pp. 288–89.

32. FDR to Eleanor Roosevelt, March 17, 1912, FDRL.

33. Nixon, *Franklin D. Roosevelt and Conservation*, Vol. 1, pp. 6–7.

34. FDR to William K. Draper, March 21, 1911, FDRL.

35. FDR to Egbert Bagg, January 31, 1912, FDRL.

36. Charles Banks Belt, "History of the Committee on Conservation of Forests and Wildlife of the Camp Fire Club of America," 1989, Camp Fire Club of America Archive, Chappaqua, NY; and Jeffrey A. Gonauer, "The Camp Fire Club of America," *Fair Chase* (Fall 2011), p. 15.

37. John Hay Jr., "George Washington, Lover of Trees," *American Forests*, Vol. 38, no. 2 (February 1932), pp. 67–75.

38. Stradling, *The Nature of New York*, pp. 76–105.

39. Frank Graham Jr., *Man's Dominion: The Story of Conservation in America* (New York: M. Evans and Company, 1971), pp. 252–53.

40. Char Miller, "Neither Crooked nor Shady: The Weeks Act, Theodore Roosevelt, and the Virtue of Eastern National Forests, 1899–1911," *Theodore Roosevelt Association Journal*, Vol. 4, no. 33 (Fall 2012), pp. 15–23.

41. Jim Robbins, "Deforestation and Drought," *New York Times*, October 11, 2015, p. 7.

42. FDR to Dexter Blagden, February 21, 1912, FDRL.

43. Timothy W. Kneeland, "Pre-Presidential Career," in William D. Pederson, ed., *A Companion to Franklin D. Roosevelt* (West Sussex: Wiley-Blackwell, 2011), pp. 40–41. See also "A Statement on the Roosevelt-Jones Conservation Bill by the Camp-Fire Club of America," in Nixon, *Franklin D. Roosevelt and Conservation*, Vol. 1, pp. 12–13.

44. FDR, "A Further Account of an Unsentimental Journey of Two Politicians and an Undergraduate," April 18, 1912, FDRL.

45. Timothy Egan, *The Worst Hard Time: The Untold Story of Those Who Survived the American Dustbowl* (New York: Houghton Mifflin, 2005), pp. 6–7.

46. Kristie Miller, "A Volume of Friendship: The Correspondence of Isabella Greenway and Eleanor Roosevelt," *Journal of Arizona History*, Vol. 40, no. 2 (Summer 1999), pp. 121–56.

47. Harold L. Ickes, *Autobiography of a Curmudgeon* (New York: Reynal and Hitchcock, 1943), pp. 6–7.

48. Arthur M. Schlesinger Jr., *The Crisis of the Old Order: 1919–1933* (Boston: Houghton Mifflin, 1957), p. 26.

49. Ickes, quoted in Tom H. Watkins, *Righteous Pilgrim: The Life and Times of Harold L. Ickes, 1874–1952* (New York: Henry Holt, 1990), p. 470.

50. Brinkley, *The Wilderness Warrior*, pp. 282–84.

51. Gifford Pinchot, *The Fight for Conservation* (New York: Doubleday, Page & Company, 1910).

52. Rachel Clothier, *Corinth* (Mount Pleasant, SC: Arcadia, 2009), p. 25.

53. Kneeland, "Pre-Presidential Career," p. 41.

CHAPTER 4: "WISE USE"

1. "Pinchot to Inspect Adirondack Forests," *New York Times*, July 21, 1911, p. 8.

2. "Protection of Adirondack Forests," *American Lumberman*, February 24, 1912, p. 39.

3. Douglas H. Strong, *Dreamers and Defenders: American Conservationists* (Lincoln: University of Nebraska Press, 1988), p. 83.

4. Eleanor Roosevelt, "My Day," October 9, 1946.

5. William Edward Coffin, "Wild Life Protection," *Proceedings of the Third National Conservation Congress, at Kansas City, Missouri, September 25–27, 1911* (Kansas City, MO: National Conservation Congress, 1912), p. 266.

6. "Pinchot to Inspect Adirondack Forests."

7. "Sees Peril to Adirondacks," *New York Times*, February 21, 1912.

8. FDR, Speech to Yale University School of Forestry, June 20, 1934.

9. FDR, "A Debt We Owe," June 1930, in Nixon, *Franklin D. Roosevelt and Conservation*, Vol. 1, p. 72.

10. FDR, Speech to Yale University School of Forestry, June 20, 1934, FDRL.

11. FDR, "Remarks at the Celebration of the Fiftieth Anniversary of State Conservation at Lake Placid," September 14, 1935. Online at American Presidency Project.

12. FDR, Speech to the Troy, New York, People's Forum, March 3, 1912, FDRL.

13. FDR to Dexter Blagden, February 21, 1912, FDRL.

14. "A Statement on the Roosevelt-Jones Conservation Bill by the Camp-Fire Club of America, February 1, 1912," in Nixon, *Franklin D. Roosevelt and Conservation*, Vol. 1, Part 1, *State Senator to the Presidency, 1911–1933*, http://www.nps.gov/parkhistory/online_books/cany/fdr/part1. htm.

15. FDR quoted in "Franklin D. Roosevelt," in Kathleen A. Brosnan, ed., *The Encyclopedia of American Environmental Policy*, Vol. 4 (New York: Facts on File, 2010), p. 1140.

16. Schlesinger, *Crisis of the Old Order*, p. 336; Jean Edward Smith, *FDR* (New York: Random House, 2007), p. 84.

17. Schlesinger, *Crisis of the Old Order*, p. 31.

18. Ibid., p. 27.

19. Cook, *Eleanor Roosevelt*, Vol. 1, p. 198.

20. Julie M. Fenster, *FDR's Shadow: Louis Howe, The Force That Shaped Franklin and Eleanor Roosevelt* (New York: Palgrave Macmillan, 2013), pp. 92–93.

21. Louis Howe, "Behind the Scenes of the National Campaign," *Jeffersonian*, Vol. 2, no. 8 (November 1932), p. 18.

22. Ottomar H. Van Norden, *Protection of Migratory Birds: Hearings before the Committee on Forest Reservations and the Protection of Game, March 6, 1912* (Washington, DC: U.S. Government Printing Office, 1912), p. 32.

23. Robert Rosenbluth, M.F., *Woodlot Forestry* (Albany, NY: J. B. Lyon, 1913), p. 9.

24. Ibid., pp. 9–10.

25. Ibid., p. 41.

26. Ward, *A First-Class Temperament*, pp. 202–3.

27. James Srodes, *On Dupont Circle: Franklin and Eleanor Roosevelt and the Progressives Who Shaped Our World* (Berkeley, CA: Counterpoint, 2012), p. 37.

28. FDR to A. S. Houghton, April 3, 1913, FDRL.

29. Fenster, *FDR's Shadow: Louis Howe.*

30. Mrs. Josephus Daniels, *Recollections of a Cabinet Minister's Wife* (Raleigh, NC: Mitchell Printing, 1945), pp. 3–5.

31. William L. Neuman, "Franklin Delano Roosevelt: A Disciple of Admiral Mahan," *U.S. Naval Proceedings*, Vol. 78 (July 1952), pp. 713–19.

32. Gifford Pinchot, *Fishing Talk* (Harrisburg: Stockpile, 1993), pp. 123–25; John F. Reiger, *Two Essays in Conservation History* (Milford, PA: Grey Towers, 1994), pp. 19–21; and Char Miller, *Gifford Pinchot and the Making of Modern Environmentalism*, pp. 241–44.

33. Elliott Roosevelt, *FDR: His Personal Letters: 1905–1928*, p. 210.

34. Susan R. Schrepfer, *The Fight to Save the Redwoods: A History of Environmental Reform, 1917–1978* (Madison: University of Wisconsin Press, 1983), p. 72.

35. J. D. Grant, "California Redwood Wonderland," *Western Woman* (December 1929–January 1930), p. 28.

36. Susan R. Schrepfer, "Sierra Club," in Brosnan, *Encyclopedia of American Environmental History*, Vol. 4, p. 1187.

37. "Timeline of the San Francisco Earthquake," Virtual Museum of San Francisco, accessed March 24, 2014, http://www.sfmuseum.net/hist10/06timeline.html.

38. Tom Turner, *Sierra Club: 100 Years of Protecting Nature* (New York: Harry Abrams, 1991), pp. 74–77.

39. John Muir, *The Yosemite* (New York: Century, 1912), pp. 255–57, 260–62. Reprinted in Roderick Nash, *The American Environment: Readings in The History of Conservation* (Reading, MA: Addison-Wesley, 1968).

40. Robert U. Johnson to FDR, October 30, 1913, FDRL

41. Michael Branch, "Robert Underwood Johnson," in George A. Cevasco and Richard P. Harmond, eds., *Modern American Environmentalists* (Baltimore, MD: Johns Hopkins University Press, 2009), pp. 229–31.

42. Ibid.

43. FDR to Robert U. Johnson, October 31, 1913, FDRL.

44. Robert U. Johnson to FDR, November 11, 1913, FDRL.

45. Ibid.

46. Johnson was right in his assertion that Wilson was no TR when it came to preserving America's wilderness. TR had established Olympic National Monument just forty-eight hours before leaving the White House in 1909 in order to protect the Roosevelt elk (*Cervus canadensis roosevelti*) from being overhunted, and to save majestic groves of Sitka spruce, red alder, and western hemlock. Six years later, President Wilson, partially short-circuiting TR's legacy, reduced the size of the magnificent monument by 50 percent to appease the lumber industry.

47. Kevin Roderick, "The Waters Flowed: 'There It Is. Take It,' " *Los Angeles Times*, October 3, 1999.

48. Robert W. Righter, *The Battle of Hetch Hetchy: America's Most Controversial Dam and the Birth of Modern Environmentalism* (New York: Oxford University Press, 2005), p. 6.

49. Horace M. Albright and Marian Albright Schenk, *Creating the National Park Service: The Missing Years* (Norman: University of Oklahoma Press, 1999), p. 23.

50. Ibid., p. 21.

51. Anne Lane and Louise Wall, eds., *The Letters of Franklin K. Lane* (Boston: Houghton Mifflin, 1922), p. 258.

52. FDR to Franklin Moon, December 20, 1915, FDRL.

53. Weintraub, *Young Mr. Roosevelt*, p. 35.

54. Harold K. Steen, *The U.S. Forest Service: A History* (Seattle: University of Washington Press, 1976), p. 123.

55. Curtis Townsend, "Flood Control of the Mississippi River," address before the National Drainage Congress in Saint Louis, Missouri, April 11, 1913.

56. FDR to John F. Coleman, January 26, 1914, FDRL.

57. Dows, *Franklin Roosevelt*, p. 19.

58. Robert Cross, *Sailor in the White House: The Seafaring Life of FDR* (Annapolis, MD: Naval Institute Press, 2003), p. 67.

59. FDR to F. F. Moon, October 22, 1915, Assistant Secretary of the Navy Papers, Box 97, FDRL.

60. Ernest Thompson Seton, *Boy Scouts of America: A Handbook of Woodcraft, Scouting, and Lifecraft* (New York: Doubleday, 1910); "Roosevelt Sees Problem of Boys Aided by Scouts," *New York City Evening World*, March 2, 1929. See also Neil M. Maher, *Nature's New Deal: The Civilian Conservation Corps and the Roots of the American Environmental Movement* (New York: Oxford University Press, 2008), p. 34.

61. Steven Beissinger, "The Next Century of 'America's Best Idea,' " *Breakthroughs* (Fall 2014), p. 23.

62. FDR, "Radio Address from Two Medicine Chalet, Glacier National Park," August 5, 1934. Online at the American Presidency Project, http://www.presidency.ucsb.edu/ws/?pid=14733.

63. Robin W. Winks, "The National Park Service Act of 1916: 'A Contradictory Mandate'?" *Denver University Law Review*, Vol. 74 (1997), p. 575.

64. Linda Flint McClelland, *Building the National Parks: The Historic Landscape Design of the National Park Service* (Baltimore, MD: Johns Hopkins University Press, 1998), pp. 1–8.

65. FDR, "Trip to Haiti and Santo Domingo, 1917," pp. 10–11, FDRL. Available online at University of Florida, http://ufdc.ufl.edu/UF00082927/00001/1x.

66. Eleanor Roosevelt, *This Is My Story*, p. 245.

67. Donald A. Ritchie, *Electing FDR: The New Deal Campaign of 1932* (Lawrence: University Press of Kansas, 2007), p. 9.

68. Gaddis Smith, "Roosevelt, the Sea, and International Security," in Douglas Brinkley and David R. Facey-Crowther, eds., *The Atlantic Charter* (New York: St. Martin's, 1994), p. 37.

69. Scott Berg, *Wilson* (New York: Putnam, 2013), p. 20.

70. Philip L. Kennicott, "World War One: An Inconsistent Memory," Amon Carter Lecture at University of Texas at Austin, April 8, 2014. Kennicott also shared with me his Commission of Fine Arts research notes (June 6, 1919) from the Commission of Fine Arts Archive in Washington, D.C.

71. FDR to Henry Heymann, December 2, 1919, Group X, FDRL.

72. FDR, speech accepting the Democratic vice presidential nomination, August 9, 1920, Hyde Park, New York.

73. Governor Thomas Riggs to FDR, August 20, 1916, in Nixon, *Franklin D. Roosevelt and Conservation*, Vol. 1, p. 41.

74. FDR to Jessie Adams (Secretary of the Lewiston Commercial Club), October 23, 1920, FDRL.

75. FDR to Henry Morgenthau, May 15, 1942, presidential diary, Henry Morgenthau Papers, FDRL. See also Warren F. Kimball, *The Juggler* (Princeton: Princeton University Press, 1991), p. 7.

76. Arthur M. Schlesinger Jr., *The Coming of the New Deal* (Boston: Houghton Mifflin, 1958).

77. Weintraub, *Young Mr. Roosevelt*, pp. 243–44.

78. John D. Leshy, "Legal Wilderness: Its Past and Some Speculations on Its Future," *Environmental Law*, Vol. 44, no. 2 (2014), pp. 554–56.

79. FDR to Eleanor Roosevelt, November 27, 1920, FDRL.

80. FDR to Eleanor Roosevelt, November 28, 1920, FDRL.

81. Weintraub, *Young Mr. Roosevelt*, p. 243.

82. Sara Delano Roosevelt, diary entry for November 2, 1920, FDRL.

83. FDR, "A Debt We Owe," *Country Home*, Vol. 54 (June 1934), pp. 12–14.

84. Norman T. Newton, *Design on the Land: The Development of Landscape Architecture* (Cambridge, MA: Harvard University Press, 1971), p. 563.

CHAPTER 5: "NOTHING LIKE MOTHER NATURE"

1. Tobin, *The Man He Became*, pp. 13–20.

2. Maher, *Nature's New Deal*, pp. 31–41.

3. David C. Scott, *My Fellow Americans: Scouting, Diversity, and the U.S. Presidency* (Dallas, TX: Windrush, 2014), pp. 84–85.

4. Ward, *A First-Class Temperament*, p. 575; and United States Geological Survey, "Geology of National Parks, 3D and Photographic Tours: Bear Mountain State Park," last modified March 6, 2014, http://3dparks.wr.usgs.gov/nyc/parks/loc9.htm.

5. Douglas Brinkley, *The Quiet World: Saving Alaska's Wilderness Kingdom, 1879–1960* (New York: HarperCollins, 2011), pp. 16–19.

6. Tobin, *The Man He Became*, p. 27.

7. FDR, "How Boy Scout Work Aids Youth," *New York Times*, August 12, 1928.

8. Scott, *My Fellow Americans*, p. 86.

9. FDR, "How Boy Scout Work Aids Youth."

10. *Annual Report of the Commissioners of the Palisades Interstate Park*, 1921 (Albany, NY: J. B. Lyon, 1922); New York State Department of Health, *Annual Report for the Year Ending December 31, 1921* (Albany, NY: J. B. Lyon, 1922), pp. 320–24.

11. Tobin, *The Man He Became*, pp. 28–29.

12. Frank Freidel, *Franklin D. Roosevelt: A Rendezvous with Destiny* (New York: Little, Brown, 1990).

13. Anna Roosevelt Boettiger, "My Life with FDR: How Polio Helped My Father," *Woman* (July 1949), pp. 53–54.

14. Tobin, *The Man He Became*, p. 48.

15. According to more contemporary medical scholarship, at his age—thirty-nine at the onset of the paralysis—it was more likely that he had contracted Guillain-Barré syndrome.

16. Ward, *First-Class Temperament*, p. 600.

17. Kenneth Davis, *FDR: The War President, 1940–1943* (New York: Random House, 2000), p. 4.

18. Eleanor Roosevelt, Foreword, in Elliott Roosevelt, *F.D.R.: His Personal Letters, 1905–1928*, p. xviii.

19. Ali Caron, "Margaret 'Daisy' Suckley," FDRL, accessed August 21, 2014, http://www.fdrlibrary.marist.edu/aboutfdr/daisysuckley.html.

20. Barbara Ireland, "At the Home of FDR's Secret Friend," *New York Times*, September 7, 2007.

21. Dawn Merritt, "The Roaring 20s: A Call to Action," *Outdoor America* (Winter 2012), pp. 24–33.

22. FDR to George Pratt, September 6, 1922, FDRL.

23. Maher, *Nature's New Deal*, p. 38.

24. Van Valkenburgh and Olney, *Catskill Park*, p. 61.

25. Adirondack Mountain Club Records, 1922 to Present, New York State Library.

26. James A. Kehl and Samuel J. Astorino, "A Bull Moose Responds to the New Deal: Pennsylvania's Gifford Pinchot," *Pennsylvania Magazine of History and Biography*, Vol. 88, no. 1 (January 1964), pp. 37–51.

27. FDR to George D. Pratt, November 25, 1922, FDRL.

28. George D. Pratt to FDR, December 1, 1922, FDRL.

29. FDR quoted in Ward, *First-Class Temperament*, p. 660.

30. FDR to Sara Roosevelt, March 5, 1923, FDRL.

31. Ward, *First-Class Temperament*, p. 660.

32. William W. Rodgers, "The Paradoxical Twenties," in Michael Gannon, ed., *New History of Florida* (Gainesville: University of Florida Press, 1996), p. 298.

33. Cross, *Sailor in the White House*.

34. FDR, "Florida Journal," February 6, 1924, FDRL.

35. Ibid., February 19, 1924.

36. FDR to Sara Roosevelt, February 22, 1924, FDRL.

37. "Robert S. Yard, 84, Once Editor Here," *New York Times*, May 19, 1945.

38. Robert Sterling Yard to FDR, February 21, 1924, FDRL. See also Nixon, *Franklin D. Roosevelt and Conservation*, Vol. 1, p. 55.

39. FDR to Maunsell Crosby, October 13, 1924, FDRL.

40. Ibid.

41. FDR to H. M. Hickeson, October 19, 1923, FDRL.

42. FDR to Charles C. Adams, Director of the Roosevelt Wild Life Forest Experiment Station at Syracuse University, December 6, 1923, FDRL.

43. Robert Sterling Yard to FDR, February 2, 1924, FDRL.

44. Donald Culross Peattie, *A Natural History of North American Trees* (San Antonio, TX: University Press, 2013), p. 3.

45. James Roosevelt, *My Parents: A Differing View* (Chicago: Playboy Press, 1976), p. 93; Fenster, *FDR's Shadow*, p. 206.

46. Schlesinger, *Crisis of the Old Order*, p. 410.

47. James Roosevelt, *My Parents*, p. 207.

48. FDR to Sara Roosevelt, October 1924, FDRL.

49. Elliott Roosevelt, *F.D.R.: His Personal Letters, 1905–1928*, p. 566.

50. FDR to Margaret Suckley, December 22, 1934. Quoted in Geoffrey C. Ward, *Closest Companion: The Unknown Story of the Intimate Friendship Between Franklin Roosevelt and Margaret Suckley* (New York: Simon and Schuster, 1995), p. 16.

51. "Franklin Roosevelt Will Swim to Health," *Atlanta Journal Sunday Magazine*, October 26, 1924.

52. Sears, *Historic Resource Study for the Roosevelt Estate*, p. 8.

53. Ruth B. Stevens, *Hi-Ya Neighbor* (New York: Atlanta, Tupper, and Love, 1947), p. 30; Hugh Gregory Gallagher, *FDR's Splendid Deception* (New York: Dodd, Mead, 1985), pp. 13–16.

54. From October 3 to 20, 1924; April 1 to May 15, 1925; March 27 to May 5, 1926; September 29 to November 10, 1926; February 11 to May 12, 1927; May 24 to June 11, 1927; June 19 to August 3, 1927; September 27 to December 5, 1927; January 20 to February 11, 1928; February 29 to May 3, 1928; mid-June (c. June 20) 1928; June 30 to July 9, 1928; and September 19 to October 5, 1928.

55. FDR to Herman Swift, October 11, 1926, Private Collection, Columbus, GA.

56. John C. Paige, *The Civilian Conservation Corps and the National Park Service, 1933–1942* (Washington, DC: U.S. Department of the Interior, 1985), p. 116.

57. "Citizens Give Land for New York Parks," *New York Times*, October 2, 1923.

58. Norman T. Newton, *Design on the Land: The Development of Landscape Architecture* (New Haven: Yale University Press, 1997), pp. 575–78.

59. Caro, *The Power Broker*, pp. 288–89.

60. Don Miller, *Supreme City: How Jazz Age Manhattan Gave Birth to Modern America* (New York: Simon and Schuster, 2014).

61. Kathleen LaFrank, "Real and Ideal Landscapes Along the Taconic State Parkway," in Alison K. Hoagland and Kenneth A. Breisch, eds., *Constructing Image, Identity, and Place: Perspectives in Vernacular Architecture*, Vol. 9 (Knoxville: University of Tennessee Press, 2003), pp. 247–62.

62. "A Brief History of Taghkanic State Park: Roosevelt's Gift to Columbia County," *Columbia County History and Heritage* (Spring 2012), pp. 33–34.

63. FDR, Introduction, in *Dutch Houses in the Hudson River Valley Before 1776.*

64. Charles C. Adams, Director of the New York State Museum in Albany, to FDR, August 1, 1928, FDRL.

65. FDR to Charles C. Adams, August 10, 1928, FDRL.

66. "Roosevelt Held Out to the Last Minute," *New York Times*, October 3, 1928.

67. Davis, *FDR: The War President*, p. 7.

68. Henry Morgenthau quoted in Herbert Levy, *Henry Morgenthau, Jr.: The Remarkable Life of FDR's Secretary of the Treasury* (New York: Skyhorses, 2010), p. 84.

69. Davis, *FDR: The New York Years*, pp. 34–35.

70. Marian Barros, "City Lawyer, Country Chickens," *New York Times*, November 14, 2007.

71. FDR to Henry Morgenthau Jr., June 13, 1928, FDRL.

72. Levy, *Henry Morgenthau, Jr.*, p. 160.

73. Harlan D. Unrau and G. Frank Williss, *Administrative History: Expansion of the National Park Service in the 1930s* (Denver, CO: Denver Service Center Publication, 1983), p. 108.

74. Graham Averill, "Appalachian Inspiration," *Nature Conservancy* (July/August 2013).

75. Davis, *FDR: The New York Years*, p. 4.

76. Dunwell, *The Hudson: America's River*, p. 264.

77. Herbert Hoover, *The Ordeal of Woodrow Wilson* (New York: McGraw-Hill, 1958), p. 297.

78. Eleanor Roosevelt, *The Autobiography of Eleanor Roosevelt* (New York: Harper & Brothers, 1961), p. 153.

79. Ritchie, *Electing FDR*, p. 10.

CHAPTER 6: "A TWICE-BORN MAN"

1. FDR, inaugural address, January 1, 1929, http://www.fdrlibrary.marist.edu/education/resources/pdfs/fdrstandard1.pdf.

2. Bernard Bellush, *Franklin D. Roosevelt as Governor of New York* (New York: Columbia University Press, 1955).

3. Stradling, *The Nature of New York*, p. 161.

4. FDR, address to New York State Forestry Association, February 27, 1929, Albany, NY.

5. Ibid.

6. Stradling, *The Nature of New York*, p. 161.

7. FDR to B. U. Hiester, June 30, 1931, FDRL.

8. FDR quoted in Schlesinger, *Crisis of the Old Order*, p. 409.

9. Harold Faber, "Savoring the Scenic Delights Along the Taconic Parkway," *New York Times*, August 14, 1987.

10. Ibid.

11. William Kennedy quoted in Mark Healy, " 'Just Drive,' Said the Road, and the Car Responded," July 5, 2002.

12. FDR to Captain Edward McCauley Jr., March 21, 1929, FDRL.

13. FDR quoted in "The Future of the Forest Preserve," in *Report of the President of the Association for the Protection of the Adirondacks*, presented at annual meeting, April 12, 1932, Paul Schaefer Wilderness Archives, Union College, Schenectady, NY.

14. Peter M. Hopsicker, "Legalizing the 1932 Lake Placid Bob-run: A Test of the Adirondack Wilderness Culture," *Olympika* 17 (2009), pp. 99–119.

15. "The Future of the Forest Preserve," pp. 2–3.

16. Frank Graham Jr., *The Adirondack Park: A Political History* (New York: Knopf, 1978), pp. 184–85.

17. "Governor Roosevelt Dedicates Whiteface, Another Similar Highway," *New York Times*, September 12, 1929.

18. Sam Rosenman, ed., *Public Papers of the President of the United States: F. D. Roosevelt*, Vol. 4 (New York: Random House, 1938), p. 361.

19. Migratory Bird Conservation Act, 45 Stat. 1222, February 18, 1929.

20. Dayton Duncan and Ken Burns, *The Dust Bowl* (San Francisco: Chronicle Books 2012), p. 33.

21. William E. Leuchtenburg, *Franklin D. Roosevelt and the New Deal* (New York: Harper & Row, 1963), pp. 2–6.

22. Adam Cohen, *Nothing to Fear: FDR's Inner Circle and the Hundred Days That Created Modern America* (New York: Penguin, 2009), p. 2.

23. Hoover quoted in Michael E. Parrish, *Anxious Decades: America in Prosperity and Desperation, 1920–1941* (New York: W. W. Norton, 1992), p. 244.

24. Roosevelt quoted in Bellush, *Franklin D. Roosevelt as Governor of New York*, p. 76.

25. Bernard Asbell, *The F. D. R. Memoirs: A Speculation on History* (New York: Doubleday, 1973), p. 105.

26. Schlesinger, *Crisis of the Old Order*, p. 390.

27. A. L. Reisch Owen, *Conservation Under F. D. R.* (New York: Praeger, 1983), p. 10.

28. FDR to Nicholas Roosevelt, February 20, 1929, FDRL.

29. Thomas W. Patton, "FDR's Trees," *Conservationist*, Vol. 49, no. 5 (April 1995), p. 26.

30. Acting Dean Nelson C. Brown Records, State University of New York College of Environmental Science and Forestry, F. Franklin Moon Library, Terence J. Hoverter College Archives and Special Collections, Syracuse.

31. Nelson Brown to Elliott Roosevelt, June 15, 1950, Brown Papers, FDRL.

32. Nelson Brown, "Governor Roosevelt's Forest," *American Forests*, Vol. 37 (May 1931), pp. 273–74.

33. Eleanor Roosevelt, preface to John Morton Blum, *From the Morgenthau Diaries*, Vol. 1 (Boston: Houghton Mifflin, 1959).

34. Elinor Morgenthau quoted in Ward, *First-Class Temperament*, p. 253.

35. FDR to Henry Morgenthau Jr., May 20, 1929, FDRL.

36. Henry Morgenthau quoted in Blum, *From the Morgenthau Diaries*, pp. 26–27.

37. FDR, speech to New York State Legislature, March 25, 1930.

38. FDR, "A Debt We Owe," *Country Home* (June 1930), pp. 12–14.

39. Egan, *The Worst Hard Time*, p. 270.

40. Florida International University Everglades Digital Library, "Reclaiming the Everglades: South Florida's Natural History, 1884 to 1934," http://everglades.fiu.edu/reclaim/timeline/.

41. Donald Worster, *Dust Bowl: The Southern Plains in the 1930s* (New York: Oxford University Press, 1979), pp. 10–11.

42. Kendrick Clements, *Hoover, Conservation, and Consumerism* (Lawrence: University Press of Kansas, 2000), p. 187.

43. Galusha, *Another Day, Another Dollar*, p. 20.

44. June Hopkins "The New York State Temporary Emergency Relief Administration," October 1, 1931, Social Welfare History Project, http://www.socialwelfarehistory.com/eras/temporary-emergency-relief-administration/.

45. Jean Christie, "Conservation," in Otis L. Graham and Meghan Robinson Wander, *Franklin D. Roosevelt: His Life and Times* (Boston: G. K. Hall, 1985), p. 77.

46. Frederic A. Delano to H. S. Hooker, October 26, 1936, FDRL.

47. Tobin, *The Man He Became*, pp. 287–88.

48. FDR, "Proclamation of Conservation Week," March 2, 1931, FDRL.

49. "Maunsell Crosby, Ornithologist, Dies," *New York Times*, February 13, 1931, p. 17.

50. Bellush, *Franklin D. Roosevelt as Governor of New York*, p. 95.

51. FDR, "A Debt We Owe."

52. Richard Davis, "National Forests of the United States," Forest History Society, September 29, 2005, http://www.foresthistory.org/ASPNET/Places/National%20Forests%20of%20the%20U.S.pdf.

53. FDR quoted in Stradling, *The Nature of New York*, pp. 162–63.

54. Maher, *Nature's New Deal*, p. 28.

55. David Gibson, "The National Adirondack Debate of 1932," http://www.adirondackwild.org/pdf/pdf_adk_almanack/post-13_national_debate_of_32.pdf.

56. Graham, *The Adirondack Park*, p. 7.

57. FDR to John G. Saxe, n.d., 1931, FDRL, quoted in Burns, *Roosevelt: The Lion and the Fox*, p. 129.

58. "Forest Measure Approved; Gov. Roosevelt Adds to His Prestige in Clash with Smith," *New York Times*, November 4, 1931.

59. Governor George H. Dern to FDR, November 10, 1931, FDRL.

60. Bellush, *Franklin D. Roosevelt as Governor of New York*, p. 98.

61. Stradling, *The Nature of New York*, p. 160.

62. Thomas B. Allen, *Guardian of the Wild: The Story of the National Wildlife Federation, 1936–1986* (Bloomington: Indiana University Press, 1987), p. 24.

63. "Hawes Will Direct New Wild Life Body," *New York Times*, September 5, 1930.

64. Hugh Bennett, "Soil Erosion: A National Menace," *USDA Circular* 33 (1928).

65. Samuel Rosenman, *Working with Roosevelt* (New York: Harper & Brothers, 1952), p. 37.

66. Fenster, *FDR's Shadow*.

67. FDR, "Address at Oglethorpe University in Atlanta, Georgia," May 22, 1932. Online at American Presidency Project, http://www.presidency.ucsb.edu/ws/?pid=88410.

68. Derek S. Hoff, "Rockefeller Family" in Brosnan, *Encyclopedia of American Environmental History*, Vol. 4, p. 1130.

69. Tom Horton, William Chesapeake Bay Foundation, *Turning the Tide: Saving the Chesapeake Bay* (Washington, DC: Island, 2013), pp. 71–72, 77–80.

70. U.S. Fish and Wildlife Service, *Okefenokee National Wildlife Refuge Comprehensive Conservation Plan*, October 2006.

71. Michael Frome, "Not Far from the Maddening Crowd," in *Wilderness* (Washington, DC: National Geographic Society, 1973), p. 250. See also Francis Harper, "The Okefinokee Wilderness," *National Geographic* (May 1934), p. 597.

72. Francis Harper and Delma E. Presley, *Okefinokee Album* (Athens: University of Georgia Press, 1981), p. 6.

73. When writing to Francis and Jean Harper, Franklin Roosevelt used the time-honored spelling "Okefinokee" that they preferred. However, the U.S. Geographical Board decided that the swamp's official name should be the more modern "Okefenokee." In the NWR declaration, the new, official spelling appeared, much to the dismay of purists like the Harpers. Composer Stephen C. Foster changed the spelling of Florida's Suwannee River for his ballad "Old Folks at Home," which he began with a longing for life "way down upon the Swanee River." Today, there is a state park dedicated to Foster, who also wrote the songs "My Old Kentucky Home" and "Beautiful Dreamer."

74. Harper and Presley, *Okefinokee Album*, p. 18.

75. Megan Kate Nelson, *Trembling Earth: A Cultural History of the Okefenokee Swamp* (Athens: University of Georgia Press, 2005), pp. 185–88.

76. Alter, *The Defining Moment*, p. 84.

77. Ibid., pp. xiv–xv.

78. John Nance Garner quoted in Jules Witcover, *The American Vice Presidency: From Irrelevance to Power* (Washington, DC: Smithsonian Books, 2014), p. 302.

79. "John Nance Garner: Out of the Vaults, onto the Screen," *Center Points* (Fall 2013), pp. 2–3.

80. John Nance Garner quoted in Witcover, *The American Vice Presidency*, p. 302.

81. Paul F. Boller Jr., *Presidential Campaigns* (New York: Oxford University Press, 1996), p. 233.

82. FDR, "Address Accepting the Presidential Nomination at the Democratic National Convention in Chicago," July 2, 1932. Online at American Presidency Project, http://www.presidency.ucsb.edu/ws/?pid=75174.

83. Ibid.

84. Eric Rutkow, *American Canopy: Trees, Forests, and the Making of a Nation* (New York: Scribner, 2012), p. 249.

85. FDR, "Address Accepting the Presidential Nomination."

86. FDR to Louis Howe, October 10, 1933, FDRL.

87. Nixon, *Franklin D. Roosevelt and Conservation*, Vol. 1, pp. 209–10.

88. FDR, "Address Accepting the Presidential Nomination."

89. Charles Lathrop Pack quoted in Rutkow, *American Canopy*, pp. 249–50.

90. Nixon, *Franklin D. Roosevelt and Conservation*, Vol. 1, p. 113.

91. "National Affairs: Hyde and Seedlings," *Time*, July 18, 1931. Also see Clements, *Hoover, Conservation, and Consumerism*, p. 192.

92. Nixon, *Franklin D. Roosevelt and Conservation*, Vol. 1, p. 113.

93. FDR to James O. Hazard, July 29, 1932, FDRL.

94. James A. Kehl and Samuel J. Astorino, "A Bull Moose Responds to the New Deal: Pennsylvania's Gifford Pinchot," *Pennsylvania Magazine of History and Biography*, Vol. 88, no. 1 (January 1964), pp. 37–51.

95. Gifford Pinchot (summer 1932, n.d.), unpublished statement, Gifford Pinchot Papers, Library of Congress, Washington, D.C.

96. "Roosevelt Leaves on Long Trip West," *Washington Post*, September 13, 1932, p. 3.

97. Worster, *Dust Bowl*, p. 12.

98. Eric Jay Dolin, *Smithsonian Book of National Wildlife Refuges* (Washington, DC: Smithsonian Institution, 2003), p. 85.

99. Margaret Bourke-White, "Dust Changes America," *The Nation*, May 22, 1935; reprinted in John R. Wunder, Francis W. Kaye, and Vernon Carstensen, eds., *Americans View Their Dust Bowl Experience* (Boulder: University of Colorado Press, 2001), pp. 90–93.

100. "Conservation Crisis Seen in Texas," *American Forests* (September 1931), p. 558.

101. Eleanor Roosevelt, *Autobiography*, p. 161.

102. Eleanor Roosevelt quoted in Thomas Patton, "Forestry and Politics: Franklin D. Roosevelt as Governor of New York," *New York History* (October 1994), pp. 397–418.

103. FDR to Lowe Shearon, in Herbert Hoover, *The Memoirs of Herbert Hoover: The Great Depression, 1929–1941* (New York: Macmillan, 1952), p. 316.

104. Hoover, *The Memoirs of Herbert Hoover*, pp. 316–17.

105. Egan, *The Worst Hard Time*, p. 104.

106. Will Rogers quoted in Cohen, *Nothing to Fear*, pp. 4–5.

CHAPTER 7: "THEY'VE MADE THE GOOD EARTH BETTER"

1. Bob Marshall, *Alaska Wilderness: Exploring the Central Brooks Range* (Berkeley: University of California Press, 2005), p. 1.

2. Harvey Manning, *Wilderness Alps: Conservation and Conflict in Washington's North Cascades* (Bellingham, WA: Northwest Wild Books, 2007), p. 76.

3. Robert Marshall, "The Problem of the Wilderness," *Scientific Monthly*, Vol. 30, no. 2 (February 1930), p. 148.

4. See Robert Marshall, "Forest Devastation Must Stop," *Nation* (August 28, 1929); "A Proposed Remedy for Our Forestry Illness," *Journal of Forestry*, Vol. 28 (March 1930).

5. Miller, *Gifford Pinchot and the Making of Modern American Environmentalism*, pp. 333–34. See also *A National Plan for American Forestry*, Senate Document no. 12, 73rd Congress, 1st Session (Washington, DC: U.S. Government Printing Office, 1933), 2 vols; Earle H. Clapp, *Major Problems and the Next Big Step in American Forestry: Summary of a Report Prepared in Response to Senate Resolution 175; Together with Table of Contents, Letters of Transmittal, and Introduction from "A National Plan for American Forestry," by the Forest Service, U.S. Department of Agriculture* (Washington, DC: U.S. Government Printing Office, 1933).

6. James M. Glover, *A Wilderness Original: The Life of Bob Marshall* (Seattle, WA: Mountaineer, 1986), pp. 146–49.

7. Bob Marshall quoted ibid., p. 150.

8. James M. Glover and Regina B. Glover, "Robert Marshall: Portrait of a Liberal Forester," *Journal of Forest History*, Vol. 30, no. 3 (July 1986), pp. 112–19. See also *Trees: The Yearbook of Agriculture* (Washington, DC: U.S. Government Printing Office, 1949), p. 713.

9. Gifford Pinchot to FDR, January 20, 1933, FDRL.

10. Conrad L. Wirth, *Parks, Politics, and the People* (Norman: University of Oklahoma Press, 1980), p. 73.

11. Gifford Pinchot to Franklin D. Roosevelt, January 20, 1933, FDRL.

12. Clements, *Hoover, Conservation, and Consumerism*, p. 179.

13. Alter, *The Defining Moment*, p. 168.

14. "Assassin Fires into Roosevelt Party at Miami; President-Elect Uninjured; Mayor Cermak and 4 Others Wounded"; "Cermak in Critical Condition at Hospital; 'Glad It Was I, Not You,' He Tells Roosevelt," *New York Times*, February 16, 1933, p. 1.

15. Philip H. Melanson, *The Secret Service* (New York: Carroll & Graf, 2002), pp. 43–44.

16. Alter, *The Defining Moment*, p. 176.

17. Black, *Franklin Delano Roosevelt*, pp. 263–64.

18. Graham White and John Maze, *Harold Ickes of the New Deal* (Cambridge, MA: Harvard University Press, 1985), p. 96.

19. Ibid., p. 98.

20. Barry Mackintosh, "Harold L. Ickes and the National Park Service," *Journal of Forest History*, Vol. 29 (April 1985), p. 78.

21. Harold Ickes, "Federal Responsibility for Planning," in Harlean James, ed., *American Planning and Civic Annual* (Washington, DC: American Planning and Civic Association, 1934), p. 3.

22. J. Leonard Bates, "Anna Wilmarth Thompson Ickes," in Edward T. James, Janet Wilson James, and Paul S. Boye, eds., *Notable American Women, 1607–1950: A Biographical Dictionary*, Vol. 1 (Cambridge, MA: Harvard University Press, 1971), pp. 251–52; Jennifer McLerran, *A New Deal for Native Art: Indian Arts and Federal Policy, 1933–1943* (Tucson: University of Arizona Press, 2009), pp. 75–76.

23. Charles F. Wilkinson, *American Indians, Time, and the Law: Native Societies in a Modern Constitutional Democracy* (New Haven, CT: Yale University Press, 1987), p. 20.

24. Elmer R. Rusco, *A Fateful Time: The Background and Legislative History of the Indian Reorganization Act* (Reno: University of Nevada Press, 2000), pp. 177–81.

25. "Today and Tomorrow," *New York Herald Tribune*, February 7, 1952; Donald C. Swain, *Wilderness Defender: Horace M. Albright and Conservation* (Chicago: University of Chicago Press, 1970), p. 218.

26. Anna Maria Gillis, "John Muir, Nature's Witness," *Humanities*, Vol. 32, no. 2 (March/April 2011), http://www.neh.gov/humanities/2011/marchapril/feature/john-muir-natures-witness.

27. Keith McClinsey, *Washington D.C.'s Mayflower Hotel* (Charleston, SC: Arcadia, 2007), p. 46.

28. Lillian Gish quoted in Alter, *The Defining Moment*, p. 239.

29. David A. Norris, "Four Terms with Franklin," *History Magazine* (October/November 2012), p. 16.

30. Wirth, *Parks, Politics, and the People*, p. 69.

31. FDR, "Inaugural Address, March 4, 1933," Washington, DC. Online at American Presidency Project, http://www.presidency.ucsb.edu/ws/?pid=14473.

32. Alter, *The Defining Moment*, p. 207.

33. Quoted in Robert A. Caro, *Master of the Senate: The Years of Lyndon Johnson* (New York: Knopf, 2002), p. 355.

34. Ellison Smith quoted in Anthony J. Badger, *FDR: The First Hundred Days* (New York: Hill and Wang, 2008), pp. 57–58.

35. FDR to Nelson C. Brown, March 8, 1933, Acting Dean Nelson C. Brown Papers, F. Franklin Moon Library of State University of New York College of Environmental Science and Forestry, Syracuse.

36. FDR, "Capable of Management for Continuous Yield," *Forestry News Digest* (January 1936).

37. FDR to F. A. Silcox, May 3, 1935, FDRL.

38. Memorandum, Roosevelt to secretaries of War, Interior, Agriculture, and Labor, March 14, 1933, in Nixon, *Franklin D. Roosevelt and Conservation*, Vol. 1.

39. William James, "The Moral Equivalent of War," *McClure's*, Vol. 35 (August 1910), pp. 463–68.

40. FDR, Address at Harvard University Tercentenary Celebration, September 18, 1936. Online at American Presidency Project, http://www.presidency.ucsb.edu/ws/?pid=15133.

41. Galusha, *Another Day, Another Dollar*, p. 33.

42. Frances Perkins, *The Roosevelt I Knew* (New York: Viking, 1946), pp. 180–81.

43. Maher, *Nature's New Deal*, pp. 11–12.

44. Jeanne Nienaber Clarke, *Roosevelt's Warrior: Harold Ickes and the New Deal* (Baltimore, MD: Johns Hopkins University Press, 1996), p. 39.

45. Stan Cohen, *The Tree Army: A Pictorial History of the Civilian Conservation Corps* (Missoula, MT: Pictorial Histories, 1980), p. 24.

46. "Fechner of CCC," *Time*, February 6, 1939, p. 11.

47. Cohen, *Nothing to Fear*, p. 214.

48. "The Text of President Roosevelt's Message to Congress Today on Unemployment," *New York Times*, March 21, 1933.

49. Cohen, *Nothing to Fear*, pp. 216–17.

50. Ibid., pp. 217–18.

51. U.S. Congress, *An Act for the Relief of Unemployment Through the Performance of Useful Public Work and for Other Purposes*, Public Law 5, 73rd Congress, 1st Session, 1933.

52. FDR, Executive Order 6101, April 5, 1933.

53. FDR, "Executive Order 6101 Starting the Civilian Conservation Corps," April 5, 1933. Online at American Presidency Project, http://www.presidency.ucsb.edu/ws/?pid=14609.

54. "Fechner Reported Resigning Because of Merging of CCC: Declines Comment but Admission Is Believed Implied by Remark Congress Hasn't Acted on Consolidation Order," *Washington Post*, May 3, 1939, p. 1.

55. John A. Salmond, *The Civilian Conservation Corps, 1933–1942: A New Deal Case Study* (Durham, NC: Duke University Press, 1967), chap. 2, http://www.nps.gov/parkhistory/online_books/ccc/salmond/chap2.htm.

56. "Quick Job Action Sought: President Asks Power to Begin Recruiting Idle in 2 Weeks," *New York Times*, March 22, 1933, 1. See also Salmond, *The Civilian Conservation Corps*, p. 14.

57. Maher, *Nature's New Deal*, p. 79.

58. Schlesinger, *The Coming of the New Deal*, p. 337.

59. "Robert Fechner, Head of CCC Dies," *New York Times*, January 1, 1940, p. 29.

60. Egan, *The Worst Hard Time*, p. 271.

61. "An Emergency Conservation Work (CCC) Chart, Prepared by Roosevelt, 3 April 1933," in Nixon, *Franklin D. Roosevelt and Conservation*, Vol. 1, p. 136.

62. Owen, *Conservation Under F.D.R.*, p. 28.

63. David M. Kennedy, *Freedom from Fear: The American People in Depression and War, 1929–1945* (Oxford: Oxford University Press, 1999), p. 118.

64. "Uniform of Spruce Green to Be Provided for CCC," *New York Times*, January 9, 1939, p. 1.

65. Galusha, *Another Day, Another Dollar*, p. 1.

66. Civilian Conservation Corps, *Forests Protected by the CCC* (Washington DC: U.S. Government Printing Office, 1941), p. 1. As Fechner describes it, " 'Three Horsemen' ride through American forests spreading destruction—Fire, Insects, and Disease. All three are deadly enemies, for they destroy timber products, wildlife, recreational and scenic values, and the forest protecting our vital watersheds."

67. Maher, *Nature's New Deal*, p. 55.

68. "Sees No Rise in Puerto Rico CCC," *New York Times*, March 5, 1939, p. 31.

69. Alfred E. Cornebise, *The CCC Chronicles; Camp Newspapers of the Civilian Conservation Corps, 1933–1942* (Jefferson, NC: McFarland, 2004), p. 7.

70. Harvey J. Kaye, *The Fight for the Four Freedoms* (New York: Simon and Schuster, 2014), p. 47.

71. Maher, *Nature's New Deal*, p. 81.

72. C. R. Hursh, Southeastern Forest Experiment Station (Asheville, NC), Civilian Conservation Corps (U.S.), "Measures for Stand Improvement in Southern Appalachian Forests," Emergency Conservation Work, Forest Publication No. 1 (Washington, DC: U.S. Government Printing Office, 1933), p. 3.

73. John A. Conners, *Shenandoah National Park: An Interpretive Guide* (Blacksburg, VA: McDonald and Woodland, 1988), p. 96.

74. Maher, *Nature's New Deal*, p. 74.

75. Dennis E. Simmons, "Conservation, Cooperation, and Controversy: The Establishment of Shenandoah National Park, 1924–1936," *Virginia Magazine of History and Biography*, Vol. 89 (October 1981), p. 401.

76. Cohen, *The Tree Army*, p. 153.

77. Ronald L. Heinemann, "Civilian Conservation Corps," in *Encyclopedia Virginia*, http://encyclopediavirginia.org.

78. Cohen, *The Tree Army*, p. 153; and Virginia State Parks, "Civilian Conservation Corps Museum," accessed August 21, 2014, http://www.dcr.virginia.gov/stateparks/ccc-museum.shtml.

79. Albert H. Good, *Park Structures and Facilities* (Washington, DC: National Park Service, 1935, 1938); and Albert H. Good, *Patterns from the Golden Age of Rustic Design: Park and Recreation Structures from the 1930s* (Lanham, MD: Roberts Rinehart, 2003).

80. Dayton Duncan and Ken Burns, *The National Parks: America's Best Idea* (New York: Knopf, Doubleday, 2009), p. 234.

81. Charles Battell Loomis, "With the Green Guard," *Liberty*, April 29, 1934.

82. Jerry J. Frank, *Making Rocky Mountain National Park* (Lawrence: University Press of Kansas, 2013), p. 99.

83. Thomas W. Patton, " 'A Forest Camp Disgrace': The Rebellion of Civilian Conservation Corps Workers at Preston, New York, July 7, 1933," *New York History* (2001), pp. 231–58.

84. Cohen, *The Tree Army*, p. 152.

85. "66 CCC Camp Sites Listed in New York," *New York Times*, March 31, 1934.

86. "18 Forest Camps for New York," *New York Times*, May 10, 1933, p. 5.

87. Henry E. Clepper, "In Penn's Woods," *American Forests* (June 1935), p. 269.

88. Olen Cole, *The African-American Experience in the Civilian Conservation Corps* (Gainesville: University of Florida Press, 1999), p. 16.

89. Gifford Pinchot to FDR, March 21, 1933, FDRL.

90. Gifford Pinchot to Louis Howe and FDR, April 17, 1933, CCC Papers, FDRL.

91. Jonathan Mitchell, "Roosevelt's Tree Army: I," *New Republic* (May 29, 1935), p. 64.

92. Rutkow, *American Canopy*, p. 250.

93. Maher, *Nature's New Deal*, p. 90.

94. Ray Condor interview by Beth Martin, September 29, 1989, Zion National Park Oral History Project.

95. "Camp Life Reader and Workbook," Language Usage Series, nos. 1–4, Federal Security Agency, U.S. Office of Education, Washington, DC.

96. Floyd Fowler oral history interview, Zion National Park Oral History Project, CCC Reunion, September 27, 1989, Southern Utah University.

97. Stradling, *The Nature of New York*, p. 164.

98. Badger, *FDR: The First Hundred Days*, p. 56.

99. Galusha, *Another Day, Another Dollar*, pp. 43–44.

100. Federal Security Agency, Civilian Conservation Corps, *The CCC at Work: A Story of 2,500,000 Young Men* (Washington, DC: U.S. Government Printing Office, 1941), p. 34.

101. "CCC: The Civilian Conservation Corps in Missouri, 1933–1942," brochure, n.d., 3 pp.

102. FDR to Robert Fechner, September 27, 1935, in "African Americans in the CCC," New Deal Network, http://newdeal.feri.org/aaccc/.

103. Cornebise, *The CCC Chronicles*, p. 5.

104. Ibid.

105. *Happy Days*, July 22, 1939.

CHAPTER 8: "HE DID NOT WAIT TO ASK QUESTIONS, BUT SIMPLY SAID THAT IT SHOULD BE DONE"

1. Richard West Sellars, "The National Park System and the Historic American Past: A Brief Overview and Reflection," *George Wright Forum*, Vol. 24, no. 1 (2007), p. 11.

2. Conners, *Shenandoah National Park*, pp. 93–94.

3. Donald C. Swain, "The National Park Service and the New Deal, 1933–1940," *Pacific Historical Review*, Vol. 41, no. 3 (August 1972), pp. 312–32.

4. Newton, *Design on the Land*, p. 538.

5. Horace Albright, *Origins of the National Park Service Administration of Historic Sites* (Philadelphia: Eastern National Park and Monuments Association, 1971), p. 19.

6. Ibid., pp. 19–22.

7. Harvey P. Benson, "The Skyline Drive: A Brief History of a Mountaintop Motorway," *Regional Review* (February 1940), p. 28.

8. See "White House Statement Summarizing Executive Order 6166, June 10, 1933." Online at American Presidency Project.

9. John T. Flynn, *Country Squire in the White House* (New York: Doubleday, Doran, 1940), p. 12.

10. Albright, *Origins of the National Park Service Administration of Historic Sites*, pp. 19–22.

11. Sellars, "The National Park System and the Historic American Past: A Brief Overview and Reflection."

12. See Gerald W. Williams, "National Monuments and the Forest Service," USDA Forest Service/National Park Service (2003).

13. Historic Sites Act of 1935, 16 USC Sec. 461–467, http://www.cr.nps.gov/local-law/hsact35.htm.

14. Eleanor Roosevelt, "My Day," April 20, 1945.

15. Phoebe Cutler, *The Public Landscape of the New Deal* (New Haven, CT: Yale University Press, 1985), p. 62.

16. Horace Albright, "Origins of National Park Service Administration of Historic Sites," http://www.nps.gov/parkhistory/online_books/albright/origins.htm.

17. Albright, *Origins of the National Park Service Administration of Historic Sites*, pp. 19–22.

18. See Lary Dilsaver, ed., *America's National Park System: The Critical Documents* (New York: Rowman and Littlefield, 1994), pp. 116–18; Char Miller, *Public Debates: A Century of Controversy* (Corvallis: Oregon State University Press, 2012), p. 71.

19. Newton, *Design on the Land*, p. 539.

20. Albright, *Origins of the National Park Service Administration of Historic Sites*, p. 23.

21. See Donald C. Swain, "Harold Ickes, Horace Albright, and the Hundred Days: A Study in Conservation Administration," *Pacific Historical Review*, Vol. 34, no. 4 (November 1965), pp. 455–65.

22. Harold L. Ickes, *The Secret Diary*, Vol. 2, *The Inside Struggle, 1936–1939* (New York: Simon and Schuster, 1954), p. 584.

23. "Frederic A. Delano," *Washington Evening Star*, March 28, 1953.

24. Philip W. Warken, "A History of the National Resources Planning Board, 1933–1945," PhD dissertation, Ohio State University, 1969, pp. 44–45.

25. Schlesinger, *The Coming of the New Deal*, p. 351.

26. Newton, *Design on the Land*, pp. 541–42.

27. "Show for Lake Placid," *New York Times*, July 16, 1933.

28. FDR, "Radio Address from Two Medicine Chalet, Glacier National Park," August 5, 1934. Online at American Presidency Project, http://www.presidency.ucsb.edu/ws/?pid=14733.

29. Elmo R. Richardson, "Federal Park Policy: The Escalante National Monument Controversy of 1935–1940," *Utah Historical Quarterly*, Vol. 33 (Spring 1965), pp. 109–33.

30. "Gov. Dern Is Urged for Vice President," *New York Times*, April 26, 1932, p. 10.

31. FDR, Campaign Address on Railroads, Salt Lake City, September 17–19, 1932.

32. Edwin G. Hill, *In the Shadow of the Mountain: The Spirit of the CCC* (Pullman: Washington State University Press, 1990), p. xvi.

33. Washington County Historical Society, "Zion National Park, Washington County, Utah."

34. See Hal Rothman, *Preserving Different Pasts: The American National Monuments* (Urbana: University of Illinois Press, 1989), p. 169.

35. *Our National Parks: America's Spectacular Wilderness Heritage* (Pleasantville, NY: Reader's Digest Association, 1985), pp. 70–73.

36. See Rothman, *Preserving Different Pasts*, p. 169.

37. *Code of Federal Regulations: The President* (Washington, DC: Office of the Federal Register, National Archives and Records Service, General Services Administration, 1968), pp. 69–70.

38. 16 U.S. Code § 346b, Consolidation of Zion National Park and Zion National Monument, https://www.law.cornell.edu/uscode/text/16/346b.

39. Richard White, *"It's Your Misfortune and None of My Own": A New History of the American West* (Norman: University of Oklahoma Press, 1991), p. 473.

40. Thomas G. Alexander, *The Rise of Multiple-Use Management in the Intermountain West: A History of Region 4 of the Forest Service* (Washington, DC: U.S. Dept. of Agriculture, Forest Service, 1988), p. 101.

41. Belden W. Lewis, "A CCC Guy's Diary," Zion National Park Archives, Springdale, UT. (Unpublished.)

42. Karl A. Larson, "Zion National Park—with Some Reminiscences Fifty Years Later," *Utah Historical Quarterly*, Vol. 37 (1969), p. 408.

43. Tony Melessa interview by Janet Seegmiller, March 8, 2006, Markaguant Plateau Oral History Project, Sherratt Library, Southern Utah University, Cedar City.

44. Quince Alvey interview by Jeff Frank, September 29, 1989, Zion National Park Oral History Project.

45. Kathleen Dalton to Douglas Brinkley, November 1, 2013.

46. Ibid.

47. FDR, "White House Statement Summarizing Executive Order 6166," June 10, 1933. Online at American Presidency Project, http://www.presidency.ucsb.edu/ws/?pid=14660.

48. Swain, *Wilderness Defender*, p. 23.

49. "Henry A. Wallace Is Dead at 77; Ex-Vice President, Plant Expert," *New York Times*, November 19, 1965.

50. Henry Wallace quoted in Clayton R. Koppes, "Environmental Policy and American Liberalism: The Department of the Interior, 1933–1953," *Environmental Review*, Vol. 7, no. 1 (Spring 1983), pp. 17–53.

51. Alter, *The Defining Moment*, pp. 292–93.

52. Hugh Bennett, *Soil Conservation* (New York: McGraw-Hill, 1939), p. 13.

53. Wellington Brink, *Big Hugh: The Father of Soil Conservation* (New York: Macmillan, 1951).

54. Hugh Bennett, *Elements of Soil Conservation* (New York: McGraw-Hill, 1947).

55. Egan, *The Worst Hard Time*, p. 134.

56. Owen, *Conservation Under FDR*.

57. J. Douglas Helms, "Hugh Hammond Bennett," in Cevasco and Harmond, *Modern American Environmentalists*, pp. 29–35; also Eagan, *The Worst Hard Time*, p. 134.

58. Bennett, *Soil Conservation*, p. v.

59. J. Douglas Helms, "Hugh Hammond Bennett and the Creation of the Soil Erosion Service," *Historical Insights*, Vol. 8 (September 2008), pp. 1–13.

60. Egan, *The Worst Hard Time*, p. 135.

61. Helms, "Hugh Hammond Bennett," p. 32.

62. American Council of Learned Societies, "Hugh Hammond Bennett," *American National Biography*, Vol. 2. (New York: Oxford University Press, 1999), pp. 582–83.

63. Arthur E. Morgan, "Tennessee Valley Becomes Laboratory for the Nation," *New York Times*, March 25, 1934.

64. Howard Zinn, *A People's History of the United States* (New York: Harper & Row, 1980), p. 393.

65. See C. Herman Pritchet, *The Tennessee Valley Authority: A Study in Public Administration* (Chapel Hill: University of North Carolina Press, 1943).

66. "New Era of Power Revolutionizes Life in the Tennessee Valley," *New York Times*, November 29, 1936.

67. Steven Solomon, *Water: The Epic Struggle for Wealth, Power, and Civilization* (New York: HarperCollins, 2010), p. 329.

68. "President Honors Norris as Liberal," *New York Times*, September 5, 1944.

69. Amity Shlaes, *The Forgotten Man* (New York: HarperCollins, 2007), p. 186.

70. David P. Billington, Donald C. Jackson, and Martin V. Melosi, *The History of Large Federal Dams: Planning, Design, and Construction in the Era of Big Dams* (Denver, CO: U.S. Department of the Interior, Bureau of Reclamation, 2005), p. 179.

71. *Page News and Courier*, August 15, 1933.

72. Cornebise, *The CCC Chronicles*, p. 105.

73. Ickes, quoted in Darwin Lambert, *Administrative History of the National Parks, 1924–1976* (Luray, VA: NPS Mid-Atlantic Region and Shenandoah Natural History Association, 1979), p. 128.

74. William Green to FDR, September 18, 1933, FDRL.

75. FDR quoted in Gaddis Smith, "Roosevelt, the Sea, and International Security," in Brinkley and Facey-Crowther, *The Atlantic Charter*, p. 35.

76. Alfred Runte, *National Parks: The American Experience* (Lanham, MD: Taylor, 2010), p. 191.

77. Quoted in Ren Davis and Helen Davis, *Our Mark on This Land* (Granville, OH: McDonald & Woodword, 2011), p. 257.

78. National Park Service, "The Civilian Conservation Corps at Platt National Park," last modified May 9, 2014, http://www.nps.gov/chic/historyculture/ccc.htm. See also Department of the Interior, *Platt National Park: Oklahoma* (Washington, DC: National Park Service, c. 1930s), http://www.nps.gov/chic/historyculture/guidebook1930s.htm.

79. James W. Cornett, *The Joshua Tree* (Palm Springs, CA: Nature Trails, 1999), p. 13.

80. Ruby DeCorsaw Culver, "The Joshua Tree: Oldest Living Thing in the California Desert," *Western World* (December 1929–January 1930), p. 26.

81. Minerva Hoyt to A. E. Dunaray, July 14, 1933, Joshua Tree National Park Archives, Twentynine Palms, CA.

82. Minerva Hoyt to Conrad Wirth, July 8, 1933, Joshua Tree National Park Archives, Twentynine Palms, CA.

83. Bertha H. Fuller, "Everything Was Peaceful," *Desert* (June 1955), p. 22.

84. Conner Sorensen, " 'Apostle of the Cacti': The Society Matron as Environmental Activist," in Doyce B. Nunis, *Women in the Life of Southern California* (Los Angeles: Historical Society of Southern California, 1996), p. 230.

85. John Steven McGroarty, "Minerva Hamilton Hoyt," in William L. Blair, *Pasadena Community Book* (Pasadena, CA: Arthur H. Cawston, 1943), pp. 752–58.

86. Edwin Way Teale, "Making the Wild Scene," *New York Times*, January 28, 1968.

87. Minerva Hoyt quoted in "Keepsake of Joshua Tree National Monument Trek," October 3, 4, and 5, 1980, Joshua Tree National Park Archives, Twentynine Palms, CA.

88. "Mrs. Hoyt Now en Route to England," April 27, 1929 [n.p.], Clippings File, Joshua Tree National Park Archives, Twentynine Palms, CA.

89. Sorensen, " 'Apostle of the Cacti,' " pp. 234–36.

90. "California Exhibit Offered by Mrs. Sherman Hoyt Made Great Impression on British," [n.d./n.p.], Clippings File, Joshua Tree National Park Archives, Twentynine Palms, CA.

91. Minerva Hoyt to A. E. Dunaray, July 14, 1933, Joshua Tree National Park Archives, Twentynine Palms, CA.

92. Samuel A. King, *A History of Joshua Tree National Monument* (Washington, DC: National Park Service, 1954), http://www.nps.gov/history/history/online_books/jotr/jotr_history.pdf.

93. Harold L. Ickes to Minerva Hoyt, November 14, 1933, Joshua Tree National Park Archives, Twentynine Palms, CA.

94. W. A. Simpson to Harold Ickes, November 20, 1933; Harold Ickes to W. A. Simpson, December 13, 1933. Both letters courtesy of Joshua Tree National Park Archives, Twentynine Palms, CA.

CHAPTER 9: "ROOSEVELT IS MY SHEPHERD"

1. R. Douglas Hurt, "Federal Land Reclamation in the Dust Bowl," *Great Plains Quarterly*, Vol. 6 (Spring 1986), p. 95.

2. Marion Clawson, *New Deal Planning: The National Resources Planning Board* (Baltimore, MD: Johns Hopkins University Press, 1981), pp. 109–11.

3. Marion Clawson, "New Deal Planning: The National Resources Planning Board," *Agricultural History*, Vol. 57, no. 1 (January 1983), pp. 116–18.

4. Clawson, *New Deal Planning*, pp. 13–14.

5. Charles E. Merriam, "The National Resources Planning Board: A Chapter in American Planning Experience," *American Political Science Review*, Vol. 38, no. 6 (December 1944), pp. 1075–88.

6. W. Dale Nelson, *The President Is at Camp David* (Syracuse, NY: Syracuse University Press, 1995), p. 4.

7. Louis Howe to Lewis Bailey, May 2, 1933, FDRL.

8. Gaddis Smith, "Roosevelt, the Sea, and International Security," in Brinkley and Facey-Crowther, *The Atlantic Charter*, pp. 39–40.

9. Cross, *Sailor in the White House*, p. 15.

10. National Industrial Recovery Act, June 16, 1933; Enrolled Acts and Resolutions of Congress, 1789–1996; General Records of the United States Government; Record Group 11, National Archives.

11. Ickes, quoted in Worster, *Dust Bowl*, p. 42.

12. *Boise City* (Oklahoma) *News*, November 2, 1933.

13. Paige, *The Civilian Conservation Corps and the National Park Service*, p. 8.

14. Newton, *Design on the Land*, pp. 577–81.

15. Wayne Franklin, Foreword, in Rebecca Conard, *Places of Quiet Beauty: Parks, Preserves, and Environmentalism* (Iowa City: University of Iowa Press, 1997), pp. xi–xv.

16. Greg Ross Harber, "How You Gonna Keep 'Em Down on the Farm? Mapping New Deal Cultural Democracy in Iowa," PhD dissertation, University of Iowa, 2007, pp. 294–95.

17. David Soll, "State Parks," in Brosnan, *Encyclopedia of American Environmental History*, Vol. 4, pp. 1230–31.

18. Cutler, *The Public Landscape of the New Deal*, p. 65.

19. Ibid.

20. Cohen, *The Tree Army*, p. 90.

21. LaFrank, "Real and Ideal Landscapes Along the Taconic State Parkway," in Hoagland and Breisch, *Constructing Image, Identity, and Place*, pp. 247–62.

22. Act of June 16, 1933 ("Industrial Recovery Act"), Public Law 73–67, 73rd Congress. NARA Online Public Access Database.

23. FDR to Henry A. Wallace, June 24, 1933, FDRL.

24. Lawrence S. Earley, *Looking for Longleaf: The Fall and Rise of an American Forest* (Chapel Hill: University of North Carolina Press, 2004), p. 4.

25. FDR to Henry A. Wallace, July 11, 1933, FDRL.

26. Charles W. Hurd, "Roosevelt Greeted Royally by Canada at Campobello Isle," *New York Times*, June 30, 1933.

27. FDR, "Greetings to the Civilian Conservation Corps, July 8, 1933," in Rosenman, *Public Papers of the President*, p. 271.

28. "Half of Bonus Army Ready to Work in Woods," *Washington Post*, May 20, 1933, p. 1.

29. "Forestry Veterans Go to New England," *Washington Post*, June 21, 1933, p. 16. There were CCC companies in Charlemont, Lee, Fall River, Hyde Park, Wrentham, Spencer, and Chicopee Falls (Massachusetts); Gardiner, Cherryfield, and Dineo Station (Maine); and Rutland (Vermont).

30. Barbara Kirkconnell, "Catoctin Mountain Park: An Administrative History," February 1988, pp. 9–10.

31. Phyllis McIntosh, "The Corps of Conservation," *National Parks* (September–October 2001), pp. 26–27.

32. Newton, *Design of the Land*, pp. 589–91.

33. Billy Townsend, "History of the Georgia State Parks and Historic Sites," http://gastateparks.org/content/georgia/parks/75th_Anniv/parks_history.pdf.

34. Davis and Davis, *Our Mark on This Land*, p. 131.

35. Newton, *Design of the Land*, pp. 502–3.

36. Richard Melzer, *Coming of Age in the Great Depression* (Las Cruces: New Mexico State University Press, 2000).

37. Edward Smith to Franklin D. Roosevelt, August 25, 1935.

38. *Woodpecker*, Vol. 1, no. 3 (August 19, 1933), Camp Cabeza de Vaca, NM.

39. FDR to Harry B. Hawes, October 5, 1939, FDRL.

40. "Hawes Will Direct New Wild Life Body," *New York Times*, September 5, 1930.

41. Harry Benton Hawes, *Fish and Game: Now or Never* (New York: D. Appleton-Century, 1935), p. 1.

42. Thomas R. Dunlap, *Saving America's Wildlife* (Princeton, NJ: Princeton University Press, 1988), p. 82.

43. Peter Carrels, "The State and Fate of the Prairie," *Outdoor America*, Issue 4 (2013), pp. 34–39.

44. Maher, *Nature's New Deal*, pp. 8–9.

45. *The CCC and Wildlife* (Washington, DC: U.S. Government Printing Office, 1938), p. 3.

46. Cohen, *Nothing to Fear*, p. 2.

47. William T. Hornaday and Harry M. Reeves, *Thirty Years War for Wild Life* (New York: Scribner, 1931), p. 191.

48. See Michael W. Giese, "A Federal Foundation for Wildlife Conservation: The Evolution of the National Wildlife Refuge System, 1920–1968," PhD dissertation, American University, 2008, pp. 97–99, 119–20.

49. Ibid., pp. 118–19.

50. Clarence Cottam, *Food Habits of North American Diving Ducks* (Washington, DC: U.S. Department of Agriculture, 1939).

51. *The CCC and Wildlife*, pp. 5–6.

52. Douglas M. Thompson, "The Cost of Trout Fishing," *New York Times*, April 11, 2015, p. A19. Thompson properly explains the environmental downside of fish hatcheries, including polluting rivers and lakes.

53. "Mr. and Mrs. Thomas H. Beck Have Good Time in Hollywood," *Wilton Bulletin*, September 19, 1940.

54. David L. Lendt, *Ding: The Life of Jay Norwood Darling* (Ames: Iowa State University Press, 1979), p. 63.

55. FDR to Henry Wallace, August 29, 1933, FDRL.

56. Paul G. Redington (Chief of the Biological Survey) to Henry Wallace, November 10, 1933, in Nixon, *Franklin D. Roosevelt and Conservation*, Vol. 1, pp. 212–24.

57. Michael V. Namorato, ed., *The Diary of Rexford G. Tugwell: The New Deal, 1932–1935* (Westport, CT: Greenwood, 1992), p. 150.

58. "Thomas H. Beck, 70, Retired Publisher," *New York Times*, October 17, 1951.

59. Kenyon B. Zahner, "Quivering Garth: The Okefenokee," *Living Wilderness*, Vol. 19, no. 50 (Autumn 1954).

60. Jean Sherwood Harper to FDR, November 25, 1933, FDRL.

61. FDR to Jean Sherwood Harper, December 19, 1933, FDRL.

62. John Wong, "FDR and the New Deal on Sport and Recreation," *Sport History Review*, Vol. 29 (1998), pp. 173–91.

63. Billy Townsend, "History of the Georgia State Parks and Historic Sites," http://gastateparks.org/content/georgia/parks/75th_Anniv/parks_history.pdf.

64. Larry I. Bland, *The Papers of George Catlett Marshall*, Vol. 1 (Baltimore, MD: Johns Hopkins University Press, 1961), p. 394.

65. *Happy Days*, April 19, 1941.

66. Paige, *The Civilian Conservation Corps and the National Park Service*, pp. 1–21.

67. William E. Leuchtenberg, *The White House Looks South* (Baton Rouge: Louisiana State University Press, 2005), pp. 29–31.

68. Ibid.

69. Ward, *Closest Companion*, p. 16.

70. Elliott Roosevelt, *F.D.R.: His Personal Letters*, Vol. 1 (New York: Duell, Sloan and Pearce, 1950), pp. 372–73.

71. Connie Huddleston, *Georgia's Civilian Conservation Corps* (Charleston, SC: Arcadia Publishing, 2009), p. 46.

72. Ibid., p. 12.

73. Ibid., p. 26.

74. FDR, Executive Order 7037, May 11, 1935. Online at American Presidency Project, http://www.presidency.ucsb.edu/ws/?pid=15057.

75. Jean Sherwood Harper to FDR, November 25, 1933, FDRL.

76. FDR to Jean Sherwood Harper, December 19, 1933, FDRL.

77. Francis Harper, "The Okefinokee Wilderness," *National Geographic* (May 1934), p. 597.

78. Jonathan Rosen, *The Life of the Skies: Birding at the End of Nature* (New York: Farrar, Straus, and Giroux, 2008), p. 267.

79. Jean Sherwood Harper to FDR, February 8, 1935, FDRL.

80. FDR to Jean Sherwood Harper, February 18, 1935, FDRL.

81. Franklin D. Roosevelt quoted in M. S. Venkataramani, *The Sunny Side of FDR* (Athens: Ohio University Press, 1973), p. 265.

82. David W. Look and Carole L. Perrault, *The Interior Building: Its Architecture and Its Art* (Washington, DC: U.S. Department of the Interior, 1986), pp. 1–27.

CHAPTER 10: "THE YEAR OF THE NATIONAL PARK"

1. Venkataramani, *The Sunny Side of FDR*, p. 37.

2. Duncan and Burns, *The National Parks*, p. 239.

3. Paige, *The Civilian Conservation Corps and National Park Service*, p. 19.

4. John Burroughs, "The Divine Abyss," in *The Writings of John Burroughs: Time and Change* (New York: Houghton Mifflin, 1912), pp. 39–45.

5. "Day by Day: March 21, 1934," Pare Lorentz Center, FDRL.

6. Stephen R. Fox, *The American Conservation Movement: John Muir and His Legacy* (Madison: University of Wisconsin Press, 1981), pp. 205–6.

7. Runte, *National Parks: The American Experience*, p. 120.

8. Harold L. Ickes to FDR, March 29, 1934, FDRL.

9. Peter Matthiessen and Patricia Caulfield, *The Everglades* (San Francisco: Sierra Club, 1970), pp. 49–63.

10. David J. Nelson, "Florida Crackers and Yankee Tourists: The Civilian Conservation Corps, the Florida Park Service, and the Emergence of Modern Florida Tourism," Ph.D. dissertation, Florida State University, 2008.

11. Cynthia Barnett, *Rain: A Natural and Cultural History* (New York: Crown, 2015), p. 232.

12. Harry A. Kersey Jr., "Seminoles and Miccosukees: A Century in Retrospective," in *Indians of the Southeastern United States in the Late 20th Century* (Tuscaloosa: University of Alabama Press, 1992), p. 102.

13. FDR to Ernest F. Coe, April 1, 1939, FDRL

14. "Roosevelt to Start Today on Fishing Trip," *New York Times*, March 27, 1934.

15. "Roosevelt Sails in Open Atlantic," *New York Times*, March 29, 1934.

16. "Roosevelt Signs Two Major Bills," *New York Times*, April 8, 1934.

17. "FDR: Day by Day for April 4, 1934," Pare Lorentz Center, FDRL.

18. "Roosevelt Starts on Way to Nassau," *New York Times*, March 30, 1934; see also "Roosevelt Sails in Open Atlantic."

19. "Roosevelt Admits Catching a Whale," *New York Times*, April 10, 1934.

20. "Big Crowd Meets Train," *New York Times*, April 14, 1934.

21. Olmsted quoted in Duncan and Burns, *The National Parks*, p. 280.

22. "Park in Everglades Voted by Congress: National Wild Life Refuge and Playground Will Be Bigger Than Rhode Island," *New York Times*, May 26, 1934.

23. Everglades National Park Enabling Act, 48 Stat. 816 (1934).

24. Marjory Stoneman Douglas, *The Everglades: River of Grass* (Sarasota, FL: Pineapple, 1997), p. 366.

25. William Sherman Jennings quoted in Michael Grunwald, *The Swamp: The Everglades, Florida, and the Politics of Paradise* (New York: Simon and Schuster, 2006), p. 210.

26. Cohen, *The Tree Army*, p. 50.

27. Florida Park Service, "Highlands Hammock State Park: History," accessed June 29, 2014, http://www.floridastateparks.org/history/parkhistory.cfm?parkid=129.

28. David Allen Sibley, *The Sibley Guide to Trees* (New York: Knopf, 2009), p. 3. See also Davis and Davis, *Our Mark on This Land*, p. 74.

29. Davis and Davis, *Our Mark on This Land*, pp. 71–72.

30. Clarke, *Roosevelt's Warrior: Harold L. Ickes and the New Deal*, pp. 100–101.

31. Harold L. Ickes, unpublished diary entry, April 29, 1945, Harold L. Ickes Papers, Manuscript Reading Room, Library of Congress, Washington, DC.

32. Mark Reisner, *Cadillac Desert: The American Desert and Its Disappearing Water* (New York: Penguin, 1993), pp. 137–47.

33. Ickes quoted in Watkins, *Righteous Pilgrim*, p. 472.

34. Anthony P. Musso, *FDR and the Post Office: A Young Boy's Fascination; A World Leader's Passion* (Bloomington, IN; Milton Keynes, UK: Author House, 2006), p. 40.

35. Ibid., p. 16.

36. Bernice L. Thomas, *The Stamp of FDR: New Deal Post Offices in the Mid-Hudson Valley* (Fleischmanns, NY: Purple Mountain, 2002), p. 7.

37. Robert U. Johnson to FDR, August 16, 1934, FDRL.

38. FDR to Robert U. Johnson, August 24, 1934, FDRL.

39. "Vermont Voters Reject Mountain Parkway Plan," *New York Times*, March 4, 1936, p. 3.

40. "Those CCC Boys," Vermont Public Radio, http://www.vpr.net/episode/43879/those-ccc-boys/.

41. See Robert W. Righter, "National Monuments to National Parks: The Use of the Antiquities Act of 1906," *Western Historical Quarterly* (1989), pp. 281–301.

42. FDR to Harold Ickes, May 29, 1934, FDRL.

43. Frederic A. Delano to Harold Ickes, May 16, 1934, FDRL.

44. Ibid.

45. Quetico-Superior Foundation, "Quetico Superior Timeline," last updated 2009, http://www.queticosuperior.org/abouttheregion/timeline.html.

46. Quetico Superior Foundation, "History of the Quetico Superior: A History of Land Use and Controversy," last updated 2009, http://www.queticosuperior.org/abouttheregion/history.html.

47. "Isle Royale Sought as National Reserve," *Washington Post*, July 10, 1942, p. 8.

48. "Lonely Isle Royale to Become a Park," *New York Times*, June 2, 1935, p. XX5.

49. James Jackson, "The Living Legacy of the CCC," *American Forests*, Vol. 94, nos. 9, 10 (1988), p. 47.

50. Cornelius M. Maher, "Planting More Than Trees: The Civilian Conservation Corps and the Roots of the American Environmental Movement, 1929–1943," PhD dissertation, New York University, January 2001, pp. 320–21.

51. Harley E. Jolly, *The CCC in the Smokies* (Gatlinburg, TN: Great Smoky Mountain National History Association, 2001), p. 12.

52. Fish and Wildlife Coordination Act of 1934, 16 U.S.C. 661–667e, 48 Stat. 401, at http://www.fws.gov/laws/lawsdigest/fwcoord.html.

53. David Wagstaff to FDR, May 30, 1934, FDRL.

54. FDR to Harold Ickes, June 6, 1934, FDRL.

55. Harold Ickes to FDR, June 19, 1934, FDRL.

56. John B. Adams interview by Don Graff, September 29, 1989, Zion National Park Oral History Project.

57. Ibid.

58. "Hawaiians Greet Roosevelt at Hilo," *New York Times*, July 26, 1934.

59. "Roosevelt to Find Rare Sea Comforts," *New York Times*, July 1, 1934.

60. Presidential log of USS *Houston*, entry for July 2, 1934, FDRL.

61. "President Speeds South on Cruiser," *New York Times*, July 3, 1934.

62. "Trip of the President: Summer 1934," entry for July 10, 1934, p. 5, FDRL.

63. "Roosevelt Watches Naval 'Hide and Seek,' " *New York Times*, July 19, 1934.

64. "Hawaiians Greet Roosevelt at Hilo," *New York Times*, July 26, 1934.

65. Summer Roper, "The Civilian Conservation Corps: An Archeological Survey of the Hilina Pali Erosion Control Project of 1940," last updated June 28, 2014, http://www.nps.gov/havo/historyculture/civilian-conservation-corps.htm.

66. Jadelyn J. Moniz, *Fire on the Rim: The Creation of Hawaii National Park* (Washington, DC: Department of the Interior, 2015), p. 11. Online at http://www.nps.gov/havo/historyculture/upload/Fire-On-The-Rim-Paper_final.pdf.

67. "Hawaiians Greet Roosevelt at Hilo."

68. National Park Service, "The Volcano House Story," last updated June 26, 2014, http://www.nps.gov/havo/parknews/history.htm.

69. "Hawaiians Greet Roosevelt at Hilo."

70. "Roosevelt Tours Hawaii Naval Base," *New York Times*, July 28, 1934.

71. "Found in the Archives," May 18, 2012, FDRL. Online at https://fdrlibrary.wordpress.com/2012/05/18/found-in-the-archives-33/.

72. "Roosevelt Arrives Off Pacific Coast," *New York Times*, August 3, 1934.

73. Eleanor Roosevelt, "By Car and Tent," *Women's Home Companion*, Vol. 61 (August 1934), p. 4.

74. See Joyce L. Kornbluh, *A New Deal for Workers' Education: The Workers' Service Program, 1933–1942* (Urbana: University of Illinois Press, 1987), p. 87; and Gwendolyn Mink, *The Wages of Motherhood: Inequality in the Welfare State, 1917–1942* (Ithaca, NY: Cornell University Press, 1996), p. 157.

75. Galusha, *Another Day, Another Dollar*, pp. 36–37.

76. Heather Van Wormer, "A New Deal for Gender: The Landscapes of the 1930s," in Deborah L. Rotman and Ellen-Rose Savulis, eds., *Divided Places: Material Dimensions of Gender Relations and the American Historical Landscape* (Knoxville: University of Tennessee Press, 2003), p. 219.

77. Nancy C. Unger, *Beyond Nature's Housekeepers: American Women in Environmental History* (New York: Oxford University Press, 2012), pp. 119–20.

78. Phyllis McIntosh, "The Corps of Conservation," *National Parks* (September/October 2001), p. 25.

79. Galusha, *Another Day, Another Dollar*, pp. 36–37.

80. Bryant Simon, "New Men in Body and Soul: The Civilian Conservation Corps and the Transformation of Male Bodies and the Male Politic," in Virginia J. Scharff, ed., *Seeing Nature Through Gender* (Lawrence: University Press of Kansas, 2003), pp. 80–102.

81. "Mrs. Roosevelt Goes On," *New York Times*, July 29, 1934.

82. "Mrs. Roosevelt Draws Crowd," *New York Times*, July 31, 1934.

83. See Billington, Jackson, and Melosi, *The History of Large Federal Dams*, pp. 193–210.

84. "Roosevelt Pledges Control of Power for Whole People," *New York Times*, August 4, 1934.

85. FDR to Harold Ickes, August 6, 1935, FDRL.

86. Gary A. Wedemeyer, "A Brief History of the Western Fisheries Research Center, 1934–2006," Western Fisheries Research Center (May 2007).

87. Dwayne Mack, "May the Work I've Done Speak for Me: African American Civilian Corps Enrollees in Montana, 1933–1934," *Western Journal of Black Studies*, Vol. 27, no. 4 (2003), p. 236.

88. "Negro Quartet at CCC Camp Gains Notice," *Western News*, September 27, 1934.

89. Donald H. Robinson, *Through the Years in Glacier National Park: An Administrative History* (West Glacier, MT: Glacier Natural History Association, May 1967), p. 22.

90. William P. Corbett, "Pipestone: The Origins and Development of a Nature Movement," *Minnesota History*, Vol. 47, no. 3 (Fall 1980), pp. 83–85.

91. For background see *Pipestone: A History of Pipestone National Monument, Minnesota*, NPS online, http://www.nps.gov/parkhistory/online_books/pipe2/sec7.htm.

92. Richard Sanderville to FDR, April 15, 1935, FDRL.

93. *Happy Days*, May 20, 1933, p. 7.

94. Calvin W. Gower, "The CCC Indian Division: Aid for Depressed Americans, 1933–1942," *Minnesota History* (Spring 1972), p. 5.

95. Ibid., p. 4.

96. Robert Fechner to John Collier, September 8, 1934, Civilian Conservation Corps Record Group 35, National Archives, Washington, DC.

97. FDR, "Radio Address from Two Medicine Chalet, Glacier National Park," August 5, 1934. Online at American Presidency Project, http://www.presidency.ucsb.edu/ws/?pid=14733.

98. Ibid.

99. Ibid.

100. Ibid.

101. Cornebise, *The CCC Chronicles*, pp. 105–6.

102. Reisner, *Cadillac Desert*, p. 158.

103. Lois Lonnquist, *Fifty Cents an Hour: The Builders and Boomtowns of Fort Peck Dam* (Helena, MT: MTSKY, 2006), pp. 78–83; "The Building of the Fort Peck Dam," *Life*, November 23, 1936.

104. Marc Johnson, "Fort Peck: Symbol of American Power," May 9, 2011. (Blog.)

105. "Roosevelt Hailed as 'Rain-Maker,' " *New York Times*, August 9, 1934.

106. Glover, *A Wilderness Original: The Life of Bob Marshall*, p. 192.

107. B. S. Yard to C. F. Truitt, January 1, 1938, Robert Marshall Papers, Bancroft Library, University of California–Berkeley.

108. Marshall quoted in Michael Frome, Foreword, in Harvey Broome, *Out Under the Sky of the Great Smokies* (Knoxville: University of Tennessee Press, 2001), p. xx.

109. Newton B. Drury, "The National Park Service: The First Thirty Years," in Harlean James, ed., *American Planning and Civic Annual* (1941), p. 34.

110. "New Park for Big Bend Region," *New York Times*, November 24, 1935, p. xxx.

111. John Jameson, *The Story of Big Bend National Park* (Austin: University of Texas Press, 2010), p. 76.

CHAPTER 11: "A DUCK FOR EVERY PUDDLE"

1. "So This Is Washington! Ducks, Geese and Other Wild Birds Play a Part in Congressional Side Shows; Bootleggers, Too; Remember Them?" *Washington Post*, January 7, 1934, p. 2.

2. Curt Meine, *Aldo Leopold: His Life and Work* (Madison: University of Wisconsin Press, 1988), pp. 314–19.

3. Ray Benson, "The President's Committee on Wildlife Restoration," *Literary Digest*, January 27, 1934.

4. Memorandum—FDR to Henry A. Wallace, Secretary of Agriculture, October 18, 1933, quoted in Allen, *Guardian of the Wild*, p. 26.

5. Lendt, *Ding*, pp. 6–12.

6. Jay Norwood Darling to Paul Errington, January 4, 1958, Darling Papers, Special Collections Department, University of Iowa Library, Iowa City.

7. Jay Norwood Darling to Juanita Lines, February 9, 1959, Darling Papers, Special Collections Department, University of Iowa Library, Iowa City.

8. Harber, "How You Gonna Keep 'Em Down on the Farm?" p. 294.

9. John M. Henry, "A Treasury of Ding," *Palimpsest*, Vol. 53 (March 1972), pp. 82–83.

10. Lendt, *Ding*, pp. 48–50.

11. Harber, "How You Gonna Keep 'Em Down on the Farm?," p. 294.

12. Eric Jay Dolin and Bob Dumaine, *The Duck Stamp Story: Art, Conservation, History* (Iola, WI: Krause Publications, 2000), p. 31.

13. Vernon Van Ness, "Rod and Gun," *New York Times*, January 23, 1934.

14. FDR to Jay N. Darling, May 9, 1935, FDRL.

15. Aldo Leopold, "Thinking Like a Mountain," in *A Sand County Almanac* (New York: Oxford University Press, 1949).

16. Ibid., p. 130.

17. Meine, *Aldo Leopold*, pp. 107–9.

18. Ibid., p. 116.

19. Aldo Leopold, *A Sand County Almanac with Essays on Conservation from Round River* (New York: Ballantine, 1990), p. 13.

20. Dolin, *Smithsonian Book of National Wildlife Refuges*, p. 87.

21. Aldo Leopold, *Game Management* (New York: Scribner, 1933; reprinted, Madison: University of Wisconsin Press, 1986).

22. Leopold, *A Sand County Almanac*, Foreword, p. xix.

23. Meine, *Aldo Leopold*, p. 319.

24. Jennifer Kobylecky, "Connecting the Starker Leopold's Legacy," *Leopold Outlook*, Vol. 12, no. 2 (Winter 2012), pp. 24–25.

25. Marc Cioc, *The Game of Conservation: International Treaties to Protect the World's Migratory Animals* (Athens: Ohio University Press, 2009), pp. 97–99.

26. Civilian Conservation Corps, *The CCC and Wildlife* (Washington, DC: U.S. Government Printing Office, 1938).

27. " 'Duck for Every Puddle' Goal in Game Restoration," *New York Times*, January 7, 1934.

28. Meine, *Aldo Leopold*, p. 315.

29. Worster, *The Dust Bowl*, p. 12.

30. Van Ness, "Rod and Gun."

31. Ding Darling to Clarence Cottam, June 25, 1959, Ding Darling Papers, Special Collections Department, University of Iowa Library, Iowa City.

32. Ibid.

33. Jay N. Darling, "The Story of the Wildlife Refuge Program, Part 1," *National Parks Magazine* (February–March 1954), pp. 6–10, 43–46.

34. Thomas Beck quoted in Meine, *Aldo Leopold*, p. 316.

35. Aldo Leopold quoted ibid., pp. 316–17.

36. Luna Leopold quoted ibid.

37. "Wild-Life Project Calls for U.S. Aid," *New York Times*, January 24, 1934.

38. Report of President's Committee on Wildlife Restoration, February 8, 1934, U.S. Fish and Wildlife Archive, Shepherdstown, WV.

39. Allen, *Guardian of the Wild*, p. 26.

40. Van Ness, "Rod and Gun."

41. "Montezuma Wetlands Complex," Friends of Montezuma Wetlands Complex, http://friendsofmontezuma.org/index.html.

42. Dolin, *Smithsonian Book of National Wildlife Refuges*, p. 88.

43. Giese, "A Federal Foundation for Wildlife Conservation," p. 139.

44. "The Reminiscences of Henry Agard Wallace," Columbia University oral history, p. 320.

45. Ding Darling to Clarence Cottam, June 25, 1959, Ding Darling Papers, Special Collections Department, University of Iowa Library, Iowa City.

46. Dolin and Dumaine, *The Duck Stamp Story*, p. 43.

47. Allen, *Guardian of the Wild*, p. 29.

48. Dolin, *Smithsonian Book of National Wildlife Refuges*, pp. 87–91.

49. Albert M. Day, "Don't Make My Duck Pond a Refuge," *Sports Afield* (February 1946).

50. Dolin and Dumaine, *The Duck Stamp Story*, p. 47.

51. Ibid., p. 48.

52. Dolin, *Smithsonian Book of National Wildlife Refuges*, p. 90.

53. George Laycock, *The Sign of the Flying Goose: A Guide to the National Wildlife Refuges* (Garden City, NY: American Museum of Natural History Press, 1965), p. 224.

54. FDR to Henry Wallace, May 24, 1934, FDRL.

55. Lendt, *Ding*, pp. 74–76. See also Frank J. Rader, "Harry L. Hopkins, the Ambitious Crusader," *Annals of Iowa*, Vol. 44 (Fall 1977), pp. 85–102.

56. *Duck Stamps and Wildlife Refuges* (Washington, DC: U.S. Government Printing Office, 1955), p. 5.

57. More Birds in America, *Small Refuges for Waterfowl* (New York: More Birds in America, 1933), pp. 35–36.

58. Giese, "A Federal Foundation for Wildlife Conservation," pp. 133–43.

59. Worster, *The Dust Bowl*, p. 13.

60. Timothy Egan, *The Worst Hard Time*, p. 7.

61. Ian Frazier, *Great Plains* (New York: Farrar, Straus and Giroux, 1989), p. 197.

62. Russell Lord, *Miscellaneous Publication No. 321: To Hold This Soil* (Washington, DC: Soil Conservation Service, 1938), pp. 72–73.

63. Craig Rupp, "History of National Grasslands," paper presented by Deputy Regional Forester Craig Rupp at National Grasslands Conference, December 10, 1975, Arlington, TX, p. 2.

64. Cutler, *The Public Landscape of the New Deal*, p. 107.

65. Wilmon H. Droze, *Trees, Prairies, and People: A History of Tree Planting in the Plain States* (Denton: Texas Woman's University, 1977), p. xxii.

66. FDR, February 10, 1937. See also Nixon, *Franklin D. Roosevelt and Conservation*, Vol. 2, pp. 3–6.

67. F. A. Silcox quoted in R. Douglas Hurt, "Forestry on the Great Plains, 1902–1942," Lecture at Kansas State University, July 1995. Online at http://www-personal.ksu.edu/~jsherow/hurt2.htm.

68. Nicolet National Forest, established by Herbert Hoover on March 2, 1933, in Presidential Proclamation 2036, 47 Stat. 2561; Chequamegon National Forest, established by Franklin Delano Roosevelt on November 13, 1933, in Presidential Proclamation 2061, 48 Stat. 1716.

69. Raphael Zon to Ovid Butler, July 18, 1934.

70. Cutler, *The Public Landscape of the New Deal*, p. 108.

71. Droze, *Trees, Prairies, and People*, pp. 63–65.

72. Joel Orth, "The Shelterbelt Project: Cooperative Conservation in 1930s America," *Agricultural History*, Vol. 81, no. 3 (Summer 2007), pp. 333–55.

73. FDR to James V. Allred, February 23, 1937, FDRL.

74. *Amarillo Globe*, July 22, 1934.

75. Orth, "The Shelterbelt Project."

76. Ibid.

77. M. G. Kains, *Five Acres and Independence: A Practical Guide to the Selection and Management of the Small Farm* (New York: Greenberg, 1935), p. 52.

78. Russell Peterson, *The Pine Tree Book* (New York: Central Park Conservancy, 1980), pp. 116–17.

79. Raphael Zon, "Shelterbelts—Futile Dream or Workable Plan," *Science*, Vol. 81 (April 26, 1935), p. 394.

80. James B. Trefethen, *An American Crusade for Wildlife* (New York: Boone and Crockett Club, 1975), pp. 235–36.

81. Paul B. Sears, "The Great American Shelterbelt," *Ecology*, Vol. 17 (October 1936), pp. 683–84.

82. FDR to Henry Wallace, January 23, 1937, FDRL.

83. "Paul Henley Roberts 1890–1871," Nebraska State Historical Society, last updated July 2011, http://nebraskahistory.org/lib-arch/research/manuscripts/family/paul-roberts.htm.

84. Agricultural Department Appropriation Bill for 1936: Hearing Before the Subcommittee of House Committee on Appropriations, United States House of Representatives, 74th Congress (1936), pp. 320–21.

85. Trefethen, *An American Crusade for Wildlife*, pp. 234–35.

86. " 'Ding' Finds Capital a Great Buck-Passer," *New York Times*, April 18, 1935.

87. Ann Vileisis, *Discovering the Unknown Landscape: A History of America's Wetlands* (Washington, DC: Island, 1997), p. 179.

88. Laycock, *The Sign of the Flying Goose*, p. 227.

89. Ibid., pp. 225–27.

90. Des Lacs: FDR, Executive Order 7154-A, August 22, 1935. Lower Souris: FDR, Executive Order 7170, September 4, 1935. Upper Souris: FDR, Executive Order 7161, August 27, 1935.

91. Author interview with Denny Holland, March 31, 2013.

92. Lendt, *Ding*, p. 72.

93. W. L. McAtee, *Local Bird Refuges* (Washington, DC: U.S. Government Printing Office, 1942), pp. 1–17.

94. Peter Matthiessen, *Wildlife in America* (New York: Viking, 1959), p. 220.

95. Giese, "A Federal Foundation for Wildlife Conservation," pp. 142–45.

96. J. C. Salyer, "Practical Waterfowl Management," in *Wildlife Restoration and Conservation; Proceedings of the North American Wildlife Conference Called by President Franklin D. Roosevelt; Connecting Wing Auditorium and the Mayflower Hotel, Washington, DC, February 3–7, 1936* (Washington, DC: U.S. Government Printing Office, 1936), p. 584.

CHAPTER 12: "SOONER OR LATER, YOU ARE LIKELY TO MEET THE SIGN OF THE FLYING GOOSE"

1. Scott Slavik, "The Blue Goose: Mythical Creature or Enduring Symbol?" Refuge Notebook, *Peninsula Clarion*, November 12, 2004.

2. Vanez T. Wilson and Rachel L. Carson, *Bear River: A National Wildlife Refuge* (Washington, DC: Government Printing Office, 1950), p. 1.

3. E. A. McIlhenny, "The Blue Goose in Its Winter Home," *Auk*, Vol. 49, no. 3 (July 1934), pp. 279–306.

4. Ira N. Gabrielson, "The Problem of Duck Conservation," speech at Illinois Sportsmen's Association, Chicago, June 29, 1936, U.S. Fish and Wildlife Archive, Shepherdstown, WV.

5. Rachel Carson, Introduction to "Conservation in Action" series, U.S. Fish and Wildlife Service, 1947.

6. Upper Souris CCC Camp Boxes, NCTS-USFWS, Shepherdstown, WV.

7. "Delivering Hope: FDR and Stamps of the Great Depression," http://postalmuseum.si.edu/deliveringhope/object_0_209045_13.html#1.

8. Tracy Casselman, "Elizabeth Beard Losey, First Female Member of the Wildlife Society," *Wildlife Society Bulletin*, Vol. 34, no. 2 (2006), p. 558.

9. Mark Madison and George Gentry, interview with Elizabeth Losey, March 15, 2003, library.fws.gov/OH/losey.elizabeth.031503.pdf.

10. Franklin S. Henika, "Sand-Hill Cranes in Wisconsin and Other Lake States," in *Wildlife Restoration and Conservation; Proceedings of the North American Wildlife Conference called by President Franklin D. Roosevelt; Connecting Wing Auditorium and the Mayflower Hotel, Washington, DC, February 3–7, 1936* (Washington, DC: U.S. Government Printing Office, 1936).

11. Tracy Casselman, *Seney National Wildlife Refuge Comprehensive Conservation Plan, 2009* (Seney, MI: U.S. Fish and Wildlife Service, 2009).

12. FDR to Henry Wallace and Harold Ickes, February 8, 1935, FDRL.

13. FDR to Henry Wallace, February 8, 1935, FDRL.

14. FDR, Address to Congress, January 24, 1935, in Nixon, *Franklin D. Roosevelt and Conservation*, Vol. 1.

15. FDR, Speech to the Society of American Foresters, January 29, 1935, FDRL.

16. Ibid.

17. Ding Darling to FDR, "Confidential Memorandum for the President and the Secretary of Agriculture; Taylor Grazing Act and Wildlife," February 4, 1935, in Nixon, *Franklin D. Roosevelt and Conservation*, Vol. 1, p. 315.

18. Skip Farrington, *The Ducks Came Back: The Story of Ducks Unlimited* (New York: Coward-McCann, 1945), pp. 3–5.

19. FDR to Joseph Pulitzer, April 19, 1935, letter 346 in Nixon, *Franklin D. Roosevelt and Conservation*, Vol. 1.

20. Jay N. Darling, "Wildlife Areas and National Land Planning," in James, *American Planning and Civic Annual*, p. 38.

21. Trefethen, *An American Crusade for Wildlife*, p. 225.

22. Ibid., p. 269.

23. J. N. Darling to FDR, February 4, 1935, FDRL.

24. "Livestock Men Meet to Discuss Program," *Bend* (Oregon) *Bulletin*, January 13, 1936.

25. Carl D. Shoemaker to FDR, March 6, 1935, FDRL.

26. FDRto John H. Baker, May 29, 1935, FDRL.

27. Ding Darling to Henry Wallace, June 8, 1935, FDRL.

28. FDR to Henry Wallace, June 17, 1935, FDRL.

29. Jason Scott Smith, *Building New Deal Liberalism: The Political Economy of Public Works, 1935–1956* (New York: Cambridge University Press, 2006), pp. 102–3.

30. Cutler, *The Public Landscape of the New Deal*, pp. 10–22.

31. Darling, "Wildlife Areas and National Land Planning," pp. 38–40.

32. Stephen J. Pyne, *Fire in America: A Cultural History of Wildland and Rural Fire* (Princeton, NJ: Princeton University Press, 1992), p. 323.

33. J. N. Darling to FDR, July 26, 1935, FDRL.

34. Ibid.

35. FDR to J. N. Darling, July 29, 1935. FDRL.

36. USDA, BBS, *Report of the Chief of the Bureau of Biological Survey, 1935* (Washington, DC: U.S. GPO, 1935), p. 23.

37. Ira N. Gabrielson, *Wildlife Refuges* (New York: Macmillan, 1943), p. 182.

38. Tugwell quoted in Leuchtenberg, *Franklin D. Roosevelt and the New Deal*, p. 136.

39. FDR quoted in Pyne, *Fire in America*, p. 322.

40. FDR to Samuel N. Spring, June 6, 1934, FDRL.

41. Asbell, *The F.D.R. Memoirs*, p. 413.

42. Davis and Davis, *Our Mark on This Land*, p. 147.

43. FDR to Scott Lord Smith, June 11, 1935, FDRL.

44. Black, *Franklin Delano Roosevelt: Champion of Freedom*, p. 4.

45. Margaret Suckley to FDR, August 3, 1935, in Ward, *Closest Companion*, p. 28.

46. Margaret Suckley to FDR, August 20, 1935, ibid., pp. 30–31.

47. Eleanor Roosevelt, "My Day," June 6, 1938.

48. Joseph P. Lash, *Eleanor and Franklin: The Story of Their Relationship* (New York: W. W. Norton & Co., 1971), p. 305.

49. Lewis, "A CCC Guy's Diary," entry for Saturday April 13, 1935.

50. Worster, *The Dust Bowl*, p. 28.

51. Douglas Brinkley and Johnny Depp, Introduction, in Woody Guthrie, *House of Earth* (Los Angeles and New York: InfinitumNihil/HarperCollins, 2013), p. xx.

52. See R. Douglas Hurt, "Federal Land Reclamation in the Dust Bowl," *Great Plains Quarterly*, Vol. 6 (Spring 1986), pp. 94–106.

53. Woody Guthrie, *Seeds of Man* (New York: Dutton, 1976), p. 122.

54. Mike Cox, *Big Bend Tales* (Charleston, SC: History Press, 2011), pp. 136–37.

55. Jameson, *The Story of Big Bend National Park*, p. 30.

56. FDR to Congressman R. Ewing Thomason, August 9, 1939, FDRL.

57. Jameson, *The Story of Big Bend*, p. xiv.

58. Harold L. Ickes, *The Secret Diary*, Vol. 1, *The First Thousand Days* (New York: Simon and Schuster, 1953), pp. 385–86, Library of Congress, Manuscript Division, Administrative Files, Box 7.

59. William Finley to FDR, December 14, 1935.

60. Kathy Durbin, "Restoring a Refuge: Cows Depart, but Can Antelope Recover?" *High Country News*, November 24, 1997.

61. William O. Douglas, *My Wilderness: The Pacific West* (New York: Doubleday, 1960), p. 74.

62. *Business Week*, May 4, 1935.

63. FDR, "Remarks at the Celebration of the Fiftieth Anniversary of State Conservation at Lake Placid," September 14, 1935, Lake Placid, NY. Online at American Presidency Project, http://www.presidency.ucsb.edu/ws/?pid=14936.

64. FDR, "Remarks at the Dedication of the White Face Memorial Highway," September 14, 1935. Online at American Presidency Project, http://www.presidency.ucsb.edu/ws/?pid=14934.

65. Benton MacKaye, "Flankline vs. Skyline," *Appalachia*, Vol. 20 (1934), pp. 104–8. See also Sutter, *Driven Wild*, p. 187.

66. Stephen Fox, *The American Conservation Movement: John Muir and His Legacy* (Madison: University of Wisconsin Press, 1981), p. 210.

67. Harold Ickes, "Twelve Years with F. D. R.," *Saturday Evening Post* (July 24, 1948), pp. 88–90.

68. Jay Norwood Darling to John Clark Salyer, November 8, 1935, Darling Papers.

69. Harold Ickes to Jay Norwood Darling, November 13, 1935, Darling Papers.

70. J. Clark Salyer and Ira N. Gabrielson, "J. Norwood Darling, 1876–1962," *Journal of Wildlife Management*, Vol. 27 (July 1963), pp. 499–502.

71. Ira N. Gabrielson, "The Problem of Duck Conservation," speech at Illinois Sportsmen's Association, Chicago, June 29, 1936, U.S. Fish and Wildlife Archive, Shepherdstown, WV.

72. Ira N. Gabrielson, unpublished memoir, U.S. Fish and Wildlife Archive, Shepherdstown, WV.

73. Henry M. Reeves and David B. Marshall, "In Memoriam: Ira Noel Gabrielson," *Auk*, Vol. 102 (October 1981), pp. 865–68.

74. *Proceedings of the North American Wildlife Conference Called by Franklin D. Roosevelt* (Washington, DC: U.S. Government Printing Office, 1936), p. 4.

75. FDR, "Greeting to the North American Wildlife Conference," February 3, 1936. Online at American Presidency Project, http://www.presidency.ucsb.edu/ws/?pid=15195.

76. "1,000 Meet Today to Save Wild Life," *New York Times*, February 3, 1936.

77. Henry Wallace, Address to North American Wildlife Conference, February 3–7, 1936, Washington, DC.

78. "Wild Life Groups United in One Body," *New York Times*, February 6, 1936.

79. Jay N. Darling, Address to North American Wildlife Conference, February 3–7, 1936, Washington, DC.

80. Harold Ickes, Address to North American Wildlife Conference, February 3–7, 1936, Washington, DC.

81. "Refuge System Celebrates Anniversary: 75 Years and Going Strong," *Fish and Wildlife News* (December 1976–January 1979), pp. 14–15.

82. George Laycock, *Wild Refuge* (Garden City, NY: Natural History Press, 1969), pp. 9–11.

83. J. C. Salyer, "Practical Waterfowl Management," *Proceedings of the North American Wildlife Conference: February 3–7, 1936* (Washington, DC: Senate Committee Printing, 1936), pp. 584–98.

84. J. Michael Scott, "Recovery of Imperiled Species Under the Endangered Species Act: The Need for a New Approach," *Frontiers in Ecology and Environment*, Vol. 3 (2005).

85. Jon Mooallem, *Wild Ones: A Sometimes Dismaying, Weirdly Reassuring Story About Looking at People Looking at Animals in America* (New York: Penguin, 2013), pp. 3–4.

86. Eleanor Roosevelt, "My Day," April 20, 1945.

87. "Bighorn History" (Arizona Desert Bighorn Sheep Society, 2012), accessed March 20, 2013, at http://www.adbss.org/bighorn_history.html.

88. William T. Hornaday, *Campfires on Desert and Lava* (Tucson: University of Arizona Press, 1985, reprint); Charles Sheldon, *The Wilderness of Desert Bighorns and Seri Indians from the Southwestern Journals of Charles Sheldon* (Tucson: American Desert Sheep Society, 1979).

89. James K. Morgan, "Slamming the Ram into Oblivion," *Audubon* (November 1975), pp. 17–19.

90. See Ralph E. Welles and Florence B. Welles, *The Bighorn of Death Valley* (Washington, DC: U.S. Government Printing Office, 1961); and *The Status of Feral Burros and Wildlife Water Sources in Death Valley National Monument* (Washington, DC: U.S. Department of the Interior, National Park Service, 1967).

91. Giese, "A Federal Foundation for Wildlife Conservation," pp. 150–51. See also Laycock, *The Sign of the Flying Goose*, pp. 210–14.

92. Renée Corona Kolvet and Victoria Ford, *The Civilian Conservation Corps in Nevada: From Boys to Men* (Reno: University of Nevada Press, 2006), pp. 93–95.

93. FDR, Executive Order 7373, Establishing the Desert Game Range in Nevada, May 20, 1936. Online at American Presidency Project, http://www.presidency.ucsb.edu/ws/?pid=61181.

94. Laycock, *Sign of the Flying Goose*, pp. 211–12.

95. U.S. Fish and Wildlife Service, "Comprehensive Conservation Plan: Desert National Wildlife Refuge Complex," August 2009, p. S-10, http://www.fws.gov/uploadedFiles/CCP%20Summary.pdf.

96. Laycock, *Sign of the Flying Goose*, p. 217.

97. Mark Madison, "Shaping the NWRs: 100 Years of History," *Fish and Wildlife News* (Spring 2003).

CHAPTER 13: "WE ARE GOING TO CONSERVE SOIL, CONSERVE WATER, AND CONSERVE LIFE"

1. FDR quoted in Arthur M. Schlesinger Jr., *The Politics of Upheaval, 1935–1936* (Boston: Houghton Mifflin, 1960), pp. 571–72.

2. "Text of Gov. Landon's Address Declaring Constitution in Peril," *Milwaukee Journal*, October 14, 1936.

3. Zachary Wimmer, "Out of the Dust: The Civilian Conservation Corps in Kansas," MA thesis, Emporia State University, 2012, pp. 3–14.

4. R. Alton Lee, "The Civilian Conservation Corps in Kansas," *Journal of the West*, Vol. 44 (2005), pp. 69–73.

5. Wimmer, "Out of the Dust," pp. 43–46.

6. Shlaes, *The Forgotten Man*, p. 278.

7. Raymond G. Carroll, "Shelterbelt," *Saturday Evening Post*, Vol. 208, no. 23 (October 5, 1935), pp. 23, 81–83, 85–86.

8. Julie Courtwright, *Prairie Fire: A Great Plains History* (Lawrence: University Press of Kansas, 2011), pp. 173–75.

9. "Rabbit Drives, 1934: Kansas Emergency Relief Committee," Kansas Memory Project, https://www.youtube.com/watch?v=YDxvc-BuS5A.

10. Craig Maier, *Next Year Country: Dust to Dust in Western Kansas, 1890–1940* (Lawrence: University Press of Kansas, 2006), p. 274.

11. "Rabbit Drives, 1934."

12. Charles W. Hurd, "President Catches a Supper of Fish," *New York Times*, March 25, 1936, p. 23; Phil Scott, *Hemingway's Hurricane: The Great Florida Keys Storm of 1935* (Camden, ME: International Marine/McGraw-Hill, 2006), pp. 187–88; Galusha, *Another Day, Another Dollar*, p. 42.

13. Les Dropkin, "Cruising with the President: An Annotated Chronology of Franklin D. Roosevelt's Cruises During the Potomac Years" (Potomac Association, March 2001), pp. 4–5, http://www.usspotomac.org/education/documents/cruising_with_the_president.pdf.

14. William Seale, *The President's House: A History* (Washington, DC: White House Historical Association, 1986), p. 984.

15. Captain Wilson Brown, USN, Log of the President's Cruise: Bahamas Waters, March 26, 1935.

16. Bernd Brunner, *The Ocean at Home: An Illustrated History of the Aquarium* (New York: Princeton Architectural Press, 2005), p. 131.

17. Eleanor Roosevelt, "My Day," February 21, 1940.

18. "1,200 Acres of Primeval Florida to Be Preserved as National Park," *New York Times*, April 25, 1937.

19. John D. Stinson, "Historical Note," National Audubon Society Records 1883–1991, New York Public Library, March 1994.

20. FDR to John H. Baker, August 12, 1939, FDRL.

21. FDR, radio address, September 6, 1936.

22. FDR, campaign speech, September 10, 1936, Charlotte, NC.

23. Look and Perrault, *The Interior Building*, pp. 11–18.

24. FDR, Address at Dedication of the New Department of the Interior Building, Washington, DC, April 16, 1936. Online at American Presidency Project, http://www.presidency.ucsb.edu/ws/?pid=15281.

25. Ibid.

26. "New Deal Friends and Foes Satirized at Gridiron Dinner," *Washington Post*, April 19, 1936, p. M1.

27. Associated Press, "New Deal Burlesqued by Gridiron Club," April 19, 1936, quoted in Fenster, *FDR's Shadow: Louis Howe*, p. 10.

28. FDR, "Statement on Signing the Soil Conservation and Domestic Allotment Act," March 1, 1936. Online at American Presidency Project, http://www.presidency.ucsb.edu/ws/?pid=15254.

29. "The President Suggests Cooperation by Farmers in the Soil Conservation Program in Their Individual and National Interest. Presidential Statement. March 19, 1936," in Sam Rosenman, ed., *The Public Papers and Addresses of Franklin D. Roosevelt*, Vol. 5, *"The People Approve," 1936* (New York: Random House, 1938), p. 136.

30. FDR to Elbert D. Thomas, March 20, 1936, FDRL.

31. Kenneth T. Walsh, *Prisoners of the White House: The Isolation of America's Presidents and the Crisis of Leadership* (Boulder, CO: Paradigm Publishers, 2013), p. 84. See also Eleanor Roosevelt, "My Day," June 15, 1936, and October 13, 1936.

32. Eleanor Roosevelt, "My Day," July 10, 1936.

33. Eleanor Roosevelt, "My Day," December 11, 1936.

34. FDR, speech at George Rogers Clark Memorial, June 14, 1936, in Nixon, *Franklin D. Roosevelt and Conservation*, Vol. 1, p. 53.

35. Eleanor Roosevelt, "My Day," July 4, 1936.

36. FDR, Address at Dedication of Shenandoah National Park, July 3, 1936. Online at American Presidency Project, http://www.presidency.ucsb.edu/ws/?pid=15316.

37. John C. Paige, *The Civilian Conservation Corps and the National Park Service, 1933–1942* (Washington, DC: Department of the Interior, 1985).

38. FDR, Address at the Dedication of Shenandoah National Park, July 3, 1936.

39. Lary M. Dilsaver, *Joshua Tree National Park: A History of Preserving the Desert* (Twentynine Palms, CA: NPS, 2015), p. 77.

40. FDR, Presidential Proclamation 2193 (50 Stat. 1760), August 10, 1936, in Thomas Alan Sullivan, comp., *Proclamations and Orders Relating to the National Park Service Up to January 1, 1945* (Washington, DC: U.S. Government Printing Office, 1947), pp. 218–19.

41. Christopher Ketcham, "The Great Republican Land Heist," *Harper's Magazine*, Vol. 330, no. 1977 (February 2015), p. 25.

42. FDR, "Remarks at Mount Rushmore National Memorial," August 30, 1936. Online at American Presidency Project, http://www.presidency.ucsb.edu/ws/?pid=15109.

43. Henry A. Wallace to FDR, May 11, 1937, in Nixon, *Franklin Roosevelt and Conservation*, Vol. 2, pp. 57–58.

44. FDR to Henry A. Wallace, June 2, 1937, FDRL.

45. FDR, "Remarks at a Luncheon in Dallas, Texas," June 12, 1936. Online at American Presidency Project, http://www.presidency.ucsb.edu/ws/?pid=15304.

46. Harold K. Steen, ed., *The Conservation Diaries of Gifford Pinchot* (Durham, NC: Forest History Society, 2001), entry from November 5, 1936, p. 169.

47. See United States Bureau of Land Management, Montana State Office, "Management Development Plan for Fort Peck Game Range," https://archive.org/details/managementdevelo29unit; and FDR, Executive Order 7562, Establishing Sacramento Migratory Waterfowl Refuge, February 27, 1937. Online at American Presidency Project, http://www.presidency.ucsb.edu/ws/?pid=61215.

48. FDR to Hendrik Willem van Loon, February 2, 1937, FDRL.

49. FDR to New York Rod and Gun Editors Association, January 25, 1937, FDRL.

50. FDR to Hendrik Willem van Loon, February 7, 1937, FDRL.

51. Clarke, *Roosevelt's Warrior*, p. 215.

52. FDR, Second Inaugural Address, January 20, 1937.

53. FDR quoted in Stefan Bechtel, *Mr. Hornaday's War* (Boston: Beacon, 2012), p. 216.

54. Hornaday quoted ibid., p. 216.

55. "Funeral Tomorrow for Dr. Hornday," *New York Times*, March 8, 1937.

56. Lee Whittlesey, *Yellowstone Place Names* (Helena, MT: Historical Society Press, 1988), p. 105.

57. Clarke, *Roosevelt's Warrior*, p. 237.

58. Jeff Shesol, *Supreme Power: Franklin Roosevelt and the Supreme Court* (New York: W. W. Norton, 2010), p. 22.

59. William O. Douglas, *Of Men and Mountains* (New York: Harper, 1950), pp. 16–17.

60. Noah Feldman, *Scorpions: The Battles and Triumphs of FDR's Great Supreme Court Justices* (New York: Twelve, 2010), pp. 60–65.

61. Jordan A. Schwartz, *The New Dealers: Power Politics in the Age of Roosevelt* (New York: Knopf, 1993), p. 173.

62. Ibid., p. 174.

63. Rosenman, *Working with Roosevelt*, p. 150.

64. William O. Douglas, "Termites of High Finance," *Vital Speeches of the Day*, Vol. 3 (November 15, 1936), pp. 186–93.

65. James F. Simon, *Independent Journey: The Life of William O. Douglas* (New York: Harper & Row, 1980), pp. 192–94.

66. Ickes, *The Secret Diary*, Vol. 2, p. 34.

67. J. Joseph Huthmacher, *Trial by War and Depression: 1917–1941* (Boston: Allyn and Bacon, 1973), p. 153.

68. Roosevelt quoted in Neil Maher, *New Deal*, p. 201.

69. Quoted in Clarke, *Roosevelt's Warrior*, p. 232.

70. Douglas quoted in James M. O'Fallon, *Nature's Justice: Writings of William O. Douglas* (Corvallis: Oregon State University Press, 2000), p. 175.

71. Robert D. Brown, "The History of Wildlife Conservation and Research in the United States—and Implications for Its Future." In H. Li, ed., *Proceedings of the Taiwan Wildlife Association* (Taipei: Taiwan National University, 2007).

72. Ecological Society of America, "About the Ecological Society of America," accessed May 1, 2013, http://www.esa.org/esa/?page_id=91aboutesa/.

73. FDR to Henry Wallace and Harold L. Ickes, February 18, 1936.

74. Edward Hoagland, *Hoagland on Nature: Essays* (Guilford, CT: Lyon, 2003), p. 295.

75. Harold Ickes to FDR, February 21, 1936, FDRL.

76. Laycock, *The Sign of the Flying Goose*, p. 35.

77. FDR to Francis and Jean Harper, September 14, 1944, FDRL.

78. FDR to Rebe E. Healtey, May 15, 1936, FDRL.

79. Numan V. Bartley, *The Creation of Modern Georgia* (Athens: University of Georgia Press, 1980), pp. 172–73.

80. "History of Piedmont National Wildlife Refuge," Piedmont National Wildlife Refuge Profile, U.S. Fish and Wildlife Service, last updated March 23, 2010, accessed January 24, 2012, at http://www.fws.gov/piedmont/history.html.

81. FDR to Harold Ickes, April 5, 1937, FDRL.

82. FDR, Presidential Proclamation 2232, Organ Pipe Cactus National Monument, April 13, 1937. Online at American Presidency Project, http://www.presidency.ucsb.edu/ws/?pid=76672. See also Natt N. Dodge, *National Park Service Natural History Handbook No. 6: Organ Pipe Cactus National Monument* (Washington, DC: U.S. Government Printing Office, 1964), http://www.nps.gov/history/history/online_books/natural/6/index.htm.

83. FDR, Proclamation 2232; also see Dodge, *National Park Service Natural History Handbook*, No. 6.

84. Fitzhugh L. Minnigerode, "Park Named for Cactus," *New York Times*, May 1, 1938.

85. FDR, Proclamation 2232; also see Dodge, *National Park Service Natural History Handbook*, No. 6.

86. Gary Snyder, *Mountains and Rivers Without End* (Berkeley, CA: Counterpoint, 1996), pp. 127–29.

87. Edward Abbey, *Abbey's Road* (New York: Plume, 1991), pp. 154–55.

88. Lynn J. Rogers, "Organ Pipe Cactus National Monument Attracts Interest of Tourists," *Los Angeles Times*, January 9, 1938.

89. Doris Evans, *Saguaro National Park* (Tucson, AZ: Western National Parks Association, 2006), pp. 12–13. A spring 2011 issue of *Park Science*, the quarterly journal of the National Park Service, ran a photo from 1935 showing a spectacular field of saguaros beside a 1998 photo of the same area. The difference was shocking, and the culprit was climate change.

90. Perkins, *The Roosevelt I Knew*, pp. 136–37.

CHAPTER 14: "WHILE YOU'RE GITTIN', GIT-A-PLENTY"

1. FDR to Henry A. Wallace, April 20, 1937, FDRL.

2. "New Orleans Hails Roosevelt Again," *New York Times*, April 30, 1937, p. 3.

3. Louisiana Federal Writers' Project, *Louisiana: A Guide to the State* (New York: Hastings House), p. 335.

4. Paul A. Keddy, *Water, Earth, Fire: Louisiana's National Heritage* (Philadelphia, PA: Xlibris, 2008), p. 159.

5. David Welky, *The Thousand-Year Flood: The Ohio-Mississippi Disaster of 1937* (Chicago: University of Chicago Press, 2011), pp. 243–46.

6. Anna C. Burns, *A History of the Louisiana Forestry Commission*, Louisiana Studies Institute Monograph Series no. 1 (1968); "Henry E. Hardtner: Louisiana's First Conservationist," *Journal of Forest History*, Vol. 22 (April 1978), pp. 78–85; "Frank B. Williams, Cypress Lumber King," *Journal of Forest History*, Vol. 24 (July 1980), pp. 127–33; "Golden Anniversary Forest Edition," *Forest and People*, Vol. 13 (First Quarter, 1963).

7. Richard C. Davis, ed., *Encyclopedia of American Forest and Conservation History* (New York: Macmillan, 1983), pp. 363–66.

8. Thanks to the New Deal, Kisatchie National Forest roared back to life. By the time Jimmy Carter was president, the Kisatchie, under proper federal regulation, generated the most revenue per acre of all the forests in the South.

9. Stephen E. Ambrose and Douglas Brinkley, *The Mississippi and the Making of a Nation: From the Louisiana Purchase to Today* (Washington, DC: National Geographic, 2002), p. 22.

10. Ross T. McIntire, *White House Physician* (New York: Putnam, 1946).

11. "William Delano, Architect, Dead," *New York Times*, January 13, 1960.

12. Matthiessen, *Wildlife in America*, p. 261.

13. Aldo Leopold, "Marshland Elegy," in *A Sand County Almanac, and Sketches Here and There* (Oxford: Oxford University Press, 1989), p. 96.

14. Log of Official Trips of the President, entry for May 1, 1937, FDRL.

15. Schwartz, *The New Dealers*, p. 270.

16. Bill Mares, *Fishing with the Presidents: An Anecdotal History* (Mechanicsburg, PA: Stackpole, 1999), p. 176.

17. Paulette Langguth, "Happy 75th Birthday to USS *Potomac*," *Currents* (Fall 2011), pp. 1–2.

18. "May 4, 1937," *Franklin D. Roosevelt: Day by Day*, Pare Lorentz Center, FDRL. Online at http://www.fdrlibrary.marist.edu/daybyday/daylog/may-4th-1937/.

19. Mike Holmes, *Fishing the Texas Gulf Coast* (Guilford, CT: Lyons, 2009), pp. 11–12.

20. Log of Official Trips of the President, entry for May 7, 1937, FDRL.

21. William O. Douglas, *Farewell to Texas: A Vanishing Wilderness* (New York: McGraw-Hill, 1967), p. 224.

22. John F. Reiger, *Escaping into Nature: The Making of a Sportsman-Conservationist and Environmental Historian* (Corvallis: Oregon State University Press, 2013), p. 80.

23. Steve Harrington, "The Silver Kings," *Texas Monthly* (May 2013), pp. 118–19.

24. "President Roosevelt Lands a Tarpon in the Gulf of Mexico," *Life*, May 24, 1937, pp. 30–31.

25. Don Farley, "President Roosevelt as I Knew Him," undated typed manuscript, FDRL.

26. Log of Official Trips of the President, entry for May 10, 1937, FDRL.

27. "Texas Acclaims Roosevelt with 21-Gun Salute," *Washington Post*, May 12, 1937, p. 2.

28. Log of Official Trips of the President, entry for May 12, 1937.

29. Carter P. Smith, Foreword, in Cynthia Brandimarte, *Texas State Parks and the CCC* (College Station: Texas A&M Press, 2013), p. x.

30. Brian Cervantez, "Lone Star Booster: The Life of Amon Carter," MA thesis, University of North Texas, Denton, December 2011, pp. 97–135.

31. Clay Reynolds, *A Hundred Years of Heroes: A History of the Southwestern Exposition and Livestock Show* (Fort Worth: Texas Christian University Press, 1995), p. 177.

32. Brandimarte, *Texas State Parks and the CCC*, pp. 3–11.

33. Cornebise, *The CCC Chronicles*, p. 36.

34. Ibid., p. 206.

35. Dan K. Utley and James Wright Steely, *Guided with a Steady Hand: The Cultural Landscape of a Rural Texas Park* (Waco: Baylor University Press, 1998), p. 68.

36. Jerry Beth Shannon, Archivist of Ropesville Farm Project, to Douglas Brinkley, February 19, 2014. See also Pam Murtha, "Growing a Community: The Ropesville Resettlement Project," *Heritage*, Vol. 2 (2007), pp. 20–23.

37. Scott C. Yaich and Greg Kernohan, "Water, Ducks, and People," *Ducks Unlimited* (November–December 2013), p. 99.

38. Mares, *Fishing with the Presidents*, pp. 78–79.

39. James O. Stevenson, "Will Bugles Blow No More?" *Audubon* (May/June 1993), p. 134; Porter Allen, *On the Trail of Vanishing Birds* (New York: National Audubon Society), p. 35.

40. "Aransas National Wildlife Refuge," U.S. Fish and Wildlife Service, February 2005.

41. James O. Stevenson and Richard E. Griffith, "Winter Life of the Whooping Crane," *Condor*, Vol. 48, no. 4 (July–August 1946), pp. 160–78.

42. R. D. W. Connor, "FDR Visits the National Archives," *American Archivist*, Vol. 12, no. 4 (January–October 1949).

43. "It's My Baby!" *Prologue*, Vol. 44, no. 2 (Summer 2012).

44. Nan Robertson, "Memorial Is Dedicated to Roosevelt in Capital," *New York Times*, April 13, 1965.

45. FDR to Daniel Beard, July 28, 1937, FDRL.

46. Irving Brant, *Adventures in Conservation with Franklin D. Roosevelt* (Flagstaff, AZ: Northland, 1989), pp. 48–49.

47. Ibid.

48. Watkins, *Righteous Pilgrim*, pp. 380–81.

49. Hugh B. Autrey (Associate Recreational Planner) to George H. Copeland, *New York Times*, June 29, 1939, Records of National Park Service, Central Classification Files, 1936, Entry 81, Box 48, File Number 0–35, 1952.

50. "First National Seashore Park," *Washington Post*, August 6, 1939, p. 48.

51. Toll quoted in Cameron Binkley, *The Creation and Establishment of Cape Hatteras National Seashore—The Great Depression Through Mission 66* (Atlanta, GA: National Park Service, Southeastern Regional Office, 2007), p. 13.

52. Douglas Caldwell, "NPS Biographical Vignette: Conrad L. Wirth," last modified December 1, 2000, http://www.nps.gov/history/history/online_books/sontag/wirt h.htm.

53. "Park Service Gets New Plane," *New York Times*, September 16, 1936.

54. "House Flouts Roosevelt on CCC," *New York Times*, May 12, 1937.

55. Delaware Federal Writers' Project, *Delaware: A Guide to the First State* (New York: Viking Press, 1938), pp. 14-16.

56. Rachel Carson, *Parker River: A National Wildlife Refuge* (Washington, DC: U.S. Government Printing Office, 1947), p. 10.

57. Binkley, *The Creation and Establishment of Cape Hatteras National Seashore*, p. 1.

58. Hamilton Gray, "First Federal Beach Mapped: North Carolina 'Banks,' Including Cape," *New York Times*, September 5, 1937.

59. Binkley, *The Creation and Establishment of Cape Hatteras National Seashore.*

60. Gray, "First Federal Beach Mapped." On June 29, 1940, Congress changed the name from Cape Hatteras National Seashore to Cape Hatteras National Seashore Recreational Area.

61. Pauline Chase-Harrell et al., *Administrative History of the Salem Maritime National Historic Site* (Boston: Boston Affiliates, 1993), pp. 3–44.

62. Ickes, *The Secret Diary*, pp. 199–204.

63. Bankhead-Jones Farm Tenant Act of 1937 (Public Law 75-210), July 22, 1937, p. 12.

64. R. Douglas Hurt, "The National Grasslands Origin and Development in the Dust Bowl," *Agricultural History*, Vol. 59 (April 1985), pp. 246–59.

65. "Pittman an Enemy of Dictatorship," *New York Times*, November 11, 1940.

66. Lonnie L. Williamson, "Evolution of a Landmark Law," in Harmon Kallman, ed., *Restoring America's Wildlife, 1937–1987: The First Fifty Years of the Federal Aid in Wildlife Restoration (Pittman-Robertson Act)* (Washington, DC: U.S. Department of the Interior, 1987), pp. 1–20.

67. U.S. Department of Agriculture, Bureau of Biological Survey, *Report of the Chief of the Bureau of Biological Survey* (Washington, DC: U.S. Government Printing Office, 1938), pp. 26–42.

68. Trefethen, *An American Crusade for Wildlife*, p. 409.

69. Walcott quoted in Nixon, *Franklin D. Roosevelt and Conservation*, Vol. 2, pp. 122–23.

70. FDR to Frederic C. Walcott, September 8, 1937, FDRL.

71. Eric Bolan, *Wildlife Ecology and Management* (Upper Saddle River, NJ: Prentice Hall, 2003), pp. 201–24.

72. FDR to A. Willis Robertson, August 23, 1937, FDRL.

73. "Gibraltar of America, Now National Monument," *New York Times*, January 22, 1935, p. 3.

74. Quoted in Barry Mackintosh, "Harold L. Ickes and the National Park Service," *Journal of Forest History*, Vol. 29, no. 2 (April 1985), pp. 78–84.

75. Robert H. Jackson, *That Man: An Insider's Portrait of Franklin D. Roosevelt* (New York: Oxford University Press, 2003).

76. Ibid., pp. 147–48.

77. See Executive Order 7764, December 6, 1937, and 2 FR 3183, December 9, 1937; Executive Order 7780, December 30, 1937; and Executive Order 7784, December 31, 1937, in Franklin Delano Roosevelt Executive Order Disposition Tables, 1937, National Archives, http://www.archives.gov/federal-register/executive-orders/1937.ht ml. In 2004 Roosevelt's Sabine and Lacassine National Wildlife Refuges became part of a larger Southwest Louisiana National Wildlife complex that also included Cameron Prairie and Shell Keys.

78. Ted Williams, "Taking a Stand," *Audubon* (July–August 2013), pp. 26–34.

79. John O. Reilly, "Wildlife Abounds in Aransas Refuge," *Washington Post*, December 12, 1948.

CHAPTER 15: "I HOPE THE SON-OF-A-BITCH WHO LOGGED THAT IS ROASTING IN HELL"

1. See Ward, *Closest Companion*.

2. FDR, "Speech to the Roosevelt Home Club, Val-Kill Farm, Hyde Park," September 11, 1937, in Nixon, *Franklin D. Roosevelt and Conservation*, Vol. 2, pp. 125–26.

3. Maher, *Nature's New Deal*, p. 22.

4. FDR to Nelson C. Brown, June 29, 1937, FDRL.

5. Nelson C. Brown to FDR, May 22, 1937, FDRL.

6. Nixon, *Franklin D. Roosevelt and Conservation*, Vol. 2, pp. 25–26.

7. William Atherton DuPuy, "Forest Service Growing Thousands of Seedlings," *Washington Post*, February 27, 1938.

8. Elmo R. Richardson, "Olympic National Park: 20 Years of Controversy," *Forest History*, Vol. 12, no. 1 (April 1968), pp. 6–15.

9. Smith quoted in Susan Ware, *Beyond Suffrage: Women in the New Deal* (Cambridge, MA: Harvard University Press, 1981), p. 114.

10. James G. K. McClure to FDR, March 3, 1937, FDRL.

11. FDR to James G. K. McClure, March 11, 1937, FDRL.

12. Maher, *Nature's New Deal*, p. 22.

13. Elmo R. Richardson, "Olympic National Park: 20 Years of Controversy," *Forest History*, Vol. 12, no. 1 (April 1968), p. 7.

14. Doug Scott, *The Enduring Wilderness* (Golden, CO: Fulcrum, 2004), pp. 44–47.

15. FDR to Nelson C. Brown, June 29, 1937, FDRL.

16. "Roosevelt Retains Popularity in Northwest, but Voters Still Oppose His Court Plan," *Washington Post*, September 26, 1938, p. B4. The *Washington Post* published a map of the United States showing the many communities the president planned to visit.

17. Eleanor Roosevelt, "My Day," September 25, 1937.

18. "Workers Start Razing Old Mammoth Hot Springs Hotel," *Helena Independent*, August 20, 1936; and Cutler, *The Public Landscape of the New Deal*, pp. 90–91.

19. Robert V. Goss, "Yellowstone Hotels and Lodges: Mammoth Hot Springs National Hotel and Mammoth Lodge," accessed June 29, 2014, http://geyserbob.org/hot-mhs.html.

20. Emergency Conservation Work (Yellowstone), "Memorandum for the Press, Release for Sunday Papers," March 18, 1936, CCC Press Releases, FDRL.

21. Cutler, *The Public Landscape of the New Deal*, pp. 90–91.

22. Franklyn Waltman, "West Puzzles as Roosevelt Hides Trip's Real Purpose," *New York Times*, September 26, 1937, p. 1.

23. Eleanor Roosevelt, "My Day," September 27, 1937.

24. Ibid.

25. Eleanor Roosevelt, "My Day," September 28, 1937.

26. Ken Robinson, *Defending Idaho's Natural Heritage* (privately published), pp. 90–114.

27. Calvin Gower, "The CCC Indian Division: Aid for Depressed Americans, 1933–1942," *Minnesota History* (Spring 1972), pp. 1–13.

28. Nick Taylor, *American Made: The Enduring Legacy of the WPA* (New York: Bantam, 2008), p. 352.

29. FDR, Remarks at Boise, ID, September 27, 1937.

30. Mike McKinley, "The Civilian Conservation Corps and Heyburn State Park Experience, 1934–1942," CCC in Idaho, http://idahoptv.org/outdoors/shows/ccc/idaho/cccheyburn.html.

31. Elizabeth Smith, *A History of the Salmon National Forest* (Washington DC: U.S. Forest Service, 1973), n.p.

32. Ibid.

33. Kenneth Hart quoted in McKinley, "The Civilian Conservation Corps and Heyburn State Park Experience, 1934–1942."

34. Vardis Fisher and Idaho Federal Writers' Project, *Idaho: A Guide in Word and Picture* (Caldwell, ID: Caxton, 1937).

35. Elers Koch, "The Passing of the Lolo Trail," *Journal of Forestry* (February 1935), pp. 98–104.

36. FDR, Remarks at Boise, ID, September 27, 1937. Eleanor Roosevelt, in coming months, traveled to Moscow, ID, to plant a Douglas fir in the Presidential Grove in front of the University of Idaho's administration building. University of Idaho Special Collections Library Assistant Julie Monroe to Douglas Brinkley, October 7, 2013.

37. Henrietta Nesbitt, *White House Diary* (New York: Doubleday, 1948), p. 71.

38. Eleanor Roosevelt, "My Day," January 23, 1937.

39. FDR to R. Walton Moore, November 21, 1937, FDRL.

40. FDR to Cordell Hull, November 22, 1937, in Elliott Roosevelt, ed., *F.D.R.: His Personal Letters*, Vol. 1, p. 728.

41. Eleanor Roosevelt, "My Day," September 29, 1937.

42. John Harrison, "Grand Coulee Dam: History and Purpose," Northwest Power and Conservation Council, last modified October 31, 2008, accessed May 21, 2013, http://www.nwcouncil.org/history/grandcouleehistory.

43. Victoria Grieve, *The Federal Art Project and the Creation of Middlebrow Culture* (Urbana: University of Illinois Press, 2009), p. 138.

44. FDR, Address at Timberline Lodge, September 28, 1937, in Sam Rosenman, ed., *The Public Papers and Addresses of Franklin D. Roosevelt* (New York: Macmillan, 1941), p. 392.

45. *The Pony Express 150th Anniversary Year* (Salt Lake City, UT: National Park Service, 2011), pp. 20–21.

46. Otis, *The Forest Service and the Civilian Conservation Corps*, pp. 68–69.

47. Richardson, "Olympic National Park: 20 Years of Controversy," pp. 6–15.

48. Doug Scott to Douglas Brinkley, October 9, 2014.

49. Scott, *The Enduring Wilderness*, p. 46.

50. Clark N. Bainbridge, "The Origins of Rosalie Edge's Emergency Conservation Committee, 1930–1962: A Historical Analysis," PhD dissertation, University of Idaho, December 2002, pp. 152–53.

51. Runte, *National Parks*, p. 127.

52. Brant, *Adventures in Conservation with Franklin D. Roosevelt*, p. 34.

53. "Biographical Note," finding aid for Irving Brant Papers, University of Iowa Libraries; "Biographical Note," finding aid for Irving Brant Papers, Library of Congress, Washington, DC.

54. "Irving Brant, 91, Writer, Expert on Constitution," *Washington Post*, September 21, 1976.

55. Brant, *Adventures in Conservation with Franklin D. Roosevelt*, pp. 2–3.

56. Bainbridge, "The Origins of Rosalie Edge's Emergency Conservation Committee, 1930–1962," pp. 158–62.

57. Irving Brant, "Protection for Petrified Forests," *Saturday Evening Post* (June 26, 1926).

58. Bainbridge, "The Origins of Rosalie Edge's Emergency Conservation Committee, 1930–1962," pp. 166–70.

59. FDR to Albert Z. Gray, n.d., FDRL.

60. Clifford Roloff and Edwin Roloff, "The Mount Olympus National Monument," *Washington Historical Quarterly*, Vol. 25, no. 3 (July 1934), pp. 214–28.

61. Homer Cummings to FDR, May 22, 1936, FDRL.

62. FDR to Henry Wallace, February 18, 1936, FDRL.

63. Ibid.

64. Henry Wallace to FDR, March 13, 1936, FDRL.

65. FDR, Executive Order 7716, September 29, 1937, Franklin D. Roosevelt Day by Day Project, entry for September 30, 1937, http://www.fdrlibrary.marist.edu/daybyday/daylog/september-30th-1937/.

66. Mary Lou Hanify, "Roosevelt Tours Olympic Peninsula," *Seattle Times Magazine*, July 14, 1968.

67. Ibid.

68. Ibid.

69. Owen A. Tomlinson to Arno B. Cammerer, October 6, 1937, RG 79, Central Classified Files, NARA, Washington, DC.

70. Richardson, "Olympic National Park: 20 Years of Controversy."

71. Irving Brant to A. E. Demaray, October 5, 1937, Record Group 79, Central Classified Files, National Archives, Washington, DC.

72. FDR, "Remarks at Grand Coulee Dam," October 2, 1937.

73. William O. Douglas, *Go East, Young Man: The Early Years—The Autobiography of William O. Douglas* (New York: Random House, 1974), p. 211.

74. National Park Service, "Places: Lake Roosevelt National Recreation Area," last updated December 10, 2013, http://www.nps.gov/laro/historyculture/old-kettle-falls.htm.

75. Aldo Leopold, "Engineering and Conservation," in Susan L. Flader and J. Baird Callicott, eds., *The River of the Mother of God* (Madison: University of Wisconsin Press, 1991), p. 254.

76. Ed Cray, *Ramblin' Man: The Life and Times of Woody Guthrie* (New York: W. W. Norton, 2004), p. 209.

77. FDR, speech, October 4, 1937, Saint Paul, MN, FDRL, Franklin D. Roosevelt, Master Speech File, 1898–1945, Ser. 2, Box 2–7.

78. Donald L. Parman, "The Indian and the Civilian Conservation Corps," *Pacific Historical Review*, Vol. 40, no. 1 (February 1971), pp. 39–56.

79. Robert Dallek, *Franklin D. Roosevelt and American Foreign Policy* (New York: Oxford University Press, 1979), p. 148.

80. FDR, "Quarantine Speech," October 5, 1937, Chicago, IL, http://millercenter.org/president/speeches/speech-3310.

81. Earl Swift, *The Big Roads: The Untold Story of the Engineers, Visionaries, and Trailblazers Who Created the American Superhighway* (New York: Houghton Mifflin Harcourt, 2011), p. 7.

82. FDR, Press Conference, October 26, 1937, FDRL.

83. Richardson, "Olympic National Park: 20 Years of Controversy."

84. FDR to Harold Ickes, December 10, 1938, FDRL.

85. Harold Ickes, diary entry, August 6, 1938, Library of Congress, Washington, DC.

86. "Last Frontier Now Olympic National Park; President Signs Bill Saving Coast Wilds," *New York Times*, June 30, 1938; Richardson, "Olympic National Park: 20 Year Controversy," pp. 6–15.

CHAPTER 16: "PERPETUATED FOR POSTERITY"

1. FDR, "Annual Message to Congress," January 3, 1938. Online at American Presidency Project, http://www.presidency.ucsb.edu/ws/?pid=15517.

2. Nixon, *Franklin D. Roosevelt and Conservation*, Vol. 2, pp. 176–77.

3. FDR to Representative J. Mark Wilcox, January 22, 1938, FDRL.

4. "Department of the Interior, Fish and Wildlife Service: Lower Florida Keys Refuges, Monroe County, FL," *Federal Register*, Vol. 73, no. 101 (May 23, 2008), http://www.gpo.gov/fdsys/pkg/FR-2008-05-23/html/E8-11617.htm.

5. United States Department of the Interior, Fish and Wildlife Service; State of Florida Department of Natural Resources, "Management Agreement for Backcountry Portions of Key West National Wildlife Refuge Great White Heron National Wildlife Refuge and National Key Deer Refuge Monroe County, Florida," September 1992, p. 2, http://floridakeys.noaa.gov/review/documents/backcountryplan.pdf.

6. T. D. A. Cockerell, "Recollections of a Naturalist: The California Islands," *Bios*, Vol. 10, no. 2 (May 1939), pp. 99–106.

7. Harold Ickes, "Should Congress Vest Ownership of the Tidelands in the States?" *Congressional Digest* (October 1, 1948), p. 255.

8. T. D. A. Cockerell, "The Botany of the California Islands," *Torreya*, Vol. 37, no. 6 (1937), pp. 117–23.

9. Richard M. Bond, "Banding Records of California Brown Pelicans," *Condor*, Vol. 44 (May 1942), p. 116.

10. FDR, Proclamation 2281, 52 Stat. 1541, April 26, 1938, quoted in National Park Service, "Establishing Channel Islands National Park," http://www.nps.gov/chis/learn/historyculture/park-history.htm.

11. Eivind T. Scoyen to Arno B. Cammerer, May 7, 1941.

12. Harry S. Truman, Proclamation 2825, Enlarging the Channel Islands National Monument, California, February 9, 1949. Online at American Presidency Project, http://www.presidency.ucsb.edu/ws/?pid=87189.

13. Jimmy Carter, National Parks and Recreation Act Amendments Statement on Signing H.R. 3757 into Law, March 5, 1980. Online at American Presidency Project, http://www.presidency.ucsb.edu/ws/?pid=33104. Also Jimmy Carter, Channel Islands Marine Sanctuary Statement by the President, September 21, 1980. Online at American Presidency Project, http://www.presidency.ucsb.edu/ws/?pid=45099.

14. Cape Meares National Wildlife Refuge, Executive Order 7957, August 19, 1938.

15. Kevin Hays, "History of Oregon's Shortest Lighthouse and the Legend of the Octopus Tree," *Salem News*, January 30, 2007.

16. U.S. Fish and Wildlife Service, "Tybee National Wildlife Refuge," http://www.fws.gov/refuges/profiles/index.cfm?id=41624.

17. Stegner, *Where the Bluebird Sings at the Lemonade Springs*, pp. 89–92.

18. "Ickes Will Seek Purification of Potomac River," *Washington Post*, July 24, 1938, p. M6.

19. "$160,000 Will Go to Purify Potomac," *Washington Post*, October 13, 1933, p. X28.

20. FDR, "Address at Oklahoma City, Oklahoma," July 9, 1938. Online at American Presidency Project, http://www.presidency.ucsb.edu/ws/?pid=15675.

21. Barbara Bryant, " 'Rivers of America': Library Celebrates 60th Anniversary of Landmark Series," *Library of Congress Information Bulletin*, Vol. 56, no. 10 (June 1997). Each title in the series, which began in 1937 with the publication of *Kennebec: Cradle of Americans* by Robert P. Tristram Coffin and ultimately produced sixty-five volumes by 1974, was written from a literary—rather than scientific or historical—perspective and contained illustrations from top talents in the field. Carmer's volume *The Hudson* (illustrated by Stow Wengenroth) was published in 1939 and was later joined by such popular entries as *The Brandywine* (by Henry Seidel Canby with illustrations by Andrew Wyeth), *The Sangamon* (by Edgar Lee Masters with illustrations by Lynd Ward), and *The Everglades: River of Grass* (by Marjory Stoneman Douglas with illustrations by Robert Fink).

22. Carl Carmer, *The Hudson* (New York: Farrar and Rinehart, 1939), pp. 362–65; chap. 32 is all about Carmer and FDR's confab of 1938 on the Hudson River.

23. Thomas Lask, "Carl Carmer, Novelist, Historian of Upstate New York, Dead at 82," *New York Times*, September 12, 1976.

24. Musso, *FDR and the Post Office*, p. 62.

25. FDR to Admiral C. J. Peoples, February 12, 1938.

26. Eleanor Roosevelt Papers Project, "Teaching Eleanor Roosevelt Glossary: Federal Project Number One," accessed October 9, 2014, http://www.gwu.edu/~erpapers/teachinger/glossary/federal-proj ect-number1.cfm. See also Don Adams and Arlene Goldbard, *New Deal Programs: Experiments in Cultural Democracy* (1995), http://www.wwcd.org/policy/US/newdeal.html#FEDONE. The list of painters employed by the Federal Art Project reads like a who's who of modern art; Jacob Lawrence, Jackson Pollock, and Mark Rothko all produced work for the government during the Depression. The Federal Writers' Project, publisher of the American Guide Series of state monographs that FDR so loved, ultimately released 800 distinct titles by a talented and diverse roster of writers that included Zora Neale Hurston, John Cheever, Ralph Ellison, and Studs Terkel.

27. USDA Forest Service, "Establishment and Modification of National Forest Boundaries: A Chronologic Record, 1891–1973" (Washington, DC: USDA Division of Engineering, 1973).

28. FDR to Daniel W. Bell, February 24, 1938, FDRL.

29. FDR to Louis B. DeKoven, June 8, 1938, FDRL.

30. Harold L. Ickes, speech to Northwest Conservation League, June 29, 1938, Ickes Papers, Library of Congress, Washington, DC.

31. Ibid.

32. Newton, *Design on the Land*, p. 544.

33. Schrepfer, *The Fight to Save the Redwoods*, p. 61.

34. Turner, *Sierra Club: 100 Years of Protecting Nature*, p. 124.

35. Tom Turner, *The Making of the Environmental Movement* (Berkeley: University of California Press, 2015), p. 32.

36. Mary Street Alinder, *Ansel Adams: A Biography* (New York: Henry Holt, 1996), pp. 105–6.

37. A. E. Demarcy to Ansel Adams, c. January 8, 1939, quoted in William A. Turnage, Introduction, in Ansel Adams, *Sierra Nevada—The John Muir Trail* (New York: Little, Brown, 2006), p. xvii.

38. Author interview with Michael Adams, April 8, 2015.

39. Alinder, *Ansel Adams*, p. 106.

40. James Glover, *A Wilderness Original* (Seattle: The Mountaineers, 1986), p. 234.

41. Black, *Franklin Delano Roosevelt: Champion of Freedom*, p. 442.

42. Waldo Schmitt, "The President's Cruise of 1938," Journal Notes, September 1938, Smithsonian Archive, Washington, DC.

43. Cordell Hull to Harold Ickes, October 28, 1933, National Park Service Files, Ickes Office Files (Record Group 48), NARA, Washington, DC.

44. Richard E. Blackwelder, *The Zest for Life, or Waldo Had a Pretty Good Run: The Life of Waldo LaSalle Schmitt* (Lawrence: University Press of Kansas, 1979), p. 123.

45. Biographical note, finding aid for Waldo Schmitt Papers, Smithsonian Institution, Washington, DC, http://siarchives.si.edu/collections/siris_arc_217388.

46. Mares, *Fishing with the Presidents*, p. 79.

47. Box 94, Schmitt Papers, Smithsonian Archive, Washington, DC.

48. Waldo Schmitt to Commander C. A. Bailey, September 9, 1938, Schmitt Papers, Smithsonian Archive, Washington, DC.

49. Waldo Schmitt, "The Presidential Cruise of 1938," unpublished essay, n.d., Schmitt Papers, Smithsonian Archive, Washington, DC.

50. See Wallace J. Nicholas, *Blue Mind* (Boston: Little, Brown, 2014); and Nicola Joyce, "Book Review: 'Blue Mind,' on the Benefits of Being Near Water, by Wallace J. Nichols," *Washington Post*, August 8, 2014.

51. U.S. Naval History and Heritage Command, photo NH 93163.

52. William K. Vanderbilt, *To Galápagos on the Ara, 1926* (Mount Vernon, NY: privately printed, 1927).

53. FDR, *Public Papers of the United States*, Vol. 7 (Washington, DC: Office of the Federal Register, 1999), p. 503.

54. FDR, *The Public Papers and Addresses of Franklin D. Roosevelt, 1938: The Continuing Struggle for Liberalism*, ed. Sam Rosenman (New York: Macmillan, 1941).

55. Mares, *Fishing with the Presidents*, p. 79.

56. Eleanor Roosevelt, "My Day," July 8, 1938.

57. "Roosevelt in Pacific Lands 45-Pound Tuna," *New York Times*, July 29, 1938.

58. FDR to Eleanor Roosevelt, July 24, 1938, FDRL.

59. Ibid.

60. FDR to Eleanor Roosevelt, July 31, 1938, FDRL.

61. FDR, Address to the Roosevelt Home Club, August 27, 1938, Hyde Park, NY.

62. "National Chiefs to Dedicate Span," *New York Times*, August 14, 1938.

63. Ibid.; and "President and Premier to Inaugurate Officially New Thousand Islands Bridge," *Ottawa Citizen*, August 17, 1938.

64. FDR, Dedication Speech of Thousand Islands Bridge near Clayton, NY, August 18, 1938.

65. Blackwelder, *The Zest for Life*, p. 123.

66. William Rigdon, *White House Sailor* (Garden City, NY: Doubleday, 1962), pp. 60–62.

67. "Scientist Roosevelt's Catch to Enrich National Museum," *Washington Post*, August 13, 1938.

68. Ibid.

69. Schmitt, "The President's Cruise of 1938."

70. Waldo L. Schmitt to Franklin D. Roosevelt, October 29, 1938, Schmitt Papers, Smithsonian Archive, Washington, DC.

71. FDR, Address to the Roosevelt Home Club, August 27, 1938.

72. Jean Sherwood Harper to Eleanor Roosevelt, January 10, 1938, Eleanor Roosevelt Papers, in Nixon, ed., *Franklin D. Roosevelt and Conservation*, Vol. 2, pp. 166–67.

73. Ibid.

74. Gabrielson, *Wildlife Refuges*, p. 17.

75. W. F. Kubichek, "The CCC Rehabilitates Waterfowl Habitat," in *Proceedings of the North American Wildlife Conference at the Mayflower Hotel, Washington, DC* (February 7, 1936).

76. Gabrielson, *Wildlife Refuges*, p. 120.

77. Author interview with Lee M. Brown, Miami, FL, January 2, 2013.

78. *Happy Days*, May 8, 1937.

79. Jimmy Carter, *An Hour Before Daylight* (New York: Simon and Schuster, 2001), pp. 104–7.

80. Author interview with former president Jimmy Carter, February 1, 2013. In 1976 Jimmy Carter launched his presidential campaign at the Little White House in honor of FDR, his boyhood hero.

81. FDR to Henry Wallace, January 19, 1938, FDRL.

82. FDR to Henry Wallace, January 25, 1938, FDRL.

83. FDR to Henry A. Wallace, January 27, 1938, FDRL.

84. Executive Order 9185, 7 C.F.R. 4713, June 25, 1942.

85. Franklin D. Roosevelt, press conference (transcript) at Hyde Park, October 7, 1938, FDRL.

86. Arthur Krock, "Taxpayers Revolt," *New York Times*, November 10, 1938.

87. Andrew E. Busch, "The New Deal Comes to a Screeching Halt in 1938," Ashbrook Center, Ashland University, May 2006, http://ashbrook.org/publications/oped-busch-06-1938/.

88. *Public Papers and Addresses of Franklin D. Roosevelt, 1938*, pp. 605–7.

89. Cissy Patterson quoted in Amanda Smith, *Newspaper Titan: The Infamous Life and Monumental Times of Cissy Patterson* (New York: Alfred A. Knopf, 2011), p. 381.

90. Ibid., p. 392.

91. Clarke, *Roosevelt's Warrior*, pp. 294–95.

92. Eleanor Roosevelt, "My Day," April 15, 1943.

93. John F. Hoffman, *Arches National Park: An Illustrated Guide* (San Diego, CA: Western Recreational Publication, 1985), p. 14.

94. Ibid.

95. Arches National Park Main Entrance Road, Moab Canyon Wash Culvert (Moab Vicinity), Grand County, UT, Historical American Engineering Record, Arches National Park Archive.

96. David R. Brower to Doug Scott (Sierra Club Conservation Director), July 13, 1989, http://www.permatopia.com/wetlands/compromise.html.

97. FDR to Henry A. Wallace, January 27, 1939, FDRL.

98. Turner, *David Brower: The Man and the Environmental Movement* (Berkeley: University of California Press, 2015), p. 35.

99. Cornebise, *The CCC Chronicles*, p. 107.

100. FDR to Angelo J. Rossi, March 24, 1939, FDRL.

101. FDR to Clyde L. Seavey, March 14, 1939, FDRL.

102. Watkins, *Righteous Pilgrim*, pp. 568–77.

103. James H. Rowe Jr., special assistant to the president, to Edwin M. Watson, secretary to the president, June 1, 1939.

104. Douglas Hillman Strong, *Trees—or Timber?* (Three Rivers, CA: Sequoia Natural History Association, 1980), p. 48.

105. William Colby quoted in Fox, *The American Conservation Movement*, pp. 216–17.

106. Glover, *A Wilderness Original*, p. 263.

107. Irving Brant, *James Madison, The Virginia Revolutionist*, 1941; *James Madison the Nationalist 1780–1787*, 1948; *James Madison: Father of the Constitution, 1787–1800*, 1950; *James Madison: Secretary of State 1800–1809*, 1953; *James Madison: The President, 1809–1812*, 1956; *James Madison: Commander in Chief, 1812–1836*, 1961 (Indianapolis: Bobbs-Merrill 1941–1961).

108. Ansel Adams, *An Autobiography* (Boston: Little, Brown, 1985), p. 343.

CHAPTER 17: "TO BENEFIT WILDLIFE"

1. Paige, *The Civilian Conservation Corps and the National Park Service*, p. 114.

2. Dolin, *Smithsonian Book of National Wildlife Refuges*, pp. 103–5.

3. Peter Van Wyk, *Burnham, King of Scouts: Baden-Powell's Secret Mentor* (Victoria, Canada: Trafford, 2003), pp. 545–46.

4. David C. Scott and Brendan Murphy, *The Scouting Party: Pioneering and Preservation, Progressivism, and Preparedness in the Making of the Boy Scouts of America* (Irving, TX: Red Honor Press, 2010), pp. 3–27.

5. Quoted in Brinkley, *The Wilderness Warrior*, p. 527.

6. Edward H. Saxton, "Saving the Desert Bighorns," *Desert* (March 1970), pp. 17–18.

7. Van Wyk, *Burnham, King of Scouts*, p. 550.

8. Leonard J. Arring, "The Sagebrush Resurrection: New Deal Expenditures in Western States, 1933–1939," *Pacific Historical Review*, Vol. 52 (1983), p. 12.

9. FDR quoted in *Boys Life* (February 1934).

10. FDR, speech to Ten Mile River Scout Camp, Narrowsburg, NY, August 23, 1933.

11. David C. Scott, *My Fellow Americans: Scouting, Diversity, and the U.S. Presidency* (Dallas, TX: WindRush, 2014), p. 93.

12. FDR to R. K. Wickstrum, January 26, 1939, FDRL.

13. Gabrielson, *Wildlife Refuges*, p. 99.

14. James K. Morgan, "Slamming the Ram into Oblivion," *Audubon*, Vol. 75, no. 6 (November 1973), p. 17.

15. Laycock, *The Sign of the Flying Goose*, pp. 251–52.

16. "Refuge System Celebrates Anniversary: 75 Years and Going Strong," *Fish and Wildlife News* (December 1978/January 1979), p. 22.

17. Harold L. Ickes to FDR, December 15, 1938, FDRL.

18. U. S. Geological Survey, "PWRC Strategic Science Plan" (Laurel, MD: USGS, 2008).

19. "Patuxent Refuge to Fashion Utopia for Wildlife," *Washington Sunday Star*, July 28, 1931. Patuxent was located twenty miles from the South Building of the Department of Agriculture in Washington, DC. To reach the refuge around the time of its dedication, biologists and visitors commonly took Maryland Avenue to Fifteenth and H streets and then Baltimore Pike to Peace Cross at Bladensburg. After making a right on Defense Highway to Lanham, they took the left-hand fork to Bowie. Once in Bowie, they made a left turn onto the Laurel-Bowie Highway. After 3.5 miles on the Laurel-Bowie Highway, a gate marked "Patuxent Research Refuge" appeared on the right side of the road.

20. Patuxent is divided into three areas. The North Tract offers opportunities for wildlife observation, hunting, and fishing. The Center Tract houses the refuge's headquarters and the Wildlife Research Center (a research study site). The South Tract contains the National Wildlife Visitor Center and various trails.

21. L. B. Morley, "Early History of Patuxent Wildlife Research Center (circa 1948)," Patuxent Archive, Laurel, MD.

22. Blair Bolles, "Wild-Life Laboratory," *New York Times*, June 18, 1939.

23. "Wallace Dedicates Wildlife Station," *Washington Evening Star*, June 5, 1939.

24. Scott Hart, "U.S. Sets Up Vast Wild Life Refuge on Historic Maryland Estate," *Washington Post*, May 28, 1939.

25. Graham, *The Land and Wildlife*, p. 11.

26. Ira N. Gabrielson, *Wildlife Conservation* (New York: Macmillan, 1941); *Wildlife Refuges* (cited above); *Wildlife Management* (New York: Macmillan, 1951).

27. David Sibley, "Birds Make Anyplace a Chance for Discovery," *Audubon* (March/April 2013), p. 18.

28. John O'Reilly, "State Wardens Inspect Federal Wildlife Station," *New York Herald Tribune*, March 22, 1940.

29. "Mr. Wallace: Please Keep It 'Experimental'!" *Modern Game Breeding* (June 1939).

30. Ira N. Gabrielson, "Waterfowl Restoration: The Plain Facts," address to Maryland State Game and Fish Protective Association, Baltimore, December 14, 1936; Ira N. Gabrielson, "The Problem of Duck Conservation," speech to Illinois Sportsman Association, Chicago, June 29, 1936, U.S. Fish and Wildlife Archive, Shepherdstown, WV.

31. "Patuxent Refuge to Fashion Utopia for Wildlife," *Washington Star*, May 28, 1939.

32. Ira M. Gabrielson, "A National Program for Wildlife Conservation," North American Wildlife Conference speech (transcript), February 7, 1937, NCTC-USFWS Archive, Shepherdstown, WV.

33. Hartley H. T. Jackson, "Conserving Endangered Wildlife Species," *Transactions of the Wisconsin Academy of Sciences, Arts, and Letters*, Vol. 35 (1944), pp. 61–69.

34. Bolles, "Wild-Life Laboratory."

35. Henry Lyon, "Historic Site Cleared for Suburban Wildlife Haven," *Washington Evening Star*, April 10, 1931.

36. Robert J. Orth and Kenneth A. Moore, "Distribution and Abundance of Submerged Aquatic Vegetation in Chesapeake Bay: An Historical Perspective," *Estuaries*, Vol. 7, no. 4 (1984), pp. 531–40.

37. Ibid.

38. Rachel Carson, "Guarding Our Wildlife Resources," *Conservation in Action*, no. 5 (Washington, DC: U.S. Government Printing Office, 1948), p. 26. See also "Mr. Wallace: Please Keep It 'Experimental'!"

39. Hoagland, *Hoagland on Nature*, p. 68.

40. Robert Bruskin, "Effort Is Made to Give Workers Better Prices," [n.d. 1939], newspaper clippings file, DuPont Papers, NCTC-USFW, Shepherdstown, WV.

41. Linda Lear, *Witness for Nature: The Life and Legacy of Rachel Carson* (New York: Henry Holt, 1997), p. 105.

42. Three of Rachel Carson's four NWR booklets are "Mattamuskeet: A National Wildlife Refuge," *Conservation in Action*, no. 4 (Washington, DC: U.S. Government Printing Office, 1947); "Parker River: A National Wildlife Refuge," *Conservation in Action*, no. 2 (Washington, DC: U.S. Government Printing Office, 1947); "Chincoteague: A National Wildlife Refuge," *Conservation in Action*, no. 1 (Washington, DC: U.S. Government Printing Office, 1947). A fourth booklet, *Guarding Our Wildlife Resources*, came out in 1948 and praised FDR's National Wildlife Refuge System to kingdom come.

43. Rachel Carson, *Under the Sea-Wind* (New York: Oxford University Press, 1941); *The Sea Around Us* (New York: Oxford University Press, 1951); *The Edge of the Sea* (Boston: Houghton Mifflin, 1955).

44. Rachel Carson, *Silent Spring* (Boston: Houghton Mifflin, 1962).

45. The refuges established by FDR in May 1939 included Appert Lake, Billings Lake, Bone Hill Creek, Buffalo Lake, Camp Lake, Canfield Lake, Charles Lake, Dakota Lake, Flickertail, Florence Lake, Half-Way, Hutchinson Lake, Johnson Lake, Lake Moraine, Lake Oliver, Little Goose, Little Lake, Lords Lake, Lost Lake, and Minnewastena.

46. FDR, speech in Grand Forks, ND, October 4, 1937, FDRL.

47. U.S. Fish and Wildlife Comprehensive Conservation Plan, *North Dakota Limited-Interest National Wildlife Refuges* (Lakewood, CO: U.S. Fish and Wildlife Service, April 2006).

48. Ibid.

49. Kathy Tandberg, "Proud to Be a CCC Boy," *The Common: Supplement to Beulah Beacon and Hazen Star*, June 17, 1999.

50. The North Dakota refuges established by FDR in 1939 as a second round include Ardoch, Brumba, Cottonwood Lake, Hiddenwood, Hobart Lake, Lake George, Lake Ilo, Lake Nettie, Lake Patricia, Lake Zahl, Lambs Lake, Maple River, McLean, Pleasant Lake, Rock Lake, Shell Lake, and Sibley Lake.

51. Laycock, *The Sign of the Flying Goose*, pp. 271–74.

52. Cohen, *The Tree Army*, p. 152.

53. Laycock, *The Sign of the Flying Goose*, pp. 271–73.

54. Elwyn B. Robinson, *History of North Dakota* (Lincoln: University of Nebraska Press, 1966), pp. 397–409.

55. National Park Service, "Federal Relief Construction in North Dakota, 1931–1943" (National Register of Historic Places Multiple Property Documentation Form, August 2010), http://history.nd.gov/hp/PDFinfo/MPDF%20Complete%20Rev%2010_2010.pdf.

56. Davis and Davis, *Our Mark on This Land*, pp. 223–24.

57. "The International Peace Garden," North American Game Warden Museum, accessed March 16, 2015, http://www.gamewardenmuseum.org/PeaceGarden.php.

58. U.S. Fish and Wildlife Service, "Waterfowl Production Areas: Prairie Jewels of the Refuge System," accessed June 22, 2014, http://nctc.fws.gov/Pubs9/NWRS_waterfowl01.pdf.

59. Dyana Z. Furmansky, *Rosalie Edge, Hawk of Mercy: The Activist Who Saved Nature from the Conservationists* (Athens: University of Georgia Press, 2010).

60. "Hawks—Good and Bad," *National Sportsman* (August 1940), p. 8.

61. Brant, *Adventures in Conservation with Franklin D. Roosevelt*, p. 49.

62. Bald Eagle Protection Act of 1940, 16 U.S.C. 668–668d, 54 Stat. 250.

63. Harpers Ferry Center, National Park Service, U.S. Department of the Interior, "Perry's Victory and International Peace Memorial, Long-Range Interpretive Plan," October 2011, pp. 9–10, http://www.nps.gov/pevi/learn/management/upload/Perrys-Victor y-LRIP_final2Update.pdf.

64. FDR, Executive Order 7937, Establishing West Sister Island Migratory Bird Refuge, August 2, 1938. Online at American Presidency Project, http://www.presidency.ucsb.edu/ws/?pid=61277.

65. Gary Wedemeyer, "A Brief History of the Western Fisheries Research Center 1934–2006," USGS Western Fisheries Research Center, last modified July 25, 2013, http://wfrc.usgs.gov/about/history.html.

66. "1,741,000,000 Trees Planted by CCC," *New York Times*, July 3, 1939.

67. Ibid.

68. Quoted in Maher, *Nature's New Deal*, p. 154.

69. Rexford Tugwell, *The Democratic Roosevelt* (New York: Doubleday, 1957), p. 331.

70. Robert Fechner, "CCC Has Conserved Youths of Nation as Well as Land," *Macon (Missouri) Chronicle-Herald*, July 25, 1939, p. 1.

71. Department of the Interior, National Register of Historic Places Multiple Property Documentation Form: "Federal Relief Construction in North Dakota, 1931–1943," http://history.nd.gov/hp/PDFinfo/MPDF%20Complete%20Rev%20 10_2010.pdf.

72. Meine, *Aldo Leopold*, p. 317.

73. "History and Culture: Mark Twain National Forest," U.S. Forest Service, accessed January 4, 2015, http://www.fs.usda.gov/main/mtnf/learning/history -culture.

74. Franklin and Eleanor, however, maintained lifetime occupancy rights. And Sara Delano Roosevelt was still living at Springwood; she would die in 1941, at the age of eighty-six.

75. FDR, "Address at the Cornerstone Laying of the Franklin D. Roosevelt Library in Hyde Park, New York," November 19, 1939, FDRL. Online at http://docs .fdrlibrary.marist.edu/php1139.html.

76. Kenneth T. Walsh, *From Mount Vernon to Crawford: A History of Presidents and Their Retreats* (New York: Hyperion, 2005), p. 93.

77. John G. Waite, *The President as Architect: Franklin D. Roosevelt's Top Cottage* (Albany, NY: Mount Ida, 2001), p. 17.

78. Tonya Bolden, *FDR's Alphabet Soup*, p. 108.

79. Leuchtenberg, *Franklin D. Roosevelt and the New Deal*, p. 133.

80. Jeanne Nienaber Clarke, *Roosevelt's Warrior: Harold L. Ickes and the New Deal* (Baltimore: Johns Hopkins University Press, 1996), p. 113.

81. Watkins, *Righteous Pilgrim*, p. 373.

82. Harold Ickes, *Back to Work: The Story of the PWA* (New York: Macmillan, 1935), p. 216.

83. Sarah T. Phillips, *This Land, This Nation: Conservation, Rural America, and the New Deal* (New York: Cambridge University Press, 2007), p. 183.

84. Fox, *The American Conservation Movement*, pp. 206–10.

85. James M. and Regina B. Glover, "Robert Marshall: Portrait of a Liberal Forester," *Journal of Forest History* (July 1986), pp. 112–19.

86. "Robert Marshall, Federal Aide, Dies," *New York Times*, November 12, 1939.

87. "CCC Director Fechner Dies Here at 63: Roosevelt Laments Loss; to Be Buried in Arlington Tomorrow," and "Fechner, Chief of CCC, Dies; Roosevelt Laments U.S. Loss; Boys of Corps Will Bear Leader's Body to Arlington Tomorrow," *Washington Post*, January 1, 1940, p. 1.

88. "J. McEntee Named to Administer CCC," *New York Times*, February 16, 1940, p. 42.

CHAPTER 18: "AN ABUNDANCE OF WILD THINGS"

1. Paul Appleby, "Roosevelt's Third-Term Decision," *American Political Science Review* (September 1952), pp. 754–65.

2. Ibid.

3. Harold L. Ickes, "Why I Want Roosevelt to Run Again," *Look*, July 4, 1939.

4. John H. Bankhead to Colonel "Pa" Watson, January 19, 1940, FDRL.

5. Gifford Pinchot to FDR, January 13, 1940, FDRL.

6. Franklin D. Roosevelt to Gifford Pinchot, January 15, 1940, in Nixon, *Franklin D. Roosevelt and Conservation*, Vol. 2, pp. 413–14.

7. Ibid.

8. Gifford Pinchot to FDR, January 17 1940, in Nixon, *Franklin D. Roosevelt and Conservation*, Vol. 2, p. 414.

9. George W. Norris to FDR, ibid., p. 417.

10. George W. Norris to FDR, January 24, 1940, FDRL.

11. Harold Ickes to FDR, February 7, 1940.

12. George McJimsey, *Harry Hopkins: Ally of the Poor and Defender of Democracy* (Cambridge: Harvard University Press), pp. 83–84.

13. Clare Boothe Luce, *Time*, quoted in the appendix of *Congressional Record: Proceedings and Debates of the 83rd Congress*, Vol. 99, part 9 (Washington, DC: U.S. Government Printing Office, 1953), p. A-1249.

14. Harold Ickes, *The Secret Diary of Harold L. Ickes: The Lowering Clouds, 1939–1941* (New York: Simon and Schuster, 1955), pp. 127–28.

15. FDR to Harold Ickes, February 7, 1940.

16. Ickes, *The Lowering Clouds*, p. 131.

17. John James Little, "Island Wilderness: A History of Isle Royale National Park," PhD dissertation, University of Toledo, p. 48.

18. "Michigan: Isle Royale Plans for the Season," *New York Times*, July 13, 1941.

19. Paul Brooks, *Roadless Area* (New York: Knopf, 1964), p. 81.

20. *Detroit Free Press*, July 2, 1935.

21. Adolph Murie, "Report on the Qualifications and Development of Isle Royale National Park," June 13, 1935, (Washington, DC: National Park Service, 1935), File 201, Box 1249, RG 79, National Archives, Washington, DC.

22. Harold L. Ickes, quoted in Watkins, *Righteous Pilgrim*, pp. 469–70.

23. "Address by Hon. Harold Ickes, Secretary of the Interior, November 20, 1934," Box 2, File Historical, Yellowstone National Park Archives, quoted in Mary Shivers Culpin, *"For the Benefit and Enjoyment of the People": A History of the Concession Development in Yellowstone National Park, 1872–1966* (Yellowstone National Park, WY: National Park Service, Yellowstone Center for Resources, YCR-CR-2003-01, 2003), p. 81.

24. Horace Albright, "The National Park Movement," in Harlean James, ed., *American Planning and Civic Annual* (1941), p. 63.

25. Eleanor Roosevelt, "My Day," December 6, 1940.

26. National Park Service, "Vanderbilt Mansion National Historic Site," file 2014, Hyde Park, NY.

27. Eleanor Roosevelt, "My Day," August 14, 1940.

28. National Park Service, "Roosevelt-Vanderbilt National Historic Sites," http://www.nps.gov/hofr/roosevelt-vanderbilt-national-historic-sites.htm (accessed April 11, 2015).

29. "Pinchot to Aid Roosevelt," *New York Times*, October 10, 1940, p. 18.

30. See "Digest of Federal Resource Laws of Interest to the Fish and Wildlife Service: Fish and Wildlife Act of 1956," U.S. Fish and Wildlife Service, http://www.fws.gov/laws/lawsdigest/fwact.html (accessed June 22, 2013); and "029 FW 4, Division of Law Enforcement," U.S. Fish and Wildlife Service, http://www.fws.gov/policy/029fw4.html (accessed June 22, 2013).

The units and responsibilities of the U.S. Fish and Wildlife Service include the National Wildlife Refuge System, the Division of Migratory Bird Management, the Federal Duck Stamp, the National Fish Hatchery System, the Endangered Species program, and the USFWS Office of Law Enforcement. By the twenty-first century, the units divisions of USFWS included an annual budget of around $2.3 billion.

31. The bureaucratic transfers of divisions and agencies dedicated to the protection of fish and wildlife is confusing. In 1939, the Biological Survey (Department of Agriculture) and the Bureau of Fisheries (Department of Commerce) were transferred to the Department of the Interior. In 1940, these two entities merged and were renamed the Fish and Wildlife Service. The 1956 Fish and Wildlife Act officially created the U.S. Fish and Wildlife Service and established the Bureau of Sport Fish and Wildlife and the Bureau of Commercial Fisheries. In 1970, the Bureau of Commercial Fisheries was transferred to the Department of Commerce and renamed the National Marine Fisheries Service. The Department of the Interior still oversees U.S. Fish and Wildlife today.

32. Rosenman, quoted in Susan Dunn, *1940: FDR, Willkie, Lindbergh, Hitler—the Election Amid the Storm* (New Haven: Yale University Press, 2013), p. 40.

33. Dunn, *1940*, p. 3.

34. James J. McEntee, "The CCC and National Defense," *American Forestry* (July 1940), pp. 1–2.

35. Eleanor Roosevelt, "My Day," June 25, 1940.

36. Eleanor Roosevelt, "My Day," July 19, 1940.

37. James Farley, *Jim Farley's Story: The Roosevelt Years* (New York: McGraw-Hill, 1948), p. 293.

38. Culver and Hyde, *American Dreamer*, p. 219.

39. Herbert Eaton, *Presidential Timber: A History of Nominating Conventions, 1868–1960* (New York: Free Press of Glencoe, 1964), p. 392.

40. Culver and Hyde, *American Dreamer*, pp. 227–28.

41. Richard Moe, *Roosevelt's Second Act: The Election of 1940 and the Politics of War* (New York: Oxford University Press), pp. 202–4.

42. Franklin D. Roosevelt, Presidential Proclamation 2416, 5 FR 2677, 54 Stat. 2717, July 25, 1940.

43. When FDR assumed the presidency in 1933, the United States had five different wildlife sanctuary designations: Reservation (e.g., Bering Sea in Alaska and

Caloosahatchee in Florida); Bird Refuge (e.g., Anaho Island in Nevada and Montezuma in New York); Migratory Waterfowl Refuge (e.g., Sand Lake in South Dakota and Lake Isom in Tennessee); Migratory Bird Refuge (e.g., West Sister Island in Ohio and Lenore Lake in Washington State); and Game Refuge (e.g., Elk Refuge in Wyoming). Roosevelt's executive action in July 1940 gathered the wild places under one rubric—NWRs—whose mandate was the protection of living things in each designated ecosystem. Before the 1940 streamlining, responsibility for overseeing federal sanctuaries was scattered among several entities from the Department of the Interior to the Light House Service to the Marine Corps to the USDA. FDR therefore considered the appellation "National Wildlife Refuge" to be more appropriate for his wide-ranging wildlife-management plan than merely "Migratory Waterfowl Refuge," which, after all, also protected the domain of the eagles, hawks, and falcons. Under Proclamation 2416, 193 reservations underwent the name change. At all of these newly dubbed refuges, it was "unlawful to hunt, trap, capture, willfully disturb, or kill any bird or wild animal . . . or to enter thereon for any purpose, except as permitted by law or by the rules and regulations of the Secretary of the Interior."

44. Trefethen, *An American Crusade for Wildlife*, p. 238.

45. *Conserving the Future: Wildlife Refuges and the Next Generation* (Washington, DC: U.S. Fish and Wildlife Service, 2011), pp. 19–20.

46. Brant, *Adventures in Conservation with Franklin D. Roosevelt*, p. 52.

47. "Ira Gabrielson, Wildlife Expert and Leading Conservationist," *National Parks and Conservation*, Vol. 51 (December 1977), pp. 18–19.

48. "Inter-American Convention on Nature Protection and Wildlife Preservation: Message from the President of the United States . . . Signed on the Part of the United States of America, on October 12, 1940" (Washington, DC: U.S. Government Printing Office, 1941).

49. Julia Craw, "U.S. Wildlife Refuges," *Travel* (December 1971), p. 50.

50. Ralph H. Imler and E. H. Kalmback, *The Bald Eagle and Its Economic Status* (Washington, DC: U.S. Government Printing Office, 1955), pp. 18–19.

51. Ibid., pp. 2–5.

52. Dunn, *1940*, pp. 1–2.

53. Maxine Newell, *The Story of the Hole N" the Rock* (privately printed, 2005).

54. FDR, "Campaign Address at Cleveland, Ohio," November 2, 1940. Online at American Presidency Project, http://www.presidency.ucsb.edu/ws/?pid=15893.

55. Philip G. Platt, "Memorandum on Water Pollution Control," in Nixon, *Franklin D. Roosevelt and Conservation*, Vol. 2, p. 453.

56. Ibid.

57. Ira N. Gabrielson, "Wildlife and the American Way of Living," in *Report of the Secretary of the Interior* (Washington, DC: U.S. Government Printing Office, 1942).

58. Harold Ickes to *Mining World* (Seattle), August 7, 1940.

59. Miller Freeman to Harold Ickes, August 20, 1940.

60. Eleanor Roosevelt, "My Day," September 4, 1940.

61. FDR, speech at the dedication of the Great Smoky Mountains National Park, Newfound Gap, Tennessee, September 2, 1940, in Nixon, *Franklin D. Roosevelt and Conservation*, Vol. 2, p. 471.

62. Richard White, *"It's Your Misfortune and None of My Own": A New History of the American West* (Norman: University of Oklahoma Press, 1991), pp. 472–502.

63. *Tamarack Times*, February, June, and July 1940.

64. Cornebise, *The CCC Chronicles*, p. 119.

65. FDR, "Campaign Address at Boston, Massachusetts," October 30, 1940. Online at American Presidency Project, http://www.presidency.ucsb.edu/ws/?pid=15887.

66. Irving Brant quoted in Fox, *The American Conservation Movement*, p. 217.

67. Dunn, *1940*, p. 1.

68. James H. Madison, *Wendell Willkie: Hoosier Internationalist* (Bloomington: Indiana University Press, 1992), pp. xiv–xvi.

69. Moe, *Roosevelt's Second Act*, p. 316.

70. FDR to John H. Baker, August 12, 1939.

71. Eleanor Roosevelt, "My Day," November 23, 1940.

72. "21,550,783 Acres in National Parks," *New York Times*, December 16, 1940, p. 28.

73. "Inaugural Crowds Demonstrate Affection for President's Mother," *Washington Star*, January 21, 1941.

74. "Burnham Hoyt, 73, Dies," *New York Times*, April 8, 1960.

75. City of Phoenix Department of Parks and Recreation, "Roosevelt's Soil Soldiers, 1933–1942," http://phoenix.gov/parks/trails/locations/south/ccc/index.html.

76. Peter MacMillan Booth, "The Civilian Conservation Corps in Arizona, 1933–1942," PhD dissertation, University of Arizona, 1991.

77. Eleanor Roosevelt, "My Day," January 24, 1941.

78. George C. Marshall to Pa Watson, January 9, 1942, CCC Files, FDRL.

79. FDR, "Fireside Chat 16: On the 'Arsenal of Democracy' (December 29, 1940)," http://millercenter.org/president/fdroosevelt/speeches/speech-3319.

80. Delbert Clark, "Mammoth Cave: New National Park," *New York Times*, July 27, 1941.

81. Ward, *Closest Companion*, p. 140.

82. Nigel Hamilton, *The Mantle of Command: FDR at War, 1941–1942* (Boston: Houghton Mifflin Harcourt, 2014), p. 38.

83. Mares, *Fishing with the Presidents*, pp. 89–93.

84. Hamilton, *The Mantle of Command*, p. 38.

85. Pottker, *Sara and Eleanor*, p. 332.

CHAPTER 19: "THE ARMY MUST FIND A DIFFERENT NESTING PLACE!"

1. "Trumpeter Swan Poster," U.S. Fish and Wildlife Archive, Shepherdstown, WV.

2. Ferdinand Augustus Silcox, U.S. Biological Survey, *Wildlife Restoration and Conservation; Proceedings of the North American Wildlife Conference Called by President Franklin D. Roosevelt; Connecting Wing Auditorium and the Mayflower Hotel, Washington, DC, February 3–7, 1936* (Washington, DC: U.S. Government Printing Office, 1936), p. 560.

3. Furmansky, *Rosalie Edge, Hawk of Mercy*, p. 236.

4. Harold L. Ickes to FDR, November 27, 1941, FDRL.

5. Ibid.

6. Irving Brant to Henry L. Stimson, November 25, 1941, FDRL.

7. FDR to Henry L. Stimson, November 28, 1941, FDRL.

8. Thomas L. Howell, "U.S. Army Winter Training Camp: West, Yellowstone, Montana," 2004, http://www.snakeriver4x4.com/armybase.php (accessed on May 17, 2013).

9. Adams quoted in Furmansky, *Rosalie Edge, Hawk of Mercy*, p. 236.

10. Dawn Merritt, "From the Jazz Age to World War II," *Outdoor America* (Spring 2012), p. 15.

11. FDR to Kenneth A. Reid, December 3, 1941, FDRL.

12. FDR quoted in Hamilton, *The Mantle of Command*, pp. 74–75.

13. FDR, "Address to Congress Requesting a Declaration of War with Japan," December 8, 1941. Online at American Presidency Project, http://www.presidency.ucsb.edu/ws/?pid=16053.

14. Hanson W. Baldwin, "Army and Navy Share Responsibility for Defense of the Island Fortress," *New York Times*, December 19, 1941.

15. Executive Order No. 8979, 6 Fed. Reg. 6471, December 16, 1941; Franklin Delano Roosevelt Executive Orders Disposition Tables, 1941, http://www.archives.gov/federal-register/executive-orders/1941.html.

16. Ibid.

17. Elaine Rhode, *National Wildlife Refuges of Alaska* (Anchorage: Alaska National History Association, 2003), p. 24; Eric Jay Dolin, *National Wildlife Refuges* (Washington, DC: Smithsonian Institution Press, 2003), p. 110. Building on FDR's action, in 1980 President Jimmy Carter, as part of his Alaska National Interest Lands Conservation Act, would change the name of the moose sanctuary to Kenai National Wildlife Refuge to reflect the diversity of wildlife in the area.

18. U.S. Fish and Wildlife Service, *Assault Cases Against Agents and Deputy Wardens* (Shepherdstown, WV: U.S. Fish and Wildlife Service, 2013), pp. 1–10.

19. FDR, "Appointment of Harold L. Ickes as Petroleum Coordinator for National Defense," May 28, 1941. Online at American Presidency Project, http://www.presidency.ucsb.edu/ws/?pid=16122.

20. Eleanor Roosevelt, "My Day," January 3, 1942.

21. Eleanor Roosevelt, "My Day," December 13, 1941.

22. Tim Smigielski, "Pioneers: Dr. John Van Oosten," *Eddies: Reflections on Fisheries Conservation* (Winter 2010–2011), pp. 8–9.

23. Harold L. Ickes to FDR, January 17, 1942, FDRL.

24. Irving Brant to Harold Ickes, January 15, 1942, Library of Congress, Washington, DC.

25. Irving Brant to Harold Ickes, January 14, 1942, Library of Congress, Washington, DC.

26. "Porcupine Mountains," Michigan Department of Natural Resources, http://www.michigan.gov/dnr/0,4570,7-153-31154_31260-54024—,00.html (accessed May 13, 2013).

27. Alan Brinkley, *The End of Reform: New Deal Liberalism in Recession and War* (New York: Random House, 1995), p. 141.

28. Robert De Vore, "McKellar Holds Scant Hopes for His Bills to Kill CCC, NYA; Death of CCC, NYA Declared Unlikely," *Washington Post*, March 25, 1942, p. 1.

29. *Report of the Secretary of War to the President, 1933* (Washington, DC: U.S. Government Printing Office, 1933), p. 8.

30. Tony Melessa interview by Janet Seegmiller, March 8, 2006, Markaguant Plateau Oral History Project, Sherratt Library, Southern Utah University, Cedar City, UT.

31. "Fechner of CCC," *Time*, February 6, 1933, p. 10.

32. Humphrey Bogart, letter published in *Happy Days*, March 7, 1942.

33. "Fechner of CCC," p. 12.

34. Wirth, *Parks, Politics, and the People*, pp. 225–26.

35. "CCC Discontinues All Nondefense Work," *Washington Post*, March 26, 1942, p. 1.

36. "Let CCC Fit Boys for Army, Chief Urges," *Washington Post*, March 31, 1942, p. 3.

37. "Ickes and Hill Recommend CCC Be Maintained to Do Spade Work for Army," *Washington Post*, April 12, 1942, p. 9.

38. FDR to Senator Elbert Thomas, March 16, 1942.

39. "Phillips' Statement on CCC Called Libel: 'Dastardly Insult' to Oklahoma Youth, McEntee Declares," *Washington Post*, April 17, 1942, p. 5.

40. John Elliott, "Roosevelt Acts to Halt Attack on CCC, NYA," *Washington Post*, March 22, 1942, p. 14.

41. "3 DC Banks Must Pay Bill for Ghost CCC," *Washington Post*, March 17, 1942, p. X23.

42. "Burning of CCC Equipment Draws Protests in House," *Washington Post*, March 12, 1942, p. 13.

43. "Internees Hired by Ickes for Farm," *New York Times*, April 16, 1943, p. 17.

44. Jacobus tenBroek, Edward N. Barnhart, and Floyd W. Matson, *Prejudice, War, and the Constitution* (Berkeley: University of California Press, 1954), pp. 100–101.

45. Rutkow, *American Canopy*, p. 262.

46. FDR to Harold D. Smith of the Bureau of the Budget, June 17, 1942.

47. John Nielsen, *Condor: To the Brink and Back—The Life and Times of One Giant Bird* (New York: HarperCollins, 2006), pp. viii–x.

48. FDR to Harold Smith, June 17, 1942, FDRL.

49. Maher, *Nature's New Deal*, p. 211.

50. R. Douglas Hurt, "Forestry on the Great Plains, 1902–1942," http://www-personal.ksu.edu/~jsherow/hurt2.htm.

51. FDR to William O. Douglas (memo), February 18, 1942; Douglas to FDR, February 2, 1942, FDRL.

52. Ellen Bidel, "Happy 65th Birthday, Smokey Bear," *New York State Conservationist* (October 2009), p. 19.

53. FDR, "A Proclamation," August 5, 1942. Reprinted in *American Forests*, Vol. 48 (1942), p. 435.

54. "Wartime Forest Fire Prevention Campaign Launched," *American Forests*, Vol. 48 (1942), p. 353. See also Rutkow, *American Canopy*, pp. 262–63.

55. Joseph M. Speakman, *At Work in Penn's Woods: The Civilian Conservation Corps in Pennsylvania* (University Park: Pennsylvania State University Press, 2006), pp. 161–62.

56. Paige, *Civilian Conservation Corps and the National Park Service*, p. 132.

57. Abraham F. Cohen, "A Tribute to the Civilian Conservation Corps," *Alaskan*, April 20, 1942.

58. Cornebise, *The CCC Chronicles*, p. 241.

59. *The Broadcast* (November 1939) and *Happy Days*, October 26, 1940.

60. J. M. Burton, M. Farrell, F. B. Lord, and R. W. Lord, "Confinement and Ethnicity: An Overview of Japanese American Relocation Sites," rev. ed., *Publications in Anthropology*, Vol. 74 (2000), n.p., http://www.nps.gov/parkhistory/online_books/anthropology74/index.htm.

61. Heidi Ridgley, "P. O. Box 1142: World War II: The Lost Chapter," *National Parks: The Magazine of the National Park Conservation Association*, Vol. 84, no. 1 (Winter 2010), pp. 42–48.

62. Tonya Bolden, *FDR's Alphabet Soup: New Deal America, 1932–1939* (New York: Alfred A. Knopf, 2010), p. 109.

63. Thomas P. Campbell, "A Best Friend in the White House," *Scouting* (March–April 2003).

64. Ibid.

65. "War Service Summary of the Boy Scouts of America, 1941 to 1945," statistics culled from the 1942 to 1946 BSA Annual Reports to Congress.

66. U.S. Department of Agriculture, *Victory Gardens: Reader's Handbook* (Washington, DC: USDA, 1944), pp. 1–4.

67. "Rediscovering the Victory Garden," *Leopold Outlook*, Vol. 2 (Summer/Fall 2011), p. 27.

68. Stan DeOrsey and Barbara A. Butler, *The Birds of Dutchess County, New York: Today and Yesterday* (Poughkeepsie, NY: The Ralph T. Waterman Bird Club, 2006), p. 10.

69. William D. Hassett, "May 31, 1943," in *Off the Record with F.D.R. 1942–1945* (New Brunswick: Rutgers University Press, 1958), pp. 172–73.

70. "J. N. 'Ding' Darling National Wildlife Refuge," J. N. Ding Darling Foundation, http://www.ding-darling.org/wildlife.html (accessed May 5, 2012).

71. U.S. Department of the Interior, Fish and Wildlife Service, Southeast Region, J. N. "Ding" Darling National Wildlife Refuge: Comprehensive Conservation Plan" (October 2010), http://www.fws.gov/uploadedFiles/Region_4/NWRS/Zone_2/JN_Ding_Darling_Complex/JN_Ding_Darling/Comprehensive%20Conservation%20Plan.pdf.

72. Lincoln A. Werden, "Wood, Field, and Stream," *New York Times*, May 7, 1943, p. 26.

73. Albert M. Day, "Control of Waterfowl Depredations," address at the North American Wildlife Conference at the LaSalle Hotel, Chicago, Illinois, April 26, 1944.

74. W. Dale Nelson, *The President Is at Camp David*, p. 6; Julie Eilperin, "For President Obama, Camp David Often Ranks as the Venue of Last Resort," *Washington Post*, March 20, 2015.

75. Nelson, *The President Is at Camp David*, p. 6.

76. Barbara M. Kirkconnell, *Catoctin Mountain Park: An Administrative History* (Washington, DC: National Park Service, 1988), pp. 74–76.

77. Rigdon, *White House Sailor*, p. 215.

78. FDR, *USS Shangri-La* logbook (1942), FDRL.

79. Nelson, *The President Is at Camp David*, pp. 4–21.

80. Winston S. Churchill, *The Hinge of Fate* (Boston: Houghton Mifflin, 1950), pp. 795–97.

81. Douglas, *Go East, Young Man*, pp. 334–35.

82. Samuel Rosenman, *Working with Roosevelt* (New York: Harper and Brothers, 1952), pp. 349–50.

83. Robert E. Sherwood, *Roosevelt and Hopkins: An Intimate History* (New York: Harper and Brothers, 1948), pp. 607–8.

84. DeOrsey and Butler, *The Birds of Dutchess County*, p. 10.

85. "Maunsell Crosby, Ornithologist Dies," *New York Times*, February 13, 1931, p. 17.

86. Philip H. Melanson with Peter Stevens, *The Secret Service* (New York: MJF Books, 2002), pp. 294–97.

87. James Whitehead, "A President Goes Birding," *Conservationist* (May–June 1977), pp. 20–23.

88. Ibid.

89. William E. Davis Jr., *Dean of the Birdwatchers: A Biography of Ludlow Griscom* (Washington, DC: Smithsonian Institution Press, 1994), p. 129.

90. Fox, *The American Conservation Movement*, p. 185.

91. Virginia Tanner, "Journeys with President Roosevelt," *Baltimore and Ohio Magazine* (April 1946).

92. Winston Churchill, *The Hinge of Fate* (Boston: Houghton Mifflin, 1950), pp. 327–28.

93. John M. Findlay and Bruce Hevly, *Atomic Frontier Days* (Seattle: University of Washington Press, 2011).

94. Rachel Carson, *Chincoteague: A National Wildlife Refuge* (Washington, DC: U.S. Fish and Wildlife Service, 1947).

95. A. C. Elmer to Ira Gabrielson, August 12, 1938, Eastern Massachusetts NWR Complex Archive, Sudbury, MA. (Thanks to Libby Herland for the document.)

96. *United States Code Congressional Service, Acts of 77th Congress, 1942* (St. Paul, MN: West Publishing Co., and Brooklyn, NY: Edward Thompson Co., 1943), p. 1238.

97. J. Clark Salyer and Francis Gillett, *Federal Refuges* (Washington, DC: U.S. Department of the Interior, 1964), p. 505.

98. Cecil Andrus and Joel Connelly, *Cecil Andrus's Western Style* (Seattle: Sasquatch Books, 1998), p. 121.

99. Nelson C. Brown, "The President Practices Forestry," *Journal of Forestry*, Vol. 41, no. 2 (February 1943), pp. 92–93.

100. "U.S. at War: Fight at Jackson Hole," *Time*, January 8, 1945.

101. Dyan Zaslowsky and T. H. Watkins, *These American Lands: Parks, Wilderness, and the Public Lands* (Washington, DC: Island Press, 1944), p. 31.

102. William Atherton DuPuy, "Jackson Hole: A Wonder in Dispute," *New York Times*, August 20, 1933.

103. John D. Rockefeller Jr. to FDR, February 10, 1943, FDRL.

104. Harold Ickes to Pa Watson and FDR, February 27, 1943, FDRL.

105. John D. Rockefeller to FDR, March 17, 1943, FDRL.

106. Richard P. Harmond, "Franklin Delano Roosevelt," in Cevasco and Harmond, *Modern American Environmentalists*, pp. 436–37.

107. Robert W. Righter, *Crucible for Conservation: The Creation of Grand Teton National Park* (Boulder: Colorado Associated University Press, 1982), pp. 113–17.

108. Ibid., pp. 114–17.

109. Zaslowsky and Watkins, *These American Lands*, p. 32.

110. George Washington Carver National Monument, H.R. 647, Public Law 148 (July 14, 1943); Harpers Ferry National Monument, Authorizing Legislation Public Law (P.L.) 78–386 (June 30, 1944).

CHAPTER 20: "CONSERVATION IS A BASIS OF PERMANENT PEACE"

1. Warren F. Kimball, *Swords or Ploughshares? The Morgenthau Plan for Defeated Nazi Germany, 1943–1946* (Philadelphia: J. B. Lippincott, 1976), p. 26.

2. Mark Robbie Eaker, "Hans Morgenthau and German Post-War Planning," MA thesis, University of Texas at Austin, 2015, p. 15.

3. FDR, Christmas Tree Folder, FDRL.

4. FDR to William Plog, October 18, 1943, Christmas Tree Folder, FDRL.

5. Grace G. Tully to William Plog, October 20, 1943, Christmas Tree Folder, FDRL.

6. Douglas, *Go East, Young Man*, p. 332.

7. Memorandum from the White House to U.S. Navy and War Production Board, January 20, 1943; Memorandum by Donald Nelson, February 26, 1943.

8. FDR to Henry Wallace, March 14, 1944.

9. FDR to Harold Ickes, March 14, 1944.

10. FDR to Harold Ickes, April 10, 1943, FDRL.

11. Eleanor Roosevelt, "My Day," March 26, 1941.

12. Jon Meacham, *Franklin and Winston*, p. 373.

13. Von Hardesty, *Air Force One: The Aircraft that Shaped the Modern Presidency* (New York: Quayside Press, 2005), p. 41.

14. FDR to Shah Mohammed Reza Pahlavi of Iran, September 2, 1944, FDRL.

15. FDR to King Ibn Saud of Saudi Arabia, February 10, 1944.

16. King Ibn Saud to FDR, April 1, 1944.

17. Elizabeth Cohen, "Getting to Know FDR Through His Books," FERI; N. H. Egleston, *Handbook of Tree Plants; or Why to Plant, Where to Plant, What to Plant, How to Plant* (New York: D. Appleton and Co., 1884), p. 124.

18. Ray Bergman, *Trout* (New York: Alfred A. Knopf, 1938); J. Fletcher Street, *Brief Bird Biographies: A Guide to Birds Through Habitat Association* (New York: Grosset and Dunlap, 1933).

19. "General Watson Dead; Roosevelt's Aide," *New York Times*, February 28, 1945.

20. Henry R. Luce, ed., *D-Day: Remembering the Battle That Won the War* (New York: Life, 2014).

21. Douglas, *Farewell to Texas*, p. 39.

22. FDR to R. Ewing Thomason, August 9, 1939.

23. Douglas, *Farewell to Texas*, p. 41.

24. Richard Phelan, *Texas Wild: The Land, Plants, and Animals of the Lone Star State* (New York: Excalibur Books, 1976), pp. 29–33.

25. FDR to President Manuel Ávila Camacho, October 24, 1944, FDRL.

26. Nancy A. Fogel, ed., *F.D.R. at Home* (New York: Dutchess County Historical Society, 2005), p. 12.

27. FDR to Democratic Party, July 11, 1944, FDRL.

28. Quoted in "Editorial: 'I Shall Accept . . . I Will Serve,' " *Life*, Vol. 17, no. 4 (July 24, 1944), p. 24.

29. Douglas, *Go East, Young Man*, p. 207.

30. Deuteronomy 20:19.

31. Zaslowsky and Watkins, *These American Lands*, p. 79.

32. Jay Winik, *1944: FDR and the Year That Changed History* (New York: Simon and Schuster, 2015), p. 496.

33. Gifford Pinchot, *Breaking New Ground* (New York: Harcourt Brace, 1947), p. 370.

34. Steen, *The Conservation Diaries of Gifford Pinchot*, p. 166.

35. Conrad Black, *Franklin D. Roosevelt: Champion of Freedom*, pp. 11–30.

36. Pinchot, *Breaking New Ground*, p. 370.

37. FDR to Gifford Pinchot, October 24, 1944.

38. U.S. Department of State, *The United Nations: Dumbarton Oaks Proposals for a General International Organization to Be the Subject of the United Nations Conference at San Francisco, April 25, 1945* (Washington, DC: U.S. Government Printing Office, 1945).

39. FDR to Cordell Hull, quoted in Pinchot, *Breaking New Ground*, 371.

40. FDR to Sam Rayburn, June 2, 1944, in Nixon, *Franklin D. Roosevelt and Conservation*, Vol. 2.

41. Saylor, *Jackson Hole*, p. 202.

42. "G.O.P. Wars on Collectivism, Dewey Asserts," *Chicago Tribune*, September 15, 1944, p. 11.

43. "Dewey Strikes at Land Grabs in the West," *Washington Post*, September 15, 1944.

44. Newton B. Drury, "The National Park Service: The First Thirty Years," in Harlean James, ed., *American Planning and Civic Annual* (1946), pp. 32–33.

45. "Ickes Credits GOP for Jackson Hole," *New York Times*, September 16, 1944. On September 14, 1950, the original 1929 Grand Teton National Park and the 1943 Jackson Hole National Monument, including Rockefeller's donation, were united into the new Grand Teton National Park, under its present-day boundaries.

46. "President Vetoes Jackson Hole Bill," *New York Times*, December 30, 1944, p. 1.

47. Eleanor Roosevelt, "My Day," June 10, 1948.

48. FDR, remarks at Clarksburg, West Virginia, October 29, 1944.

49. Eleanor Roosevelt, "My Day," October 28, 1944.

50. David B. Roosevelt, *Grandmère, A Personal History of Eleanor Roosevelt* (New York: Warner Books, 1982), p. 151.

51. Asbell, *The F. D. R. Memoirs*, p. 413.

52. David McCullough, *Truman* (New York: Simon & Schuster, 1992), p. 259.

53. Harold K. Steen, *The U.S. Forest Service: A History* (Seattle: University of Washington Press, 1976), p. 255.

54. FDR to Edward R. Stettinius, November 22, 1944, FDRL.

55. Ibid.

56. Edward Stettinius to FDR, December 16, 1944, FDRL.

57. FDR to Edward R. Stettinius Jr., January 2, 1945, FDRL.

58. Johnna Rizzo, "Japan's Secret WWII Weapon: Balloon Bombs," *National Geographic*, May 27, 2013, http://news.nationalgeographic.com/news/2013/05/130527-map-video-balloon-bomb-wwii-japanese-air-current-jet-stream/.

59. Lee Juillerat, "Japanese Balloon Bomb Killed 60 Years Ago Today," *Klamath Herald and News*, May 5, 2005, http://www.heraldandnews.com/news/top_stories/article_3b8041b 6-5bed-5c1b-9b8c-a12baba734fd.html?mode=jqm; "War Memorial, Lake County," The Oregon History Project, Oregon Historical Society, http://www.ohs.org/education/oregonhistory/historical_records/dspDocument.cfm?doc_ID=148D9401-9F23-2285-AB2059A24134757B.

60. Hoai-Tran Bui, "Be Sure the Birthday Cake's Candles Are Out," *USA Today*, August 8, 2014.

61. Kristin Fawcett, "This Just In," *Smithsonian* (July–August 2014), p. 122. In 1947 Smokey's motto became "Only YOU Can Prevent Forest Fires!" Then, in the spring of 1950, in the Capitan Mountains of New Mexico, a young cub was found after getting caught in a burning forest. He took refuge in a tree and ultimately escaped the blaze, but was badly burned. The firefighters were so moved by what the bear had been through that they named him Smokey. News about the real, live Smokey's heroism spread across America. He was soon given a home at the National Zoo in Washington, DC. The living Smokey became the symbol of woodlands conservation. Smokey died in 1976 and was buried in the Capitan Mountains in what is now known as Smokey Bear Historical Park.

62. Pinchot, *Breaking New Ground*, p. 368.

63. Harold L. Ickes, diary, April 29, 1942.

64. Gifford Pinchot to FDR, January 2, 1945, FDRL.

65. Forest History Society, "American Tree Farm History Timeline," http://foresthistory.org/atfs/.

66. Eleanor Roosevelt, "My Day," February 2, 1945.

67. William Souder, *On a Farther Shore: The Life and Legacy of Rachel Carson* (New York: Crown Publishers, 2012), pp. 7–10.

68. Clarence Cottam and Elmer Higgins, "DDT: Its Effect on Fish and Wildlife," *U.S. Fish and Wildlife Circular 11* (Washington, DC: U.S. Government Printing Office, 1946).

69. Carson quoted in Souder, *On a Farther Shore*, p. 9.

70. Steen, *Conservation Diaries of Gifford Pinchot*, entry for January 20, 1945.

71. FDR, Inaugural Address, January 20, 1945. Online by Gerhard Peters and John T. Woolley, *The American Presidency Project*, http://www.presidency.ucsb.edu/ws/?pid=16607.

72. Goodwin, *No Ordinary Time*, p. 574.

73. Kennedy, *Freedom from Fear*, p. 800.

74. FDR, press conference aboard the USS *Quincy*, February 23, 1945, FDRL.

75. Goodwin, *No Ordinary Time*, p. 584.

76. Gifford Pinchot to FDR, March 28, 1945, FDRL.

77. Douglas, *Go East, Young Man*, p. 254.

78. Quoted in Arthur M. Schlesinger, *A Life in the Twentieth Century: Innocent Beginnings, 1917–1950* (New York: Houghton Mifflin Harcourt, 2002), p. 430.

79. Robert H. Ferrell, *The Dying President: Franklin D. Roosevelt, 1944–1945* (Columbia: University of Missouri Press, 1998), p. 119.

80. Chris West, *A History of America in Thirty-Six Postage Stamps* (New York: Picador, 2014), p. 179.

81. Eleanor Roosevelt, "My Day," April 20, 1945.

82. Steen, *Conservation Diaries of Gifford Pinchot*, entry for April 12, 1945.

83. Ralph McGill, "FDR," *Atlanta Constitution*, April 14, 1945.

EPILOGUE: "WHERE THE SUNDIAL STANDS"

1. Merriman Smith, "Funeral Train Seen on Way by Silent Crowds," *New York Herald Tribune*, April 14, 1945, p. 2.

2. Harold L. Ickes, unpublished diary, April 29, 1945.

3. Robert Klara, *FDR's Funeral Train: A Betrayed Widow, A Soviet Spy, and a Presidency in the Balance* (New York: Palgrave Macmillan, 2010), pp. 152–53.

4. Harold L. Ickes, "Should Congress Vest Ownership of the Tidelands in the States?" *Congressional Digest*, October 1, 1948, p. 255.

5. Jill York O'Bright, *The Perpetual March: An Administrative History of the Effigy Mounds National Monument* (Omaha, NE: National Park Service, Midwest Regional Office, 1990).

6. Eleanor Roosevelt, "Speech at the Opening of the Home of Franklin D. Roosevelt National Historic Site," April 12, 1946, FDRL.

7. Dayton Duncan and Ken Burns, *The National Parks: America's Best Idea* (New York: Alfred A. Knopf, 2009), p. 290.

8. Watkins, *Righteous Pilgrim*, p. 951.

9. Newton B. Drury, "The National Park Service: The First Thirty Years," in Harlean James, ed., *American Planning and Civic Annual* (1946), pp. 29–35.

10. FDR, "A Message to the Congress on the Use of Our National Resources," January 24, 1935, reprinted in *The Public Papers and Addresses of Franklin D. Roosevelt*, Vol. 4, *1935*, (New York: Random House, 1938), p. 59.

11. The reforestation work the CCC did here led, in the end, to this picturesque part of Greater Cleveland becoming a national park—Cuyahoga Valley National Park—in 2000.

12. J. McEntee, *Final Report of the Director of the Civilian Conservation Corps, April 1933 Through June 30, 1942* (Washington, DC: U.S. Government Printing Office, 1942), pp. 104–9.

13. Nonprofit president Joan Sharpe provided the author with files about the statue program honoring CCC veterans. CCC Legacy Archives, Edinburg, VA.

14. Lyndon B. Johnson, Johnson White House Tapes, August 7, 1964, LBJ Presidential Library, Austin, TX.

15. Douglas Brinkley, "Rachel Carson and JFK, an Environmental Tag Team," *Audubon* (May–June 2012), p. 2.

16. Norman B. Lehde, "JFK's Visit Thrills Thousands," *Union-Gazette* (Port Jervis, NY), September 25, 1963, p. A1.

17. Douglas, *My Wilderness.*

18. Kate Raftery quoted in Harvey Chipkin, "Built to Last: Parks and Youth Conservation Corps," *Parks & Recreation*, Vol. 46, no. 7 (July 2011), p. 47.

19. Salazar, quoted in "Federal Agencies Announce National Council to Build 21st Century Conservation Corps," news release, U.S. Department of the Interior, January 10, 2013.

20. Bill McKibben, "A Green Corps," *Nation*, April 7, 2008; also published as the foreword to Galusha, *Another Day, Another Dollar*, p. ix.

21. Owen A. Tomlinson, memorandum to custodian Muir Woods, May 5, 1945, Golden Gate National Recreation Area, Park Archives and Records Center, Presidio, San Francisco.

22. Press release, quoted in John Auwaerter and John F. Sears, *Historic Resource Study for Muir Woods National Monument: Golden Gate National Recreation Area* (Boston: Olmsted Center for Landscape Conservation, 2006), p. 341.

23. California Department of Parks and Recreation, "Honoring Those Who Built the Foundation of California's State Park System," news release, April 10, 2008.

24. "Redwoods Gain Company of Another Giant; Plaque to Roosevelt is Placed in Grove," *New York Times* May 20, 1945, p. 24.

25. Ibid.

26. Neal Maher, "The New Deal and Climate Change?" *Solutions: For a Sustainable and Desirable Future*, Vol. 1, no. 5 (October 7, 2010), pp. 72–75.

27. "Roosevelt Paid Honor by Delegates," *Los Angeles Times*, May 20, 1945, p. 3.

28. Gifford Pinchot and Harold K. Steen, "Conservation as the Foundation for Permanent Peace," *Forest History Today* (Spring–Fall 2001), p. 5.

29. John Auwaerter and John F. Sears, "Muir Woods, William Kent, and the American Conservation Movement," in *Historic Resource Study for Muir Woods National Monument; Golden Gate National Recreation Area* (Boston: Olmstead Center for Landscape Preservation, National Park Service, 2006), pp. 341–42.

30. "Stories—Muir Woods," National Park Service website, http://www.nps.gov/muwo/learn/historyculture/stories.htm.

INDEX

Page numbers of photographs appear in italics.
Abbreviatons used: ER = Eleanor Roosevelt; FDR = Franklin Delano Roosevelt; TR = Theodore Roosevelt; CCC = Civilian Conservation Corps; FWS = U.S. Fish and Wildlife Service; NPS = National Park Service; NWR = National Wildlife Refuge; RDA = Recreation Demonstration Area; WPA = Works Progress Administration

IMAGES AND CREDITS

Roosevelt touching tree, Cover: Roosevelt's Little White House/Georgia State Parks and Historic Sites Division

Aerial Springwood, page 5: Franklin D. Roosevelt Presidential Library & Museum (FDRL)

FDR bow and arrow 1892, page 20: Franklin D. Roosevelt Presidential Library & Museum (FDRL)

FDR and father 1895, page 24: Franklin D. Roosevelt Presidential Library & Museum (FDRL)

FDR with camera 1897, page 27: Franklin D. Roosevelt Presidential Library & Museum (FDRL)

FDR ER Sara 1905, page 41: Franklin D. Roosevelt Presidential Library & Museum (FDRL)

Roosevelts Campobello picnic 1906, page 50: Franklin D. Roosevelt Presidential Library & Museum (FDRL)

FDR campaign 1910, page 61: Franklin D. Roosevelt Presidential Library & Museum (FDRL)

FDR Pinchot 1931, page 65: National Park Service

FDR Franklin Lane hunting, page 76: National Archives and Records Administration

FDR grouper Florida, page 102: Franklin D. Roosevelt Presidential Library & Museum (FDRL)

FDR Crosby Florida, page 103: Franklin D. Roosevelt Presidential Library & Museum (FDRL)

FDR Warm Springs Pool 1930, page 111: Franklin D. Roosevelt Presidential Library & Museum (FDRL)

FDR Morgenthau 1931, page 119: Franklin D. Roosevelt Presidential Library & Museum (FDRL)

FDR sworn in as NY governer 1930, page 137: Franklin D. Roosevelt Presidential Library and Museum (FDRL)

Francis Jean Harper Okefenokee 1930, page 146: Delma Presley South Georgia History & Culture Collection at Georgia Southern University

Robert Marshall standing, page 160: U.S. Forest Service

Ickes Native American chiefs, page 165: Library of Congress, Prints and Photographs Division

CCC map 1933, page 177: National Park Service

Henry Wallace, page 199: Library of Congress, Prints and Photographs Division

Tennessee Valley map, page 204: Franklin D. Roosevelt Presidential Library & Museum (FDRL)

FDR Ickes Big Meadows Camp August 1933, page 206: National Park Service

Minerva Hoyt, page 210: National Park Service

Frederic A Delano, page 213: Library of Congress, Prints and Photographs Division

FDR shaking hands with farmer near Warm Springs, page 220: Franklin D. Roosevelt Presidential Library & Museum (FDRL)

FDR sailfish Florida 1933, page 241: Franklin D. Roosevelt Presidential Library & Museum (FDRL)

FDR in Hawaii, page 253: Franklin D. Roosevelt Presidential Library & Museum (FDRL)

Bonneville Project Columbia River Oregon, page 257: Franklin D. Roosevelt Presidential Library & Museum (FDRL)

FDR at Glacier National Park, page 262: Associated Press

Aldo Leopold with birds, page 272: Aldo Leopold Archives, University of Wisconsin

Thomas Beck, page 279: Library of Congress, Prints and Photographs Division

Ding Darling Duck Stamp, page 282: U.S. Fish and Wildlife Service

Ding Darling at Post Office, page 283: U.S. Fish and Wildlife Service

Dust Storm Colorado, page 287: Library of Congress, Prints and Photographs Division

John Calrk Salyer, II, page 294: U.S. Fish and Wildlife Service

Blue Goose Work Sacramento Refuge, page 299: Living New Deal

Gabrielson releasing duck, page 321: Library of Congress, Prints and Photographs Division

FDR pruning tree Warm Springs, page 330: Atlanta History Center

FDR at Shenandoah dedication 1936, page 339: National Park Service

William O. Douglas 1937, page 350: Yakima Valley Museum

Organ Pipe National Monument 1944, page 355: National Park Service

FDR Port Aransas Tarpon, page 368: Port Aransas Museum

Franklin Roosevelt, Yellowstone National Park, 1937, page 396: National Archives and Records Administration

Fishermen at Olympic National Forest by Irving Brant, page 407: National Park Service

ER visiting proposed waterfront park site former dump, page 428: Franklin D. Roosevelt Presidential Library & Museum (FDRL)

Waldo Schmitt iguana, page 435: Smithsonian Institution

FDR shark Galapagos July 1938, page 439: U.S. Naval History and Heritage Command

Self-Portrait in Victorian Mirror, Atherton, California Center for Creative Photography, page 453: Center for Creative Photography, The Ansel Adams Publishing Rights Trust

Frederick Burnham Carlsbad 1941, page 456: Hoover Institution Archives, Stanford University

Rosalie Edge Hawk Mountain, page 472: Hawk Mountain Sanctuary

FDR speech Great Smoky dedication 1940, page 502

Red Rocks Amphitheater, page 508: Western History/Genealogy Department, Denver Public Library

Roosevelt Churchill fishing Shangri-La, page 533: Franklin D. Roosevelt Presidential Library & Museum (FDRL)

FDR Suckley Griscom Thompson Pond, page 537: Franklin D. Roosevelt Presidential Library & Museum (FDRL)

John D. Rockefeller Jr. and wife Jenny Lake 1931, page 545: National Park Service

FDR and globe 1944, page 549: Franklin D. Roosevelt Presidential Library & Museum (FDRL)

FDR Amon Carter 1944, page 556: National Park Service

Smokey the Bear poster 1944, page 568: U.S. Department of Agriculture, National Agricultural Library

FDR King Ibn Saud, Col. W. A. Eddy and Admiral William Leahy aboard the USS Quincy. *February 14, 1945*, page 573: Franklin D. Roosevelt Presidential Library & Museum (FDRL)

Franklin D. Roosevelt's funeral in Hyde Park, New York, page 578: Franklin D. Roosevelt Presidential Library & Museum (FDRL)

Grant Everglades 1937, Nature Insert page 1 top: National Park Service

Old Faithful Geyser, Yellowstone National Park, erupting, against dark sky, Nature Insert page 1 middle: National Archives and Records Administration

Malheur Bird Refuge 1937, Nature Insert page 1 bottom: Oregon Historical Society

Sacramento Migratory Bird Refuge 1938-refuge-entrance, Nature Insert page 1 bottom: Living New Deal

Glacier National Park, Nature Insert page 2 top: National Archives and Records Administration

Grand Canyon, Nature Insert page 2 middle: National Archives and Records Administration

Timberline Lodge Mt Hood, Nature Insert page 2 bottom: National Archives and Records Administration

Grant Grand Gorge Capitol Reef 1935, Nature Insert page 3 top right: National Park Service

Cedar Breaks National Monument, Utah, c. 1947 by Ansel Adams, Nature Insert page 3 bottom: The Ansel Adams Publishing Rights Trust

Grant Olympic Mts. from Hurricane Ridge Olympic National Park 1938, Nature Insert page 4 top: National Park Service

Grant Giant Dome Carlsbad Caverns 1934, Nature Insert page 4 middle: National Park Service

Grant Cedar Breaks NM 1935, Nature Insert page 4 bottom: National Park Service

Untitled Mesa Verde, Nature Insert page 5 top: National Archives and Records Administration

Burro Mesa and the Chisos Mountains, Big Bend National Park, Texas by Ansel Adams, Nature Insert page 6 top: The Ansel Adams Publishing Rights Trust

Moonrise, Joshua Tree National Monument, California, 1948 by Ansel Adams, Nature Insert page 6 bottom: The Ansel Adams Publishing Rights Trust

Tetons Snake River, Nature Insert page 6 middle: National Archives and Records Administration

Saguaros, Saguaro National Monument, vertical full view of cactus with others surrounding, Nature Insert page 7 top: National Archives and Records Administration

Rocky Mountain, Nature Insert page 7 middle: National Archives and Records Administration

Kings Canyon Rac River, Nature Insert page 7 bottom: National Archives and Records Administration

Zion National Park, Nature Insert page 8 top: National Archives and Records Administration

Grant General Grant Tree Sequoia and Kings Canyon National Parks 1936, Nature Insert page 8 bottom: National Park Service

FDR, Howe, Ickes, Fechner, Wallace, Tugwell, CCC, Virginia, CCC Insert page 1 top: Franklin D. Roosevelt Presidential Library & Museum (FDRL)

ER CCC Yosemite, CCC Insert page 1 middle: Franklin D. Roosevelt Presidential Library & Museum (FDRL)

CCC soil conservation, CCC Insert page 1 bottom: Franklin D. Roosevelt Presidential Library & Museum (FDRL)

CCC Bandelier National Monument, CCC Insert page 2 top: Living New Deal

CCC men Ludington State Park barracks, CCC Insert page 2 middle: Living New Deal

Indian CCC Mobile Unit in 1938, CCC Insert page 2 bottom: National Park Service

CCC workers clearing snow Rocky Mountain NP, CCC Insert page 3 top: National Park Service

Queen Elizabeth of England looking at the CCC exhibit at Fort Hunt Virginia; L to R-Mackenzie King, Prime Minister of Canada, Queen Elizabeth and Eleanor Roosevelt. June 9, 1939, CCC Insert page 3 middle: National Park Service

CCC camp Chisos, CCC Insert page 3 bottom: National Park Service

CCC Camp from Trail Ridge Road Rocky Mt. Natl. Park, Colo, CCC Insert page 4 top: University of North Texas Portal to Texas History, Marfa Public Library

CCC wash room, CCC Insert page 4 middle: Pomona Public Library - The Frasher Foto Postcard Collection

CCC Jackson camp man with sign Indiana, CCC Insert page 4 bottom: Franklin D. Roosevelt Presidential Library & Museum (FDRL)

CCC men in library, CCC Insert page 5 top: Franklin D. Roosevelt Presidential Library & Museum (FDRL)

CCC plant nursery, CCC Insert page 5 middle: Franklin D. Roosevelt Presidential Library & Museum (FDRL)

CCC 2 men with shovels experimental farm Maryland, CCC Insert page 5 bottom: Franklin D. Roosevelt Presidential Library & Museum (FDRL)

CCC night school chair making, CCC Insert page 6 top: Franklin D. Roosevelt Presidential Library & Museum (FDRL)

CCC Boxing Team at Marsh Field, San Diego, Calif, CCC Insert page 6 middle: Franklin D. Roosevelt Presidential Library & Museum (FDRL)

CCC splitting firewood California, CCC Insert page 6 bottom: Franklin D. Roosevelt Presidential Library & Museum (FDRL)

CCC tree planting, CCC Insert page 7 top: Franklin D. Roosevelt Presidential Library & Museum (FDRL)

CCC beavers Salmon National Forest Idaho, CCC Insert page 7 middle: Franklin D. Roosevelt Presidential Library & Museum (FDRL)

CCC surveying Black Hills National Forest, CCC Insert page 7 bottom: Franklin D. Roosevelt Presidential Library & Museum (FDRL)

CCC Training, CCC Insert page 8 top: Franklin D. Roosevelt Presidential Library & Museum (FDRL)

CCC tree planting, CCC Insert page 8 bottom: Franklin D. Roosevelt Presidential Library & Museum (FDRL)

ABOUT THE AUTHOR

DOUGLAS BRINKLEY is a professor of history at Rice University, the CNN Presidential Historian, and a contributing editor at *Vanity Fair* and *Audubon*. The *Chicago Tribune* has dubbed him "America's new past master." His recent book *Cronkite* won the Sperber Prize for Best Book in Journalism and was a *Washington Post* Notable Book of the Year. *The Great Deluge* won the Robert F. Kennedy Book Award. He is a member of the Society of American Historians and the Council on Foreign Relations.

BOOKS BY DOUGLAS BRINKLEY

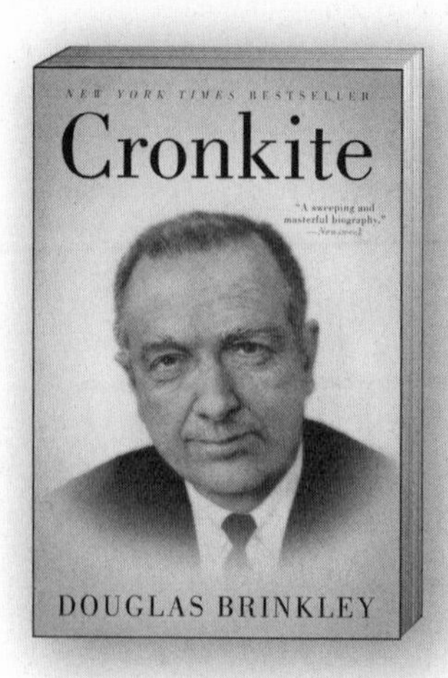

CRONKITE

Available in Paperback and eBook

"A majestic biography. . . . *Cronkite* will endure not for what it tells us about broadcast media but for what it reveals about the man. . . evidence that a job can be done just about perfectly. That goes for the man and this exceptional biography."

—*The New York Times Book Review*

THE QUIET WORLD

Saving Alaska's Wilderness Kingdom, 1879-1960

Available in Paperback and eBook

Brinkley traces the Wilderness Movement, documenting the heroic fight to save wild Alaska—Mount McKinley, the Tongass and Chugach National Forests, and Glacier Bay, among other treasured landscapes—from despoilers.

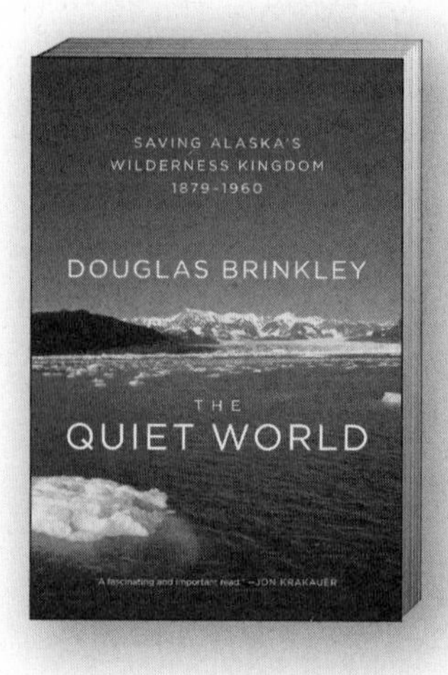

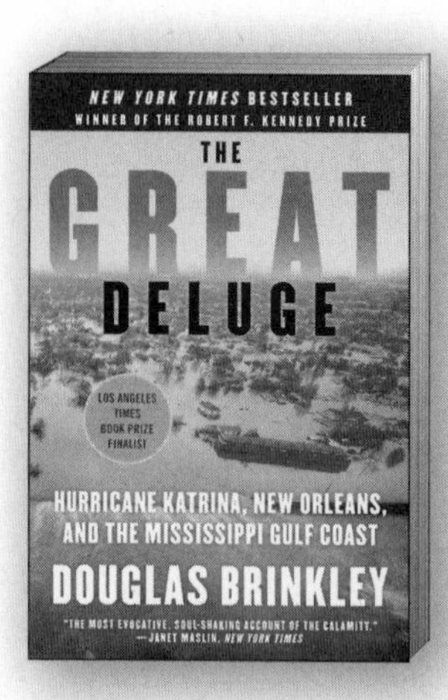

THE GREAT DELUGE

Hurricane Katrina, New Orleans, and the Mississippi Gulf Coast

Available in Paperback and eBook

The complete tale of the terrible storm, offering a piercing analysis of the ongoing crisis, its historical roots, and its repercussions for America.

THE WILDERNESS WARRIOR

Theodore Roosevelt and the Crusade for America

Available in Paperback and eBook

A historical narrative and eye-opening look at the pioneering environmental policies of President Theodore Roosevelt.

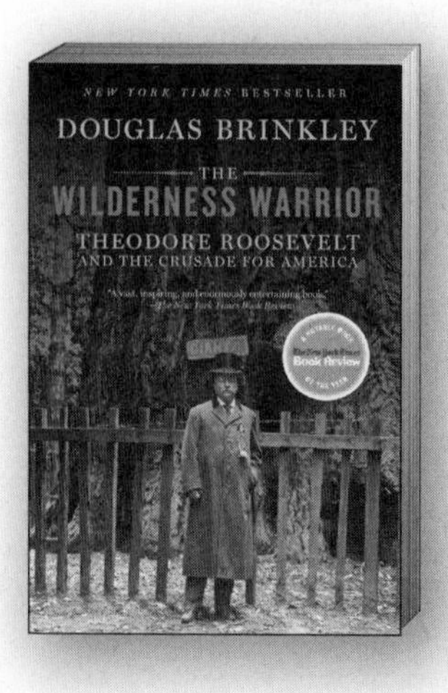

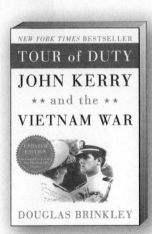

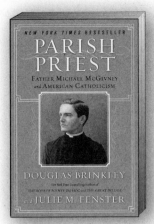

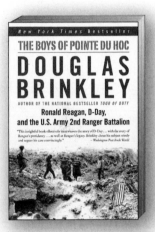